Sustainable Shipping

Harilaos N. Psaraftis
Editor

Sustainable Shipping

A Cross-Disciplinary View

Springer

Editor
Harilaos N. Psaraftis
DTU Management Engineering
Technical University of Denmark
Kongens Lyngby, Denmark

ISBN 978-3-030-04329-2 ISBN 978-3-030-04330-8 (eBook)
https://doi.org/10.1007/978-3-030-04330-8

Library of Congress Control Number: 2018968417

This Springer imprint is published by the registered company Springer Nature Switzerland AG
The registered company address is: Gewerbestrasse 11, 6330 Cham, Switzerland

To Aleka, Anastasia and Nikos

Foreword and Acknowledgments

This book is a compilation of material on sustainable shipping. The material comes from various sources, mostly from work of several external invited colleagues, but also from research projects that my associates and I have been involved in. An outline of the scope and contents of the book is presented in the Preface.

The idea for this book originated in October 2016, a few months after my previous book, *Green Transportation Logistics: The Quest for Win-Win Solutions,* also published by Springer, came out.

As developments at the International Maritime Organization (IMO) and the European Union (EU) on how to reduce maritime greenhouse gas (GHG) emissions looked interesting at the time, it occurred to me that a book focusing on sustainable shipping may be of interest to maritime stakeholders. The basic foundation for the book was research that originated circa 2008, when the Hellenic Chamber of Shipping (HCS) awarded to the National Technical University of Athens (NTUA), my former affiliation, a small study on ship emissions. The study included an analysis of CO_2 emissions statistics for the world fleet and the development of a rudimentary online ship emissions calculator, which is actually still in place.[1] The HCS study was the first among several larger projects dealing with the interface of ship emissions and logistics, either as a central subject or as part of a wider context. Additional sponsors at NTUA included the European Commission, Det Norske Veritas, the American Bureau of Shipping, the Lloyd's Register Foundation and the General Secretariat for Research and Technology (Greece).

Momentum to the book idea was added by my parallel involvement in IMO matters on both the GHG emissions subject (initially in the period 2010–2013 as an adviser to HCS and more recently after mid-2017 as an adviser with the Intercargo delegation) and on the subject of environmental risk evaluation criteria as applied to oil pollution (period 2007–2011 as an adviser to HCS). In addition, in the period 2014–2018, I was involved in the European Sustainable Shipping Forum (ESSF) subgroup on competitiveness (established by the European Commission,

[1] http://www.martrans.org/emis/

DG MOVE) as an invited expert on matters mostly pertaining to the impact of sulfur regulations on European Short Sea Shipping. Several new projects at the Technical University of Denmark (DTU) that relate to maritime emissions (period 2015 on) and several DTU MSc theses provided further interesting material for the book.

Relevant research conducted at DTU and reported here was funded in part by the RoRo SECA project (supported by the Danish Maritime Fund and the Orients Fund), the Blue SIROS project (supported by the European Space Agency, DTU space leader) and the ShipCLEAN project (supported by the Swedish Energy Agency, Chalmers University leader). Three recent DTU MSc theses, by Juan Morales, Massimo Giovannini and Fabio Vilas, have also contributed some material to Chap. 10 of the book. My own time in editing the book and in writing Chap. 10 and parts of Chaps. 7, 11 and 13 was covered in part by an internal grant by the president of DTU and by an internal grant at the DTU Department of Management Engineering, Management Science Division.

Still, most of the material of this book came from invited colleagues, who in fact have written fully 9 of the book's 13 chapters and have contributed to 3 more. So obviously my biggest thanks have to go to all these (23) authors, who kindly accepted my invitation to contribute to the book and without whom the book would be impossible. Among them, I would especially like to thank Poul Woodall, director, Environment & Sustainability at DFDS, and Thalis Zis, postdoc at DTU, for their constructive review of some of the chapters of the book. I want also to thank Kostas Gkonis, secretary general of Intercargo, for giving me the opportunity to attend some of the recent IMO/MEPC meetings on the subject of reducing maritime GHG emissions, as an adviser with the Intercargo delegation. Naturally, any opinions that I express in this book are only my own and do not necessarily represent Intercargo's position or anybody else's position for that matter.

Last but not least, I am grateful to Springer for kindly accepting my proposal to be the editor of this book and, in particular, to Matthew Amboy, Faith Su and Kalaiselvi Ramalingam for their excellent administrative and technical support during book production.

Kongens Lyngby, Denmark Harilaos N. Psaraftis
October 2018

Preface

Scope of the Book

International shipping is currently at a crossroads. The decision of the 72*nd* session of the Marine Environment Protection Committee (MEPC 72) of the International Maritime Organization (IMO) in April 2018 to achieve by 2050 a reduction of at least 50% in maritime greenhouse gas (GHG) emissions vis-à-vis 2008 levels epitomizes the last among a series of recent developments as regards sustainable shipping. It also sets the scene on what may happen in the future. Even though many experts and industry circles believe that the MEPC 72 decision is in line with the COP21 climate change agreement in Paris in 2015, others disagree, either on the ground that the target is not ambitious enough or on the ground that no clear pathway to reach the target is currently visible.

This debate actually goes even further and transcends maritime transportation. The COP21 climate change agreement itself was hailed by many as a most significant achievement, but others were not equally enthusiastic. The decision of American President Trump to steer the United States away from COP21 has caused disappointment or even consternation to the broad spectrum of nations that endorsed the Paris Agreement and has injected a new dose of uncertainty as to what may happen to climate change. Irrespective of the US path, the COP21 Agreement upheld the noninclusion of international shipping (as well as aviation) within its mandate, something that has received mixed reviews by the international community. The rationale for the noninclusion has been that action in these two sectors is within the mandate of the IMO for international shipping and of the International Civil Aviation Organization (ICAO) for aviation. Some industry circles think this is correct; however, environmental groups perceive this as a sign of inability or unwillingness to act and are not happy about it.

Before COP21, the most sweeping piece of regulation pertaining to maritime GHG emissions reduction was the adoption of the so-called Energy Efficiency Design Index (EEDI) by the IMO. This was agreed upon at MEPC 62 in July 2011. This was a no-consensus decision, as adoption was put to a vote in which a group of developing countries (such as China, India, Brazil, Saudi Arabia, South Africa and others) were firmly against the agreement. During the same session, the Ship Energy Efficiency Management Plan (SEEMP) was also adopted.

2011 was also the year the European Union (EU) adopted the new Transport White Paper, which targets drastic reductions in GHG emissions from all modes of transport in the EU by 2050. An aggregate 60% reduction vis-à-vis 1990 levels is stipulated. The target for maritime transportation GHG emission reductions is 40% and if possible 50%. Such targets are highly ambitious because the stipulated reductions are nontrivial. However, and even though a detailed implementation plan has also been proposed in the White Paper, at least for maritime transportation, it is not clear how or if the above reduction targets can be realized.

There have also been some setbacks. For instance, the discussion on a possible adoption of market-based measures (MBMs) for GHGs, initiated in 2010 at the IMO and entailing a comprehensive review of some 11 MBM proposals, was finally suspended in 2013. Relevant discussion was rechanneled toward a system for monitoring, reporting and verification (MRV) of CO_2 emissions. Progress after COP21 was equally mixed. At the IMO, a roadmap was agreed in October 2016. The roadmap foresaw the adoption of an Initial Strategy in 2018 to meet the targets of COP21, which entered into force in November 2016. The strategy will be validated by actual emission figures gathered through the IMO's fuel data collection system (DCS) as of 2019. This would then lead to a final agreement on targets and measures, including an implementation plan, by 2023. The April 2018 IMO decision was an important link in the chain of events that will lead to 2023.

On the more controversial side, perhaps the most significant development has been the February 2017 vote of the European Parliament (EP) to include shipping into the EU Emissions Trading System (ETS) as of 2023, in case no global agreement is reached by 2021, and the subsequent (November 2017) alignment of the EU process with that of the IMO. The EP vote had raised extensive voices of protest from industry circles such as ECSA (European Community Shipowners Associations), ICS (International Chamber of Shipping) and many national ship owner associations. The shipping industry is concerned that an EU ETS may create significant distortions and obstacles for efficient trade, may not be compatible with the IMO roadmap and in fact may not be a good instrument for reducing GHG emissions.

This book is an attempt to shed some light on these and other related developments. To do so, it tries to answer the following questions: where does shipping currently stand as regards sustainability, and what are the prospects for the future? At first glance, these questions may look easy to pose. However, we shall see that they are not so easy to address.

To define "sustainable shipping", we reproduce Fig. 1 of Chap. 1 of this book.

One can see from Fig. 1 that a great number of factors are at play as regards sustainable shipping. In addition, and aside from the considerations of Fig. 1, it should also be realized that sustainability in general, at least as reflected in the United Nations Sustainable Development Goals (UN SDGs), includes additional

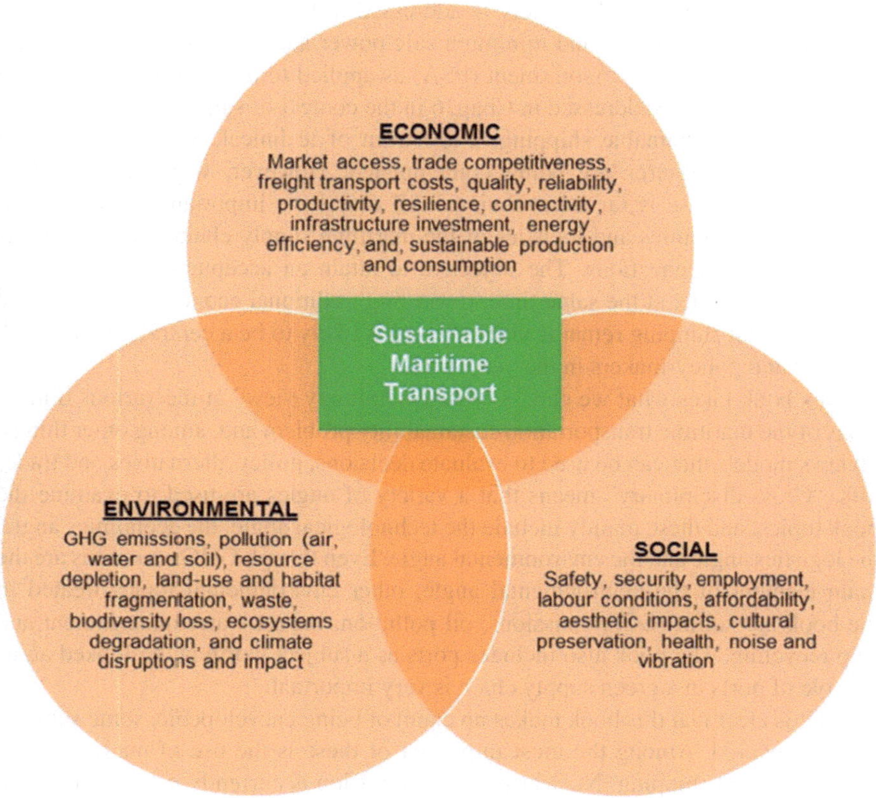

Fig. 1 Defining sustainable maritime transport. Source: UNCTAD. (Reproduced from Chap. 1 of the book)

issues such as poverty, hunger, gender equality, education and several others.[2] So we need to clarify that for the specific object of study of this book (maritime transportation), the distinct perspective of the book is the interaction between the *environmental* and the *economic* dimensions. This means that the central issue of this book is how one can achieve a balance between environmental and economic objectives. Achieving such a balance is important as it would make no sense for a ship, a shipping operation or a maritime supply chain system to be performing well environmentally but be non-viable from an economic perspective. Achieving "win-win" solutions is therefore an indispensable prerequisite for sustainable shipping.

Social criteria such as safety, security, employment, labor conditions, health and others are also important in their own right. But aside from the general sustainability discussion of Chap. 1, their *in-depth* consideration is outside the main scope of the

---

[2]For a list of the UN SDGs, see https://www.un.org/sustainabledevelopment/sustainable-development-goals/

book. We note however that safety is addressed in Chap. 3 in the context of the interaction between EEDI and minimum safe power and in Chap. 5 in the context of IMO's Formal Safety Assessment (FSA) as applied to oil pollution. Health and safety issues are also addressed in Chap. 6 in the context of ship recycling.

To achieve sustainable shipping, a spectrum of technical, logistics-based and market-based measures are being contemplated. However, which are the best measures to choose is far from obvious. All may have important side effects as regards the economics and logistics of the maritime supply chain, including ports and hinterland connections. The objective to attain an acceptable environmental performance while, at the same time, respecting traditional economic performance criteria so that shipping remains viable, is and is likely to be a central goal for both industry and policy-makers in the years ahead.

This book takes what we call "a cross-disciplinary view" at the various dimensions of the maritime transportation sustainability problem and, among other things, reviews models that can be used to evaluate decisions, policy alternatives and trade-offs. "Cross-disciplinary" means that a variety of angles are used to examine the book topics, and these mainly include the technological angle, the economics angle, the logistics angle and the environmental angle. Even though GHG emissions are the main element of the environmental angle, other environmental topics treated in the book are sulfur (SO_x) emissions, oil pollution, ballast water management and ship recycling. The book also includes ports as a subject worth being looked at, as the role of ports in a green supply chain is very important.

As it is clear that this book makes no claim of being encyclopedic, some subjects are *not* covered. Among the most important of these is the use of nuclear power for commercial shipping. Nuclear marine propulsion is currently confined to naval vessels and ice-breakers, and there is nothing a priori obvious that would preclude its consideration in commercial shipping. In fact a distinct advantage of nuclear propulsion is the complete elimination of GHG and other operational emissions. However, issues such as safety, disposal of radioactive waste and economic viability are also important. Proponents of the nuclear option argue that such issues have been resolved.

Even though the exclusion of this topic from the book does not imply a judgment against nuclear propulsion, it is noted that the nuclear option is not included (at least as of yet) as one of the candidate measures postulated by the IMO in their April 2018 decision. To be more precise, the nuclear option is not explicitly excluded by the IMO as a potential measure; however, this option is not visible in any of the current discussions on alternative (low-carbon or zero-carbon) fuels to reduce maritime GHG emissions. This is so in spite of the fact that a GHG emissions reduction goal of at least 50% would seem to encourage a stance not to exclude any solutions, however radical these solutions may seem. However, political considerations, especially after the Fukushima Accident in Japan, seem to be a factor that currently weighs against the use of this option in commercial shipping. Whether or not this exclusion continues as we move toward 2050 is not clear at this point in time. Readers interested in the topic are referred to the work of the Royal Academy of Engineering,[3] of Lloyd's Register[4] and of Hirdaris et al. (2014), among others.

Another issue which this book does not attempt to address is how various emission reduction options may impact climate change and more specifically the mean temperature of the planet. This is true not only as regards GHGs but also as regards other emissions. For instance, the anticipated drastic (but largely unknown) reduction of maritime SO_x emissions as soon as the global 0.50% sulfur cap kicks in as of January 1, 2020, will reduce the "radiative cooling" effect caused by SO_x emissions in the atmosphere. As such, it may increase global warming. But what the increase will be is basically unknown (for a discussion of the relevant issues, see Eyring et al. (2010) and more recently Gratsos (2018), among others). In addition, producing vast quantities of low-sulfur fuels would certainly require some energy, which, if not from renewable or nuclear sources, would also increase global CO_2. Again, the impact of this development on climate change is by and large unknown. Realizing that these issues are very important, this book makes no attempt to estimate their impact.

The above disclaimers notwithstanding, it is hoped that the material assembled in this book is of interest and that it will help clarify some of the important issues that are at stake as regards sustainable shipping, as we move toward 2050. Assuming this is the case, the book may eventually contribute toward the identification of solutions that are feasible and viable, both for private maritime enterprises and for society as a whole.

Book Organization

The rest of the book is organized as follows:

Chapter 1 by Benamara, Hoffmann and Youssef sets the stage for the rest of the book, by identifying the linkages between shipping and sustainable development and highlighting what are the stakes in sustainable shipping, who are the stakeholders, what are the trade-offs, what are the policy issues, what may be the obstacles and enablers of sustainable shipping and the role of international institutions including the IMO and UNCTAD (the UN Conference on Trade and Development).

Chapter 2 by de Kat and Mouawad looks at the topic of technological solutions for sustainable shipping. These include air lubrication, wind-assisted propulsion and solar power, waste heat recovery systems, ballast water management systems, more efficient (energy-saving) engines, more efficient ship hulls and designs, more efficient propellers, hybrid systems and others, both for the main engine and the auxiliary engines.

---

[3]https://www.raeng.org.uk/news/news-releases/2013/July/a-sea-of-options-for-future-ship-propulsion

[4]https://www.lr.org/en/nuclear-power/

Chapter 3 by Polakis, de Kat and Zachariadis looks at the only regulatory measure thus far in place to reduce GHG maritime emissions, EEDI. Chapter 3 goes over the rationale for EEDI and the factors that are important and also goes over related concepts, such as the SEEMP plan and the Energy Efficiency Operational Indicator (EEOI) and the Existing Vessel Design Index (EVDI). Possible weaknesses of EEDI and how to improve the EEDI are also presented.

Chapter 4 by Fjørtoft and Berge is complementary to Chap. 2 and looks into ICT (information and communication technologies). These do not lead to direct environmental benefits, but their smart use can definitely do so, by increasing the efficiency of the maritime supply chain, improving safety, by improving the load factor, etc. This chapter reviews relevant ICT systems in shipping and considers their impact on improving environmental performance.

Chapter 5 by Ventikos, Louzis and Sotiralis highlights the most significant attributes of oil pollution in the context of the sustainable shipping. The chapter presents the current legislative framework for the environmental protection against oil pollution and depicts the utility of the implementation of risk control options (RCOs). Furthermore, the measures of containment of the oil pollution cost are illustrated along with the incorporation of the environmental risk evaluation criteria in IMO's Formal Safety Assessment (FSA). Finally, the chapter discusses feasible ways of achieving a sustainable future without undermining the environmental integrity.

Chapter 6 by Mikelis addresses the recycling of ships, otherwise known as dismantling, ship breaking, scrapping and demolition. The chapter outlines the efforts to implement existing international legislation to ship recycling and the development of the Hong Kong Convention and provides a critical analysis of the development of regional legislation by the EU. It also discusses the combination of voluntary and legislative mechanisms that will secure the global implementation of minimum standards for safe and environmentally sound ship recycling.

Chapter 7 by Zis and Psaraftis presents an overview of the main issues of sulfur emissions and the legislative framework that seeks to reduce the sulfur footprint of the maritime sector. It also analyses potential modal shifts toward less efficient land-based modes which may happen as a result of sulfur regulations and investigates the related potential economic damage to ship operators. To that effect, the chapter presents a methodological framework that can be used to estimate such modal shifts as well as to measure the efficacy of possible measures to reverse such shifts.

Chapter 8 by Wang, Norstad, Fagerholt and Christiansen examines, from a tramp ship operator's point of view, how potential CO_2 emission reduction measures impact operational decisions, and their economic and environmental consequences. Two MBMs are discussed, the bunker levy scheme and the emission trading scheme, and it is shown that both can be incorporated in a similar way into a typical tramp ship routing and scheduling model.

Chapter 9 by Hellsten, Pisinger, Sacramento and Vilhelmsen looks into green liner shipping network design problems, these being defined as problems in green logistics related to the design of maritime services in liner shipping with focus on reducing the environmental impact. The chapter discusses how to more efficiently plan the vessel services with the use of mathematical optimization models.

Chapter 10 by Psaraftis focuses on speed optimization. This involves the selection of an appropriate speed by the vessel so as to optimize a certain objective. As ship speed is not fixed, depressed shipping markets and/or high fuel prices induce slow steaming which is being practiced in many sectors of the shipping industry. The chapter presents some basics, discusses the main trade-offs and also examines combined speed and route optimization problems. Some examples are presented so as to highlight the main issues that are at play, and the regulatory dimension of speed reduction via speed limits is also discussed.

Chapter 11 by Psaraftis and Woodall focuses on the concept of MBMs to reduce GHG emissions from ships and reviews several distinct MBM proposals that were under consideration by the IMO. The chapter then moves on to discuss the concept of MRV of CO_2 emissions and the distinct mechanisms set up the European Union (EU) and the IMO for MRV. The two issues are connected as a next possible step after MRV can be an MBM.

Chapter 12 by Zis examines the issues associated with a green port operation. These include technologies such as cold ironing, market-based practices such as differentiated fairway dues, speed reduction, noise and dust abatement and others. The legislative framework in various countries is explained and various environmental scorecards are discussed. The chapter emphasizes on the implementation of speed reduction programmes near the port, use of cold ironing at berth and the effects of fuel quality regulation, considering the perspectives of the port authority and the ship operator.

Last but not least, Chap. 13 by Psaraftis and Zachariadis looks at the way ahead, with a focus on the April 2018 IMO Initial Strategy on how to reduce maritime GHG emissions. The chapter includes a section on alternative fuels, these figuring centrally among candidate measures included in the IMO Initial Strategy.

All chapters are to a great extent self-contained, with cross-referencing among them wherever appropriate.

Intended Audience

The intended audience of the book includes:

- Faculty, students and researchers active in maritime transportation and interested in the environmental dimension of shipping
- Carriers, shippers, infrastructure managers and other logistics providers who aim at improving their environmental performance while staying in business
- Technology designers and providers
- Policy-makers at the national and international level, including the IMO and the EU
- Other stakeholders, environmental or other

- Industry associations such as ICS,[5] IAPH,[6] Intertanko,[7] Intercargo,[8] BIMCO,[9] IACS,[10] WSC,[11] ECSA,[12] ESPO,[13] ECS,[14] etc.
- Academic associations such as IAME,[15] SNAME,[16] INFORMS,[17] etc.

Kongens Lyngby, Denmark Harilaos N. Psaraftis

References

Eyring, V., Isaksen, I. S. A., Berntsen, T., Collins, W. J., Corbett, J. J., Endresen, Ø., et al. (2010). Transport impacts on atmosphere and climate: Shipping, *Atmospheric Environment*, *44*, 4735–4771.

Gratsos, G. (2018). Global warming, mankind, the environment, transport, shipping and clear thinking. Accessible at https://www.academia.edu/.

Hirdaris, S. E., Cheng, Y. F., Shallcross, P., Bonafoux, J., Carlson, D., Prince, B., et al. (2014). Considerations on the potential use of nuclear Small Modular Reactor (SMR) technology for merchant marine propulsion. *Ocean Engineering*, *79*, 101–130.

[5]International Chamber of Shipping

[6]International Association of Ports and Harbours

[7]International Association of Independent Tanker Owners

[8]International Association of Dry Cargo Shipowners

[9]Baltic International Maritime Council

[10]International Association of Classification Societies

[11]World Shipping Council

[12]European Community Shipowners Associations

[13]European Sea Ports Organisation

[14]European Shippers Council

[15]International Association of Maritime Economists

[16]Society of Naval Architects and Marine Engineers

[17]The Institute for Operations Research and the Management Sciences

Contents

About the Editor

Harilaos N. Psaraftis is a professor at the Technical University of Denmark (DTU), Department of Management Engineering, Management Science Division. He has a diploma from the National Technical University of Athens (NTUA, 1974), two MSc degrees (1977) and a PhD (1979) from the Massachusetts Institute of Technology (MIT). He was an assistant and associate professor at MIT from 1979 to 1989 (Department of Ocean Engineering) and professor and director of the Laboratory for Maritime Transport (LMT) at NTUA from 1989 to 2013 (School of Naval Architecture and Marine Engineering). Harilaos' main interests are in transport optimization and maritime and intermodal transport, with a recent focus on green transport, and maritime transport policy and regulations. He has been a principal investigator (PI) or co-PI in some 50 research projects on various transportation subjects. Among the 24 EU projects he was involved in, he has coordinated three multi-partner EU consortia, including project SUPERGREEN on European Green Corridors (2010–2013). His published work includes 4 books, over 150 refereed articles in journals and conferences, over 130 other publications (including 17 book chapters) and over 180 lectures in various other conferences. His previous book was *Green Transportation Logistics: the Quest for Win-Win Solutions* (Springer, 2016). He has been on the editorial board of several journals, including *Transportation Science* (1987–2018), *WMU Journal of Maritime Affairs* (since 2002), *Maritime Economics and Logistics* (since 2016) and *Networks* (since 2016). Since 2006, Harilaos has participated in many meetings of the International Maritime Organization (IMO), both at the Maritime Safety Committee (MSC) and the Marine Environment Protection Committee (MEPC). He has served as the chairman of various MEPC correspondence and working groups. Harilaos and his group have received various academic and industry awards, including two from Lloyd's List. In addition to his academic duties, Harilaos has served as the CEO of the Piraeus Port Authority (OLP) for close to 5.5 years (1996–2002). His tenure has been linked with developments that include (a) Piraeus making the list of the top 50 world container ports (No. 41 in 1998) with traffic more than doubling from 575.000 TEU (1996) to 1.160.000 TEU (2001) and (b) the port being transformed from a public law undertaking into a corporation (1999). Harilaos is the third longest serving CEO since OLP's inception in 1930, having served under four Ministers of Shipping in Greece.

About the Authors

Hassiba Benamara is a maritime transport and trade specialist with a postgraduate degree in economics from the University of Ottawa (Canada) and over 20 years professional experience in transport and trade logistics. She joined UNCTAD in 2005 and has been working on various transport issues. She is a co-author of the UNCTAD annual Review of Maritime Transport focusing particularly on seaborne trade and port issues. Other areas of work include climate change, energy market developments, transport costs, supply-chain security and sustainable transport. Before joining UNCTAD, she has been working for the Canadian Ministry of Transportation in the international shipping and trade divisions, respectively. Areas of work included marine insurance and liability, arrest of ships, maritime liens and mortgages, maritime security, antitrust immunity and liner conferences, cabotage as well as transport and logistic services trade liberalization. She represented the Ministry at relevant meetings of the International Maritime Organization and the World Trade Organization.

Svein Peder Berge is a senior business developer at SINTEF Ocean AS, that changed name from MARINTEK in 2017, and holds a PhD in engineering cybernetics from the Norwegian University of Science and Technology (NTNU). He has 25 years of experience from both maritime industry and research companies. He has been working with project management, research and product development for both energy and maritime industries. Areas of expertise are software development, mathematical modeling, simulation technology and design and development of energy and maritime applications.

Marielle Christiansen works as a professor of operations research at the Norwegian University of Science and Technology. In addition, she is the head of the Department of Industrial Economics and Technology Management. Her primary research interests concern development and implementation of optimization models and methods for industry-related planning problems as regards transportation, logistics and production. She is particularly interested in applications where maritime transportation and supply chain challenges are considered and has been involved in many shipping industry-sponsored projects. Her research has resulted in numerous papers in journals like *Computers & Operations Research*, *European Journal of Operational Research*, *Journal of the Operational Research Society*, *Naval Research Logistics*, *OMEGA*, *Networks*, *Transportation Research Part C* as well as *Transportation Science*. Furthermore, she has contributed with several surveys within maritime transport optimization in general and within combined inventory management and routing and fleet composition and routing in particular.

Jan Otto de Kat has worked as director in ABS since 2013 in Copenhagen. He is responsible for global coordination of activities related to containerships; his previous areas of responsibility included energy efficiency and operational and environmental performance. He has worked as a senior director and head of innovation in AP Moller-Maersk between 2007 and 2012. From 1989 through 2006, he worked at the Maritime Research Institute Netherlands (MARIN), Wageningen, the last 7 years as director of research and development. He received his MSc and PhD in naval architecture and offshore engineering from the University of California at Berkeley and his BEng from the University of New South Wales, Sydney.

Kjetil Fagerholt is a professor in operations research at the Norwegian University of Science and Technology, Department of Industrial Economics and Technology Management. He also holds positions as an adjunct professor in the Department of Marine Technology at the same university and as a scientific adviser at SINTEF Ocean. His main research interest is on developing and applying operations research models and methods on planning problems in maritime transportation. He has published around 100 papers in international peer-reviewed journals, such as *Transportation Science*; *European Journal of Operational Research*; *OMEGA*; *Transportation Research Part C, D and E*; *INFORMS Journal on Computing*; *Computers & Operations Research*; and *Maritime Policy & Management*. He is ssociate editor for *Transportation Science*, *Flexible Services and Manufacturing* and *Asia-Pacific Journal of Operational Research* and deputy editor for *Journal of Marine Science and Technology*.

Kay E. Fjørtoft is a senior research scientist at SINTEF Ocean AS. He holds an MSc and has more than 23 years of experiences from maritime research. Kay has been involved in several research projects with thematic covering maritime logistics and operation, software architecture and development, integrated operations and planning, maritime safety operations, Arctic maritime operations, freight transport, port community systems and communication (telecom) as examples. He also held a position for ESA as an ambassador for Norway within the Business Applications programme, where the mission is to assist and promote space technology for the maritime sector and for the Norwegian industry.

Erik Hellsten is a PhD student at the Technical University of Denmark. His work revolves around applying mathematical optimization to improve operations in the feeder line industry. He is currently working on vessel scheduling and network design. Erik received his BSc in engineering physics and MSc in engineering mathematics at Chalmers University of Technology.

Jan Hoffmann joined UNCTAD in 2003 and is currently chief of the organization's Trade Logistics Branch, responsible for research and technical assistance programmes in international transport and trade facilitation. Jan is co-author and coordinator of the *Review of Maritime Transport*, initiated the UNCTAD *Maritime Country Profiles* and created the annual *Liner Shipping Connectivity Index*. Previously, he spent 6 years with the UN Economic Commission for Latin America and the Caribbean in Santiago de Chile and 2 years with the IMO in London and Santiago. Prior to this, he held part-time positions as assistant professor, import-export agent, consultant and seafarer on an Antigua and Barbuda flagged tweendecker. Jan has studied in Germany, the United Kingdom and Spain and holds a PhD in economics from the University of Hamburg. He is member of the boards of various journals and associations and past president of the International Association of Maritime Economists (IAME).

Konstantinos Louzis has a master's in naval architecture and marine engineering from the School of Naval Architecture and Marine Engineering (NA&ME) of the National Technical University of Athens (NTUA), Greece. He is currently a PhD candidate at the Laboratory for Maritime Transport of the same school conducting research in the fields of life cycle risk assessment and management in the maritime domain. His expertise is in the following scientific areas: marine safety, risk analysis/assessment and risk engineering, maritime accident investigation and modeling, human factors and human reliability, resilience engineering and environmental risk aspects of shipwrecks. Konstantinos Louzis has published parts of his work in international peer-reviewed journals and has participated as a speaker at various international scientific conferences with topics relevant to his expertise.

Nikos Mikelis graduated in naval architecture from the University of Newcastle and obtained master's and doctorate degrees from London University. He has worked in ship classification and for a shipping company as a superintendent, technical manager and then director. In 2006, he joined the International Maritime Organization, from where he retired as head of Marine Pollution Prevention and Ship Recycling Section. Currently he is a consultant; a non-executive director of GMS, the world's largest buyer of ships for recycling; and a director of QACE, the entity that has been set for the quality assessment and certification of classification societies recognized by the European Union. He has served in senior positions in shipping industry bodies as council member of INTERTANKO, chairman of the "INTERTANKO Safety, Technical & Environmental Committee" and member of the Safety Committee of the Union of Greek Shipowners. He has written over 50 learned papers and numerous articles in the maritime press and is a freeman of the city of London.

Jad Mouawad is a naval architect, graduated in 2005 from the Norwegian University of Science and Technology (NTNU). He has since had several positions in the maritime industry, first at the Norwegian Maritime Administration (NMA), before joining DNV GL where he started and developed their class services for Ballast Water Management. In 2013, Jad Mouawad founded Mouawad Consulting which has become a major engineering and consultancy firm specialized in Ballast Water Management, with offices in Norway, China, Korea and Lebanon. Jad Mouawad issued type approval certificates for more than 11 Ballast Water Management Systems (BWMS) and initiated the development of the first DNV GL rules for installation of BWMS onboard ships. He also contributes heavily to the development of the guidelines of the BWM Convention, developed at the International Maritime Organization (IMO) as an advisor to NMA, and represents Norway in the ISO Technical Committee dealing with standards for BWM.

Inge Norstad is a senior business consultant at Tieto Norway AS in Trondheim, Norway. He works with optimization and decision support in the midstream part of the supply chain in the oil and gas industry. He has a PhD in operations research from the Norwegian University of Science and Technology. Prior to working at Tieto, he has been a research scientist at SINTEF Ocean in Trondheim, Norway, for 12 years. His research interests are in the field of optimization of maritime transportation and maritime supply chains.

David Pisinger is a professor in operations research at DTU Management Engineering and adjoint professor at DIKU, University of Copenhagen. His research interests include maritime optimization, vehicle routing, railway optimization, energy models and loading problems. He has been leading several research projects in maritime logistics and railway optimization. David received the Hedorfs Fonds Prize for Transport Research 2013 and received the Teaching Prize 2016 at DTU Management. He has supervised more than 20 PhD students: two of these have received the VeRoLog Dissertation Prize and one has received the TSL dissertation prize.

Maria Polakis is a technical professional with over 20 years of experience in the maritime industry. She graduated from the National Technical University of Athens (NTUA) with an MSc in naval architecture and marine engineering. She joined the American Bureau of Shipping (ABS) Operational and Environmental Performance group in 2013 where she acquired extensive experience through research projects focused on environmental technologies and product development. Current role includes cooperation with IACS Correspondence Group on EEDI on regulatory developments beyond phase 2 and projects with marine and offshore clients to evaluate the energy efficiency and operational performance of new and existing marine assets in view of current and future regulations. Key competencies include fuel efficiency of ships in service, reduction of ship emissions and hydrodynamic performance of ships. She is a senior trainer on EEDI verification and environmental regulatory requirements for ABS engineering offices globally. Her previous roles involved naval architecture consultancy and technical ship management in the Greek shipping industry.

David Sacramento received his BSc in mathematics at the University of La Laguna (ULL) and his MSc in mathematical modeling and computation with focus area in operations research at the Technical University of Denmark (DTU), where he is currently a PhD student. His research is focused on shipping network design and vessel scheduling under a feeder line structure.

Panagiotis Sotiralis holds a university degree in naval architecture and marine engineering from the School of Naval Architecture and Marine Engineering (NA&ME) of the National Technical University of Athens (NTUA), Greece. His diploma thesis, titled "Probabilistic Study Spill Oil Tankers from: Quantities and Costs", was awarded by the Greek section of SNAME for the best diploma thesis of the year 2011. He has a master's degree in techno-economic systems (MBA) from National Technical University of Athens (School of Electrical and Computer Engineering) – University of Piraeus (Department of Industrial Management and Technology). He is currently a PhD candidate at the Laboratory for Maritime Transport of the same school in the scientific areas of dynamic risk analysis/assessment and maritime safety. Panagiotis Sotiralis' expertise is in maritime safety, maritime accident investigation and modeling, human factors, resilience engineering, statistical modeling and optimization of the operational and environmental performance of different ship types.

Nikolaos P. Ventikos is an associate professor in the Laboratory for Maritime Transport at the School of Naval Architecture and Marine Engineering of the National Technical University of Athens, Greece. Dr. Ventikos' expertise is in marine safety, risk engineering, (oil) marine pollution, human reliability (including human-centered design), marine casualty investigation, resilience engineering, systems engineering, offshore and environmental risk, maritime sustainability, blue growth and statistical and probabilistic modeling. He has published and presented his work in many highly esteemed scientific journals such as *Ocean Engineering*, *Reliability Engineering & System Safety*, *Marine Pollution Bulletin*, *Journal of Risk Research*, *Journal of Risk and Reliability*, etc. and in more than 100 scientific conferences. Dr. Ventikos was awarded the GP Livanos Grand Prize on Environmental Issues sponsored by the Hellenic Chamber of Shipping (2000).

Charlotte Vilhelmsen is an assistant professor in operations research at DTU Management Engineering. Her research interests include maritime optimization, health-care planning and optimization under uncertainty. Charlotte received DTU Management's Teacher of the Year Award in 2018.

Xin Wang is a business consultant at Tieto Corporation in Oslo, Norway. His work focuses on optimization and decision support in the oil and gas industry. During 2015–2018, he was a postdoctoral fellow in the Department of Industrial

Economics and Technology Management at Norwegian University of Science and Technology. His research interests lie in the area of operations research, particularly decision-making under uncertainty and optimization in transportation and logistics. He completed his undergraduate studies at Shanghai Jiao Tong University and his PhD at Lancaster University. During his PhD study, he was also visiting researcher at Norwegian School of Economics in Bergen, Norway, and CIRRELT in Montreal, Canada.

Poul Woodall is director of environment and sustainability at DFDS A/S. He has over 40 years of experience within the maritime and transport industry. His career includes 17 years' overseas experience with postings in Asia, the Middle East and Europe. Since 2013 his focus has been maritime environmental issues. Poul Woodall has a degree from Copenhagen Business School supplemented with management education at Insead and Stanford Universities. Poul is also involved with the Trident Alliance, Green Ship of the Future, Interferry and various European Sustainable Shipping Forum (ESSF) working groups in Brussels. In July 2018, Poul joined the new Connect Europe Facility (CEF) Transport Advisory Group of the Innovation and Networks Executive Agency (INEA).

Frida Youssef joined UNCTAD in 1997 and is currently the chief of the Transport Section of the Trade Logistics Branch at UNCTAD. She has more than 20 years of professional experience in broad range of issues including transport, commodity supply chain, energy, structured and project finance. She is a co-author of the UNCTAD's annual Review of Maritime Transport focusing particularly on freight rates, maritime transport costs and PPPs in ports. Other areas of work include sustainable freight transport, transport and transit corridors and finance. Ms. Youssef is also member of the Global network of Export-Import Bank (G-NEXID) steering committee. She holds an MBA from Boston University and an MA in International Affairs from the Institut des hautes études internationales (France).

Panos Zachariadis, mechanical engineer (BSc) and naval architect (MSE) from the University of Michigan, has a 35-year experience in shipping, having served for years as marine superintendent in New York, after periods of shipbuilding supervision in Japan and sea service in bulk carriers and oil tankers. He is a technical director of Atlantic Bulk Carriers since 1997 and a member of the Greek Delegation to IMO since 2004, having contributed to the development of several well-known regulations (e.g. GBS, PSPC, FSA) and the prevention of others (e.g. double hull bulk carriers). In cooperation with major Korean shipyards, he has applied numerous energy-saving ideas, some of which have become industry standards. He is a member of technical committees of several classification societies, UGS and Hellenic Chamber of Shipping, BIMCO Marine Committee and BoD HELMEPA and MARTECMA. He has been honored with the 2011 Efkranti Award for his contribution to Greek shipping, the Hellenic Shortsea Shipowners Association 2017

Maritime Personality Award and the 2017 Lloyd's List "Achievement in Safety or Environmental Protection" Award.

Thalis P. V. Zis is a postdoctoral research associate at the Technical University of Denmark (DTU), Department of Management Engineering, Management Science Division. He holds a diploma in mechanical engineering from the National Technical University of Athens (NTUA), an MSc in transport from Imperial College London and a PhD in port operations from Imperial College London. He has conducted significant research on improving the environmental performance of port operations and also on the issues of sulfur emissions from shipping, focusing on economic and logistical issues arising from the stricter sulfur legislation. He currently focuses on applying game theory to improve compliance rates with the sulfur regulation among ship operators. He serves as an international member of the Standing Committee on Marine Environment and the Standing Committee on Ports and Channels of the Transportation Research Board (USA).

Abbreviations

4C	Commitment, Competence, Continuous Learning, Collaboration
AER	Annual Efficiency Ratio
AGV	Automated Guided Vehicles
AIS	Automatic Identification Systems
ALARP	As Low As Reasonably Practicable
AMP	Alternative Marine Power
AMVER	Automated Mutual Assistance Vessel Rescue System
BAF	Bunker Adjustment Factor
BAU	Business As Usual
BDN	Bunker Delivery Note
BIMCO	Baltic and International Maritime Council
BMEP	Brake Mean Effective Pressure
BRI	Belt and Road Initiative
BWM	Ballast Water Management
CAF	Cost of Averting a Fatality
CATS	Cost of Averting a Tonne of Spilt oil
CBDR-RC	Common But Differentiated Responsibilities and Respective Capabilities
CFD	Computational Fluid Dynamics
CH_4	Methane
CO_2	Carbon Dioxide
COP	Conference of the Parties [of the United Nations Framework Convention on Climate Change]
COW	Crude Oil Washing
CPP	Controllable Pitch Propeller
CSI	Clean Shipping Index
CSR	Corporate Social Responsibility
CV	Contingent Valuation
DCS	Data Collection System
DNC	Digital Nautical Charts
DNV	Det Norske Veritas

DoC	Document of Compliance
DSC	Digital Selective Calling
EC	European Commission
ECA	Emission Control Area
ECDIS	Electronic Chart Display and Information System
ECSA	European Community Shipowners Associations
EESH	Energy Efficiency Per Service Hour
EGR	Exhaust Gas Recirculation
EIS	Efficiency Incentive Scheme
EIV	Estimated Index Value
EMSA	European Maritime Safety Agency
ENC	Electronic Navigational Charts
EPA	Environmental Protection Agency (US)
ESD	Energy-Saving Device
ESPO	European Seaport Organization
ESSF	European Sustainable Shipping Forum
ETS	Emission Trading System
EU	European Union
EU SRR	European Union Ship Recycling Regulation
EVDI	Existing Vessel Design Index
FAL	The Convention on Facilitation of International Maritime Traffic
FORS	Fuel Oil Reduction Strategy
FPP	Fixed Pitch Propeller
FSA	Formal Safety Assessment
GHG	Green House Gas
GMDSS	Global Maritime Distress and Safety System
GWP	Global Warming Potential
HEA	Habitat Equivalent Analysis
HKC	Hong Kong International Convention for the Safe and Environmentally Sound Recycling of Ships, 2009
IACS	International Association of Classification Societies
IAMS	Integrated Alarm and Monitoring Systems
IAPH	International Association of Ports and Harbours
IAS	Integrated Automation System
ICAF	Implied Cost of Averting a Fatality
ICS	International Chamber of Shipping
ICT	Information and Communication Technologies
IEE	International Energy Efficiency
IMO	International Maritime Organization
INTERTANKO	International Association of Dependent Tanker Owners
IOT	Internet of Things
IPCC	Intergovernmental Panel on Climate Change
IPL	Integrated Planning and Logistics
IPTA	International Parcel Tanker Association
IRRC	International Ready for Recycling Certificate

ISM	International Ship Management
ISO	International Organization for Standardization
ISPI	Individual Ship Performance Indicator
IT	Information Technology
ITCP	Integrated Technical Cooperation Programme
ITOPF	International Tanker Owners Pollution Federation Limited
ITS	Intelligent Transport Systems
ITTC	International Towing Tank Conference
IUCN	International Union for the Conservation of Nature
LBSI	Lean Burn Spark Ignition
LDC	Least Developed Country
LIS	Leveraged Incentive Scheme
LNG	Liquefied Natural Gas
LPDF	Low Pressure Dual Fuel
LPG	Liquefied Petroleum Gas
LSNDP	Liner Shipping Network Design Problem
LSP	Liner Service Planning
MAC	Marginal Abatement Cost
MARPOL	International Convention for the Prevention of Pollution from Ships
MBM	Market-Based Measure
MCFP	Multi-Commodity Flow Problem
MCR	Maximum Continuous Rating
MDO	Marine Diesel Oil
MEPC	IMO Marine Environment Protection Committee
MIMS	Maritime Information Management System
MPP	Minimum Propulsion Power
MRV	Monitoring, Reporting and Verification
MSW	Maritime Single Window
MTO	Man, Technology and Organization
NGO	Nongovernmental Organization
NH_3	Ammonia
NOAA	National Oceanic and Atmospheric Administration
NOx	Nitrogen Oxides
OBC	Oxygen Blown Converter
OECD	Organisation for Economic Co-operation and Development
OILPOL	International Convention for the Prevention of Pollution of the Sea by Oil
OPA	Oil Pollution Act
OPRC	International Convention on Oil Pollution Preparedness, Response and Co-operation
OPS	Onshore Power Supply
OSIR	Oil Spill Intelligence Report
OT	Operational Technology
P&I	Protection & Indemnity

PCS	Port Community System
PM	Particulate Matter
POLA	Port of Los Angeles
POLB	Port of Long Beach
PSC	Port State Control
RCOs	Risk Control Options
RMG	Rail Mounted Gantry
RPM	Revolution Per Minute
RTG	Rubber Tired Gantry
S-AIS	Satellite-AIS
SCR	Single Catalytic Reduction
SDGs	Sustainable Development Goals
SECA	Sulfur Emission Control Area
SECT	Ship Efficiency and Credit Trading
SEEMP	Ship Energy Efficiency Management Plan
SFOC	Specific Fuel Oil Consumption
SIDS	Small Island Developing State
SO	Sulfur Monoxide
SO_2	Sulfur Dioxide
SO_3	Sulfur Trioxide
SO_x	Sulfur Oxides
SOLAS	International Convention for the Safety of Life at Sea
SRFP	Ship Recycling Facility Plan
SRP	Ship Recycling Plan
SSS	Short-Sea Shipping
SSSCRP	Simultaneous Ship Scheduling and Cargo Routing Problem
SW	Single Window
TOS	Terminal Operation System
TSP	Traveling Salesman Problem
TSW	Trade Single Window
UNCTAD	United Nations Conference on Trade and Development
UNFCCC	United Nations Framework Convention on Climate Change
VES	Vessel Efficiency System
VHL	Value of Human Life
VLCC	Very Large Crude Carrier
VNS	Variable Neighborhood Search
VR	Virtual Reality
VRP	Vehicle Routing Problem
VSRP	Vessel Speed Reduction Programme
VTMIS	Vessel Traffic Management and Information Services
VTS	Vessel Traffic Services
WHO	World Health Organization
WSC	World Shipping Council
WTO	World Trade Organization

Chapter 1
Maritime Transport: The Sustainability Imperative

Hassiba Benamara, Jan Hoffmann, and Frida Youssef

Abstract The role of maritime transport in addressing the global sustainability imperative is increasingly recognized. Safe, secure, energy-efficient, affordable, reliable, low-carbon, climate-resilient and rule-based maritime transport systems contribute to achieving an economically efficient, socially equitable and environmentally sound development. However, for this role to effectively materialize, unsustainable maritime transport practices that result in wide-ranging external costs need to be addressed.

In the context of the ongoing implementation of the 2030 Agenda for Sustainable Development, the sustainable development goals (SDGs) and the Paris Agreement on climate change, there is a renewed opportunity to tap the sustainability potential of the maritime transport sector. Leveraging maritime transport in support of the sustainable development agenda requires that economic, social and environmental sustainability criteria be fully integrated and mainstreamed into relevant maritime transport planning processes, policies and investment decisions.

This chapter highlights key issues lying at the interface of maritime transport and sustainable development while emphasizing the role of the sector as a catalyst of a development path that promotes people, prosperity, environment and effective and relevant partnerships.

Abbreviations

ASD	Agenda for Sustainable Development
BIMCO	Baltic and International Maritime Council
CO_2	Carbon dioxide
COP	Conference of the Parties [of the United Nations Framework Convention on Climate Change]

H. Benamara (✉) · J. Hoffmann · F. Youssef
Trade Logistics Branch, Division on Technology and Logistics, UNCTAD, Geneva, Switzerland
e-mail: Hassiba.Benamara@UN.org

H. N. Psaraftis (ed.), *Sustainable Shipping*,
https://doi.org/10.1007/978-3-030-04330-8_1

DME	Dimethyl ether
DOALOS	Division for Ocean Affairs and the Law of the Sea
EU	European Union
IACS	International Association of Classification Societies
IAPH	International Association of Ports and Harbors
ICS	International Chamber of Shipping
INTERTANKO	International Association of Independent Tanker Owners
ITOPF	International Tanker Owners Pollution Federation
LNG	Liquefied natural gas
LPG	Liquefied petroleum gas
MEPC	IMO Marine Environment Protection Committee
NDCs	Nationally determined contributions
P&I Clubs	Protection and Indemnity Clubs
SDGs	Sustainable development goals
TEU	Twenty-foot equivalent unit
UNCTAD	United Nations Conference on Trade and Development
UNFCCC	United Nations Framework Convention on Climate Change

1 Introduction

In 2015, the world community strengthened its commitment to sustainable development and reaffirmed its pledge to global climate action. Two landmark agreements were adopted during the year: (i) the 2030 Agenda for Sustainable Development (ASD), under the framework of the United Nations Sustainable Development Summit, and (ii) the Paris Agreement on climate change (the Paris Agreement), adopted under the auspices of 21st Conference of the Parties (COP21) of the United Nations Framework Convention on Climate Change (UNFCCC).

Articulated around 17 sustainable development goals (SDGs) and 169 targets, the ASD sets a development path that integrates the economic, social and environmental dimensions of sustainability.[1] The agenda encompasses sectoral and cross-cutting priority areas that extend beyond basic poverty reduction, health and education aspirations. New areas of focus span among other issues, economic growth, decent jobs, cities and human settlements, energy and climate change. The parameters of the ASD were further defined by the Paris Agreement which established an action plan to limit global warming to below 2 °C above pre-industrial levels and to pursue efforts to limit the temperature increase even further to 1.5 °C.[2]

[1] See http://www.un.org/sustainabledevelopment/development-agenda.

[2] See http://unfccc.int/paris_agreement/items/9485.php.

Governments have the primary responsibility to follow-up and review progress made in implementing the ASD and the Paris Agreement. Under the voluntary national reviews' (VNRs) mechanism,[3] member states are encouraged to submit to the United Nations High-Level Political Forum (HLPF) their ASD implementation progress reports (paragraph 79). These national reviews are expected to serve as a basis for the regular reviews by the HLPF meetings under the auspices of ECOSOC (paragraph 84). Meanwhile, Article 4, Paragraph 2 of the Paris Agreement requires each party to prepare, communicate and maintain successive nationally determined contributions (NDCs) that it intends to achieve.[4] The NDCs indicate countries' levels of ambition and assess their efforts to curb national greenhouse gas (GHG) and adapt to the impacts of climate change.

No specific SDG was expressly and exclusively dedicated to the thematic area of transport. When elaborating the goals, the international community recognized that the cross-sectoral nature of transport and its role as a critical enabling factor of several SDGs were better served by integrating and mainstreaming transport considerations into a range of SDGs (see Box 1.1).

Box 1.1 Transport and the SDGs

Transport has been mainstreamed into many SDGs, including as follows:

Poverty (Goal 1); hunger, food security, nutrition and sustainable agriculture (Goal 2); health (Goal 3); access to affordable, reliable, sustainable and modern energy for all (Goal 7); sustained, inclusive and sustainable economic growth, full and productive employment and decent work for all (Goal 8); infrastructure resilience building, inclusive and sustainable industrialization and innovation (Goal 9); inequality within and among countries (Goal 10); inclusive, safe, resilient and sustainable cities and human settlements (Goal 11); sustainable consumption and production patterns (Goal 12); climate change and its impacts (Goal 13); conservation and sustainability of oceans, seas and marine resources for sustainable development (Goal 14); protection, restoration and promotion of sustainable use of terrestrial ecosystems, sustainable management of forests, desertification and land degradation and biodiversity loss (Goal 15); peace and inclusive societies for sustainable development, access to justice for all and the building of effective, accountable and inclusive institutions at all levels (Goal 16); and strengthening the means of implementation and revitalizing the global partnership for sustainable development (Goal 17).

---

[3] See https://sustainabledevelopment.un.org/vnrs.

[4] See http://unfccc.int/focus/items/10240.php.

Sustainable transport has long been recognized as a development objective as illustrated by relevant policy processes such as the 1992 Earth Summit,[5] the United Nations Conference on Sustainable Development (RIO+20),[6] UNCTAD Quadrennial Conferences (XIII and XIV),[7] the Third International Conference on Small Island Developing States (SIDS),[8] the Second United Nations Conference on the Landlocked Developing Countries (LLDCs),[9] the United Nations General Assembly Resolution on the "Role of transport and transit corridors in ensuring international cooperation, stability and sustainable development" (A/RES/69/213)[10] and the work carried out under the framework of the United Nations Secretary-General's High-Level Advisory Group on Sustainable Transport (United Nations 2016).

Maritime transport is core to the sustainability debate. An economic sector in its own right, the sector is often considered the lifeline to the rest of the world. Almost all countries, developed and developing alike, depend on maritime transport to link relevant supply chains, support international production processes, carry international trade and provide access to the global marketplace. In 2017, over 80% of international merchandise trade by volume and over two-thirds by value were carried by sea (UNCTAD 2018). In addition to enabling trade, maritime transport is a critical input that sustains the productivity of other sectors and industries such as marine equipment manufacturing, maritime auxiliary services (e.g. insurance, banking, brokering, classification and consultancy), fisheries, tourism, offshore energy sector, shipbuilding and ship demolition.

Given its strategic economic and social importance as well as its relative environmental-friendliness, when measured in tonne-miles, maritime transport could emerge as a sustainable development enabler. However, this is not a straightforward exercise as maritime transport could also erode some of its own benefits if unsustainable transport patterns are maintained. These patterns often result in external costs particularly in the form of pollution (marine, air), GHG emissions, infrastructure degradation, resource depletion and biodiversity loss, to name but a few. In view of their growth aspirations, developing countries have generally emphasized the economic dimensions of a sustainable development over the social and environmental aspects. In a maritime transport context, this translates into higher priority given to, for example, shipping connectivity, access to markets, participation in value chains, and infrastructure development as opposed to controlling air pollution or achieving gender balance.

In this context and against the backdrop of the ongoing implementation of the ASD and the Paris Agreement, there is a renewed opportunity to tap the

[5]See http://www.un.org/geninfo/bp/enviro.html.

[6]See https://sustainabledevelopment.un.org/rio20.html.

[7]See http://unctad.org/en/Pages/Meetings/UNCTAD-Conferences.aspx.

[8]See https://sustainabledevelopment.un.org/sids2014.

[9]See http://www.lldc2conference.org.

[10]See http://unctad.org/meetings/en/SessionalDocuments/ares69d213_en.pdf.

sustainability potential of the maritime transport sector. Leveraging this potential will help generate added economic, social and environmental value and effectively respond to the sustainability imperative.

Drawing largely upon UNCTAD's work in the field of sustainable transport[11] (UNCTAD 2015, 2017), this chapter highlights some relevant issues lying at the interface of maritime transport and sustainable development while emphasizing the role of the sector in promoting a development path centered around people, prosperity, environment and partnership-building (United Nations 2015).[12] The chapter focuses on maritime transport and its strong nexus to trade, growth and development.

Section 1 sets the scene and introduces issues at stake. Section 2 defines the concept of sustainable maritime transport and provides an overview of selected trends shaping the sustainability agenda in maritime transport. Section 3 highlights some of the challenges that could undermine shipping and ports sustainability. Section 4 considers examples of sustainability initiatives applied in maritime transport. It also identifies key players and stakeholders whose involvement is critical to enforcing sustainability criteria in the sector. Finally, Sect. 5 offers some concluding remarks.

2 Relevant Issues at the Interface of Maritime Transport and the Sustainability Imperative

2.1 Sustainable Maritime Transport: Defining the Concept

Promoting sustainability in transport entails striking the right balance between varied and potentially competing economic, social and environmental objectives. The aim is to generate more value while building on synergies and complementarities, ensuring optimal use of resources and promoting system coherence. The application of the sustainability concept may vary depending on the country, the stakeholder, the sector and the activity.

While not intended as an exhaustive list, sustainability in maritime transport entails, among other features, the ability to provide transportation infrastructure and services that are safe, socially inclusive, accessible, reliable, affordable, fuel-efficient, environmentally friendly, low-carbon and resilient to shocks and disruptions including those caused by climate change and natural disasters.[13]

[11] See http://unctad.org/TLB.

[12] See also http://unctad.org/en/Pages/DTL/TTL/Infrastructure-and-Services/Sustainable-Transport.aspx and http://unctad.org/en/Pages/DTL/TTL/Infrastructure-and-Services/SFTF-Toolkit. aspx.

[13] For additional information on UNCTAD's work on Maritime Transport and Climate Change Impacts and Adaptation, see also http://unctad.org/en/Pages/DTL/TTL/Legal/Climate-Change-and-Maritime-Transport.aspx.

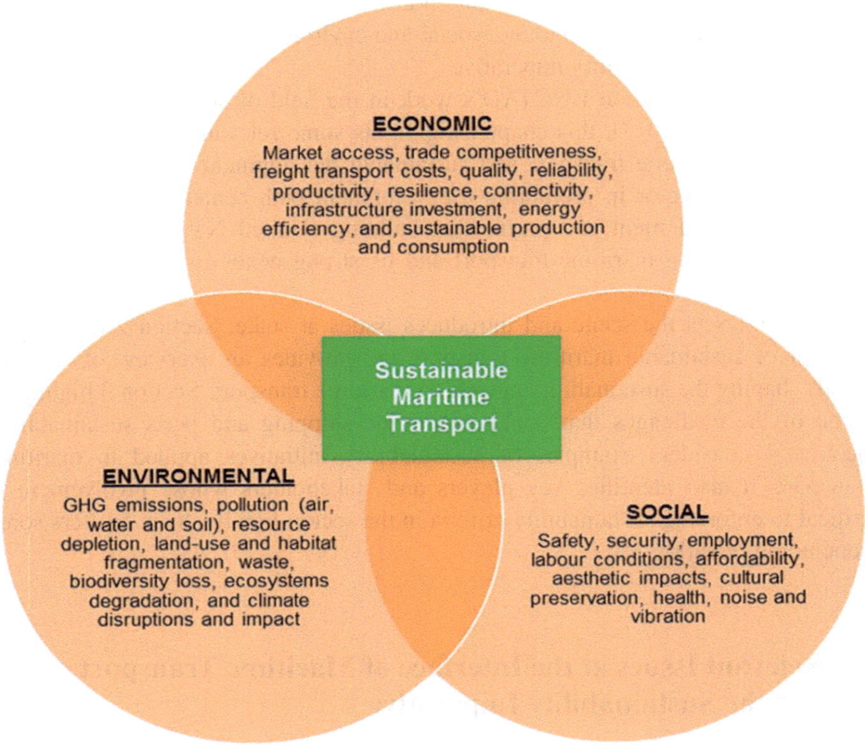

Fig. 1.1 Defining sustainable maritime transport. (Source: UNCTAD 2015)

Figure 1.1 illustrates the intersection between the economic, social and environmental dimensions of sustainable maritime transport. Economic parameters may include, among others, market access, connectivity, infrastructure capacity, trade competitiveness and transport costs. As to the social dimension, relevant considerations include, for example, safety, health, security, employment and working conditions. Common environmental concerns relate to, inter alia, air emissions (pollutants and GHGs), waste control, spills and pollution (e.g. oil and other substances), climate change impacts, biodiversity loss as well as resource and energy depletion.

2.2 *Key Trends Shaping the Sustainability Agenda in Maritime Transport*

The ability of the maritime transport sector to effectively deliver on the sustainability imperative is heavily influenced by developments shaping the sector's operational and regulatory framework. Some trends unfolding over recent years relate in

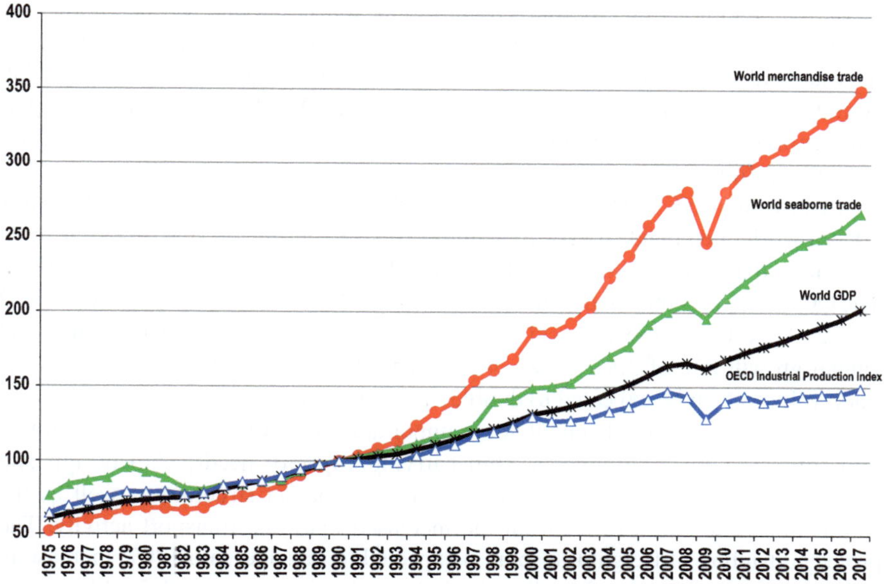

Fig. 1.2 Organization for Economic Co-operation and Development index of industrial production and world indices: gross domestic product, merchandise trade and seaborne shipments, 1975–2017 (1990 = 100) (Sources: UNCTAD secretariat calculations, based on data from OECD 2018; UNCTAD Review of Maritime Transport, various issues; World Trade Organization, appendix tables, table A1a; WTO press release 820, 12 April 2018)

particular to the slow-moving global economic and trade growth; developments in energy markets and bunker fuel consumption patterns; the volatility of energy prices and the sharp decline in oil price levels since mid-2014; the oversupply of ship carrying capacity and related implications for freight markets, profitability, investments, market structure and ports; the emergence of new technologies, automation and big data; piracy and security concerns; and vulnerability to external shocks and natural disasters. Depending on how these trends further evolve, their impact could potentially be significant and affect the ability of the maritime transport sector to enhance its sustainability.

2.2.1 Economic Growth and Demand for Maritime Transport

Demand for maritime transport grows in tandem with the growing world population, consumption needs, industrial activity, urbanization, trade and economic growth. The close statistical association between some of these variables has long been established (UNCTAD 2016). Figure 1.2 shows the historical positive correlation that prevailed between gross domestic product (GDP), industrial production, merchandise trade and seaborne cargo shipments.

In its baseline scenario, the Organization for Economic Co-operation and Development (OECD) projects that the total freight transport demand (domestic and international as measured in billion tonne-kilometers) will triple over the 2015–2050 period, primarily driven by economic growth (OECD 2017c). Maritime transport will continue to account for the largest share of demand with a contribution of 75% in 2050, up from 71% in 2015.

Meanwhile, UNCTAD estimated world seaborne trade to have accounted for over 80% of world merchandise trade by volume in 2017. During the year, seaborne trade volumes expanded by 4%, taking the total to 10.7 billion tonnes (UNCTAD 2018). The OCED expects this modal share to remain steady at around 80% in 2050 (OECD 2017c). UNCTAD projects world seaborne trade volumes to expand at a compound annual growth rate of 3.8% between 2018 and 2023. At this rate, world seaborne trade volumes can be expected to double in about two decades.

Combined, these trends entail some important implications for maritime transport infrastructure requirements, ship carrying capacity needs, ship design and technology, port developments and performance, market structure as well as the externalities that could result from the increased maritime transport activity. The projected tripling of freight transport is expected to translate into 120% increase in freight transport-related carbon dioxide (CO_2) emissions over the 2015–2050 period (OECD 2017c).

2.2.2 Shift in the Geography of Economic Influence and Trade

The incremental shift of economic influence and geography of trade observed over recent years and whereby developing countries have been contributing larger shares to global economic growth and trade, can influence the sustainability-building efforts in maritime transport. In 2017, a total of 60% of world seaborne trade volumes originated in developing countries, and 63% of this trade were delivered on their territories (UNCTAD 2018). Developing countries have become active players both as exporters and importers and are no longer only a source of supply for raw materials and fossil fuel energy. They contribute significantly to relevant globalized manufacturing processes and constitute a growing source of consumption import demand, including of raw materials, such as oil. In terms of geographical influence, Asia remains the main cargo loading and unloading area globally (UNCTAD 2018). These trends are further reflected in port activity shares as shown in Fig. 1.3. In 2017, world container port throughput was estimated at over 750 million 20-foot equivalent units (TEUs), with Asia accounting for the lion's share (64%). Together, these trends are generating new patterns of geographical distribution of production and consumption centers (e.g. changes in distances traveled by cargoes) and altering cargo flows and direction. These entail some implications for maritime transport, including for shipping networks and configuration, fuel consumption, transport costs, ship emissions and climate change.

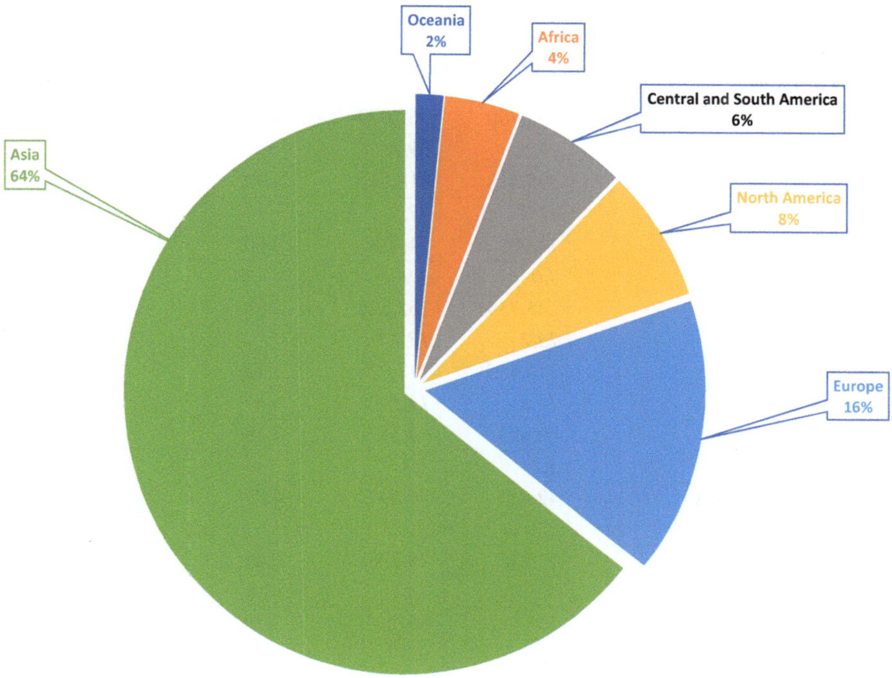

Fig. 1.3 World container port volumes by region, 2017 (Percentage shares). (Source: UNCTAD secretariat, Review of Maritime Transport 2018)

2.2.3 Ship Supply Capacity and Market Structure

On 1 January 2018, the world commercial fleet consisted of 94,171 vessels, with a combined tonnage of 1.92 billion dwt. After 5 years of decelerating growth, 2017 saw a slight upturn in the growth rate. The deadweight tonnage of the commercial shipping fleet grew by 3.31% in the 12 months leading up to 1 January 2018, up from 3.15% recorded in the previous year leading up to 1 January 2017 (Table 1.1).

The container shipping segment, which carries around 16% of world trade by volume and over half of its value, has been particularly affected by developments in the world fleet capacity and size (UNCTAD 2018). In an oversupplied market characterized by mega containerships and overall weak growth in global demand, the container shipping industry has turned to consolidation and rationalization to optimize capacity utilization and reduce costs. Over the 2016–2017 period, the industry intensified its consolidation efforts which have altered the overall liner shipping dynamics and forces. The arrival of megaships intensified the consolidation activity (e.g. mergers and acquisitions) and the formation of new and larger shipping alliances.

These developments have some important implications for ship deployment, market access and connectivity through changes to the number and frequency

Table 1.1 World fleet by principal vessel type, 2017–2018 (thousands of deadweight tonnes and percentage shares)

Principal types	2017	2018	Percentage change 2018/2017
Oil tankers	535,700	561,079	4.74
	28.8	*29.2*	
Bulk carriers	795,518	818,612	2.90
	42.7	*42.5*	
General cargo ships	74,908	74,458	−0.60
	4.0	*3.9*	
Container ships	245,759	252,825	2.88
	13.2	*13.1*	
Other	210,455	217,028	3.12
	11.3	*11.3*	
Gas carriers	60,003	64,317	7.19
	3.2	*3.3*	
Chemical tankers	42,853	44,597	4.07
	2.3	*2.3*	
Offshore	77,845	78,228	0.49
	4.2	*4.1*	
Ferries and passenger ships	5944	6075	2.20
	0.3	*0.3*	
Other/not available	23,810	23,811	0.01
	1.3	*1.2*	
World total	**1,862,340**	**1,924,002**	**3.31**
	100.0	*100.0*	

Source: UNCTAD secretariat calculations, based on data from Clarksons Research Services
Notes: Propelled seagoing merchant vessels of 100 gross tonnes and above, beginning-of-year figures and percentage share in italics

of services, routes, trade networks, competition levels as well as freight rates, earnings and profitability. They could lead to market power concentration and push the alliances to reduce port calls and service offerings to enhance network efficiency. For example, alliances may decide to focus on reducing transit times and enhance reliability at the expense of service frequency. Additionally, shippers could face higher prices for shipping services, which in turn can undermine their competitiveness in the global marketplace. They may also be required to redefine their supply chains because of changes or reductions in port calls. Depending on how these trends continue to unfold, they clearly have the potential to significantly shape the sustainability agenda of maritime transport, especially in terms of the economic dimension of sustainability (e.g. costs, financial solvability, competitive advantages, etc.). Chapters 8 and 9 of this book look at tramp and liner shipping (respectively) from an optimization/supply chain/sustainability perspective.

2.2.4 Megaships, Shipping Services and Ports

The deployment of megaships affects port terminals across the ship-port interface, the yard and terminal activities, as well as the gate and hinterland operations. As maritime access may be limited by draft restrictions, larger container ships will normally call at fewer ports. The physical features of such ships and handling requirements add pressure to berth and crane operations. Larger ship calls are often associated with lower service frequency and periods of peak volumes at port terminals. Less frequent calls with greater cargo volumes being handled per call create surges and pressure on yard operations, given the ensuing peaks. As more equipment is required to move containers to and from stacking areas, additional equipment and labor are necessary. Pressure is also imposed on the restacking of containers through increased requirements for gantry cranes of shipyards and stacking density. For specialized cargo such as refrigerated goods, larger port call volumes exert pressure on the usage of reefer slots. Sharp increases in cargo volumes also create greater demands on gate access, with more trucks arriving and leaving with larger numbers of containers. This creates local congestion as more trucks are waiting to enter the port.

Thus, large container ships provide economies of scale at sea, but these economies do not necessarily extend to ports. As ships and alliances become larger, the number of ports and terminals that can accommodate their ship calls becomes limited. Ports will be increasingly required to increase productivity. Enhancing port efficiency and reducing port dwell time will become even more important to cut costs and enhance trade competitiveness.

Ports sustainability objectives will be defined by these additional pressures. Port operators and administrators and port community stakeholders will need to improve their performance across the economic, social and environmental dimensions of sustainability. Measuring performance and setting up indicators with multidimensional metrics spanning a range of factors (e.g. efficiency, cost-effectiveness, productivity, profitability, connectivity, access, social inclusiveness and environmental sustainability) are increasingly important for ports and their users and customers.[14]

Relevant initiatives seeking to advance the work on port performance measurement include the Portopia project, which brings together an international consortium of academic, research and industrial partners with experience in port performance management.[15] Portopia aims to support the European port industry with performance data to inform policy formulation and monitor implementation. Another example is the work carried out under the Joint Working Group 174 on Sustainability Reporting for Ports of the International Association of Ports and

[14]For additional information see the proceedings of UNCTAD's "Ad Hoc Expert Meeting on Measuring Shipping Connectivity and Performance: The Need for Statistics and Data". 15 May 2017 http://unctad.org/en/pages/MeetingDetails.aspx?meetingid=1364.

[15]See http://www.portopia.eu.

Harbors (IAPH) and the World Association for Waterborne Transport Infrastructure (PIANC).[16] A key goal of this working group is to develop guidance relating to sustainability reporting for ports. Issues pertaining to green ports are discussed in Chap. 12 of this book.

3 Challenges to Sustainable Maritime Transport: An Overview

The following section highlights, in overview, some of the more pressing or topical issues that are arising as important in the current context of growing international momentum relating to the ASD, global climate action and ocean governance.[17]

3.1 Energy Consumption and Heavy Reliance on Oil for Propulsion

More than 50% of global oil demand today is concentrated in the transport sector. According to the US Energy Information Administration (US EIA), global transport accounted for around 55% of total end-use sector liquid fuels in 2015 (EIA 2017). During the same year, the share of refined petroleum and other liquid fuels in the transportation sector amounted to 95%. Projections for 2040 indicate that over half of the increase in freight transport energy use can be attributed to shipping, reflecting developing countries' growing demand for goods and services and further integration of local producers into global supply chains.

Energy demand from international shipping increased at an annual rate of 1.6% from 2000 to 2014. In 2012, shipping was estimated to have used 300 million tonnes of bunker fuel per year with international shipping accounting for 86% of this total. Main fuel consumers were, respectively, container ships, bulk carriers and oil tanker (IMO 2014). While some decoupling between maritime transport activity and marine bunker fuel consumption may have been observed over the past few years, this trend does not necessarily reflect energy efficiency improvements or reduced dependence on oil. Instead, it probably reflects factors such as the constrained activity that followed the 2009 downturn, the upgrading of the global container fleet to larger and more efficient ships, the scrapping of older ships that were inefficient and the slow steaming, which helped absorb excess capacity.

Energy consumption and heavy reliance on oil for propulsion are real challenges to sustainable maritime transportation. Marine bunker fuels have a high carbon

[16] See https://www.pianc.org.

[17] See relevant work by UN-Oceans. http://www.unoceans.org/activities/en.

intensity and are very polluting. At the same time, the sector is not yet in a position to fully switch to alternative fuels or widely adopt energy efficiency technologies. Apart from pollution and carbon emissions, heavy reliance on oil could undermine energy conservation objectives, challenge affordability of maritime transport services due to oil price volatility and exacerbate environmental degradation and climate change. Consequently, reducing the overdependence of the sector on fossil fuels is a necessary condition to achieving greater sustainability in maritime transport.

3.2 Infrastructure Needs, Access and Connectivity

Inadequate and poor conditions of maritime transport infrastructure can undermine the sector's role as an engine of trade, a driver of global economic integration and a sustainable development enabler. Poor and insufficiently maintained port infrastructure, inadequate cargo handling equipment, limited and constrained physical access to ports and inadequate hinterland connections often combine and result in increased costs, extended delays, reduced reliability of service, limited access and low shipping connectivity. These factors are contributing to marginalizing countries and preventing their effective participation in relevant transport networks and supply chains.

Shipping connectivity is crucial for countries' economic sustainability. Shipping connectivity is even more important for sea-locked countries such as SIDS where maritime transport connections are the lifelines linking local populations and economies to the regional and global market place and providing access to resources and services (UNCTAD 2014).

Closing the maritime transport infrastructure gap is therefore a precondition to sustainable shipping as is a well-articulated transport infrastructure vision that builds on a careful coordination of the social, economic and physical development of maritime transport systems. However, to close the gap on the infrastructure deficit in developing countries, including in transportation, spending must reach \$1.8 trillion–\$2.3 trillion per year by 2020 compared with the current levels of \$0.8 trillion–\$0.9 trillion a year (UNDP 2013). Given the magnitude of the required investments and in view of the long-life cycle of maritime transport infrastructure, not accounting for the longer-term sustainability requirements at the inception phase may involve costly retrofitting of infrastructure and equipment and adjustment of operations and services. Maritime transport infrastructure developers, investors and managers should mainstream sustainability criteria into their infrastructure development plans at the early stages of relevant decision-making and investment processes.

3.3 Affordability and Transport Costs

Sustainable maritime transportation entails affordable and reasonably priced shipping and port services which, at the same time, generate value for the service provider. This requires effective control of factors driving maritime transport costs, including infrastructure, trade (volumes, economies of scale, directional imbalances), competition, type of products shipped and the position within relevant shipping networks (e.g. center/periphery, hub/feeder ports/services). With maritime transportation being heavily dependent on oil for propulsion, oil price volatility and developments in the energy markets are also relevant for maritime transport costs and can influence efforts to achieve sustainable maritime transportation.

The sector's oil dependency enhances the exposure of freight rates and transport costs to oil price volatility. While since mid-2014, oil prices have generally trended downward, there remain concerns that any rise in oil prices as experienced in the first half of 2018 combined with the continued reliance on oil for propulsion may drive up rates and costs. As developing countries are already facing disproportionately higher transport costs, the negative implications of rising fuel costs for their sustainable development cannot overemphasized.

On average, geographically disadvantaged countries, namely, LLDCs and SIDS, face relatively higher transport costs than other economic groupings. In 2016, average transport costs represented about 17% of the value of imports for developing countries, 19% for LLDCs and almost 22% for SIDS, compared with a world average of 15% (see Fig. 1.4). Commercially unaffordable and therefore, unsustainable, maritime transport services may require governments to intervene and subsidize certain services which may impose further pressure on already constrained national budgets, especially in vulnerable economies such as SIDS.

In this context, greater sustainability in maritime transport requires as a matter of priority that the determinants of maritime transport costs be better understood and controlled and the overdependence on oil-based propulsion systems be effectively reduced. Limiting exposure to volatile oil/fuel prices and costs through investment in energy efficiency measures and alternative energy sources as well as the adoption of more sustainable operational and management practices can help ensure that fuel and transport costs are effectively controlled and also derive efficiency gains to enable improved access to markets and promote trade competitiveness.

Although lower oil prices may be a welcome development from the perspective of trade and shippers, they could nevertheless undermine other sustainability measures. Lower oil price levels could reverse some existing trends introduced over recent years and undermine the sustainability gains generated by measures such as slow steaming, investment in more efficient ships and equipment, speed management and the selection of economical routing options. Lower oil price levels may also prompt some operators to sail at faster speeds and therefore increase air emissions as well as create excess tonnage capacity at a time when the shipping market remains generally oversupplied.

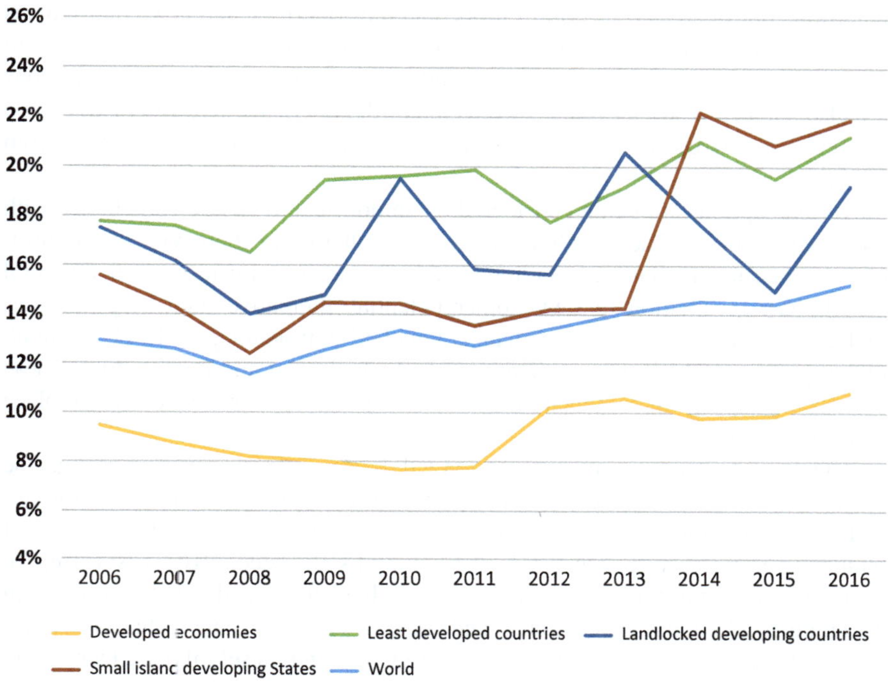

Fig. 1.4 Transport and insurance costs of international trade (2006–2016). Percentage share of value of imports. (Source: UNCTAD secretariat calculations. Note: All modes of transport)

3.4 Air Pollution

The emission of air pollutants from maritime transport including emissions of sulfur oxide (SOx) and nitrogen oxide (NOx) is closely correlated with the use of dirty heavy bunker fuels. International ship emissions of NOx and SOx were estimated at about 13% and 12% of global NOx and SOx total, respectively, over the 2007–2012 period (IMO 2014). One large container ship visiting a port is estimated to produce the same amount of NOx as 12,500 cars (over the same period) (McKinnon 2016). These emissions are causing serious problems to human health and can cause death. In 2010, China saw an estimated 1.2 million premature deaths caused by ambient air pollution (McKinnon 2016). Locking-in fossil fuels and related technologies in maritime transport systems will perpetuate unsustainable transport patterns and undermine efforts to reduce the sector's over dependence on fossil fuels, control its air pollution and maintain its carbon emissions at manageable levels.

Air pollution from shipping is regulated by the International Maritime Organization (IMO) through the International Convention for the Prevention of Pollution from Ships, 1973, as modified by the Protocol 1978 (MARPOL). Airborne emissions from ships are specifically regulated under Annex VI of MARPOL which entered into force in 2005. Stricter limits on SOx emissions are already in place in emission control areas (ECAs) in Europe and the Americas, and new control areas are being established in port areas in China. By 2020, shipping will be required to meet the global sulfur cap of 0.5%.[18] In addition to SOx emissions, NOx emissions are also subject to stricter limits such as in North America where NOx ECA came into effect in 2016. The North European NOx ECA will apply to ships built from 2021.[19]

Various options are being considered by shipowners to ensure compliance with these requirements. These include switching to alternative fuels with lower sulfur content (marine distillates), installing scrubber systems and using liquefied natural gas (LNG). In addition to LNG, other potential alternative fuel/energy sources include electricity, biodiesel, methanol, liquefied petroleum gas (LPG), ethanol, dimethyl ether (DME), biogas, synthetic fuels, hydrogen and nuclear fuel. These options can be used in combination with conventional bunker fuels or fully replace conventional fuels (DNV GL 2017).

The decision to adopt any given option will be determined by factors such as the availability of the fuels and their costs, uncertainty relating to other technologies and their maturation levels as well as the investment requirements in terms of bunkering infrastructure (DNV GL 2017; Lloyd's Register 2016). LNG use, for example, will require important investments in bunkering facilities. Scrubbers will also require additional expenditures, and some of the underlying technology is yet to be tested and proven. Distillates (e.g. marine gas and diesel oils) are technically feasible. However, an increased demand for distillates will likely widen the cost differential with the conventional bunker fuels. The adoption of some alternative fuels can also be challenged by the physical features of the fuels (e.g. levels of flashpoints and toxicity). The International Code of Safety for Ships using Gases or other Low-flashpoint Fuels (IGF Code) could potentially help address some of these challenges. Chapters 2 and 4 of this book look into the subject of technologies for greener shipping, including ICT, and a discussion on alternative fuels is provided in Chap. 13.

Energy efficiency is also important for the realization of the twin objective of reducing oil dependency and harmful air emissions. The International Energy Agency (IEA) considers energy efficiency as the world's "first fuel" and estimates global investment in energy efficiency to have increased by 9% to USD 231 billion in 2016 (IEA 2017). In addition to the environmental benefits, energy efficiency promotes energy security. Efficiency improvements since 2000 avoided additional spending on energy imports in many countries (OECD 2017b). Looking at 22 potential ship efficiency measures and calculating their aggregated cost-effectiveness and reduction potential, one study finds that, by 2020, the industry's growing fleet could reduce annual CO_2 emissions by 33% of the projected annual total (ICCT 2011).

[18]See http://www.imo.org/en/OurWork/Environment/PollutionPrevention/AirPollution/Pages/Air-Pollution.aspx.

[19]See http://ec.europa.eu/environment/air/sources/maritime.htm.

Another study investigated 28 energy-saving options and estimated a reduction of CO_2 emissions of more than 50% in shipping by 2030 (Alvik et al. 2010).

Key regulatory measures promoting maritime transport energy efficiency are the IMO's Energy Efficiency Design Index (EEDI), Energy Efficiency Operational Indicator (EEOI) and Ship Energy Efficiency Management Plan (SEEMP), in force since 2013 (IMO 2017). By addressing energy efficiency, these regulations help reduce maritime transport energy intensity as well as air emissions. The EEDI aims to increase the energy efficiency of new ships over time. Intended as a nonprescriptive, performance-based mechanism, ship designers and builders can select the most cost-efficient solutions in complying with the regulations (IMO 2017). It is complemented by the EEOI and the SEEMP dealing the operational efficiency of ships. Chapter 3 of this book provides an in-depth discussion on EEDI.

To support this work, the European Commission (EC) recently contributed €10 million funding to an EC-IMO energy efficiency project that aims to establish Maritime Technology Cooperation Centres in five regions: Africa, Asia, the Caribbean, Latin America and the Pacific.[20] The centers will promote the uptake of energy-efficient and low-carbon technologies and operations in maritime transport in the less developed countries of the regions identified.

Sustainability gains generated by energy-efficient ships and clean fuels underscore the need to support the development and implementation of energy-efficient shipping systems, including by scaling up existing finance levels, diversifying sources of finance and addressing barriers to investments such as the split incentives involving shipowners and charterers. On the positive side, banks are increasingly accounting for sustainability criteria and ship energy efficiency performances, when making financing decisions. With energy-efficient ships being more likely to have higher asset values and a longer lifespan, banks are reported to be increasingly favoring investments in sustainable ships such as "eco-ships". The latter offer low-risk financing opportunities given the associated improved chartering potential and lower fuel costs (The Marine Professional 2015).

3.5 *Greenhouse Gas Emissions (GHGs)*

Transport including all modes and both passenger and freight was the second largest CO_2 emitter in 2015, accounting for 24% of the world CO_2 emissions from fuel combustion. Emissions from marine bunkers expanded rapidly over the 1990–2015 period, at a rate of 77% and faster than road transport (see Fig. 1.5) (OECD 2017a). IMO estimates total shipping emissions to have reached approximately 938 million tonnes CO_2 in 2012 (IMO 2014). International shipping emissions accounted for 85% of this total with 796 million tonnes CO_2. This value represents

[20] See http://www.imo.org/en/MediaCentre/PressBriefings/Pages/01-2016-MTCC-.aspx (accessed on 28 April 2018).

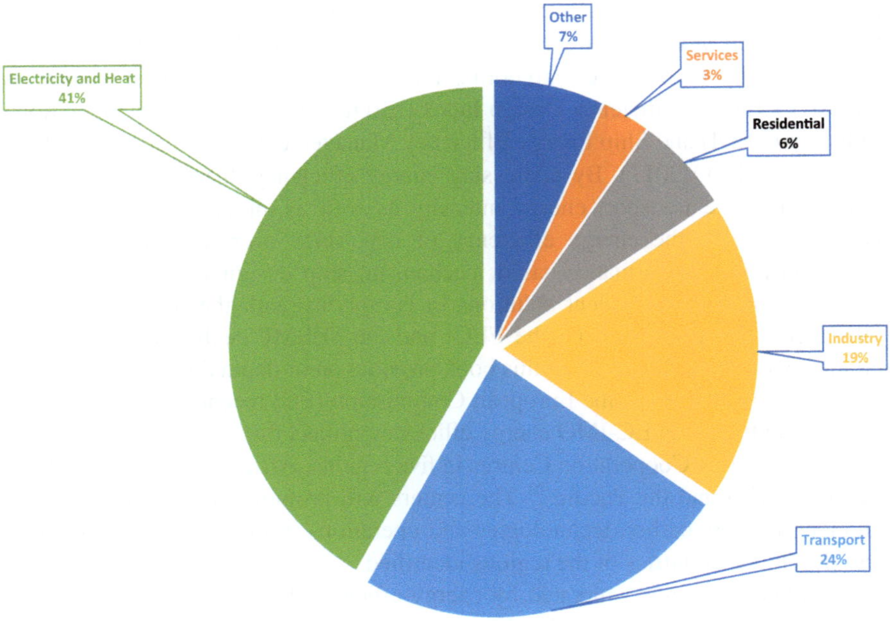

Fig. 1.5 World CO2 emissions from fuel combustion by sector, 2015. (Source: IEA 2017). CO2 Emissions from Fuel Combustion: Overview (2017 edition), Paris)

approximately 2.2% of global total CO_2. While shipping is considered one of the most efficient modes of transport in terms of CO_2 emissions per cargo carried and distance traveled, without any mitigation action, carbon emissions from the sector are expected to rise.

Forecast scenarios for the medium term suggest that international carbon emissions could increase by 50–250% by 2050, depending on economic growth and global energy demand (IMO 2014). Tapping the transport sector's mitigation potential is required to achieve the 2 °C limit and even more so to reach the 1.5 DC aspiration. Interestingly and despite this urgency, freight transport of which maritime transport is a dominant component and which contributes some 40% of global CO_2 emissions, is mentioned in only 29% of the UNFCCCs' NDCs that propose transport measures (Gota et al. 2016). Moreover, a limited number of VNR submitted by countries to the 2016 and 2017 HLPF to review SDGs include measures on transport, including freight transport.

Article 2.2 of the Kyoto Protocol states that the parties included in Annex I shall pursue limitation or reduction of emissions of GHGs not controlled by the Montreal Protocol from marine bunker fuels, working through the IMO.[21] While the 2015 Paris Agreement made no express reference to IMO, some progress has

[21] For additional information, see http://unfccc.int/methods/emissions_from_intl_transport/items/1057.php (accessed 28 April 2018).

been achieved. In 2016, IMO adopted a mandatory data collection system for fuel consumption of ship and approved a roadmap for developing a comprehensive IMO strategy on the reduction of GHG emissions from ships (IMO 2016).

In April 2018, the IMO Marine Environment Protection Committee (MEPC) adopted an initial strategy on the reduction of GHG emissions from ships (IMO 2018). This strategy, which represents the first global climate framework for shipping, includes quantitative GHG reduction targets through 2050 and a list of candidate short-, medium- and long-term policy measures to help achieve these targets. Among other targets, the strategy sets out to reduce CO_2 emissions per transport work, as an average across international shipping, by at least 50% by 2050, compared to 2008, while at the same time, pursuing efforts toward phasing them out entirely.[22] The strategy envisages potential market-based measures (MBMs) as potential midterm measures to be agreed and decided upon between 2023 and 2030. The initial strategy is intended as a stepping stone to the final IMO's GHG strategy expected by 2023 (Psaraftis 2018).

Meanwhile, apart from technical and operational measures focusing on improving energy efficiency to reduce air emissions, currently the focus, including at the European Union (EU) level, seems to be on monitoring, reporting and verifying fuel consumptions and emissions. Although the international community, under the auspices of IMO/UNFCCC, is considering several MBMs such as levies/taxes and emission trading mechanisms, at present, an agreement on any international market-based instrument to regulate carbon emissions from international shipping is yet to be achieved. Proposals made tend to focus on incentivizing shipping companies to reduce carbon through operational changes and/or accelerate the adoption of new, more carbon-efficient vessels. Others prefer the introduction of a carbon tax on shipping, including in the form of a levy on bunker fuel with revenues generated being potentially used to fund research, and/or early adoption of new low-carbon technologies, pay for carbon offsets or contribute to the Green Climate Fund established under the UNFFCCC framework (GSF 2015). These considerations, including the potential implications of MBMs for transport costs, trade and production, are part of the international policy agenda. More on MBMs to reduce GHG emissions can be found in Chap. 11 of this book.

3.6 Resilience: Adapting to Climate Change Impacts and Enhancing Resilience

While curbing GHG emissions remains an urgent imperative to ensure manageable global warming levels, the effects of climate variability and change, irrespective of its cause, are already being felt in different parts of the world, often in the poorest countries with low adaptive capacity. Transport networks and seaports in

[22]See http://www.imo.org/en/mediacentre/pressbriefings/pages/06ghginitialstrategy.aspx.

particular are likely to be highly affected by climate change factors given port location and vulnerability (Asariotis et al. 2017). Climatic factors such as rising water levels, floods, storms, precipitation, extreme weather events and associated risks such as coastal erosion, inundation and deterioration of hinterland connections have implications for shipping volumes and costs, cargo loading and capacity, sailing and/or loading schedules, storage and warehousing. With international trade being increasingly multimodal and requiring the use of rail, road and waterway transport, these impacts will also affect transport corridors above and beyond the ports acting as gateways. Mapping out the vulnerabilities the maritime transport sector; conducting risk assessments, especially in ports; and identifying adequate adaptation measures to build the climate resilience of maritime transport systems are therefore necessary sustainability pre-conditions.[23]

3.7 *Ship Recycling*

A large share of world ship recycling activity is taking place in South and East Asia under unsustainable conditions that pose serious risks to human health, environment and society and populations, including of children working in ship breaking activities. South Asia accounted for 77% of ship scrapping in 2017. Turkey maintained a market niche for scrapping some gas carriers, as well as some ferries and passenger ships (see Table 1.2).

The human and environmental dangers generated by unsustainable patterns of ship recycling relate to the hazardous and oily materials (e.g. asbestos, polychlorinated biphenyls, oils, oil sludge) contained in many of the old ships being scrapped. To address this problem, the IMO adopted the 2009 Hong Kong International Convention for the Safe and Environmental Sound Recycling of Ships (HKC). The convention seeks to ensure that the design, construction, operation and preparation of ships enable their sustainable recycling and safeguard the safety, the environment, and the operational efficiency of ships. An enforcement mechanism for ship recycling, incorporating certification and reporting requirements is provided for under the convention. At the EU level, a specific regulation on ship recycling (Regulation 1257/2013) sets out the requirements for EU-flagged ships. These requirements are expected to come into force no later than the end of 2018 (European Commission 2017). More on ship recycling can be found in Chap. 6 of this book.

[23]For additional information about the science of climate change and the impacts of climate change on transport, including coastal transport infrastructure, see, for example, relevant documentation about UNCTAD's work carried out in the field and available for downloading at http://unctad.org/en/Pages/DTL/TTL/Legal/Climate-Change-and-Maritime-Transport.aspx (accessed 7 December 2018).

Table 1.2 Reported tonnage sold for demolition, major vessel types and countries where demolished, 2017 (thousands of gross tonnes)

	China	India	Bangladesh	Pakistan	Unknown Indian subcontinent	Turkey	Others/ unknown	World total
Oil tankers	1	1935	3245	0	749	12	40	5982
Bulk carriers	2464	1062	1460	2527	470	139	0	8123
General cargo	82	420	155	102	0	312	108	1178
Containerships	650	1755	892	748	140	309	3	4498
Gas carriers	4	145	59	0	0	173	5	387
Chemical tankers	2	109	35	0	44	0	6	196
Offshore	90	318	57	77	157	128	404	1230
Ferries and passenger ships	0	165	35	5	0	51	21	277
Other	152	415	321	0	0	133	23	1044
Total	**3445**	**6323**	**6260**	**3459**	**1560**	**1257**	**611**	**22,916**

Source: UNCTAD secretariat estimates, based on data from Clarksons Research Services
Notes: Propelled seagoing merchant vessels of 100 gross tonnes and above

3.8 *Waste Discharge by Ships*

Ships generate various types of wastes (oily wastes, drainage from the bilges, sewage and garbage, cargo residues). Damping these wastes in a marine environment results in negative impacts spanning chemical pollution and nondegradable waste affecting marine life as well as degradation to the natural and economic value of the coastal areas. These concerns have been on the agenda of the international community for many years as reflected in the MARPOL Convention. Relevant obligations are mirrored at the EU level, for example, with Directive 2000/59/EC on port reception facilities for ship-generated waste and cargo residues (European Commission 2000). The directive requires that member states implement a cost recovery system to cover the costs of planning for collecting and disposal of waste.

3.9 *Ballast Water*

The transfer of harmful aquatic organisms and pathogens between marine ecosystems through ships' ballast water and sediments is a huge environmental challenge as it can significantly damage coastal and marine environments and ecosystems. Shipping is a major channel introducing invasive aquatic species when ballast water is discharged without treatment. Consequently, the IMO adopted the International Convention for the Control and Management of Ships' Ballast Water and Sediments

(BWM) which came into force in September 2017. The convention requires ships to have a ballast water treatment system to remove alien species and invasive organisms from the ballast water.

The Global Environment Facility (GEF), the United Nations Development Programme (UNDP) and the IMO are seeking further solutions by identifying ship hull fouling as another priority action area. Building on the approach and experience of GloBallast,[24] a new project entitled GloFouling,[25] has recently been approved to assist developing countries and engage private sector actors in reducing the transfer of invasive species by way of "biofouling" on ship hulls and other mobile marine infrastructure.[26] More on ballast water can be found in Chap. 2 of this book.

3.10 Ship-Source Oil Pollution

As about half of global crude oil production is carried by sea, the significant pollution risks associated with an oil spill cannot be overemphasized. Ship-source oil pollution may arise from accidents (collisions, groundings) and during ordinary cargo operations. Consequently, and following several important oil spills from ships, the IMO focused its environmental work on regulating the prevention of marine pollution by oil. This work has led to the adoption of the International Convention for the Prevention of Pollution from Ships (MARPOL) in 1973, which was later amended to include additional measures to prevent marine pollution and to cover pollution from other types of materials such as chemicals. In addition to prevention efforts, other instruments have contributed to an effective oil pollution regulatory regime including the International Convention on Oil Pollution Preparedness, Response and Co-operation (OPRC), the 1969 International Convention Relating to Intervention on the High Seas in Cases of Oil Pollution Casualties and the 1973 Protocol.

Mandatory standards of liability and compensation for victims have also been adopted under the IMO framework. Relevant conventions include the 1969 and 1992 Civil Liability Conventions, the 1992 Fund Convention and the 2003 Supplementary Fund Protocol (CLC-Fund regime).[27] In assessing the CLC-Fund regime, UNCTAD finds that a considerable number of coastal states, including developing countries that are potentially exposed to ship-source oil pollution incidents, were not yet contracting parties to the latest legal instruments in the field and, as a result, would not be benefitting from significant compensation in the event of a major oil spill

[24] See http://archive.iwlearn.net/globallast.imo.org (accessed 7 December 2018).

[25] See https://www.thegef.org/project/building-partnerships-assist-developing-countries-minimize-impacts-aquatic-biofouling (accessed 7 December 2018).

[26] See http://www.imo.org/en/MediaCentre/PressBriefings/Pages/21-BWM-EIF.aspx (accessed 7 December 2018).

[27] See http://www.iopcfunds.org/about-us (accessed 7 December).

affecting their coasts or other areas under their marine jurisdiction (UNCTAD 2012). Adopting and implementing relevant oil pollution legal instruments are important policy considerations, especially in many coastal and small island developing countries.

Overall and while oil trade by sea has grown over the years, the number of oil spills from tankers had dropped reflecting in large part the effectiveness of the international oil pollution regulatory framework under the IMO. However, concerns remain about other types of pollution, including from spills of hazardous and noxious substances. Continued efforts to address these concerns are reflected in the 1996 Hazardous and Noxious Substances Convention (HNS Convention and the 2010 HNS Protocol). More on ship-sourced marine pollution can be found in Chap. 5 of this book.

4 Selected Maritime Transport Sustainability Initiatives and Key Players

The sustainability concept is bound to take further hold and become a strategic and critical consideration for key players involved in the maritime transport sector, including transport policymakers and regulators, port administrations, operators and terminals, as well as shipping companies. This is reflected not only in the number and scope of regulatory measures adopted over the years at country, regional and international level such as under the auspices of the IMO and at the EU level, but also in the widespread use of voluntary self-regulation by shipping and ports. Self-regulation allows for greater participation of all stakeholders while allowing for some flexibility through the choice of approaches and solutions adopted. Table 1.3 presents selected examples of intervention measures that could be applied in the maritime transport sector irrespective of whether these are mandated by governmental authorities and regulation or self-imposed by industry.

4.1 Examples of Government-/Country-Led Initiatives

Governments can improve sustainability in maritime transport through a "comprehensive and integrated approach". Such an approach should be based on a considered cost-benefit analysis and assessment of trade-offs that may prevail between competing sustainability priorities (energy efficiency gains, transport cost reduction, speed and reliability of services) as well as sectors (e.g. transport, energy, mining, environment, labor) and modes of transport. An example of an integrated transport planning approach is the 2011 EC White Paper on Transport that defines a strategy toward competitive and resource-efficient transport systems and sets clear objectives and targets such as (i) optimizing the performance of multimodal

Table 1.3 Examples of relevant sustainability-motivated intervention measures (Source: UNCTAD based on review of literature, including Kusumal Ruamsook and Evelyn A. Thomchick (2012): Sustainable Freight Transportation: A Review of Strategies)

Type of intervention measures	Example of measures and actions
Technology and innovation	Intelligent transportation systems (ITS), electronic devices for monitoring engines, computers to measure fuel efficiency, computerized routing and scheduling software with global positioning system (GPS), software to alert drivers to the most cost-effective fuelling locations, devices that automatically switch off idling engines, energy-saving technologies (including low friction under water paint, high efficiency propeller and rudder)
Fuel-related measures	Cleaner fuels, cleaner-burning engines, improved vehicle and propulsion technology, better aerodynamics to improve fuel efficiency, efficient routing practices, speed reduction
Economic measures	Controlling determinants of transport costs, reforming maritime transport prices and investment practices, applying full-cost pricing, congestion pricing, carbon pricing, taxation, grants and subsidies (e.g. to speed up old engine turnover)
Strategic and operational	Plan and organize routings and scheduling to reduce empty mileage and optimize operations, slow steaming
Regulatory	Emission standards, design of vessels and infrastructure, speed limits, targets for use of renewable energy sources, targets for energy efficiency, emission and noise standards for ships and ports
Other/soft measures	Awareness-raising activities enable greater access to information and communication technology

logistics chains; (ii) promoting more energy-efficient modes of transport at a larger scale, facilitated by efficient and environmentally friendly freight corridors; (iii) instigating a 50% shift in longer-distance freight journeys from road to other modes; (iv) instigating a 40% use of sustainable low-carbon fuels in aviation; and (v) achieving at least a 40% cut in shipping emissions. The ultimate objective is to achieve a 60% reduction in both CO_2 emissions and oil dependency (European Commission 2011).

The 2012 China Green Freight Initiative (CGFI) is another example which seeks to improve fuel efficiency, reduce CO_2 and air pollutant emissions from road freight transport and adopt cleaner technologies and smarter freight management practices.[28] Indonesia has also introduced relevant policies to promote sustainable freight transport systems through improved fuel efficiency, enhanced maritime shipping, better use of integrated transport networks systems and reduced road transport work (and therefore related emissions), which account for about 70% of freight tonne-kilometers (Susantono 2012). In the United States, the Washington State launched its Maritime Blue 2050 initiative in 2017 to increase technology, reduce

[28]See http://cleanairasia.org (accessed 7 December 2018).

pollution, update workforce training and ensure sustainability for the industry.[29] This was the first effort of the kind in the United States. Singapore created a focus group between port authorities and maritime administrations from Asia, Europe and the United States to promote LNG bunkering. This resulted in the world's first set of harmonized safety standards for LNG bunkering.[30]

4.2 Examples of Industry-Led Initiatives

Both industry and governments are leveraging sustainability strategies in maritime transport to derive economic, social and environmental gains. Integrated public and private initiatives that promote greater policy coherence and synergies are essential to achieving an optimum outcome. Box 1.2 illustrates, with few examples, actions taken at the industry level both in response to as well as in anticipation of greater demands for improved performances in terms of sustainability.

Self-regulation in the maritime transport sector often reflects growing pressure from customers, rising demands from supply chain partners and greater public scrutiny. Also, for the industry, investing in sustainability makes a good business case given the co-benefits and the competitive advantages that may arise from sustainability-minded measures. Over recent years, the industry has made considerable progress integrating sustainability into their business processes and systems.

Industry-led voluntary actions include, among others, (i) CMA CGM taking delivery of the Antoine de Saint-Exupery, a 20,600 TEU container vessel that includes IMO-required ballast water treatment system to mitigate the transport of marine invasive species. The vessel benefits from the best technologies like a Becker Twisted Fin allowing to improve the propeller's performance and helping to reduce significantly the energy expenditure for a 4% reduction in CO_2 emissions. It also benefits from a new-generation engine that significantly reduces oil consumption (25%) for a 3% average reduction of CO_2 emissions, (ii) the decision by CMA CGM to equip its future giant containerships with engines using LNG, a technological innovation for a large reduction of pollutant emissions; (iii) Maersk's "eco voyage" maritime software tool, which can help cut fuel costs and make a voyage plan resulting in minimum fuel consumption; and (iv) Mitsui O.S.K. Lines (MOL) eco-sailing programme that includes the new 20,000 TEU-class containerships equipped with new sustainable technologies.

With shipping being inherently an international industry and involving a complex web of linkages and multiple stakeholders, many bodies are involved in enforcing international commitments and standards. While not intended as an exhaustive list

[29] See https://medium.com/wagovernor/leading-in-the-maritime-sector-washington-launches-maritime-blue-2050-initiative-d54f7d5730cc (accessed 7 December 2018).

[30] See https://www.businesstimes.com.sg/transport/singapore-maritime-week-2018-singapore-shares-its-collaboration-initiatives-to-promote.

but rather as an indication of the varied perspectives that are specific to maritime transport, relevant industry players with a role in promoting sustainability include entities as diverse as the International Chamber of Shipping (ICS), the International Association of dependent Tanker Owners (INTERTANKO), the International Tanker Owners Pollution Federation (ITOPF), the International Group of Protection and Indemnity (P&I) Clubs, the International Association of Classification Societies (IACS), the Baltic and International Maritime Council (BIMCO), the International Association of Ports and Harbors (IAPH), regional port associations and individual shipping companies. In enforcing the mandated top-down regulatory initiatives, controls imposed by coastal, ship flag and port states are equally important.

Box 1.2 Examples of Voluntary Self-Regulation in Maritime Transport

The *Clean Cargo Working Group* provides tools and methodologies to help understand and manage sustainability impacts. Relevant measures include the average trade lane emissions data that can be used for a benchmarking of carrier's performance based on their carbon emissions as well as for more informed decisions by both carriers and shippers.[31]

The *World Ports Climate Initiative (WPCI)* under the International Association of Ports and Harbors (IAPH). The 50 participating ports in the WCPI are engaged in reducing GHG emissions from their activities, including by influencing the sustainability of supply chains.[32]

IAPH Air quality and Greenhouse Gas Tool Box and work relating to climate adaptation in ports such as the climate protection plan development.[33]

The *Sustainable Shipping Initiative (SSI)* brings together leading industry companies and around the world to promote a sustainable future. Relevant activities include the launch of the Case for Action report in 2011 and efforts to promote greater uptake of sustainable shipping rating schemes to provide transparency and comparability and to enable cargo owners, charters and shipowners to integrate sustainability into commercial decisions. SSI developed its Roadmap to a Sustainable Shipping Industry by 2040. Seventeen shipping firms and suppliers aims to achieve up to 90% reduction in carbon emissions through various methods including installing advanced power management systems on ships and reducing speed to decrease fuel consumption.[34]

(continued)

[31] https://www.bsr.org/collaboration/groups/clean-cargo-working-group

[32] http://wpci.iaphworldports.org

[33] http://wpci.iaphworldports.org/iaphtoolbox

[34] http://www.ssi2040.org

Box 1.2 (continued)

Eco-Ships: many shipping companies have been investing and ordering eco-ships that are generally referred to as a new generation of vessels that are eco-friendly and at the same time fuel efficient.[35]

Charterers representing 20% of global shipped tonnage are adopting policies to avoid using inefficient ships based on their GHG emission performance.[36]

Environmental charging initiatives (e.g. Environmental Ship Index, Clean Baltic Sea Shipping and Green Award). For example, 30 ports applying an environmental charging scheme in Europe.[37]

Global Industry Alliance (GIA): a new public-private partnership initiative of the IMO under the framework of the GEF-UNDP-IMO GloMEEP Project that aims to bring together maritime industry leaders to support an energy-efficient and low-carbon maritime transport system.[38]

Task Force on Decarbonizing Shipping: launched by the Global Maritime Forum, Carbon War Room, the Carbon Pricing Leadership Coalition (CPLC) and University College London (UCL) in 2017, this industry-led initiative aims to develop tangible pathways for shipping's decarbonization through dedicated working groups, including a working group carbon pricing, finance and technology.[39]

Green Ports Initiatives[40]: promoted in various ports to reduce air emissions from ships (e.g. the United States, Europe and to some extent Asia[41]). Activities could include, for example, reducing port dues/tax, reducing or exemption for cleaner ships and investment in port infrastructure.

5 Concluding Remarks

Maritime transport is at the heart of sustainable development as it enables economic growth, promotes trade, improves market access and links communities and societies. Safe, secure, energy-efficient, affordable, reliable, low-carbon, environmentally friendly, climate-resilient and rule-based maritime transport systems

[35] https://worldmaritimenews.com/archives/tag/eco-ships

[36] https://www.sea-technology.com/news/archives/2015/

[37] https://ec.europa.eu/transport/sites/transport/files/2017-06-differentiated-port-infrastructure-charges-exec-summary.pdf

[38] http://glomeep.imo.org/global-industry-alliance-gia

[39] https://www.carbonpricingleadership.org/news/2017/10/12/industry-led-task-force-pushes-for-decarbonization-of-maritime-industry

[40] http://www.greenport.com/congressamerica

[41] http://www.bureauveritas.com/home/about-us/our-business/commodities/news-and-media/oil-and-gas-news/

contribute to achieving an economically efficient, socially equitable and environmentally sound development. Thus, maritime transport has an important role to play in addressing the sustainability imperative. However, for this role to effectively materialize, unsustainable maritime transport practices and related external costs need to be addressed with relevant sustainability criteria being fully integrated and mainstreamed into key maritime transport planning, policies and investment decisions. Tailored and targeted policies, regulations, incentives and enabling programmes will be required to promote more efficient, competitive, less energy-intensive and more environmentally friendly maritime transportation systems.

In view of the multi-faceted nature of maritime transport and the complex web of stakeholders and players involved in maritime business, enhancing the sustainability of the sector calls for a multi-stakeholder approach involving governments, transport industry, financial institutions and other relevant partners. Furthermore, as maritime transport is inherently international and implies far-reaching implications that extend beyond national borders, enhancing communication and coherence in policies and institutions and improving global coordination are equally important. Inter-agency coordination among relevant United Nations agencies which mandates cover maritime transport and its sustainability, should be furthered. Examples include the IMO, the Division for Ocean Affairs and the Law of the Sea (DOALOS), United Nations Environment as well as the five UN regional commissions.

A successful transition to a more sustainable maritime transport paradigm requires scaling up financial resources and investments. In this respect, worth noting is the growing role of the regional, subregional and national development banks as important enablers of finance and facilitators of access to sustainable transport finance. At the Rio+20 Conference, eight development and multilateral banks committed to providing over $175 billion of loans and grants for transport in developing countries over 2012–2022. At the same time, in February 2018, the European Investment Bank (EIB) and ING signed an agreement to support the European shipping market by providing a €300 million worth of green investments.[42] Equally important is the need to improve access to relevant technologies, through inter alia cooperation in development of technology and operational procedures and the carrying out of pilot and demonstration projects to validate and prove relevant new technologies.

The involvement of the private sector, academia as well as the scientific and research community is key. Working with governments and public authorities through effective public-private partnerships that are underpinned by sustainability criteria and that promote constructive dialogues, synergies, joint research and development can speed up and facilitate sustainable maritime transport efforts.

Managing the sustainability performance of maritime transport systems requires the ability to measure parameters such as impacts, emissions and externalities as well as data transparency (e.g. fuel consumption and emissions). However, global standards for relevant measurements to be carried out on GHG emissions, energy

[42]See https://www.governmenteuropa.eu/green-future-sustainable-shipping/85728/.

efficiency and fuel consumption, for example, are lacking. This may work against a level playing field. Greater standardization, coherence and harmonization of standards and methods should be promoted at the global and multilateral levels.

In addition to finance, capacity building through technical assistance (e.g. technology transfer, advisory services and best practice sharing) will also shape the sector's efforts to build its sustainability. Capacity building in the field of maritime transport is imperative especially in developing regions, notably in SIDS. In this respect, technical assistance programmes such as those delivered by UNCTAD and IMO can help build and strengthen the institutional and human capacity in the field of sustainable maritime transportation. UNCTAD is currently implementing a technical assistance project aimed at building the capacity of developing countries to shift toward sustainable transport, including sustainable shipping. This work is also considering ways in which such a shift could be financed in a sustainable manner.[43] Relevant outcomes resulting from this work include the UNCTAD Sustainable Freight Transport and Finance Toolikit[44] and UNCTAD Sustainable Freight Transport Portal.[45] UNCTAD has also implemented a technical assistance project to help build the adaptive capacity of SIDS in the Caribbean region in the face of climate change impacts on coastal transport infrastructure.[46] Main outputs under the project is the methodology used as a tool to assist transport infrastructure managers and other relevant entities in SIDS in assessing climate-related impacts and adaptation options regarding coastal transport infrastructure.

Acknowledgment The views represented in this chapter are those of the authors and do not necessarily reflect the views of the UNCTAD secretariat.

References

Alvik, S., Eide, M. S., Endresen, Ø., Hoffmann, P., Longva, T. (2010). *Pathways to low carbon shipping. Abatement potential towards 2030.* DNV.

Asariotis, R., Benamara, H., Mohos-Naray, V. (2017). *Port industry survey on climate change impacts and adaptation.* UNCTAD Research Paper No. 18. UNCTAD/SER.RP/2017/18. December.

DNV GL. (2017). *Low carbon shipping towards 2050.* Norway.

EIA. (2017). *International energy outlook 2017.* U.S. Energy Information Administration. Norway.

European Commission. (2000). *Directive 2000/59/EC of the European parliament and of the council on port reception facilities for ship-generated waste and cargo residues.* 27 November. https://eur-lex.europa.eu/legal-content/EN/TXT/?uri=celex:32000L0059.

European Commission. (2011). *Roadmap to a single European transport area – towards a competitive and resource-efficient transport system.* Brussels.

[43] See https://unctad.org/en/PublicationChapters/tc2015d1rev1_S02_P06.pdf.

[44] See http://unctad.org/en/Pages/DTL/TTL/Infrastructure-and-Services/SFTF-Toolkit.aspx.

[45] See https://unctadsftportal.org/.

[46] See https://sidsport-climateadapt.unctad.org/.

European Commission. (2017). *On the feasibility of a financial instrument that would facilitate safe and sound ship recycling*. Report from the Commission of the European Parliament and the council. COM (2017) 420 final. Brussels.

Gota, S., Peet, K., Windisch, E., Bongardt, D., Eichhorst, U. (2016). *Proposed avenues for NDCs*. Increasing the potential of Nationally Determined Contributions (NDCs) for ambitious action on transport and climate change. November. http://www.ppmc-transport.org/wp-content/uploads/2015/06/Proposed-Avenues-for-NDCs-November-2016version.pdf. Accessed 28 Apr 2018

GSF. (2015). *CO2 emissions from the maritime sector. GSF policy statement for COP21 Paris*. Global Shippers Forum. December 2015. https://www.globalshippersforum.com/media/1217/gsf_maritime_policy_statement_cop21_paris_2015.pdf. Accessed 28 Apr 2018

ICCT. (2011). *Reducing greenhouse gas emissions from ships cost effectiveness of available options*. White Paper. International Council for Clean Transportation. July. https://www.theicct.org/sites/default/files/publications/ICCT_GHGfromships_jun2011.pdf. Accessed 28 Apr 2018.

IEA, Energy Efficiency. (2017). Market Report Series. https://www.iea.org/publications/freepublications/publication/Energy_Efficiency_2017.pdf.

IMO. (2014). *Third IMO GHG study 2014. Executive summary and final report*. London: International Maritime Organization.

IMO. (2016). Amendments to the annex of the protocol of 1997 to amend the international convention for the prevention of pollution from ships, 1973, as modified by the protocol of 1978 relating thereto, annex 3, resolution mepc. 278(70) (MEPC 70/18/Add.1) Adopted on 28 October 2016.

IMO. (2017). *Considerations of how to progress the matter of reduction of GHG emissions from ships*. Note by the Secretariat. ISWG-GHG1/2. International Maritime Organization. London, 21 February.

IMO. (2018). Report of the Marine Environment Protection Committee on it its Seventieth Session, Road Map for Developing a Comprehensive IMO Strategy on, Reduction of GHG Emissions from Ships, MEPC 70/18/Add.1,Annex 11, page 1, https://www.mardep.gov.hk/en/msnote/pdf/msin1707anx3.pdf

Lloyd's Register. (2016). *Low carbon pathways 2050*. https://www.lr.org/en/insights/sustainability/low-carbon-pathways. Accessed 28 Apr 2018.

McKinnon, A. (2016). *UNCTAD workshop on sustainable freight transport and finance*. Nairobi.

OECD. (2017a). *CO2 emissions from fuel combustion: Overview*. Paris: Organisation for Economic Co-operation and Development/International Energy Agency.

OECD. (2017b). *Energy efficiency market 2017*. Paris: Organisation for Economic Co-operation and Development/International Energy Agency (2017).

OECD. (2017c). *ITF transport outlook*. Paris: OCED Publishing.

Organization for Economic Cooperation and Development. (2018). Main economic indicators. Available at http://www.oecd.org/std/oecdmaineconomicindicatorsmei.htm. Accessed April 2018.

Psaraftis, H. N. (2018). Decarbonization of maritime transport: to be or not to be? *Maritime Economics and Logistics*. https://doi.org/10.1057/s41278-018-0098-8.

Susantono, B. (2012). *Sustainable freight transport policy in Indonesia*. Presentation of the Vice Minister for Transportation Republic of Indonesia, at the UNCTAD XIII side event: Paving the Way for Sustainable Freight Transport, Doha, 25 April 2012.

The Marine Professional. (2015). *Banks more likely to finance efficient ships*. 22 April.

UNCTAD. (2012). *Liability and compensation for ship-source oil pollution: An overview of the international legal framework for oil pollution damage from tankers*. Studies in Transport Law and Policy – 2012 No. 1. New York.

UNCTAD. (2014). *Closing the distance: Partnerships for sustainable and resilient transport systems in SIDS*. New York. 2014. https://unctad.org/en/PublicationsLibrary/dtltlb2014d2_en.pdf.

UNCTAD. (2015). *Sustainable freight transport systems: Opportunities for developing countries. Note by the UNCTAD secretariat*. Geneva. TD/B/C.I/MEM.7/11.

UNCTAD. (2016). *Review of maritime transport 2016*. New York. October 2016.
UNCTAD. (2017). *Trade logistics and the 2030 agenda for sustainable development, note by the UNCTAD secretariat*. Geneva. TD/B/C.I/MEM.7/14.
UNCTAD. (2018). *Review of maritime transport 2018*. New York. October 2018. https://unctad.org/en/PublicationsLibrary/rmt2018_en.pdf
UNDP. (2013). *Human development report 2013. The rise of the South: Human progress in a diverse world*. New York.
United Nations. (2015). *The road to dignity by 2030: ending poverty, transforming all lives and protecting the planet*. Synthesis report of the Secretary-General on the post-2015 sustainable development agenda. New York. December 2014, A/69/700.
United Nations. (2016). *Mobilizing sustainable transport for development*. Secretary-General's High-Level Advisory Group on Sustainable Transport. New York. 2016.
World Trade Statistical Review and WTO. (2018). *Trade statistics and outlook strong trade growth in 2018 rests on policy choices*, Press Release 820, 12 April 2018.

Chapter 2
Green Ship Technologies

Jan Otto de Kat and Jad Mouawad

Abstract This chapter provides information on green ship technology measures. Included are background information, descriptions of the technologies, explanation of key issues, general pros and cons of each measure, and limits of applicability or effectiveness, as well as practical issues related to implementation. The technical measures described here include the design of energy-efficient ships using hull form optimization, efficient propellers, energy-saving devices, and other novel technologies; attention is paid also to air lubrication, wind-assisted propulsion, and solar power. A subsequent section on machinery systems covers key areas for machinery technology efficiency improvements including the main and auxiliary engines, waste heat recovery systems, auxiliary machinery, and hybrid power storage/production equipment. The last section on ballast water management addresses regulations and provides an overview of ballast water treatment systems and related issues.

Abbreviations

ABS	American Bureau of Shipping
AC	Alternating current
B	Ship beam
BMEP	Brake mean effective pressure
BWM	Ballast water management
BWMS	Ballast water management system
C_B	Block coefficient
CFD	Computational fluid dynamics

J. O. de Kat (✉)
American Bureau of Shipping, Houston, TX, USA
e-mail: jdekat@eagle.org

J. Mouawad
Mouawad Consulting, Hamar, Norway
e-mail: jad.mouawad@bwm.no

H. N. Psaraftis (ed.), *Sustainable Shipping*,
https://doi.org/10.1007/978-3-030-04330-8_2

CO_2	Carbon dioxide
Cp	Prismatic coefficient
CPP	Controllable pitch propeller
DC	Direct current
ECA	Emission control area
EGR	Exhaust gas recirculation
ESD	Energy-saving device
FOC	Fuel oil consumption per 24 h
FPP	Fixed pitch propeller
IMO	International Maritime Organization
L	Ship length
LCB	Longitudinal center of buoyancy
MCR	Maximum continuous rating
NOx	Nitrogen oxides
PM	Particulate matter
PTI/PTO	Power take in/power take out
PV	Photo voltaic
RANS	Reynolds-averaged Navier-Stokes
SCR	Selected catalytic reduction
SFOC	Specific fuel oil consumption
T	Ship draft
UV	Ultraviolet
VFD	Variable frequency drive
WHR	Waste heat recovery

1 Introduction

This chapter has been compiled to provide information on the current state of the art of green ship technology measures. Included are background information, descriptions of the technologies, explanation of key issues, general pros/cons of each measure, and limits of applicability or effectiveness, as well as practical issues related to implementation. Treatment does not include information and communication technologies, which are covered in Chap. 4 of this book.

The rest of this chapter comprises the following sections: Sect. 2 ("Design of Energy-Efficient Ships") addresses issues related to the basic hull form design including selecting proper proportions and reducing resistance by optimizing the hull form and appendage design. Topics covered include hull optimization, efficient propellers, energy-saving devices, and other novel technologies, including air lubrication, wind-assisted propulsion, and solar power. Section 3 ("Machinery Systems") covers the key areas for machinery technology efficiency improvements that can be applied to support sustainable shipping. The section is divided into four main subsections covering main and auxiliary engines, waste heat recovery systems, auxiliary machinery, and hybrid power storage/production equipment. Finally, Sect. 4 ("Ballast Water Management") addresses regulations and provides an overview of ballast water treatment systems and related issues. Ballast water is essential to the safe and efficient operation of shipping, but it also poses a serious ecological, economic, and health threat through the transfer of invasive aquatic species inadvertently carried in the ships. Living organisms can be eliminated from ballast water using a variety of technologies, which are summarized in this section. In addition, the section addresses some key issues associated with the installation and operation of these treatment systems.

2 Design of Energy-Efficient Ships

2.1 Hull Optimization: General Consideration

Hull form and propulsion optimization provide an effective means to improve the energy efficiency of ships. When assessing hull form optimization, the owner has several options available for consideration:

(a) Accept a standard, readily available hull form and propulsion system offered by the shipyard.
(b) Modify an existing and preferably well-optimized hull form to address the expected operating profile.
(c) Develop a new hull form design based on expected operational profile.

Option (a) involves the least capital expense – substantive savings in vessel construction costs are often realized by adopting the standard design offered by a shipyard. Due to the need to improve fuel efficiency, many of these standard ships have well-optimized hull forms and propulsors, albeit usually only optimized at the design condition and to a lesser extent at the normal ballast condition or other service conditions. Hydrodynamic performance varies significantly with changes in draft and ship speed; however these operating conditions may not be fully considered.

Option (b) enables optimization of the design for specific service conditions (e.g., a number of expected operating draft, trim, and speed combinations with their associated service durations). This optimization process generally involves modifications to the forebody design (the bulb and transition into the forward shoulder) but may also involve modifications to the propeller.

Option (c) enables optimization of vessel hull particulars to be in concert with the propeller and power plant for the relevant service conditions as mentioned under (b). This option is usually justified when a larger series of ships is being ordered, or when the shipyard under consideration does not offer a suitable standard design.

This section provides an overview of the key process elements that lead to an optimized hull design. Further details can be found in the American Bureau of Shipping (2013). Fathom Shipping (2013) provides another useful overview on ship efficiency features.

2.2 *Main Considerations Prior to Detailed Optimization of Vessels*

Before starting the optimization of a vessel design or retrofit, it is important to look at the vessel and its main parameters. It is recommended to evaluate the current trends of vessels in the same class. After the design evaluation against peers, it is important to take into consideration the operational profile, area of operation (trade route), principal dimensions, constraints, and hard points before embarking on the detailed optimization of the vessel. These elements are described below.

2.2.1 Vessel Operational Profile

Until recently all optimization for any vessel was done for a single design point, at service speed and design draught. After the rise in fuel prices, many owners found that the single design point rarely, if ever, occurred during service. Thus, vessels were overpowered and not operated at the design point for which hull and propeller had been designed.

Therefore, when embarking on a new vessel design project, it is important to take the anticipated operational profile into consideration. As a starting point, data from existing vessels (noon reports, AIS data or similar) can be used, or alternatively an operational profile can be determined based on the anticipated route network, vessel carrying capacity, etc. The operational profile is a matrix of speeds and draughts (and trim) where the vessel will operate with a percentage of time attached to each point.

The impact of optimizing the vessel over the operational profile is largest on ships with pronounced bulbous bows designed for higher speeds, but also tankers and bulk carriers can gain several percent points of efficiency over the operational profile when properly optimized.

2.2.2 Area of Operation

The vessel design/retrofit design can depend on the planned area of operation for the vessel, as design will affect the vessels motions and added resistance in waves. Whereas ship motions are related primarily to safety of ship and cargo and crew comfort, added resistance in waves can have a significant impact on the fuel consumption of the vessel.

The impact is more significant on routes with higher waves like the north Atlantic and less on routes in more calm weather conditions such as Indian Ocean or Mediterranean Sea.

Recent trends with slower speeds have resulted in vessel designs with vertical stems and no pronounced bulbous bows – these designs have shown merit in waves and in varying loading conditions for ships like container vessels and smaller tankers.

2.2.3 Principal Dimensions Study

Once the operational profile has been defined, the next step is to consider the principal dimensions. The main dimensions of the vessel are typically limited by numerous constraints. For a modern tanker or bulk carrier, typical constraints are:

- Beam may be restricted due to the port limitations.
- Draught may be restricted due to water depth at the berths or channels/seaways leading into the ports.
- Length may be restricted due to port constraints and/or lock constraints.

Typically, tankers and bulk carriers are built to fixed dimensions, but there is a relatively large savings potential in the selection of the main dimensions. For this reason, the option of altering the main dimensions should be considered in close dialogue with the ship yard/designer of the vessel at a very early design stage.

For container vessels and Ro-Ro or RoPax vessels, there are constraints such as:

- Beam variation is restricted due to number of rows of containers (both in hold and on deck) or lanes on trailer vessels.
- Draught may be restricted due to water depth at the berths or channels/seaways leading into the ports.
- Length variation is restricted due to number of bays of containers or number on trailers in trailer decks.

At this stage, semiempirical models/databases (possibly supported by computational fluid dynamics (CFD)) can be used to predict the preliminary powering requirements of the design variants.

Once the main dimensions have been decided, the final powering prediction for the vessel can be calculated, and an initial selection of the propeller diameter and other characteristics can be made for use in the subsequent process. The final propeller design and diameter will be revisited once the lines have been optimized. The matching of the propeller and engine is very important to ensure that the necessary power is available with the lowest fuel consumption.

The potential for savings in a main dimensions study depends highly on the starting point, conditions, and constraints. But variation of main dimensions can easily decrease or increase the fuel consumption by 2–5% between the variants investigated.

Increasing the length/beam ratio and/or increasing length and reducing the block coefficient can provide reductions in propulsion fuel consumption up to 3–5%. Increasing the length while reducing the beam and maintaining the draft, displacement and block coefficient (C_B) constant normally yields improvements in hull efficiency, provided additional ballast is not needed to maintain adequate stability. A higher length/beam ratio tends to reduce wave-making resistance, while the reduced beam/draft ratio tends to reduce wetted surface and therefore the frictional resistance.

Increasing draft by reducing C_B and/or beam results in improvements to hull efficiency and may provide the additional advantage of allowing for a larger propeller to be fitted. Increasing length while reducing C_B reduces the required power. Reducing beam while increasing C_B also tends to reduce required power.

The longitudinal prismatic coefficient (Cp) is a commonly applied indicator of the longitudinal distribution of displacement. A lower Cp, favored for faster ships, implies a greater concentration of displacement amidships and a finer entrance angle. Tankers and bulk carriers with fuller (bluff) bow shapes will have a higher Cp.

Of course, main particulars and hull form coefficients cannot be selected based on hydrodynamic principles alone. The accommodation of the cargo block and main propulsion units, minimization of ballast, and restrictions from port and canal infrastructure are some of the factors that must be accounted for. Such design constraints are assessed against economic factors, including fuel consumption and construction cost. Other factors must be taken into consideration such as berth availability for the longer ships and structural reliability as the length/depth ratio increases. Nevertheless, driven by rising fuel costs, the longer-term trend will be toward increasing the length/beam ratio and reducing the block coefficient or reducing the design speed.

It is important that studies to determine optimal dimensions consider the effects of speed loss in waves. For early-stage analysis, a semiempirical approach such as Townsin and Kwon (1993) is adequate for estimating speed loss. As the design progresses, model tests in waves and numerical analysis provide a more accurate behavior of the specific hull form in waves.

2.2.4 Hard Points and Constraints Evaluation

Once the main dimensions have been determined, the detailed optimization of the hull form can begin. It is important to have a close and open dialogue between the ship yard/designers and the ship owner. Especially the discussion of the constraints/hard points on the vessel is critical, and it is important that the effect of these points is discussed on the basis of the preliminary general arrangement and preliminary lines plan. Constraints and hard points on tankers and bulk carriers are typically:

- Displacement and cargo intake
- Cargo holds/tanks layout

- Engine position in the hull in relation to the hull surface
- Rudder head box design
- Sea chest position and extension on the hull

For a container ship, it is also important to evaluate the container bay positions in relation to the hull surface. For Ro-Ro/Ro-Ro passenger vessels, equipment such as internal ramps and external doors and rams are often hard points that need to be included in the design process.

The following section describes the methods available to today's naval architect for optimizing hull form and propeller and outlines some of the issues that vessel owners should consider in the assessment of the hull form aiming to enhance vessel fuel efficiency.

2.3 *Hull Form Optimization*

Computational fluid dynamics (CFD) methods have reached the stage in which they can predict resistance and propulsion characteristics in calm water conditions with sufficiently high accuracy. With the advent of powerful computers, it is not necessary anymore to assume inviscid fluid conditions for hull flow calculations, and instead one can model viscous fluid effects using the Reynolds-averaged Navier-Stokes (RANS) equations. With CFD-RANS tools, it is possible to consider free surface effects in combination with fluid viscosity, including flow interaction with rudder and propeller. CFD-RANS is useful in assessing the influence of changes to the entrance angle, optimizing the location and shape of the fore and aft shoulders, as well as shape of fore and ship. CFD calculations are to be employed sequentially, allowing for refinement of shape and elimination of less favorable variations (see, e.g., Larsson et al. 2010).

There is substantive potential for fuel savings by optimizing for the off-design conditions where the expected operating profile differs from a single design draft and design speed. Changes in draft, trim, and speed can dramatically change the wave profile and overall resistance. Therefore, the owner and designer should prepare a clear specification of the different operating drafts and speeds on different legs of the expected voyages. Numerical analysis and model tests should then cover all operating conditions at which the vessel may spend a significant portion of its time at sea.

While designers are comfortable using CFD for quantitative assessment of required power, model tests are recommended for confirmation of the numerical results and for final power prediction. When developing lines, numerous trade-offs are considered. Although considerable progress has been made in numerical hull form shape optimization tools, the creation of lines remains part art and part science, and there is still no substitute for the experienced designer. There is considerable advantage in beginning with a good parent hull of similar proportions and in having an extensive database for benchmarking purposes. Therefore, many of the

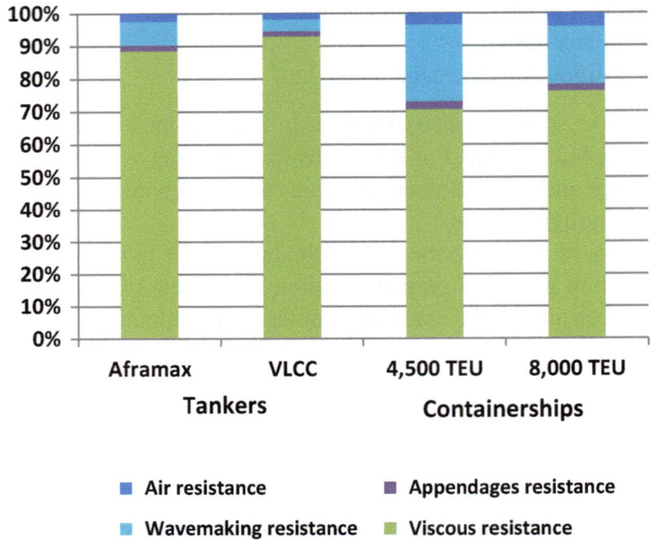

Fig. 2.1 Components of hull resistance in calm water conditions. (Source: ABS 2013)

best performing hull forms have been developed by the major model basins or yards with their own proven testing facilities, well validated through full-scale trial comparisons.

2.3.1 Approach to Improving Key Elements of Resistance

As shown in Fig. 2.1, viscous (frictional) resistance is the major component of overall resistance, accounting for between 70% and 93% of total resistance. The percentage of total resistance attributed to viscous (frictional) resistance is greatest for slower, larger ships. Wave-making resistance increases with ship speed and is a larger component of overall resistance for high-speed fine, form ships than it is for slower, full form ships.

When developing a full body hull form such as a tanker, emphasis is placed on reducing wetted surface as viscous resistance is a major component of overall resistance. Another important consideration is to provide a smooth and gradual transition to the propeller, to avoid separation of flow at the stern, and to provide for a uniform wake field (i.e., constant axial velocities at each radius). This encourages the LCB to be as far forward as practical, although care must be taken to avoid a harsh shoulder forward. Mitigating wave propagation at the forward shoulder is more important than reducing wave making by fining up the entrance angle, encouraging a blunter bow shape to accommodate smoother transitions through the forebody. The blunter bow shape allows a shift in volume from the midship region into the forebody region, resulting in better overall resistance performance for full body ships.

For higher speed and therefore finer hull forms typical for larger containerships, wave making is more significant (18% of total resistance for a standard 8000 TEU containership shown above). Such a vessel will have more slender proportions as compared with a tanker, with a higher L/B ratio. In this case, the more slender and finer hull allows the LCB to be moved aft while still maintaining good flow into the propeller. This enables a reduced entrance angle and softer forward shoulders. The bulb on a containership will be elongated with finer shape to reduce wave-making resistance.

2.3.2 Forebody Optimization

Forebody optimization includes consideration of the bulb design, waterline entrance, forward shoulder, and transition to the turn of the bilge. A properly designed bulbous bow reduces wave-making resistance by producing its own wave system that is out of phase with the bow wave from the hull, creating a canceling effect and overall reduction in wave-making resistance. Physical factors considered in bulb optimization include volume, vertical extension of the center of volume, longitudinal extension, and shape. Further details on bulbous bow and forebody design can be found in Larsen et al. (2010).

The characteristics of the bulbous bow must be carefully balanced with the shape of the entrance and the transition toward the forward shoulder and bilge. Bulbs are most effective at certain Froude number (speed-length ratio) and draft. Changes in speed and draft significantly change the wave created, such that reductions in draft or speed can lead to increases in wave-making resistance. As few commercial vessels operate solely at one design draft, compromises in the bulb design are needed to provide good performance over the expected range of operating drafts and speeds. For a container vessel fuel savings of over 5% were reported by modifying the bulbous bow of a shipyard design that was optimized to the design draft, so that it provided more favorable performance over the anticipated operating profile of drafts and ship speeds (De Kat et al. 2009).

2.3.3 Aftbody Optimization

Aftbody optimization includes efforts to mitigate stern waves, improve flow into the propeller, and avoid eddy effects. A properly designed stern can reduce the aft shoulder crest wave as well as the deep wave trough and stern waves. Improving the nature of the stern flow can lead to improved propulsive efficiency. Viscous flow calculations are needed to evaluate aftbody flow through the propeller and wetted transom flows in way of a submerged transom because these are dominated by viscous effects.

Single screw sterns forward of the propeller may be V-shaped, U-shaped, or bulb types. The tendency is toward the bulb shape, as the improved wake reduces cavitation and vibration. Asymmetrical sterns can be designed to improve propulsive efficiency through pre-rotation of the flow to the propeller and to some extent by reducing the thrust deduction. The pre-rotation of the flow into the propeller helps reduce the separation of flow in the stern aft of the propeller. To date, these enhancements have not been proven to be sufficiently effective to offset the extra complexity involved in construction, apart from some twin-skeg designs.

Twin-screw propulsion arrangements offer enhanced maneuverability and redundancy and are also adopted when the power required for a single propeller is excessive. Propulsion power may exceed what can be handled reasonably by a single propeller if, for example, the vessel design is draft limited, and the propeller diameter is correspondingly reduced. For a twin-screw design, there is the choice of open shafts with struts or twin skegs (or gondolas).

For twin-screw propulsion with open shafts, efficiency is generally compromised when compared to a single screw design, in part due to the high appendage resistance from struts and bearings. The introduction of the twin gondola-type skeg design eliminates the need for these appendages and can provide favorable hydrodynamic performance, especially for full-bodied ships and those with wide beams and/or shallow drafts. For slender, higher-powered ships, the open shaft twin-screw design may be more favorable when two propellers are required because the open stern shape provides lower wake variation, resulting in less cavitation and vibration.

For full hull form ships, it has been found that twin skegs may provide a 2–3% efficiency improvement over well-optimized single screw designs with corresponding characteristics (SSPA 2009). If the propeller diameter on a single screw design is suboptimal due to draft restrictions, unloading of the propellers in twin-skeg arrangements can lead to significant propulsion efficiency improvements. While there may be improvements in the overall efficiency of the vessel, in relation to fuel consumption, the fitting of twin skegs does have disadvantages that should be evaluated, including:

- The wetted surface is typically about 4–5% higher for a twin skeg vs. a single screw design. The lower the C_B, the more pronounced the effect on wetted surface.
- The hull steel weight is increased (by roughly 4–5% for tankers).
- Twin skeg arrangements are more expensive to build.

As there are numerous design and installation arrangements for twin skegs, each unique to the specific vessel design, it is essential that an optimization effort consisting of CFD and model testing be employed to achieve the desired results.

2.3.4 Appendage Resistance

For cargo vessels in calm water conditions, appendage resistance is about 2–3%. Roughly half of the appendage resistance is attributable to the rudder and half to bilge keels. Rudder resistance can increase substantially in severe wind/weather conditions or for directionally unstable ships as noted below.

Added resistance from a bow thruster tunnel can be significant (in the range of 1–2% of calm water resistance). Grid bars are frequently placed over the opening perpendicular to the flow direction. They serve to break up laminar flow and reduce vortices. Anti-suction tunnels can be used to reduce the pressure variation across the bow thruster tunnel.

2.3.5 Maneuvering and Course-Keeping Considerations

A high block coefficient, forward LCB, lower length to beam ratio, and open stern are factors that can lead to reduced directional stability. Accordingly, performance should be assessed through CFD or by model tests, either through captive tests in a towing tank or by free running models testing in an open basin. Where the vessel's mission requirements necessitate the use of a hull form with reduced directional stability, effective course keeping can be provided by larger rudders, high-performance rudders, or skegs, which will induce a penalty in overall efficiency when compared to vessels not provided with such rudders or skegs. In such cases, viscous flow CFD assessment and model tests are recommended as the drag and added resistance resulting from the larger, high-performance rudders and skegs can vary substantially.

2.4 Propulsion Arrangement and Propeller Selection

Once the resistance and propulsion characteristics of the hull have been optimized, the propeller designers can start their work and optimize the final propeller(s) for the vessel. This should be done in close dialogue with the engine manufacturer to ensure the best possible match. It is well known that a larger diameter propeller in general gives better efficiency, but when the propeller diameter is limited by other factors, advanced propeller design can still help to increase the propulsive efficiency.

Important aspects to consider during the design process of the propeller include:

- Engine layout
- Sea margin
- Light running margin

A range of independent propeller vendors are offering a variety of modern propeller designs for commercial vessels, which have the potential of increasing the propulsive efficiency.

Often the designs from different vendors are tested in a model basin for the same range of speed and draughts used in the tests with stock propellers. To ensure best possible relative comparison, all propellers should be tested during the same test session, including a retest of stock propeller. It is important to note that the operational profile might also affect the selection of final propeller design, as some of the design variants will be less efficient under certain operational conditions.

Differences between new propeller designs can be as high as 2–5% on a modern tanker or bulk carrier, so there is potential for significant savings with minimum extra new building costs. The value of savings naturally depends on the constraints put on the design space for the propeller (operating RPM, maximum propeller diameter, minimum pressure pulses, etc.).

2.4.1 Single Screw Vessels

For single screw vessels such as tankers, bulk carriers, container vessels, and some Ro-Ro vessels, a fixed pitch propeller (FPP) tends to be most appropriate. Commonly used FPP vendors include:

- Stone Marine Group Ltd., UK
- MAN Kappel Propellers, Denmark
- Wärtsilä Propellers, The Netherlands
- MMG, Germany

Figure 2.2 shows an example.

2.4.2 Twin-Screw Open Shaft

For Ro-Ro vessels, Ro-Ro passenger vessels, and cruise vessels, the most common solution is controllable pitch propellers (CPP) fitted on open shafts supported by bossings and brackets. This solution is also used on smaller container vessels and smaller tanker vessels. Commonly used vendors include:

- Rolls Royce (KaMeWa), Sweden
- MAN Propellers, Denmark
- Wärtsilä Propellers, The Netherlands
- Nakashima Propellers, Japan

The CP propellers allow the ship crew to optimize the pitch setting through the combinatory to match pitch to the present speed and loading of the vessel, resulting in minimum fuel consumption.

Fig. 2.2 Three-bladed Kappel propeller. (Courtesy MAN)

Fig. 2.3 Rolls Royce CPP. (Courtesy Rolls Royce – Commercial Marine)

The open shaft systems require attention to the design and optimization of the appendages, as the bracket systems can contribute largely to the resistance of the vessel. Figure 2.3 shows an example of CPP.

2.4.3 Azimuthing Propulsion and Pod Propulsion

For offshore vessels the twin open shaft propulsion system is often replaced with twin azimuthing thruster units to ensure high maneuverability and DP capability. This arrangement can also give less resistance and better propulsion efficiency. Commonly used vendors include:

- Rolls Royce (KaMeWa), Sweden
- Wärtsilä Propellers, The Netherlands
- Nakashima Propellers, Japan

On offshore vessels many of the azimuthing thruster units are fitted with ducts to ensure high bollard pull for towing operation and DP application.

For larger passenger vessels and cruise vessels with diesel electric configuration, POD units are often seen as an alternative to open shaft arrangements. The POD units are typically configured with electrical motor in the gondola, but some smaller units are equipped with mechanical connections of the propeller shafts. Commonly used vendors include:

- ABB, Finland (Azipods)
- Rolls Royce (KaMeWa), Sweden

2.5 *Energy-Saving Devices*

2.5.1 Overview

During the years, many different devices have been studied to either correct the energy performance of suboptimal ship designs or to improve an already optimized design by exploiting physical phenomena usually regarded as secondary in the normal design process. The final objective is to reduce fuel consumption related to propulsion. It should be noted the devices described here are not necessarily compatible, and their combined effects could be less than the sum of the savings of the individual components.

This section explores a range of energy-saving devices (ESDs), most of which historically concentrate on the improvement of propeller propulsion effectiveness. However, the industry has also seen the recent development of a series of devices aimed at either reducing the hull frictional resistance or exploiting readily available natural resources, such as solar and wind energy. Energy-saving technologies such as air lubrication are examined in Sect. 2.6.

The following propulsion efficiency related ESDs are described in this section:

(a) Wake Equalizing and Flow Separation Alleviating Devices

    (i) Wake Equalizing Ducts
    (ii) Vortex Generators of fins

(b) Pre-swirl Devices

  (i) Pre-swirl Fins and Stators
  (ii) Pre-swirl Stators with Wake Equalizing Ducts

(c) Post-swirl Devices

  (i) Asymmetric Rudders
  (ii) Rudder Bulbs
  (iii) Propeller Boss Cap Fin types
  (iv) Rudder Thrust Fins
  (v) Post-swirl Stators

It is important to note that the operational profile might also affect the selection of ESDs, as some of them may be less efficient under certain operational conditions.

2.5.2 Evaluation and Analysis of Energy-Saving Devices (ESDs)

Once the design (hull, propeller and engine) of the base vessel has been finalized, an analysis of feasible ESDs for a modern vessel can be initiated. But before starting to delve into the details of which ESDs are feasible for full form vessels, it is instructive to review the different types of propulsion losses that ESDs are supposed to help reduce and/or recover.

In the quest to maximize fuel efficiency, it is important to understand the origins of the energy consumption of the vessel in question.

In the process of converting the shaft rotation to a longitudinal force that can propel the vessel forward, energy can be saved by:

(a) Reducing the required propulsion power (i.e., *optimize hull* for resistance)
(b) Reducing energy losses (*optimize wake field* and *optimize propeller*)
(c) Recovering energy losses

Item (a) has already been discussed in Sect. 2.3, so the following section is focused on how the flow into the propeller can be further improved and how some of the energy losses can be recovered.

The total propulsion efficiency of a propeller varies typically between 50% and 70%. The losses for an average propeller (with an efficiency of 60%) can be attributed to three primary physical phenomena:

Axial losses – A propeller generates thrust, due to the acceleration of the incoming water. Behind the vessel, the outcoming flow mixes with the surrounding flow. Due to this turbulence, energy will be lost. Typically, the axial loss amounts to approximately 20%.

Frictional losses – When the propeller rotates, water in contact with the propeller blade surface causes friction and thus losses. The total blade surface, speed of rotation, and surface roughness are the primary factors affecting frictional losses of a propeller. The frictional losses can primarily be reduced by reducing the number of blades and reducing the blade area ratio within the limitation of risk of cavitation. Typically, the frictional losses amount to approximately 13%.

Rotational losses – Rotation of the propeller blades causes not only the generation of a longitudinal acceleration of the water, which generates thrust, but also an unwanted rotational acceleration, which generates swirl. The energy that goes into swirl is a loss. Typically, the rotational loss amounts to approximately 7%. This is an important number to remember because this means that if we were able to remove all rotational losses, then we would save at most 7%. Some ESDs introduce a pre-swirl that also improves the propeller inflow (wake field), and by combining these effects, they can thus result in higher savings.

The rotational losses can be approached in two ways:

- In front of the propeller (Pre-swirl – MEWIS ducts or similar, Schneekluth ducts with Grothues spoilers, vortex generators, stator fins, and similar)
- Behind the propeller (Post-swirl – Rudder position, rudder bulb, Propeller Boss Cap Fin (PBCF), thrust fins, post stator, and similar).

2.5.3 Wake Equalizing Duct and/or Flow Guide Fins

For many full form ships, the wake field is not very even, and the flow into the propeller is retarded in the upper half of the propeller disk. In general, wake equalization and flow separation alleviating devices are features to improve the flow around the hull that were developed to obviate propeller problems and/or added ship resistance caused by suboptimal aft hull forms. As such, they are less effective when the ship geometry has been designed correctly, with an eye at optimizing the flow to the propeller and avoiding the generation of detrimental hydrodynamic effects such as bilge vortices.

An example solution is a wake equalizing duct (WED) or to install flow guiding fins (also referred to as vortex generators). The concept for both solutions is to condition the flow in front of the propeller. This is done by guiding water from regions with high flow velocity into regions of low velocity and thus making the wake field more even. Some examples of such systems include, with applicability to tankers, bulk carriers and containerships:

- Schneekluth Duct – savings potential 2–4%, and additional 1–2% in combination with Grothues spoilers
- Vortex generators from SHI (SAVER fins) – savings potential 1–2%

2.5.4 Pre-swirl Devices

To further improve the propeller inflow, a contrarotation in the flow can be introduced in front of the propeller – this has the effect of reducing the rotational losses behind the propeller. For full form, low-speed ships, there are several vendors, but the main supplier is Becker Marine System with their MEWIS duct, where a

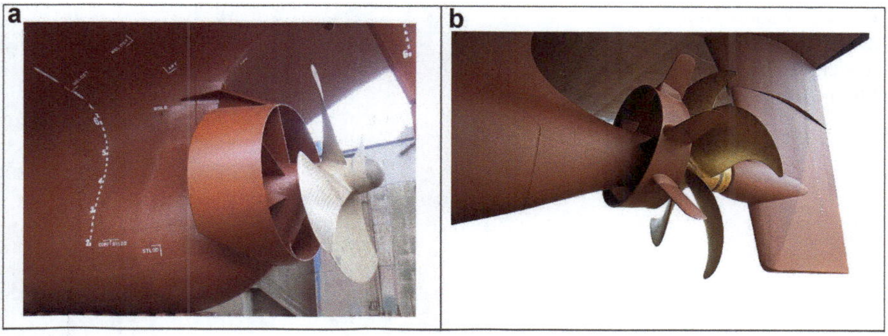

Fig. 2.4 Example of MEWIS duct and Becker Twisted Fin. (**a**) MEWIS Pre-Swirl duct on a tanker vessel. (**b**) Becker Twisted fin on a large container vessel. The vessel is also fitted with a twisted rudder and integrated rudder bulb. (Courtesy Becker Marine Systems)

wake equalizing duct and pre-swirl stators are combined in one ESD and pre-swirl stators system (PSS) developed by DSME.

Savings for MEWIS duct lie typically in the range of 3–6% for a tanker or bulk carrier application (Mewis and Guiard 2011), whereas savings with PSS typically are in the range from 2% to 4% for a typical tanker or bulk carrier application (Simonsen et al. 2011). Savings will vary depending on the actual vessel design and its characteristics (hull form, propeller loading, rudder design, etc.).

Pre-swirl systems have been developed for medium-/high-speed vessels with slender hull forms by Becker Marine Systems with the Becker Twisted Fin, where a pre-swirl stators and a structural ring/duct are combined, and by DSME with the pre-swirl stators system (PSS). Initially the Becker Twisted Fin systems included a full ring, but the newest designs include a partial ring. Savings for Becker Twisted Fin lie typically in the range of 2.5–3.5% for container vessel applications and similar numbers apply for pre-swirl stators systems. Savings may vary depending on the actual vessel design and its characteristics (hull form, propeller loading, rudder design, etc.). Figure 2.4 shows an example of two pre-swirl devices.

2.5.5 Rudder Position

In the optimization of the vessels propulsion, it is important to investigate the longitudinal position of the rudder as this has an impact on the recovery of the rotational losses. Several studies have been performed on this topic and presented at conferences worldwide. There are obviously some limitations to the position of the rudder in relation to the hull, which must be considered in an early stage of the design.

Potential savings may be in the order of 1–2% on tankers and bulk carriers from best to worst position (relatively small range of variation in position). More information can be found in Reichel (2009) and in Minchev et al. (2013).

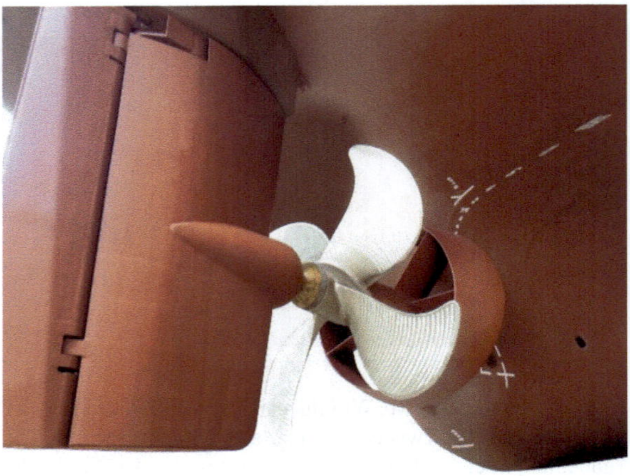

Fig. 2.5 Rudder bulb mounted on twisted flap rudder. (Courtesy Becker Marine Systems)

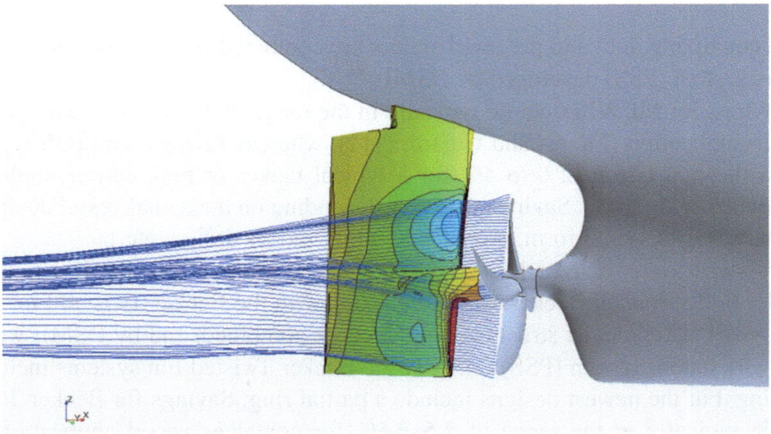

Fig. 2.6 CFD flow simulation of rudder bulb. (Source: ABS graphics)

2.5.6 Rudder Bulb

Once the rudder position is fixed, the implementation of a rudder bulb on the rudder can be addressed. The aim is to remove the hub vortex (high radial distribution in the flow near the propeller hub) and thus recover some of the rotational losses resulting in reduced fuel consumption.

Several designs exist from a range of vendors. The concept has been widely used and has been implemented on to all types of vessels including full form tankers and bulk carriers (see Fig. 2.5).

Potential savings are in the order of 1–2% for full form tankers and bulk carriers. The final rudder bulb can be optimized using advanced CFD analysis taking the actual flow behind the vessel and operational profile into account (see Fig. 2.6).

Fig. 2.7 Twisted flap rudder. (Courtesy Becker Marine Systems)

Savings may vary depending on the actual design, propeller loading, propeller hub diameter, distance between propeller hub, and rudder leading edge.

2.5.7 Twisted Rudder

A twisted rudder (twisted above and below the propeller center line) can also reduce the fuel consumption. The effect from the twisted rudder is not to regain loss but to reduce the drag on the rudder due to angled flow (due to propeller rotation) over the rudder (see Fig. 2.7).

Several designs exist from a range of vendors with some variation in designs. The concept has been widely used and implemented on all types of vessels; the biggest gains have been observed on faster vessels. Potential saving is a reduction of power in the order of 1–2%. A twisted rudder is also often seen in a combination with a rudder bulb where the gains can be compounded.

2.6 Novel Technologies

2.6.1 Air Lubrication

In ship resistance, the three main components are friction resistance, form resistance, and wave resistance. The dominant component is the skin friction resistance, which can make up 60% or more of the total resistance. In the past three decades, there has been continuous interest in air lubrication as a method to reduce the skin friction drag of the ship's hull.

Three categories of air lubrication methods can be distinguished:

- Bubble drag reduction
- Air layer drag reduction
- Air cavity drag reduction

In bubble drag reduction, small bubbles are generated by compressor or blower and injected into the turbulent boundary layer of the ship's hull. When very small bubbles are generated, this drag reduction method is referred to microbubble drag reduction. The size of such microbubbles is generally less than 0.1 mm and in the order of microns. Typically, bubbles are injected at the forward end of the flat part of the ship's bottom.

In the UK, Silverstream Technologies has developed a technology to reduce the frictional resistance of a vessel by injecting air into cavities on the bottom of the ship. Conceivably the interface between the air cavities and water creates microbubbles that follow the streamlines beyond the injection point. The system has been installed and tested on a 40 k DWT product tanker in cooperation with Shell (Silberschmidt et al. 2016). It has been reported that the net amount of power saving from the air lubrication system was measured to be about 4%, as mentioned in below press release statements:

http://www.shell.com/business-customers/trading-and-supply/trading/news-and-media-releases/silverstream-air-lubrication-technology.html
https://www.marineinsight.com/shipping-news/silverstream-air-lubrication-technology-proven-to-deliver-significant-long-term-energy-savings/

The system has been installed on some recently built cruise vessels, but no performance results have been reported yet.

In air lubrication, with sufficient air injected into a turbulent boundary layer, air layer drag reduction occurs when the injected air bubbles coalesce into a continuous or nearly continuous layer (film) of air separating the solid surface from the water flow and subsequently result in a skin friction drag reduction.

The following applications use the air layer drag reduction concept:

- Mitsubishi Air Lubrication System (MALS)
- Samsung Heavy Industries SAVER system

The Mitsubishi Air Lubrication System (MALS) is a patented air lubrication system using the drag reduction method developed by the Japanese shipbuilder Mitsubishi Heavy Industries Ltd. (MHI). Frictional drag reduction is achieved using an air injector device to deliver air through injection holes at the ship's bottom to generate air bubbles and form a layer to separate the surrounding water from the hull surface (Mizokami et al. 2011). The locations of the air injection outlets are designed to allow the air bubbles to cover the ship bottom as widely as possible. It is understood that the MALS can be applied to different types and sizes of seagoing vessels. Typically, for low-speed full form ships such as tankers and bulk carriers that have a large, flat bottom, one spanwise air injection outlet at the bottom forward near the bow might be adequate. For fine form ships such as ferries and containerships with the flat bottom being narrow near the bow and at stern, a triple outlet scheme (e.g., one centerline and two side injection outlets) at the forward of the flat bottom might be more appropriate. Calm water trials resulted in net power savings of around 10%, but the performance in waves during regular service conditions has not been publicized.

Samsung Heavy Industries (SHI) has been developing an air lubrication system, referred to as the SAVER system (Lee et al. 2017; Jang et al. 2014). Model tests and full-scale trials have been carried out on several vessels. In 2014, SHI designed an air lubrication system for a Heavy Cargo Carrier (HCC, $L = 165$ m, $B = 42$ m, $T = 5.25$ m), which was retrofitted, and conducted sea trials to measure fuel-saving effects. Subsequently in 2015, a joint development project to develop an air lubrication system for an LNG carrier retrofit was set up and conducted in cooperation with BG Group (now Shell), Gaslog, and ABS. The LNGC has a cargo capacity of 170,000 m^3 and a length of 290 m. Lee et al. (2017) describe the design and testing of the SHI system for the two full-scale ships, along with model test results. The power savings of the systems have been evaluated through sea trials and in-service voyages, and the full-scale results have been compared with model tests. Generally it was observed that the model test would overestimate the propulsion power savings when compared with the sea trial results. The full-scale performance data suggest that for the HCC, the system can lead to an average power saving of 8.8% on actual voyages; for the LNG carrier net savings of about 4–5% saving were observed on the basis of the full-scale measurements (Lee et al. 2017).

The air cavity concept is based on the usage of a recess (or several recesses) in the bottom of a ship, where air is supplied to it so that an artificially inflated air cavity is formed and separates a part of the bottom from the contact with the water, therefore reducing the frictional resistance. Here, the air layer in the cavitating flow is much thicker than the turbulent boundary layer on the ship hull. Air is continuously injected into the cavity to make up for air dissipation into the surrounding fluid.

For ocean-going ships either the bubble drag or air layer drag reduction technology seems to be most suited. To date net savings have been documented to be in the order of 5%, which typically applies at the higher operational speed range for the vessels fitted with the systems.

2.6.2 Renewable Energy

The utilization of renewable energy sources is currently benefiting from a vast international attention in all industrial fields, including shipping. In our industry, attempts in this direction are naturally concentrating on wind power, since this is readily available at sea and has a long history of successful exploitation. However, photovoltaic (PV) solar panels are also being considered in specific fields such as the generation of auxiliary power.

Wind Propulsion

Wind has been used to propel ships for millennia, but the vast practical benefits of modern propulsion systems have meant the progressive decline and disappearance of sails from all merchant vessels. In many ways, it is hard to imagine a return to sails and the complexity of operation imposed by this type of propulsion. However, the large fuel-saving benefits that wind power can provide should not be underestimated. Even if the need to compromise between optimization of shipping operations and minimization of fuel consumption will imply only a partial reduction of the latter, it is reasonable to expect this to be easier to achieve and to offer a greater potential with the use of wind power, than with the adoption of most other energy-saving measures.

In recent decades we have seen the application of Flettner rotors, sails, kites, and wing foils on different ship types. Flettner rotors are vertical cylinders spinning around their axis. A propulsive force is generated in the direction perpendicular to that of the wind hitting the rotor as a result of the Magnus effect. For this reason, rotor sails offer maximum efficiency near apparent beam wind conditions. The rotors need to be driven by an electrical motor to achieve the necessary RPM; this power needs to be added to the propulsive power. In the applications presented so far, the Flettner rotor shall be considered as a supplement to the normal propulsion system.

Applications are still limited, but there has been some success for the Flettner rotors, especially from the Finnish Company Norsepower.[1] An issue of the Flettner rotor is the negative drag when heading into the wind; there have been designers proposing a version that can be folded away or telescopically collapsed to minimize aerodynamic drag (and air height in port) when they are not in use.

Towing kites are currently the only wind power exploitation technology commercially available to ships. The principle behind it is relatively simple, although the technology necessary to deploy, control, and recover the kite is complex. In practice, extra power is provided to propel the ship by flying a kite tethered to her bow. The kite speed through the air increases its efficiency compared to standard sails, but the setup requires a computer to control the kite. Naaijen et al. (2006) estimate that significant fuel savings are possible using these systems for slower ships (typically bulk carriers and tankers); however the envelope of operability of kites is limited to a relatively narrow range of wind conditions (essentially quartering winds), which further limits the usefulness of these systems. To evaluate the actual cost-benefit of kites, it is necessary to estimate their potential when deployed on specific routes, where wind patterns can be predicted.

[1]https://www.norsepower.com/ Accessed Oct. 2, 2018

A concern regarding towing kites is the complexity of its operation and the risk associated with the system behavior in rough weather. As the largest gains provided by towing kites are when strong tail winds are present, it is paramount that the system can be operated safely, reliably, and with no additional strain of the already limited crew resources available onboard.

Solar Power

There have been attempts to use photovoltaic (PV) panels to power small craft, such as the 30-m-long catamaran Planet Solar, designed to circumnavigate the world on a 500 m^2 array. However, because of the low electrical output per unit surface, PV solar panels are better suited as an additional source of auxiliary power. In this role they have already been utilized on commercial vessels such as the NYK car carrier Auriga Leader, equipped with 328 solar panels. The energy generated by the 40 kW solar arrays on this ship is used to power lighting and other applications in the crew's living quarters.

The drawback of PV solar power is the high capital cost and required surface area.

3 Machinery Technology

This section covers the key areas for machinery technology efficiency improvements that can be applied to support sustainable shipping. The section is divided into four main subsections covering main and auxiliary engines, waste heat recovery systems, auxiliary machinery, and hybrid power storage/production equipment.

While there is efficiency improvements that can be applied individually to each type of installed equipment or system, the biggest efficiency gains may be achieved where installed machinery is considered in a more holistic approach for the entire ship and ship operating profiles. With this approach it may be possible, for example, to make the best use of emerging advanced medium and high-speed engine designs coupled with electric drive, high-voltage or direct current (DC) power distribution systems, and energy storage devices such as battery packs or capacitors. While these techniques may be best suited to ship applications with high transient power demands and short mission profiles, the principles are equally applicable to all ship types.

To fully realize this efficiency potential will of course require bold challenges to established machinery and propulsion arrangements in what is traditionally a very conservative market. The use of modern simulation and modeling techniques can help support this process at the concept and detail design stages. The knowledge base can be further improved through the life of the ship by comprehensive data collection, analytics, and machinery optimization.

The decoupling of machinery and vessel speeds by the use of the electric propulsion systems mentioned above therefore suggests the advent of electric drive, energy storage, and hybrid propulsion systems may sound the demise of the direct-drive slow-speed diesel engine. However, the additional system losses associated with these propulsion systems may negate the advantages; at present we see no significant emerging trend away from direct-drive slow-speed propulsion systems for the bulk of the commercial deep sea fleet. Determining the most fuel-efficient ship design and operational practices remains a very ship-specific process.

The slow-speed two-stroke engine has long had the highest thermal efficiency of any prime mover and hence by selection alone provides a fuel-efficient solution. Although fuel prices are notoriously volatile, there has always been an operational incentive to operate engines as fuel efficiently as possible and hence reduce operational costs. Some quirks of the marine industry with regard to who actually pays for the fuel of course can obscure this objective in certain cases, but when combined with the IMO statutory design and operational energy efficiency regulations that are now in place, the general trend is clearly to continue to minimize fuel consumption in the years ahead. The marine efforts to reduce CO_2 emissions are consistent with the regulatory regimes in other transport sectors.

In support of these objectives, the long-term predominant marine fuel and prime mover choice will likely emerge in the 2025–2030 time frame through a combination of market success with technologies currently being trialed and other market and political forces. Liquefied natural gas (LNG) is one of those emerging for non-gas carriers since around 2000. In the meantime traditional marine fuels and diesel engines will dominate the shipping sector with a wide variety of other potential solutions, which may be very ship type, size and trading route specific, emerging in increasing quantities. Gas turbines are expected to continue to find niche marine propulsion and power generation applications, and fuel cells may play an increasing role in ship power generation in the next 15 years. However, the internal combustion engine will continue to be the dominant marine prime mover in years to come, and the main steps to be taken for sustainable shipping, from the efficiency and emissions viewpoint, will continue to be reducing fuel consumption and improving overall ship efficiency. In view of the 2018 IMO decision to work toward a 50% reduction in GHG emissions by 2050 from the shipping industry, the emphasis will shift to the additional benefits that can be gained from replacing traditional fossil fuels with bio or carbon neutral fuels, such as hydrogen and ammonia. Such fuels can be burned in internal combustion engines without drastic engine modifications. For a discussion on alternative fuels, see Chap. 13 of the book.

This section presents the most practical and widely available energy efficiency measures that may be applied to ship machinery and a brief look at emerging technologies.

3.1 *Main and Auxiliary Internal Combustion Engines*

3.1.1 Propulsion and Power Generation Arrangements

The traditional propulsion and power generation arrangement for large deep sea commercial shipping has generally been a single slow-speed two-stroke main propulsion engine direct coupled to a fixed pitch propeller with three medium-speed four-stroke auxiliary engines driving generator sets. A simple, reliable, and cost-effective solution with minimum system losses has been the de facto propulsion solution since the marine industry made the transition from coal to oil as the primary energy source.

The slow-speed two-stroke engine is typically defined as one with a rated speed of less than 400 rpm and which includes a long-stroke design with the piston rod connected to the connecting rod using a crosshead construction, as illustrated in Fig. 2.8. This crosshead construction supports very long-stroke designs and enables the cylinder lubrication to be separated from the bearing lubrication systems. This feature supports the use of specialized alkaline oil lubrication of the piston/liner interface and enables accurate control of corrosion caused by high sulfur fuels. More information on the two-stroke slow-speed cylinder lubrication systems is given further below. The suitability to burn high-sulfur residual fuel oils has long been one of the key features of the slow-speed engines and has enabled the supply of cheap refinery residue fuels to the marine market. The slow-speed design has evolved to ever larger piston strokes, and stroke/bore ratios in excess of 4.5 are now common. There has also been a trend for improved fuel consumption through

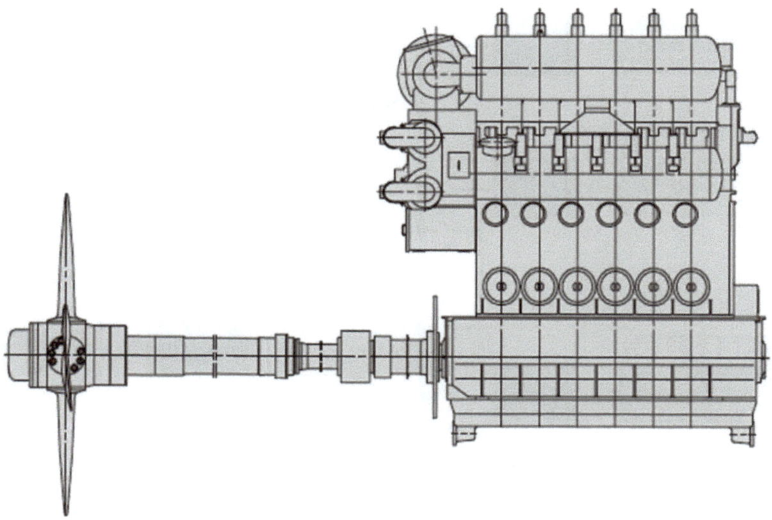

Fig. 2.8 Direct-coupled slow-speed two-stroke propulsion arrangement with CPP. (Courtesy MAN)

increased brake mean effective pressure (BMEP). Typical BMEPs have risen from 18 to 22 bar and typical firing pressures increased from 140 to 180 bar since 2005. This has supported the trend in reducing rated engine speed and the use of ever larger diameter propellers for increased propulsion efficiency.

The medium-speed four-stroke diesel engines are of a trunk piston design and are defined as engines with a rated speed between 400 and 1400 rpm. These are the predominant auxiliary engine design used for power generation on deep sea fleets and usually have higher specific fuel oil consumption (SFOC) characteristics than the slow-speed two-strokes. However, sophisticated modern medium-speed engine designs that incorporate, for example, two-stage turbocharging systems with intercooling and aftercooling, electronic fuel injection, common rail fuel supply, miller timing, etc. are now achieving SFOC values as low as the two-stroke slow-speed designs under certain conditions.

High-speed diesel engines are defined as engines with rated speeds over 1400 rpm and are also of a trunk piston design. These may be utilized as auxiliary or emergency generator sets on larger vessels and as propulsion and auxiliary engines on smaller vessels such as ferries or patrol craft where the high power to weight and power to volume metrics are a requisite. These engines have long been closer to large automotive and off-road engines and hence have included the advanced design features mentioned above for medium-speed engines for many years.

An example of a typical main and auxiliary engine arrangement is shown in Fig. 2.9. The advent of electric propulsion and hybrid systems means there are many potential variants now emerging.

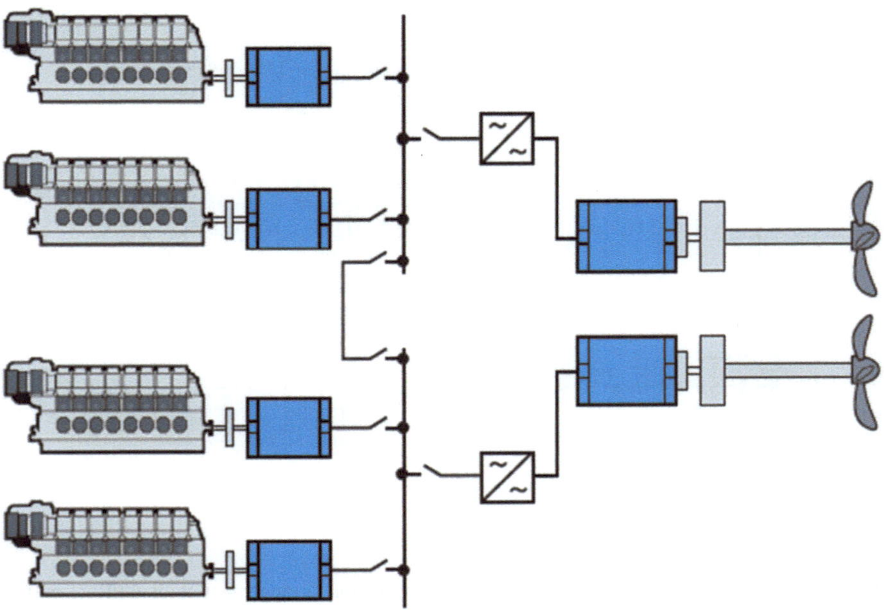

Fig. 2.9 Twin-screw medium-speed diesel electric drive and power generation. (Courtesy MAN)

3.1.2 Propulsion Engines

The slow-speed two-stroke engine is the most efficient prime mover available and can achieve fuel efficiency in excess of 52% from the base engine. This compares, for example, to an efficiency of approximately 42% from an advanced road car diesel engine. With minimal losses from the direct-coupled engine to propeller configuration, the two-stroke slow-speed propulsion arrangement, particularly with a fixed pitch propeller (FPP), provides the simplest and most fuel-efficient propulsion option.

Medium-speed two-stroke diesel engines have slightly higher SFOC, typically 3–4% higher than a two-stroke slow-speed design at similar power levels. Similarly, high-speed four-stroke engines may have SFOC levels 4–5% higher than the medium-speed designs. Since the propulsion shaft speed of medium- and high-speed engines needs to be reduced significantly to match an efficient propeller speed, these engines must be connected to the propeller through a speed reduction transmission system. This can be either through a mechanical reduction gear unit or an electric drive system. These transmission systems introduce additional losses, approximately 2–4% from gear units and 10% from an electric drive system. Hence there can be significant fuel penalties for medium- and high-speed installations compared to slow speed. Figure 2.10 shows a comparison on typical SFOC curves for slow-, medium-, and high-speed engines.

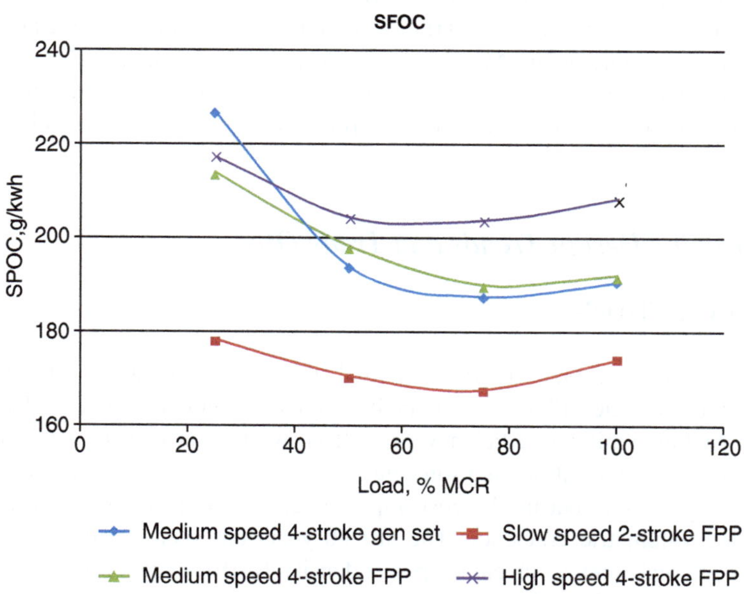

Fig. 2.10 Comparison of SFOC curves

The use of controllable pitch propellers (CPP) with constant speed engine operation and the potential for engine operation in the fuel efficiency "sweet spot" that electric drive provides are ways that some of these penalties can be reduced. The potential to apply even slower propeller speeds, or operate the propeller at its most efficient operating point by changing gear ratios, and through the delinked engine and propeller speed feature of electric drive arrangements are further ways these penalties can be reduced. This perhaps can best be demonstrated by the emerging use of high-torque, high-speed engines in variable speed and variable load electric drive arrangements, particularly in combination with hybrid features, to give potential overall ship fuel oil consumption (FOC) reductions of up to 20%.

3.1.3 Power Generation Engines

The main electrical power onboard ship is generated by the auxiliary diesel generators, which may also be supplemented by a shaft generator driven from the two-stroke slow-speed main propulsion engine. This shaft generator provides the opportunity to provide electrical power at the lower SFOC applicable to the two-stroke slow-speed engine. Ship electrical systems typically utilize alternating current (AC) architectures at a frequency of 50 or 60 Hz. This require the generators to be driven at a constant synchronous speed, which can be determined by dividing 7200 (for 60 Hz) or 6000 (for 50 Hz) by the number of generator poles (only an even number of poles are used). The larger the number of poles, the slower the generator speed, and generally the higher its cost. Medium-speed generator engines typically operate at 720, 750, 900, and 1000 rpm with high-speed engines running at 1500 or 1800 rpm, depending on the AC frequency selected and generator design.

Where engines are arranged in a diesel electric drive arrangement, the power generation engines provide power for both propulsion and ship electrical loads.

3.2 Engine Design Trends and Trade Offs

3.2.1 Design Trends

Overall engine efficiency, or brake thermal efficiency, is made up of a number of individual engine efficiencies, notably volumetric and mechanical. There are some potential significant efficiency gains from waste heat recovery systems (see below for more information) and some gains from improving mechanical efficiency by reducing friction, but the biggest improvements have generally been made by improving volumetric efficiency. The last 60 years has seen significant developments in internal combustion engine power density, BMEP, cylinder firing pressures, and efficiency. Much of this has been enabled by the increasing use and level of turbocharging technologies. A downside of the high BMEP, highly turbocharged, modern internal combustion engines is a reduction in transient response. This can be improved by the use of turbocharging and air management techniques such as exhaust waste gates/bypass, variable geometry turbochargers, and air bypass features, but as with all adjustable engine parameters, it is a compromise to achieve optimum performance. The increasing trend in required charge air boost pressures has until now largely been accommodated by single-stage turbocharging systems. Future engine designs will see an increase in the use of two-stage and sequential turbocharging systems to support high BMEP, low-emission designs incorporating features such as Miller timing.

3.2.2 Trade-Offs

An expansive discussion of engine design fundamentals and features is beyond the scope of this publication, but it is worth noting that for any particular engine design, there are a number of significant trade-offs to be reconciled between design features and settings to achieve low fuel consumption together with low exhaust emissions. Achieving lower CO_2 emissions by reducing fuel consumption often directly conflicts with achieving low emissions of other emissions species that may be regulated, for example, the nitrogen oxides (NOx) versus SFOC trade-off, the particulate matter (PM) versus SFOC trade-off, and the NOx versus PM trade-off. Figure 2.11 shows an example of these trade-off curves. Balancing these conflicting characteristics result in a compromise of settings to achieve the optimum within a particular engine.

Figure 2.11 shows a typical SFOC, or brake-specific fuel consumption (BSFC), versus NOx trade-off curve, in this instance a high-speed truck diesel engine. This curve is typical of all engines, and the lowest fuel consumptions have a tendency toward the highest NOx emissions. A step change in engine technology, such as common rail fuel injection, or the so-called second-generation common rail systems, with higher injection pressures will shift the curve toward the plot origin.

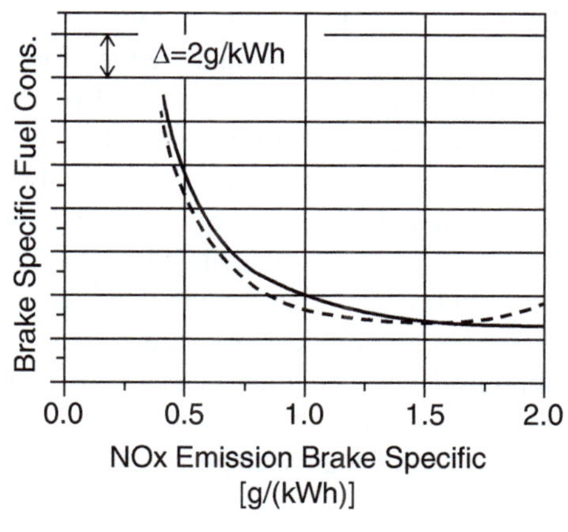

Fig. 2.11 Typical diesel engine BSFC vs NOx trade-off characteristic. (Source: IMechE)

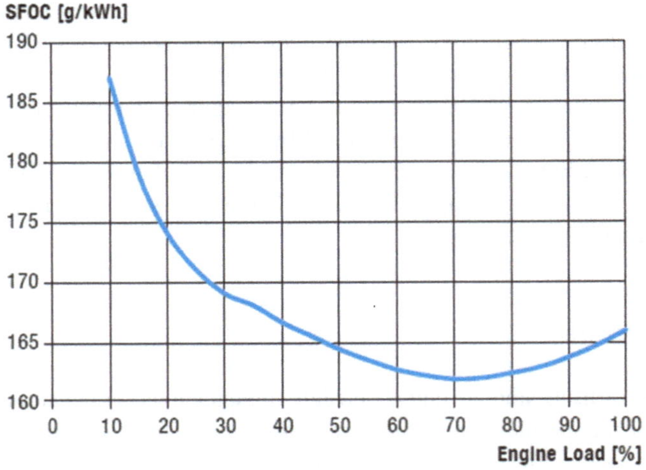

Fig. 2.12 Typical slow-speed SFOC vs engine load characteristic. (Courtesy MAN)

3.2.3 Fuel Consumption Characteristics

In addition to understanding the fuel consumption versus emissions trade-offs, to understand how to implement efficiency improvements effectively, it is also important to recognize the typical fuel consumption characteristics of internal combustion engines.

Figure 2.12 shows typical two-stroke slow-speed fuel consumption versus engine load curve, and optimum fuel consumption is in the 60–80% load range with significant increases in SFOC at lower engine loads. This plot is generated from the typical propeller curve based on the engine maximum continuous rating (MCR).

Seeing how the fuel consumption changes across the engine speed versus torque or BMEP map requires a much greater number of fuel consumption test points to be measured. These “ISO” SFOC maps are more readily available for medium- and high-speed engines and are therefore particularly useful for variable speed and variable load applications but are also important for direct-drive applications where the effects of heavy or light propellers, or propeller or hull fouling, can significantly shift the engine load away from the nominal propeller curve. This highlights the importance of accurate ship model and tank test results for the hull and propeller to enable accurate power demand estimation and therefore correct engine matching. The importance of understanding engine manufacturers’ recommended selection processes, in particular for direct-drive propulsion arrangements, and guidance on appropriate propeller and sea and power margins is critical to obtaining an efficient design that is also fit for purpose.

The use of high-voltage or DC power distribution systems, variable frequency drives (VFD), and inverters and the removal of the requirements for synchronous power generation speeds open the door for electric propulsion systems that incorporate a variety of fixed and renewable energy production and storage equipment.

However, the internal combustion engine is expected to remain at the heart of this mixed propulsion arrangement system for the time being and a key feature for targeting an overall fuel-efficient system.

3.2.4 Air Pollution Considerations

In 1997 IMO adopted a new protocol to amend the MARPOL Convention and adopt a new Annex VI "Regulations for the Prevention of Air Pollution from Ships." To balance all of the design variables and trade-offs for a particular slow-, medium-, or high-speed engine design to achieve improved performance and reduced fuel consumption, while meeting these air pollution limits, has emerged as one of the biggest challenges facing engine designers and the marine industry.

NOx formation is linked to peak combustion temperatures; therefore engine changes to optimize for fuel efficiency, for example, by increasing BMEP and maximum firing pressures (and hence combustion temperatures), can increase NOx. To shift the characteristic trade-off curves and reduce SFOC, at the same time reducing NOx, requires a step change in engine complexity or features, for example, mechanical to electronic fuel injection equipment (FIE), adoption of common rail system, or higher injection pressures.

The IMO NOx certification process incorporates steady-state testing undertaken at test-bed under reference conditions in accordance with the requirements detailed in the IMO NOx Technical Code, which was adopted by IMO at the same time as Annex VI and which is itself based on the ISO 8178 standard series for exhaust emission measurement of internal combustion engines. Each certified NOx emission value for a particular engine type is a cycle weighted value determined from the test-bed testing at discrete engine load points and the applicable weighting factor. These engine loads, or mode points, are weighted in accordance with the applicable duty cycle appropriate for that application.

The IMO NOx limits are based on engine rated speed, with the lowest limits applicable to medium- and high-speed engines, and these IMO Tier I, II, and III limits are shown in Fig. 2.13. The figure includes an example of typical NOx emissions from a Tier II slow-speed engine together with some example NOx emissions from low-pressure Otto combustion cycle dual fuel (DF) and gas engines running on gas and meeting the Tier III limit. The Tier I NOx limit was retrospectively applicable to engines fitted to ships with keels laid after 1 January 2000 once Annex VI entered into force on 19 May 2005. Once the Annex entered into force, steps were taken to progressively reduce the NOx limits, and the Tier II limit entered into force on 1 January 2011. The Tier III limit is only applicable in Emission Control Areas (ECA) and represents a NOx reduction of 80% from the Tier I limits. Currently the only NOx ECA in force is the North American ECA, which entered into force on 1 January 2016. The existing Baltic and North Sea SOx ECAs will also become NOx ECAs from 1 January 2021.

The IMO Tier III NOx emissions limits are now driving the use of new technologies and alternative fuels, such as DF engines and exhaust emission abatement

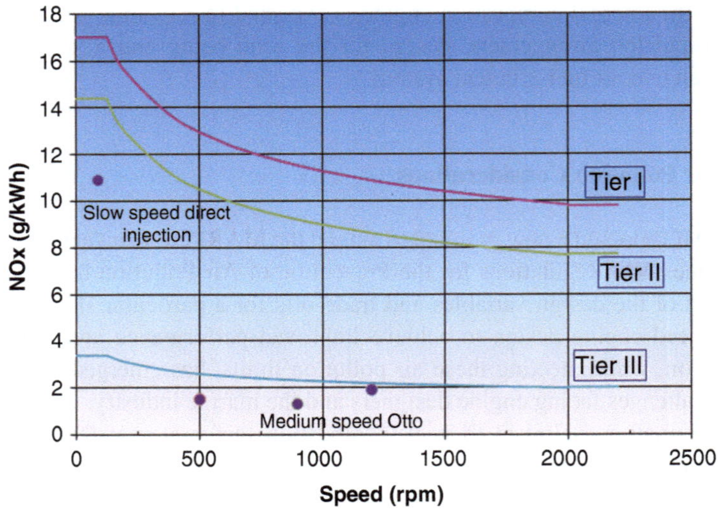

Fig. 2.13 IMO MARPOL annex VI regulation 13 engine speed-related NOx test limit

equipment. The most likely exhaust emission abatement systems to be used to meet the IMO Tier III limits are EGR and SCR. The challenge for the marine industry is the development of EGR and SCR systems suitable to the high sulfur and residual fuels currently prevalent in the marine sector.

3.3 Internal Combustion Engine Efficiency Improvements

From the ship efficiency perspective, improvements can be achieved via installation of new equipment and systems, via upgrades or modifications to existing machinery, by improved operating procedures, or a mixed combination of all three. Instrumentation and data collection equipment and analysis are essential additional requirements to verify the impact of any implemented efficiency reduction measures. This is a topic in its own right, but it is worth noting that the two most important parameters to validate any efficiency improvements, power output, and FOC are among the most difficult to measure accurately in a marine environment.

We can see from above that diesel engines generally have only one operating point in the speed versus power curve at the highest efficiency. To further understand how improvements to the main and auxiliary engine efficiency can support the ship system efficiency improvements, some of the background to engine characteristics, propulsion arrangements, and the techniques and equipment used to improve the fuel efficiency are further expanded below. Figure 2.14 shows the energy balance for a MAN 12S90ME-C9.2 two-stroke slow-speed engine design in standard configuration. As can be seen, extracting waste energy from the exhaust is the

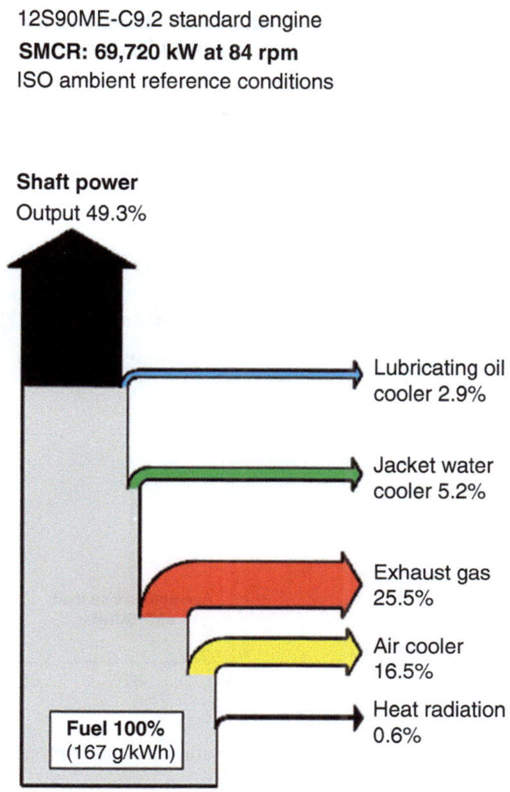

Fig. 2.14 Energy balance for MAN 12S90ME-C9.2. (Courtesy MAN)

obvious target for improving overall efficiency of the propulsion system, but there are also potential gains to be made from other areas, notably jacket and air cooler circuits.

3.3.1 Propulsion Engine Derating

One of the trends in recent years to reduce the main propulsion engine fuel consumption is by the selection of a derated engine. Engine selection is one of the critical factors to ensure acceptable vessel performance, but the overlap in possible engine types, bore sizes, and number of cylinders offered by manufacturers to deliver the required vessel design power provides the scope for a derated engine selection. Slow-speed two-stroke engine designs are typically offered with a wide range of potential engine ratings in the rating layout, with the normal MCR offered at the highest engine speed and power density.

Generally, selecting an engine type with a larger stroke/bore ratio, higher BMEP, and lower design speed provides improved fuel efficiency and gives improved propulsion system efficiency through the use of a larger-diameter, more efficient

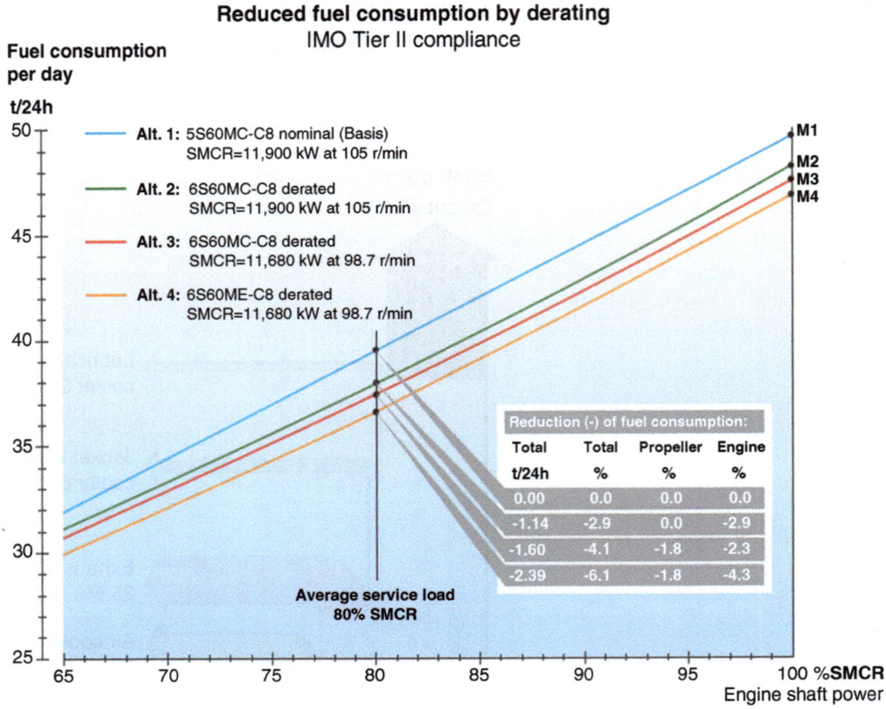

Fig. 2.15 Sample effects of derating and larger propeller on fuel consumption. (Courtesy MAN)

propeller. This is a trend supported by the latest slow-speed two-stroke ultra-long-stroke engine designs from both MDT ("G" series) and WGD ("X" type). Selecting a contract rating, sometimes referred to as NCR (nominal contract rating), CMCR (contract maximum continuous rating), or SMCR (service maximum continuous rating), that is in the lower range of the layout map provides the opportunity to run the engine at a lower engine speed and BMEP, hence a lower SFOC. An example of how this engine type and rating selection can be applied to MDT engine designs is shown in Fig. 2.15. In this example savings of 2.9–6.1% are possible with a combination of alternative engine selections and propeller optimization at lower speeds for the same power demand. The main advantages of each example are achieved by adding an extra cylinder, selecting a lower operating speed and selecting the electronically controlled version of the engine. Since operating and maintenance costs can be increased by the number of cylinders, as with all modifications, any savings by selection of a derated engine need to be weighed against the total cost of ownership through a life cycle analysis for the specific ship design and operating profile.

3.3.2 Slow Steaming

While "slow steaming" and "super slow steaming" are not design changes to improve efficiency and hence would fall under operation practices and route optimization (for which more in Chap. 10 of the book), it is worth briefly putting into context how these savings are achieved with respect to engine fuel consumption characteristics and engine modifications. The fundamental of this is the generic ship propulsion power demand of the nominal propeller curve, which can be represented typically by a cube law relationship to speed. As we have seen from previous subsections, all engine design systems, parameters, and settings are a compromise to balance the engine within the engine thermodynamic and mechanical limits while remaining compliant within statutory air emissions requirements. The engine is optimized toward the vessel design point.

For the direct-drive slow-speed main engine arrangements, reducing engine speed will obviously reduce vessel speed, but the potential fuel savings are very significant if the operator can commercially accept this and the engine is not put into a barred speed position. The cubic propeller law relationship means that only 12.5% power is needed to deliver 50% engine speed. This would put the engine into a part of the fuel consumption map that has approximately 10% worse SFOC. Although engine efficiency will vary a little across the speed range, total FOC is approximately proportional to power. Therefore, the total fuel consumption can be reduced by approximately 80% by slow steaming and reducing ship operation speed, for example, from 25 to 15 knots.

In times of high fuel prices and overcapacity, this is the low-hanging fruit and quick fix, but it may have other longer-term impacts on the engine. Since the engine is not running at the design point, there will be increased smoke and increased fouling, which would need to be managed with operational practices such as increasing engine speed for a short period of time and/or managing this with additional maintenance. In many respects the costs of these actions may be more than compensated from the savings in reduced overall maintenance costs from running the engine at much lower BMEPs, where the engine is less stressed. If slow steaming becomes a permanent mode of operation, then the efficiency can be further optimized by making engine changes, such as turbocharger cut out, to bring the engine back toward optimum operating performance at the new operating point. More permanent ship changes such as changing propellers and modifying ship bows can also be applied to optimize for the new operating speed. Figure 2.16 shows a turbocharger cut out upgrade kit from MAN where gate valves are installed to reduce the number of turbochargers in operation during slow steaming and the potential SFOC benefits, depending on the number of installed turbochargers that are cut out during slow steaming, for each engine type.

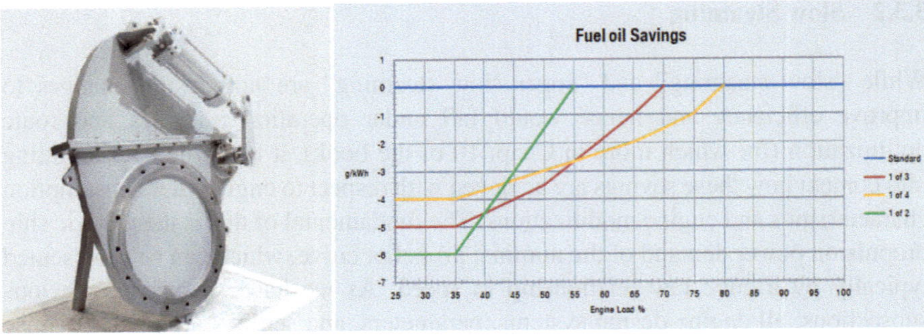

Fig. 2.16 Turbocharger cut out upgrade kit and SFOC reductions. (Courtesy MAN)

3.3.3 Electronic Engine Control and Common Rail

The adoption of electronically controlled engines represents the most important step change in engine technology that opens the door for flexibility on control of many engine related parameters, ancillary systems and the ability to achieve significant fuel consumption reductions. In a similar way to the introduction of electronic fuel injection controls to the automotive and off-road diesel engines in the 1980s, the advent of reliable microprocessors and computer controls has enabled the same transition for marine engines since around 2000. Primarily aimed at providing electronic control of fuel injection timing and fuel quantity, it is now possible, and indeed a necessity, to also control many other engine components and systems. For two-stroke slow-speed engines, the control of exhaust valve timing and lift is an additional key feature, but electronic control of turbocharger waste gates, variable geometry turbochargers, sequential turbocharging, turbocharger cut out, air management bypass valves, variable valve timing, and emission control features are just some examples of electronically controllable engine features that can be adjusted to provide the optimum fuel efficiency, transient response, and exhaust emissions settings. The balancing of these settings for any given point in the engine speed versus power map is a complex compromise to achieve the optimum fuel efficiency within the mechanical, thermal, and air fuel ratio limits and exhaust emissions limits for any particular engine type. This "calibration" of the engine electronic control unit (ECU), or "map" settings, is now perhaps the most important aspect of modern electronically controlled engines and is one that can have significant statutory air emissions implications.

From the fuel efficiency perspective, the switch from pure mechanical drive and control of fuel injection systems to electronic control can achieve fuel efficiency improvements of approximately 5%. This optimization is maximized when the electronic control is combined with application of the so-called "common rail" principles. While the maximum fuel injection pressures and phasing thereof are historically limited by the mechanical camshaft drive in conventional internal combustion engines with mechanical fuel injection, in common rail engines, the available fuel pressure curve is delinked from the engine speed. This gives the opportunity for much higher fuel injection pressures, particularly at lower engine loads, then achievable with camshaft-driven conventional mechanical drives. This enables fuel efficiency improvements and smoke-free operation.

Electronically controlled engines provide an opportunity to have accurate control of the fuel injection timing and quantity; common rail engines give the opportunity for much higher injection pressures at lower loads, and for the two-stroke designs, the exhaust control provides an opportunity to adjust compression pressures and cylinder scavenging. Medium-speed designs are also adding electronic capability for air management with turbocharger waste gates, air bypass valves, and inlet valve camshaft phasing or variable valve timing (VVT) units. The range of electronic "calibration" settings is almost limitless, but from the above trade-off characteristics, we can understand that any given "calibration map" is targeting a robust all-round performance but may be optimized for a particular application. Electronic engine control therefore represents the key engine technology shift to enable improvements in fuel efficiency and perhaps more importantly enable control of many other engine and system design features to achieve the optimum performance and efficiency while meeting the exhaust emissions regulations.

Figure 2.17 shows an example of the various electronic fuel tuning maps offered that are targeting the lowest possible SFOC in certain parts of the load curve: "delta tuning," "low-load tuning," "part load optimized." SFOC savings of 2–7% are possible compared to standard tuning and are well suited to support low-load ship operation, such as slow steaming or high-load applications.

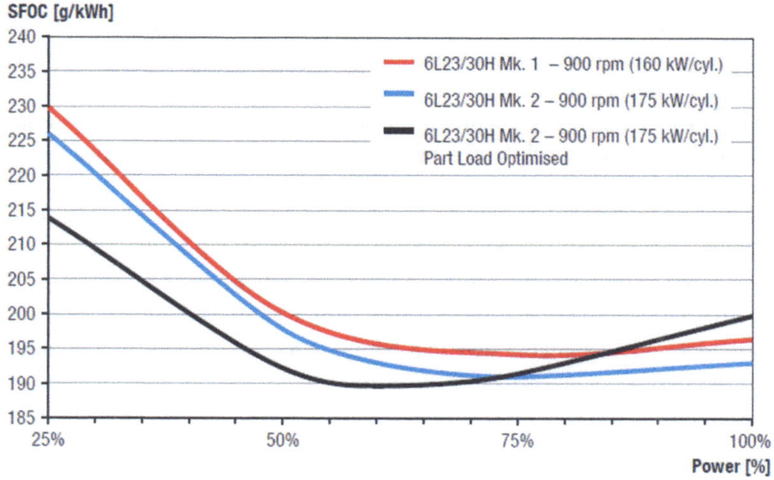

Fig. 2.17 "Part Load Optimized" electronic engine control. (Courtesy MAN)

3.3.4 Engine Instrumentation, Monitoring, and Control

Assessment of the total ship operational efficiency and machinery condition is always likely to include some element of manual data collection; however, the increase of electronic control systems enables the automation of much of this data collection. An electronically controlled engine, propulsion system, cargo control system, or other such system must be built around the capability to effect a change of settings or parameters, such as pump speed, valve position, etc. These changes are typically effected through actuators and speed control and are built around continual monitoring of basic parameters such as temperature, pressure, and position with more complex instrumentation measuring flow, vibration, or utilizing strain gauge instrumentation techniques. Complex machinery, such as engines, monitors such functions at a very high frequency that is linked to engine speed on a combustion cycle basis.

The installation of comprehensive instrumentation and the collection of such data therefore become essential to ensuring ship machinery systems, equipment, and components continue to operate in the most efficient way in service. Much of this instrumentation will be installed as part of the base machinery instrumentation, some can be added for additional capability and some would need to be added to provide a more total picture of ship operational efficiency.

To evaluate the energy efficiency of a ship's propulsion system, it is necessary to accurately measure and track fuel consumption and power. The installation of accurate power or torque measuring instrumentation and fuel consumption equipment are examples of additional ship instrumentation that may be fitted. Alternatively, these may be fairly accurately predicted from the data collected from the engine instrumentation and known as "calibration" map data. The addition of supplementary condition monitoring equipment for engines, such as bearing wear monitoring, cylinder drain oil analysis, and water in oil sensing, is a further example of additional instrumentation that can be added to improve condition monitoring and efficient operation. Full details on specific machinery or ship instrumentation are beyond the scope of this publication.

The collection of data or the use of this for optimization, condition-based monitoring, and predictive analytics is not a new technology but is an important emerging trend in the marine industry. Concerns such as cyber security, intellectual property on data, and how any service provision to help shipowners operate in the most efficient and sustainable way can be delivered in a practical way are current industry themes.

3.3.5 Energy Efficiency Optimization

When considering the total ship efficiency, it is important to consider parasitic loads as part of the operation of the main engine. The number of pumps, compressors, and other items of equipment installed is determined by classification society, IMO, and flag state regulations, based on the need for redundancy in case of failure of a running unit and to provide operational flexibility. Unit size/capacity and the number of units installed are selected to meet the most severe design conditions. These coolant, oil, and fuel pumps are typically driven by electric motors, the power for which is provided by the auxiliary generators running at four-stroke SFOC values. With these pumps driven at constant speed, the delivery rates are typically set for maximum load, and hence there is a lot of waste energy from throttling or spilling the pump output. For example, often three sea water cooling pumps are provided, each rated for 50% of the maximum sea water demand when the sea water is at the maximum design temperature. In service, the sea water temperature is often significantly below the maximum design temperature, some cooling loads may not be in operation, heat exchangers may not be fouled to the extent assumed in the design specifications, and the main engine is operating at less than its maximum continuous rating. The result is that the system's cooling requirements may be served by only one pump, thus providing the potential of saving the energy required for running a second pump. The use of variable speed motors and VFD drives can also recover some of these losses and operate the systems in a more efficient manner.

With the main engine parasitic and supplementary system loads increasing with engine complexity and after treatment demands, for example, hydraulic servo systems, DF fuel supply systems, EGR blowers, SCR heating, reductant dosing and soot blowing, etc., these parasitic loads can be significant and need to be considered within the total ship optimization plans.

3.3.6 Exhaust Emission Abatement Equipment

Exhaust emission abatement equipment is covered briefly here to highlight the impact on the base engine considerations. As discussed above, the primary international air emissions control regulations are those detailed in MARPOL Annex VI Regulation 13 for NOx and 14 for SOx. Most shipowners are expected to comply with Regulation 14 using sulfur compliant fuels and a number using exhaust gas cleaning technologies. For NOx compliance, engines are expected to install either SCR or EGR systems or to apply Otto-cycle process gas as fuel operation, to comply with the Tier III limits.

All of these Tier III technologies involve significant additional equipment and a change in the operational mode and settings of the engine. There are also associated supplementary support systems that impact the machinery space arrangements and the dosing of additional consumables, together with additional pump, compressor, and heating loads. At present there are still a small number of Tier III installations in operation, so real-world experience is limited; obtaining the optimum efficiency will require careful management of engine and Tier III technologies to ensure both environmental compliance and efficient operation.

For EGR systems there are the additional electrical loads associated with operation of the EGR blowers, operation of scrubbing water systems, the additional NaOH consumable for neutralizing acid formations in the wash water systems, and additional compressed air consumption for sealing of the EGR blower together with

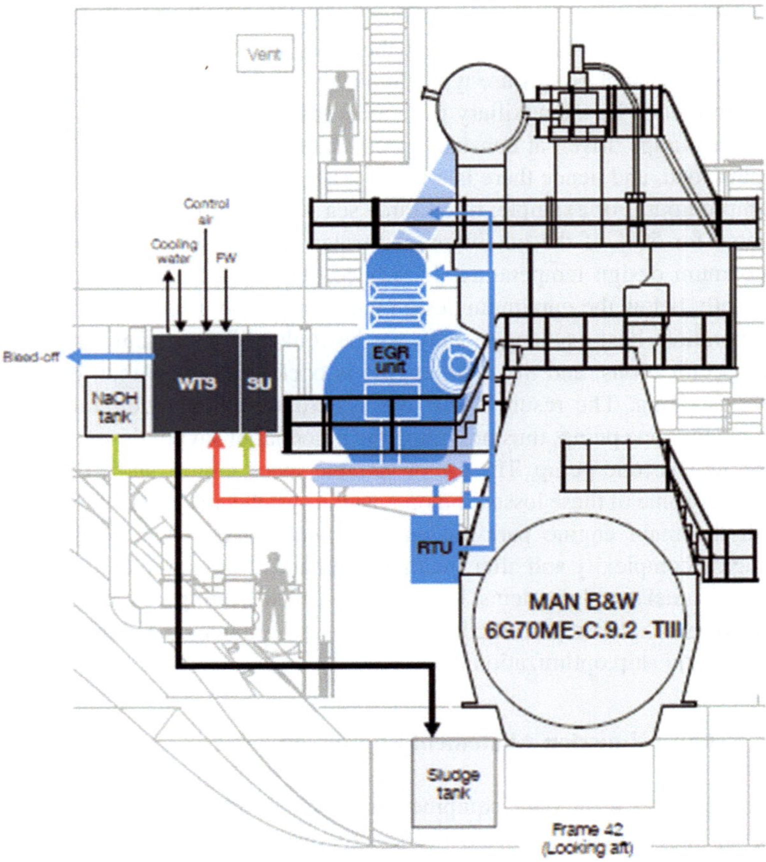

Fig. 2.18 Example of an EGR installation. (Courtesy MAN)

the increase in SFOC associated with EGR combustion. For Tier III operation at 100% engine load, the increase in SFOC can be 5 g/kWh (from the Tier II only engine); across the load range, it varies between 2 and 5 g/kWh. There is also an increase in cooling water flow required for the charge air cooling system to accommodate the higher heat load from the recirculated exhaust gases. Depending on the concentration of NaOH solution used in the water treatment system (WTS), there may be additional loads for the heating of the NaOH tank. There is the collection and disposal of the residues collected by the WTS to be considered. Figure 2.18 shows a schematic of an EGR installation, and Table 2.1 gives an example of the additional fuel consumption (compared to Tier II only engine), loads, and consumables for a slow-speed MDT 6G60ME-C9 engine with an MCR of 16,080 kW at 97 rpm operating in Tier III EGR mode.

For SCR systems the additional considerations depend on whether a before turbine, high-pressure (HP) SCR is installed or an after turbine, low-pressure (LP)

Table 2.1 Example MAN 6G60ME-C9 Tier III EGR data

Load, %	Additional fuel, kg/day	Power EGR blower, kW	Power WTS, kW	NaOH, liter/day
100	1929.6	88.4	64.3	108.1
75	1157.8	67.5	54.7	96.5
50	578.9	69.1	46.6	69.5
25	0	45.0	37.0	46.3

system is installed. In both cases there are the costs associated with the dosing of the reductant (typically UREA) together with air supplies for reductant injection and soot blowing of the SCR reactor. For HP systems there is the additional heat load that will be necessary to heat the SCR reactor, probably through electrical trace heating, and for LP systems, this will be seen through the additional cost of supplying fuel to the exhaust gas burner fitted upstream of the SCR reactor to raise exhaust gas temperatures. These reactor heating loads will increase dramatically with extended low-load operation. SCR operation will increase SFOC compared to a low-load tuned engine across the load range but may actually have lower SFOC (compared to a high load tuned engine) at the 50% and 25% load points; an increase of 2 g/kWh at 100% load is typical. The auxiliary blowers will need to be upgraded from the standard arrangement since they will need to be capable of being operated across the whole engine load range and require approximately 2.2 times the capacity of standard blowers. It is also important to note that even though there are control systems and operation strategies in place for reductant dosing and to minimize ammonia slip, it is likely that ammonium bisulfates would form in the exhaust gas boiler or economizer if operated at the same time as the SCR. Therefore it is strongly recommended to install a bypass of the boiler for when the SCR is in operation. SCR catalysts have a finite life which will depend on many factors; the catalysts may be considered consumables. Figure 2.19 shows a schematic of an HP SCR installation, and Table 2.2 gives an example of the additional (and reduced) fuel consumption (compared to Tier II only engine), loads, and consumables for a slow-speed MAN 6G60ME-C9 engine with an MCR of 16,080 kW at 97 rpm operating in Tier III mode with a HP SCR.

The above illustrates that there are considerable additional equipment, electrical loads, and costs associated with the operation of Tier III technologies to be considered. Tier III operation may also have indirect impacts on operation of other waste heat recovery systems and hence impact the total ship efficiency. The actual costs will be very dependent on the specific operational profile, time in ECAs, and engine load profile when operating in the ECA. The advent of Tier III technologies is challenging the status quo of traditional ship and machinery space designs, and highlighting that a ship-specific assessment for total cost of ownership is the only way to determine the most fuel efficient and sustainable way to operate ships of the future.

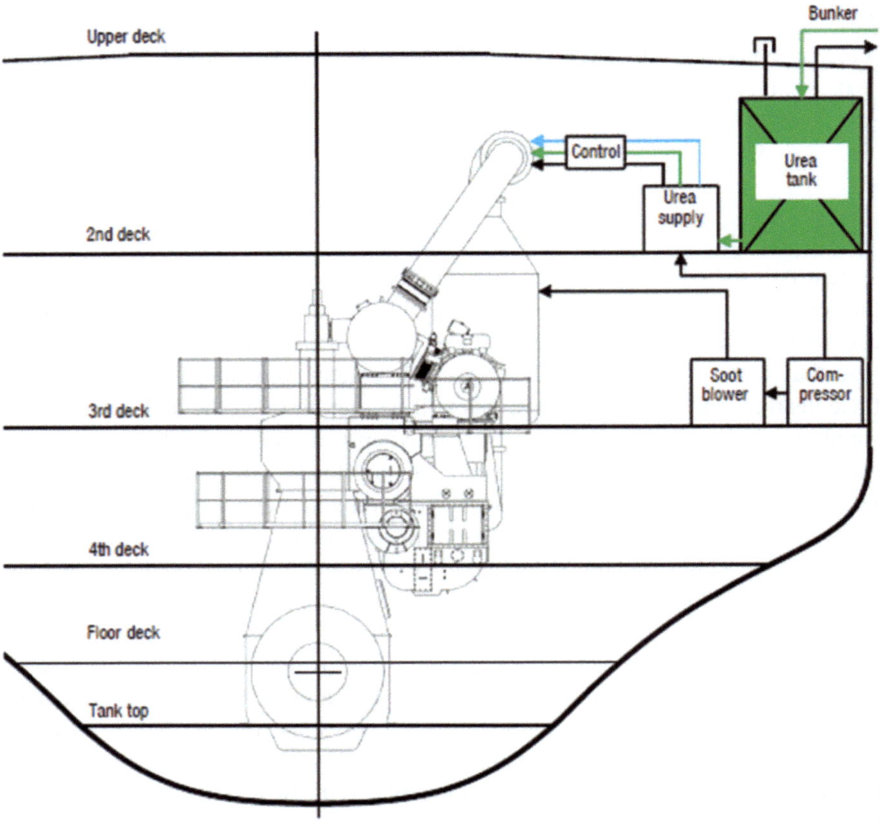

Fig. 2.19 Example of an HP SCR installation. (Courtesy MAN)

Table 2.2 Example MAN 6G60ME-C9 Tier III HP SCR data

Load, %	Additional fuel, kg/day	Power, kW	UREA, liter/day
100	771.84	80.4	6560.6
75	144.72	80.4	4920.5
50	−675.36	80.4	3280.3
25	−964.8	80.4	1640.2

3.4 Waste Heat Recovery

While modern diesel engines are very efficient, they still generate a large amount of waste heat when running at full load which can be utilized to improve the overall propulsion system efficiency. Figure 2.20 shows an example of the MAN 12S90ME-C9.2 (previously shown in Fig. 2.14) increasing the plant efficiency from 50% to 55% by the use of waste heat recovery (WHR) techniques.

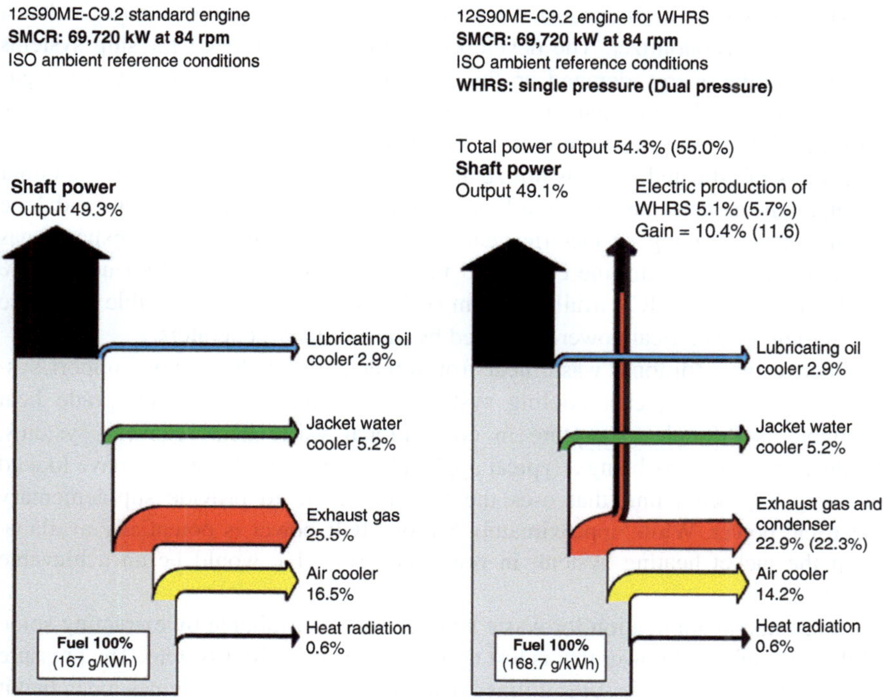

Fig. 2.20 Example of increase in overall propulsion engine efficiency by WHR. (Courtesy MAN)

As can be seen, about 5% of the fuel energy goes to the engine jacket cooling water system, and about 25% is contained in the exhaust gas. For many years it has been common to use the heat from the main engine jacket cooling system to generate fresh water and the heat in the exhaust gas to generate steam for heating. As the size of the ship and its engine increase, the amount of exhaust heat available increases much more rapidly than the demand for steam for heating. This is because the primary uses for the steam are heating oil tanks and accommodation spaces. For most commercial ships, the total size of the accommodations is about the same, and the amount of steam for oil heating grows only slightly with the engine size. This results in a surplus of heat available on ships with large engines after the more traditional services have been fulfilled. The 15% waste energy in the air cooling circuit is another potential source of WHR. In all cases, the actual quantities of WHR available and the efficiencies of the WHR systems need to be carefully considered with respect to the available waste heat at any particular engine load point. For example, the slow steaming operation mentioned above may impact the available waste heat to the extent that it is just not efficient to extract by the WHR plant. However, the right systems and operation modes can significantly improve overall engine efficiency. A few of the common WHR systems are discussed below.

The simplest form of exhaust WHR is by the use of a steam generating exhaust gas boiler or economizer. The developed steam can be utilized for ship systems and reduce the energy demand on the ships auxiliary boiler. Typical exhaust gas boilers are available in ranges from 0.1 to 21 MW with 0.2–17 t/h steam capacity and specifically designed to minimize soot build up.

More sophisticated systems utilize an off-engine skid unit that comprises either a standalone exhaust gas-driven turbine driving an electrical generator, a standalone steam turbine driving an electric generator, or a unit containing an exhaust gas turbine and a steam turbine connected to a common generator. With up to 10% of the main engine MCR available from such a WHR unit, it is possible to reduce the amount of electrical power generated by the auxiliary generators.

Recovering additional waste heat from other parts of the engine support systems, such as the jacket cooling system, is possible, but the low-grade heat available is difficult to capture in conventional waste heat recovery systems, freshwater generators being a typical application. Several pilot studies have looked at using a process unit that uses the Rankine cycle to provide supplementary electrical power. While approximately 5% of MCR power is potentially available from the jacket heating system, in practical terms, 1% would be an achievable amount.

A promising application for waste heat recovery is available by extracting some of the 15% of MCR power that is lost to the charge air cooling system. The pressure ratios and high boost pressures utilized in modern turbocharged engines mean that it is not unusual to need to cool charge air from temperatures over 150 °C at full load, which represents a potential higher grade heat source. However, the quality of the available waste heat is very dependent on engine load. The greatest benefits would perhaps come where the recovered energy could be used in association with existing steam turbine waste heat recovery units, in a feedwater preheater arrangement. The use of two or multistage air cooler units would be necessary and would add to engine complexity and cost but can contribute to obtaining the maximum achievable WHR from the installed systems.

3.5 Auxiliary Equipment

Adopting a complete ship system approach to energy efficiency means any assessment needs to consider the potential improvements to be gained from the auxiliary equipment; this section looks at some of that auxiliary equipment.

3.5.1 Shaft Generator

The addition of a shaft generator powered by a two-stroke slow-speed main engine gives the potential to generate electric power at low SFOC but under certain conditions. There are several different types of shaft generators in common use on ships. The simplest type is a shaft generator connected to the main engine by a gearbox with a fixed gear ratio. To obtain constant frequency electric power, the main engine must operate at constant RPM, which requires the use of a CPP. Such a shaft generator cannot operate in parallel with the ship's auxiliary generators since the main engine speed variation will vary more than the diesel generators speed, particularly when the ship is pitching in waves. The transient response of the two different engine types is also very different which makes load sharing at constant frequency difficult. The losses from such gear-driven shaft generators reduce efficiency to approximately 92%, and operating the CPP anywhere other than full load will put the propeller into a less efficient point of operation. This type of shaft generator will therefore only offer overall fuel-saving benefits where the engine and CPP are operated near full load for long periods.

Alternative shaft generators are available that have either variable ratio gears or frequency control. Both of these types can work with a fixed pitch propeller over a range of RPM (usually 75–100% RPM), alleviating some of the issues with the constant gear ratio shaft generator. However, these shaft generators are more expensive and less efficient. Typical efficiency for a variable speed gear drive is 88–91% and for the variable frequency shaft generator 81–88%. If the incorporation of a shaft generator can enable a reduction in the number of auxiliary generators, then it is a viable option for improving overall efficiency and maintenance costs. The greatest benefits however perhaps would come where the unit is a combined generator/motor and used in a hybrid PTO/PTI configuration – see below.

3.5.2 Number/Size of Ships Auxiliary Generators and Power Management Systems

The number and size of installed auxiliary generators is chosen to provide sufficient power for the electrical loads for various modes of operation of the vessel, with sufficient standby power to meet SOLAS requirements and replace the largest generator in operation should a failure occur. For some ships the use of a shaft generator can be sized to provide all hotel loads during ship voyages and hence avoid operation of the higher SFOC generator sets. However, in most cases, the generator sets will be in operation, and as we have seen from previous sections, optimum efficiency only occurs in a small part of the engine speed/load map. So the target loading of the generators should be to keep the engines within this maximum efficiency operating point. The use of Power Management Systems (PMS) to automatically determine how many of the installed generators should be in operation simultaneously and how each of those is loaded therefore becomes an essential tool in obtaining maximum ship efficiency.

3.5.3 Heating, Ventilation, and Air Conditioning (HVAC)

While heating, ventilation, and air conditioning (HVAC) systems are typically not large consumers of power on commercial cargo ships, a holistic approach to the total ship systems should assess the systems for potential improvements in design and/or operation. For example, modern heating and air conditioning systems incorporate preheating and recirculation features that reduce the required energy input. Upgrading to this type of system can offer significant reductions in operational energy demands. Similarly, ventilation fans may frequently be sized for the maximum air change requirements and controlled with crude speed controls. The use of variable speed motors and the use of automated control systems can significantly reduce the energy demands and should be considered along with any changes to operation that can be implemented to encourage low energy usage.

3.5.4 Variable Speed Motors: Pumps and Fans

In a similar manner to the reductions achievable with the engine parasitic loads detailed above, the use of variable speed motors and VFDs can improve the operating efficiency of pumps and fans that operate at variable loads in other ship systems. With a variable speed pump, the required flow rate can be achieved at a reduced head by slowing the pump down. Although the variable speed system consumes slightly more power at full load, pumps are rarely operated at maximum demand. Therefore, there are significant savings to be gained over the range of flow rates that the pump would typically operate. Similar benefits can be obtained from variable speed control of all other auxiliary equipment onboard.

3.6 Hybrid Systems and Equipment

One of the promising areas for future ship propulsion system developments, and the potential for significant ship efficiency improvements combined with lower emissions, comes from the adoption of the so-called hybrid technologies. One of the key enablers for this is a switch to electric propulsion systems. However, the additional complexity and inherent system losses do pose challenges, while the overall potential gains can be significant. The connected equipment and system nodes, such as the use of DC systems, increased control electronics, system integration, data logging, online optimization, data collection, etc., all feed into a future for ships and ship propulsion systems, where connectivity and ship power grids allow the integration of unconventional energy production and storage systems with conventional power generation and propulsion systems. Hybrid propulsion and hybrid ships: an evolutionary rather than revolutionary approach, but a key enabler to more efficient and sustainable shipping in the future.

The internal combustion engine will remain at the heart of these hybrid ships for many years due to its high power density but to a different level of integration and complexity than has been traditional. The propulsion norm may no longer be large, slow-speed main engines directly coupled to propellers but may shift toward medium- and high-speed power units supplementing the power in energy storage systems for use in electric propulsion units and for hotel loads. The use of electric propulsion pods and azimuthing thrusters will become more prevalent and are particularly suited to applications requiring accurate station keeping. Eventually the internal combustion (IC) engine may be completely replaced by fuel cells, but the difficulties with hydrogen as a fuel, the poor transient response characteristics of fuel cells, and the challenges for development of the hydrogen economy may mean that fuel reforming remains a part of the fuel cell deployment strategy for many years and fuel cells will only form part of the total power system. The types of fuel supplied to the marine industry in the years ahead will have a large influence on how the details of the hybrid systems will evolve, including the continued use of IC engines.

Though some have seen a shift to gas as the next big step in ship propulsion, i.e., from sail to coal and from coal to oil and from oil to gas, the shift is likely to be to electric propulsion using a variety of "fuel" sources. Sooner, rather than later, all energy and transport infrastructure will need to shift from fossil fuels for climate change reasons. If bio fuels, or perhaps more accurately, carbon free or carbon neutral fuels, can be developed and supplied in sufficient quantities and at competitive prices, then the ships of the future will still look much like they are now. It is just there will be increased use of hybrid power generation and storage systems, and ships will be thought of as total "electric" systems rather than just by a handle linked to propulsion method or fuel. To achieve this will require a multilateral approach to how primary energy is produced. It makes sense to generate hydrogen from a "clean" source, such as land-based solar or wind and deliver this as a clean fuel for use in the transport sector, either directly as hydrogen or in a hydrogen carrier fuel.

The early marine hybrid adopters will be the local and short sea shipping sectors where limited range and frequent refueling will not hinder performance and operation. The most suited applications are those with large transient power requirements and where continuous operation at high load is not a dominant part of the ship operation profile; but transient operation is a factor for all ships, and all ships can benefit from some form of energy storage system to smooth out the transient demands and improve overall efficiency. This energy storage approach enables equipment such as fuel cells or DF engines that have reduced transient capability to be efficiently deployed. It is no longer important for the engine to meet traditional transient response requirements or be type tested across a wide speed/load map that is not appropriate for its use but merely that the engine is capable of delivering power reliably and efficiently at a constant speed within a small-speed/load window. It is therefore the ship power generation and storage system, and how that system is managed, that must meet the transient performance that a particular ship type needs.

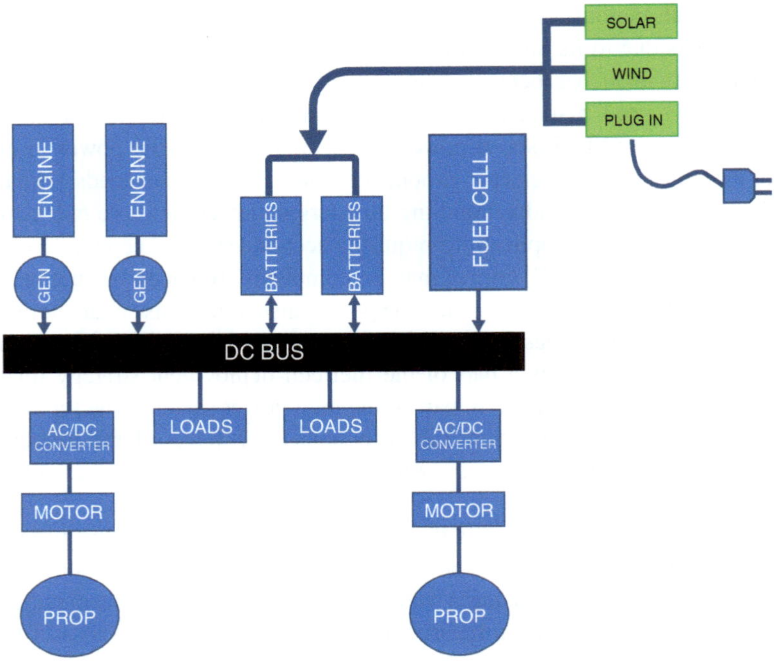

Fig. 2.21 Hybrid power system example

While wind and solar may find some small niche supplementary power generation capability for certain ships, the dominant energy storage technology being deployed will be batteries. The ratio of batteries to engines and fuel tanks will be what changes between a given ship type based on peak power demand and vessel range requirements. The use of batteries enables "cold ironing," or perhaps a more appropriate name would be "plug in hybrid," when at berth so that power generation requirements for the ship at berth can be delivered from clean shoreside sources and the power used to fully charge the batteries for use in the next sea deployment. The combination of energy storage and electric drive also enables ships to be operated with no exhaust emissions in sensitive air quality areas such as ports, rivers, and estuaries or even completely within ECAs.

Figure 2.21 shows how a hybrid ship with a variety of energy storage and production equipment is integrated by connection to a DC grid.

3.6.1 Batteries

Battery technology has advanced quite rapidly over the past few decades with significant steps being made away from the traditional lead/acid battery chemistry through nickel cadmium (NiCd) and nickel-metal hydride (NiMH) chemistries to lithium ion. These advances have obviously been seen in the development

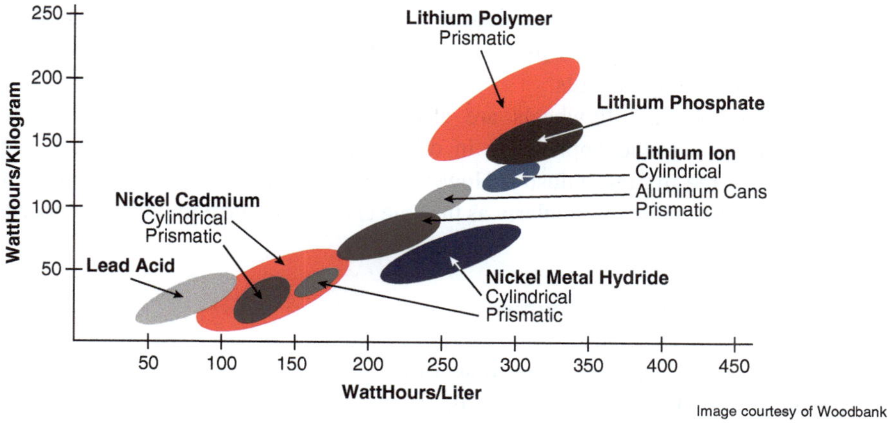

Fig. 2.22 Battery development (ABS 2017)

of batteries for mobile phones and laptops, but the energy density and weight advantages of the lithium-ion batteries have now enabled significant performance increases in the range of electric and hybrid vehicles. While still facing challenges on energy density and cost, lithium-ion batteries represent the most practical energy storage unit for marine applications. Figure 2.22 shows the development in energy density of the various battery chemistries.

One of the critical factors for lithium-ion batteries is thermal management. Battery cooling requirements and the prevention of thermal runaway are perhaps the most significant practical and safety issues to be considered when integrating batteries into a hybrid ship. The battery is no longer a simple cell, or group of cells within a casing, but is typically a power unit with its own battery management system (BMS) measuring cell temperatures and controlling charge and discharge rates. The cells, BMS, and sensors may be referred to as a module and the battery pack comprised of a number of modules. The battery packs may themselves be grouped in an array or system to complete the battery power storage and supply unit. Safety aspects and battery system application issues are addressed in, for example (ABS 2017).

The main application for battery systems will be for balancing loads and peak-shaving where they can act as the transient buffer in the system both to supply and absorb energy when there is excess production. This can allow generator sets to be operated at a near constant load at the most efficient load point. This capability also has the potential to improve operational efficiency by reducing maintenance on engines due to optimal loading and reduced engine running hours. Battery power also enables a vessel to operate in electric mode in port or during transit to give a zero emission operation mode or can be used to supplement propeller power when high speed is required.

Dependent on the type of ship, significant savings are also possible if the battery system is large enough to be considered a standby power source and may mean fewer generators need to be installed or in operation. The ability to be used to prevent ship blackout and act as an emergency power source provide additional redundancy and safety benefits.

Supercapacitors and flywheels are also potential alternative energy storage devices that have found application in other industries and are therefore being considered for the marine industry. However, both are unlikely to find significant marine application soon, particularly as the sole energy storage equipment on board.

3.6.2 Alternative Energy Sources

As indicated above, wind and solar power may provide some niche supplementary power on certain ship types. However the energy density is very low, and the area and volume requirements on board the ship make these technologies difficult to implement in a practical manner. For example, utilizing all available surface area for photovoltaic (PV) installations on a bulk or oil carrier may enable generation of 2–10% of main engine power but places sensitive equipment in cargo areas. The costs for PV cells have dropped dramatically in recent years, so this may provide a viable payback in certain circumstances; however the most promising alternative energy source in the long term is the fuel cell.

The fuel cell concept can be traced to the 1830s but did not find commercial application until deployment in the space program. A fuel cell is an electrochemical cell that produces electricity by a chemical process reaction from hydrogen rich fuel and air supplies. Ideally the only emission from a fuel cell is water. As indicated above, a pure hydrogen fuel supply provides the simplest fuel cell arrangement, and this has found limited application in the automotive car and bus sectors. Fuel cell power densities and cost are approaching levels where they can be considered a viable alternative to the internal combustion engine.

There are however still significant challenges with the use of hydrogen as a fuel, and commercial marine application is likely to come where the fuel cell is close coupled with a reformer to produce hydrogen rich fuel from a fuel source such as natural gas or methanol. There are still challenges with the packaging of this type of fuel cell power system, as well as the inherent poor transient performance, fuel sensitivity of the fuel cell stack, and issues with excess fuel or fuel slip in the exhaust stream. When looking at the full hybrid system approach, fuel cells could be incorporated in the power generation system as an alternative auxiliary power source and potentially replace one, more or all of the diesel generator sets. When combined with a battery storage energy unit, the fuel cell represents a viable part of the hybrid power mix. The marine regulations for fuel cells are still under development but will form part of the IMO International Code of Safety for Ships Using Gases or Other Low-Flashpoint Fuels (IGF Code).

There are many different types of fuel cell available that are largely characterized by the type of electrolyte: proton-exchange membrane (PEM), alkaline, phosphoric acid, molten carbonate, and solid oxide fuel cells. The high-temperature PEM fuel cell is emerging as one of the most suitable for marine applications.

4 Ballast Water Management

Shipping moves over 80% of the world's commodities and transfers approximately 3–5 billion tons of ballast water internationally every year. Ballast water is essential to the safe and efficient operation of shipping, but it also poses a serious ecological, economic, and health threat through the transfer of invasive aquatic species inadvertently carried in it.

The transfer of invasive marine species into new environments via ballast water has been identified as one of the major threats to the world's oceans. In response, the International Maritime Organization (IMO) adopted the Ballast Water Management Convention (BWM Convention) in 2004, which later entered into force 8 September 2017.

The BWM Convention includes two-tiered steps to comply with its requirements, which apply to all vessels irrespective of age, size, type, or trade, unless trading in domestic waters, naval ships, or for ships that do not discharge ballast water.

While the IMO aims at regulating ballast water in a similar manner worldwide, individual countries have the right to enforce their own domestic regulations. The most important of those countries that have local ballast water requirements that are different than those of the IMO is the United States. Australia is another example, but the majority of the Australian regulations are similar to those of the IMO.

4.1 Requirements Under the BWM Convention

Regulation B-3 of the BWM Convention stipulates the dates at which ships flying the flag of a Party or discharging ballast water in the waters of a Party must comply with the D-1 standard or the D-2 standard.

The D-1 standard applies as of 8 September 2017 and requires ships to perform mid-ocean exchange of their ballast water. The exchange must ensure that at least 95% of the water is exchanged and can only be done following one of the three methods:

Flow-through: which means the water is pumped into a full tank and out on deck through adequate openings, long enough to ensure exchange of three times the volume of each ballast tank

Sequential: which means the water is emptied and refilled

Dilution: which is similar to flow-through, only the ship ensures that the ballast tank level is kept constant until three times the volume exchanged. Dilution applies to ships with ballast tanks partially filled.

In addition to the above, ballast water exchange is to take place as follows:

Whenever possible, at least 200 nm from the nearest land and at least 200 m in depth.
In cases where the above is not possible, at least 50 nm from the nearest land and at least 200 m in depth.
When the above is not possible, designated areas or ballast water exchange must be used. In all cases, ships are never required to deviate from their original routes to meet the requirements for exchange stipulated above.

The D-2 standard applies to new ships keel-laid after 8 September 2017 and to existing ships mainly at their first IOPP Renewal Survey after 8 September 2019. The D-2 standard is a biological performance standard that requires all ballast water discharge to not exceed:

Ten organisms/m^3 for organisms with size larger than 50 μm
Ten organisms/mL for organisms with size between 10 and 50 μm
Toxicogenic *Vibrio cholerae* (O1 and O139) with less than 1 colony forming unit (cfu) per 100 ml or less than 1 cfu per 1 g (wet weight) zooplankton samples
Escherichia coli less than 250 cfu per 100 ml
Intestinal *Enterococci* less than 100 cfu per 100 ml

To meet the requirements above, ships have several options, including non-discharge of ballast water, discharge to reception facility, or treatment onboard. Treatment onboard is required to be done by a type-approved ballast water management system (BWMS), following the Code for Type Approval of BWMS.

4.2 *Requirements in the United States*

For ships trading in the United States (US), different sets of requirements are applicable when discharging ballast water in the United States.

4.2.1 Federal Regulations Under the US Coast Guard

The Federal Regulations falling under the USCG require exchange or treatment (by a USCG Type Approved BWMS, not only IMO Type Approved). The compliance dates for treatment are the first scheduled dry-dock after 1 January 2016 or 1 January 2014 (depending on the ballast capacity of the ships), with extensions to those dates issued by the USCG in case the ship cannot find suitable USCG Type Approved BWMS.

4.2.2 Federal Regulations Under the US Environmental Protection Agency

The US EPA regulates ballast water through the Vessel General Permit (VGP), following in general the same standards as the USCG Regulations, but requiring annual testing of the Ballast Water and reporting back to the EPA.

4.2.3 State Regulations

Individual states in the US are allowed to have their own additional requirements to the discharge of ballast water under the VGP. The State of California is the most active state, requiring additional measures on top of the USCG and the EPA.

4.3 Ballast Water Management Systems

Ballast Water Management Systems (BWMS) are the most common way for ships to comply with the D-2 standard and the US Regulations. In order to do so, BWMS must be type approved by an Administration and the USCG, following the BWMS Code requirements of the IMO and the §162.060 requirements of the USCG.

Type Approval consists of three parts, (a) Readiness Evaluation where the system's ability to meet the requirements, its documentation, and test plans are evaluated. (b) Once satisfied that the BWMS is ready to be tested, a series of land-based tests (five tests for each salinity: fresh-, brackish-, and marine water) is conducted with challenge water to verify the efficacy of the BWMS. A series of three shipboard tests (IMO) or five (USCG) is also required onboard a commercial vessel. The electric and electronic components of the BWMS are tested for environmental compatibility. (c) When the test reports show that water treated by the BWMS managed to pass the D-2 standard, a type approval application is submitted to the IMO Administration and the USCG for issuance of the type approval certificate.

4.4 Technologies Used in BWMS

Treatment technologies can be divided into three bulk areas: mechanical, chemical, and physical. In those main categories, it is possible to identify 12 main processes divided in turn into some 23 specific types that are used by the industry today. Table 2.3 is a summary of technologies used by BWMS.

We will briefly introduce the technologies that are mainly used by ships, which are filtration, UV, and electrolysis.

Table 2.3 Overview of technologies used in BWTS

Main technology	Technology	Sub-technology
Physical	Ultraviolet (UV)	Low pressure
		Medium pressure
	Ultrasound (US)	
	Cavitation	
	Deoxygenation	Inert gas stripping
		Nitrogen injection
	Heat treatment	
Mechanical	Filtration	Screen filters
		Disk filters
		Hydrocyclones
		Magnetic separation and coagulation
	Pressure drop	
Chemical	Electrolysis	Electrolysis
		Electrocatalysis
		Electrochlorination
	Ozonation	
	Chemical injection	Sodium hypochlorite
		Chlorine dioxide
		Other
	High energy plasma	
	Advanced oxidization	Titanium dioxide
		AOP: Ozone + UV
		AOP: Other

4.4.1 Filtration

The aim of filtration is the separation of larger organisms and solids from ballast water. Most BWMS using mechanical processes use screen filters. Filters are always used as a pre-step to another technology, for example, UV or electrolysis.

Screen filters range from 10 to 50 μm screens weaved in many ways and according to different standards. Even references to screen sizes are not standardized so knowing exactly what is meant with a 50 μm screen can be a challenge and can differ from one vendor to the other. All screen filters in the market are of a self-backwashing design, creating a challenge related to the ballast water pumping capacity of ships as filters typically use the same ballast water to backwash, reducing significantly the flow rate during the backwashing period. Up to 30% of flow rate loss can be expected during the backwashing period, which is significant in the cases where filters backwash continuously depending on the conditions of the water being filtered. The installation of a backwash pump will increase the volume of water being backwashed and so increase the loss in ballasting capacity.

Screen filters are currently used together with most other processes in ballast water treatment with the main aim to remove larger organisms (pore sizes of 25–50 μm are mainly used) and reduce the number of solids in ballast water.

Screen filters will generally not reduce the amount of sediments in the ballast tanks as most sediment in the seas where ships take on ballast water are fine silt and clay with nominal pore sizes between 2 and 10 μm. However, caking is a known phenomenon where small-sized TSS can clog a filter pore.

4.4.2 UV Technologies

All BWTS using UV use amalgam lamps surrounded by quartz sleeves to produce UV light. Generally, at doses used for disinfection of water, UV light changes the molecular structure of DNA in organisms and thereby prevents them from reproducing. New interpretations of the regulations by the USCG have led the industry to increase the dose significantly in order to kill the organisms directly, not only damaging their DNA.

The majority of UV-based BWMS use medium-pressure amalgam lamps. UV efficiency depends on five main parameters:

The type of lamp used (low pressure or medium pressure)
The length of the lamp being used (the arc length)
The physical design of the UV's water exposure chamber
The water flow rate through the UV's exposure chamber
The condition of the water being treated

With items 1–3 being fixed by the design of the BWMS without the possibility to change, and the flow rate (item 4) being tested at its continuous maximum (Treatment Rated Capacity or TRC), the only variable affecting the efficiency of UV lamps is the condition of the water being treated, which will also affect the amount of energy needed to clean the ballast water.

Of all water quality parameters, ultraviolet transmittance (UV-T) is the most important. This is because the UV-T of the water will determine how well the UV light will penetrate the water in order that the pathogens in the water may be exposed to sufficient UV light to be inactivated. Although parameters such as POC, DOC, and turbidity all influence the extent to which UV light penetrates the water, they are all effectively accounted for by the UV-T reading. Total suspended solids (TSS) is also important. TSS is important because of the phenomenon known as "shielding" whereby the pathogens can be "shielded" from the UV light by the particles suspended within the water.

4.4.3 Electrolysis

Electrolysis is the process of oxidizing seawater through an electrolytic process using all or part of the seawater as the source of the ions. Electrolysis is by far the most used in situ process in ballast water treatment.

Both temperature and salinity are critical parameters affecting the efficiency of electrolysis.

In general terms, and common to all electrolysis processes used by BWMS, the lower the temperature, the higher energy you need to produce hypochlorite and disinfect the water. The increase in energy need by the BWMS follows an increasing exponential curve. Normal lower temperatures for operation of electrolysis processes in BWMS range between 10 and 17 °C, although some manufactures claim that their electrodes would still be efficient at 1 °C. Low salinity makes it difficult for those processes to generate disinfectants.

The last common issue to all electrolysis processes is the generation of hydrogen and chlorine gases that are explosive and toxic. High temperatures and high salinity of water are ideal for the generation of large volumes of hydrogen. Mixture of hydrogen and chlorine has a wider range of flammability than mixtures of hydrogen in air and so must be avoided. Management of dangerous gases is an important parameter to consider when installing BWMS using electrolysis on ships.

In ballast water management, electrolysis has been applied in two ways:

1. Side stream where a small percentage of water is taken from the main stream of ballast water and stimulated by a certain voltage difference to create the hypochlorite and other chemicals needed to disinfect the main stream, once injected back into it.
2. Full stream where the complete flow of water is stimulated by the voltage difference.

The side-stream solution is by far the most common when applying electrolysis as a process in the BWMS in the market. Some advantages of side-stream injection of in situ generated hypochlorite are the ability to overcome the temperature challenge by applying heating jackets to the side-stream pipe and to overcome the salinity challenge by using a storage tank with adequate water (salinity and temperature) to drive the treatment process through at least one ballasting sequence.

4.5 Compliance Challenges and Alternatives

As the BWM Convention's aim is to reduce the risk of spread of non-indigenous species through ships' ballast water, many questions started popping up regarding the usefulness of the D-1 and D-2 discharge in certain trades and areas:

- Short sea shipping including especially ferries like in the North Sea; Baltic Sea; the area around Singapore, Indonesia, and Malaysia; the area between China, Korea, and Japan; the intra-Great Lakes trade; etc.
- The effect of biofouling on spread of invasive species

We will shortly discuss the problematic aspects of the BWM Convention, although those issues deserve their own book digging deep into the technical and economic aspects of this regulation.

4.5.1 Short Sea Shipping

While the most known aspects of the BWM Convention are its requirements to exchange ballast water or treat ballast water, other less widely applied or discussed alternatives include the use of freshwater as ballast water, exemptions from the requirements and exceptions to the requirements.

Freshwater as Ballast Water

It is a common misconception that using freshwater generators onboard ships should be good enough for that ship to meet the D-2 standard when it uses that water as ballast water. The IMO through the course of many years had long and detailed discussions about this issue where it was concluded that while fresh or potable water generated onboard might meet the D-2 standard, those generators must go through a type approval process like any other BWMS to prove their ability to consistently meet the D-2 standard under the challenging conditions the type approval process presents.

The BWM Convention does not allow use of fresh, municipal water taken from shore as being ballast water meeting the D-1 or D-2 standard, so this is not an option for ships under the BWM Convention. However, the USCG opens up for such a possibility by allowing ships to take US municipal water and discharge it in the sea in the United States.

While at first sight, the IMO regulations seem unreasonable as water suitable for drinking should be good enough to discharge in the sea, a closer look at the different standards applied around the world on drinking water, the experience of algae growing in still fresh, drinkable water onboard vessels (like Offshore Supply Vessels when those fail to deliver the water to the platforms due to weather), as well as the problematic of access to freshwater in certain parts of the world may shed a new light on this aspect of the BWM Convention.

Exemptions and the Application of Same Risk Area

Regulation A-4 of the BWM Convention opens up for allowing ships to be exempted from the requirements in Regulations D-1 and D-2, when trading between specific ports where a risk assessment applied in accordance with Guidelines G7, have concluded that there is no risk of spread of invasive species between those ports.

This regulation was further expanded to introduce the concept of Same Risk Area, where such risk assessments, always done in accordance with Guidelines G7, apply to a region or area with multiple ports. This concept is especially useful in heavily trafficked seas like the Southeast Asian passages around Singapore/Malaysia and Indonesia, the Great Lakes, the North Sea, and other local areas and ports.

Furthermore, such exemptions are very much applicable to ferries and passenger ships going on shuttle traffic between two and three ports or on very short voyages (e.g., ferries going between Norway, Denmark, and Sweden with less than 6–12 h sailing route).

The challenge with applying those exemption guidelines, according to several sources in the shipping industry, is that the exemption work is done by the ship owner but is then applicable to all other ships in the same route, so that the cost of such exemption is carried by one owner, for the benefit of all. This has shown to be challenging for shipping companies to apply. This fact, in addition to the complex sampling and analysis procedure for establishing the noninvasive nature of organisms between two ports, has led so far to very little, if any, such analysis taking place.

Exceptions from the Requirements of the BWM Convention

Regulation A-3 of the BWM Convention allows ships, in certain circumstances, to discharge unmanaged ballast water. Those circumstances are:

- The uptake or discharge of Ballast Water and Sediments necessary for the purpose of ensuring the safety of a ship in emergency situations or saving life at sea
- The accidental discharge or ingress of Ballast Water and Sediments resulting from damage to a ship or its equipment
- Ballast operations for the purpose of avoiding or minimizing pollution incidents from the ship
- The uptake and subsequent discharge on the high seas of the same Ballast Water and Sediments
- The discharge of Ballast Water and Sediments from a ship at the same location where the whole of that Ballast Water and those Sediments originated and provided that no mixing with unmanaged Ballast Water and Sediments from other areas has occurred

Exceptions are operational situations that, on a case by case basis, ships do not need to manage their ballast water. However, those ships must still be able to discharge ballast water compliant with the D-1 or D-2 standard, as applicable, including having onboard a BWM Plan, an International BWM Certificate and a Record Book.

4.5.2 Biofouling

Biofouling is also considered one of the main vectors for bioinvasions and is described as the undesirable accumulation of microorganisms, plants, algae, and animals on submerged structures (especially ships' hulls). Studies have shown that biofouling can be a significant vector for the transfer of invasive aquatic species. Biofouling on ships entering the waters of States may result in the establishment of invasive aquatic species which may pose threats to human, animal, and plant life, economic and cultural activities, and the aquatic environment.[2]

The IMO adopted in 2011 the Biofouling Guidelines, which are voluntary guidance for ship owners on how to avoid this important vector of spread of invasive species. In 2012, the IMO expanded those guidelines to include recreational craft with length less than 24 m through *Guidance for minimizing the transfer of invasive aquatic species as biofouling for recreational craft*.

Biofouling is also recognized by a large number of coastal states as a threat to their environment, which forced them to regulate how often ships clean their hulls, propellers, and other submerged parts. In the United States, biofouling is regulated through the EPA's VGP, and the USCG requires Biofouling Management Plans for ships.

Means to control biofouling include mainly routine cleaning of the hull of the ship, anchor chains, and niche areas like thruster tunnels, rudders, propellers, and sea chests, to name a few. The VGP includes a detailed list of actions required to do proper biofouling management.

Acknowledgment The corresponding author would like to thank the American Bureau of Shipping for their permission to use ABS references and graphics, and he would like to extend his gratitude to Mark Penfold for his significant contribution to Sect. 3 ("Machinery Technology").

Disclaimer The views and opinions expressed in this chapter are those of the authors and do not necessarily reflect the position or views of the American Bureau of Shipping.

[2]From the IMO website www.imo.org.

References

American Bureau of Shipping. (2013). *Ship energy efficiency measures: Status and guidance*. https://ww2.eagle.org/content/dam/eagle/advisories-and-debriefs/ABS_Energy_Efficiency_Advisory.pdf

American Bureau of Shipping. (2017). *ABS advisory on hybrid electric power systems*. https://ww2.eagle.org/content/dam/eagle/advisories-and-debriefs/ABS_Hybrid_Advisory_17033.pdf

De Kat, J. O., Cerup-Simonsen, B., Jakobsen, O. G., Pedersen, L. R. Petersen, J. B., & Posborg, T. (2009). *An integrated approach towards cost-effective operation of ships with reduced GHG emissions*. 2009 SNAME annual meeting proceedings.

Fathom Shipping. (2013). Ship efficiency – The guide. In *A comprehensive guide to ship eco-efficiency technologies and measures* (2nd ed.). London, UK: Fathom Eco-Efficiency Consultants Ltd.

Jang, J. H., Choi, S. H., Ahn, S. M., Kim, B. K., & Seo, J. S. (2014). Experimental investigation of frictional resistance reduction with air layer on the hull bottom of a ship. *IJNAOE, 2*(6), 363–379.

Larsson, L., et al. (2010). *Principles of naval architecture series: Ship resistance and flow* (pp. 196–201). Alexandria, VA, USA: Society of Naval Architects and Marine Engineers.

Lee, J., Kim, J., Kim, B., Jang, J., McStay, P., Raptakis, G., & Fitzpatrick, P. (2017). *Full scale applications of air lubrication for reduction of ship frictional resistance*. Houston: SNAME Maritime Convention.

Mewis, F., & Guiard, T. (2011). Mewis Duct® – new developments, solutions and conclusions. In *Second international symposium on marine propulsors, SMP'11*. Hamburg.

Minchev, A., Schmidt, M., & Schnack, S. (2013). Contemporary bulk carrier design to meet IMO EEDI requirements. In *Proceedings of the third international symposium on marine propulsors SMP'13*. Launceston.

Mizokami, S., Kawakita, C., Kodan, Y., Takano, S., Higasa, S., & Shigenaga, R. (2011). Development of air lubrication system and verification by the full scale ship test. *Journal Japan Society of Naval Architects and Ocean Engineers, 12*, 69–77.

Naaijen, P., Koster, V., & Dallinga, R. P. (2006). On the power savings by an auxiliary kite propulsion system. *International Shipbuilding Progress, 53*, 255–279.

Reichel, M. (2009). Influence of rudder location on propulsive characteristics of a single screw container ship. In *Proceedings of the first international symposium on marine propulsors, SMP'09*. Trondheim.

Silberschmidt, N., Tasker, D., Pappas, T., & Johannesson, J. (2016). *Silverstream® system – air lubrication performance verification and design development*. Conference of Shipping in Changing Climate.

Simonsen, C. D., Nielsen, C., Klimt-Møllenbach, C., Holm, C. R., & Minchev, A. (2011). *CFD based investigation of potential power saving for different rudder types, positions and pre-swirl fins, DCMT*. Copenhagen: FORCE Technology.

SSPA. (2009). Newsletter 46/2009: Green hydrodynamic design.

Townsin, R. L., & Kwon, Y. J. (1993). Estimating the influence of weather on ship performance. *RINA Transactions, 135*, 191–209.

Chapter 3
The Energy Efficiency Design Index (EEDI)

Maria Polakis, Panos Zachariadis, and Jan Otto de Kat

Abstract Thus far the only regulatory measure to reduce greenhouse gases (GHGs) from ships is the adoption of the Energy Efficiency Design Index (EEDI) by the IMO in 2011. This chapter will go over the rationale behind EEDI and the important factors that influence compliance of a vessel's Attained EEDI with the regulatory limit of ship-type specific reference lines (Required EEDI) set by the IMO. This chapter will also go over related concepts and requirements, such as the Ship Energy Efficiency Management Plan (SEEMP) and the Energy Efficiency Operational Indicator (EEOI). Concerns around possible implications directly linked or relevant to the EEDI framework will be outlined, including EEDI vs minimum propulsion power. The Existing Vessel Design Index (EVDI) rating of RightShip will also be presented. Last but not least, a discussion of the weaknesses of EEDI will be provided.

Abbreviations

AER	Annual Efficiency Ratio
AHEWG-TT	Ad Hoc Expert Working Group on Facilitation of Transfer of Technology for Ships
BIMCO	Baltic and International Maritime Council
CO_2	Carbon dioxide
CSI	Clean Shipping Index
DCS	Data Collection System
DWT	Deadweight
EEDI	Energy Efficiency Design Index

M. Polakis (✉) · J. O. de Kat
American Bureau of Shipping, Houston, TX, USA
e-mail: MPolakis@eagle.org

P. Zachariadis
Atlantic Bulk Carriers Management, Ltd., Piraeus, Greece

H. N. Psaraftis (ed.), *Sustainable Shipping*,
https://doi.org/10.1007/978-3-030-04330-8_3

EEOI	Energy Efficiency Operational Indicator
EIAPP	Engine International Air Pollution Prevention
ETS	Emission trading system
EVDI	Existing Vessel Design Index
FORS	Fuel Oil Reduction Strategy
FPSO	Floating production storage and offloading
FSU	Floating storage unit
GHG	Greenhouse gas
GT	Gross tonnage
IACS	International Association of Classification Societies
IEE	International Energy Efficiency
IMO	International Maritime Organization
ISO	International Organization for Standardization
ITTC	International Towing Tank Conference
LNG	Liquefied natural gas
MARPOL	The International Convention for the Prevention of Pollution from Ships
MCR	Maximum continuous rating
MEPC	Marine Environment Protection Committee
MPP	Minimum propulsion power
PP	Propulsion power
PSC	Port State Control
RO	Recognized organization
SEEMP	Ship Energy Efficiency Management Plan
SFOC	Specific fuel oil consumption
SMS	Safety management system
VLCC	Very large crude carrier

1 Introduction

For the past decade, energy regulations and global demand for reducing international shipping's greenhouse gas (GHG) emissions have progressively stimulated innovation and targeted technology readiness of all components influencing the performance of a ship from its design phase.

The Energy Efficiency Design Index (EEDI) was established as part of the International Maritime Organization's (IMO) strategy to reduce carbon dioxide (CO_2) emissions from shipping and provides a benchmark for comparing the energy efficiency of vessels, while setting a minimum required level of efficiency for different ship type and size segments.

The EEDI was the first legally binding climate change treaty to be adopted since the Kyoto Protocol and made mandatory for new ships at the 62nd session of IMO's Marine Environment Protection Committee (MEPC 62) with the adoption of amendments to MARPOL Annex VI, IMO (2011a). Following this breakthrough, the IMO MEPC, at its 63rd session of March 2012, adopted four important guidelines, IMO (2012a, b, c, d) aimed at assisting the implementation of the mandatory regulations on Energy Efficiency for Ships formally introduced into Annex VI of the International Convention for the Prevention of Pollution from Ships (MARPOL).

The EEDI for new ships aims at promoting the use of more energy-efficient (less polluting) equipment and engines. The EEDI requires a minimum energy efficiency level per capacity mile (e.g., tonne mile) for different ship type and size segments. From 1 January 2013, following an initial 2-year Phase 0 when new ship design will need to meet the reference level for their ship type, the level is to be tightened incrementally by 10% every 5 years. Therefore, regulations on EEDI are intended to stimulate continued innovation and technical development of all the components influencing the fuel efficiency of a ship from its design phase.

The EEDI is a non-prescriptive, performance-based mechanism that leaves the choice of technologies to use in a specific ship design to the industry. As long as the required energy efficiency level is attained, ship designers and builders are free to use the most cost-efficient solutions for the ship to comply with the regulations. EEDI is thus a goal-based technical standard intended to encourage improvements in ship design and to promote the use of less polluting equipment and engines.

The EEDI provides a specific numerical figure for an individual ship design, expressed in grams of CO_2 per ship's capacity mile (the smaller the EEDI, the more energy efficient is the ship's design) and is calculated by the formula below which is based on the technical design parameters for a given ship:

$$\frac{\left(\prod_{j=1}^{n} f_j\right)\left(\sum_{i=1}^{n\mathrm{ME}} P_{\mathrm{ME}(i)} \cdot C_{\mathrm{FME}(i)} \cdot \mathrm{SFC}_{\mathrm{ME}(i)}\right) + (P_{\mathrm{AE}} \cdot C_{\mathrm{FAE}} \cdot \mathrm{SFC}_{\mathrm{AE}}) + \left(\left(\prod_{j=1}^{n} f_j \cdot \sum_{i=1}^{n\mathrm{PTI}} P_{\mathrm{PTI}(i)} - \sum_{i=1}^{n\mathrm{eff}} f_{\mathrm{eff}(i)} \cdot P_{\mathrm{AEeff}(i)}\right) C_{\mathrm{FAE}} \cdot \mathrm{SFC}_{\mathrm{AE}}\right) - \left(\sum_{i=1}^{n\mathrm{eff}} f_{\mathrm{eff}(i)} \cdot P_{\mathrm{eff}(i)} \cdot C_{\mathrm{FME}} \cdot \mathrm{SFC}_{\mathrm{ME}}\right)}{f_i \cdot f_c \cdot f_l \cdot \mathrm{Capacity} \cdot f_w \cdot V_{\mathrm{ref}}}$$

Starting on 1 January 2013, an initial 2-year "Phase 0" required new ship designs to meet the reference level for their specific ship type. From that point on, new designs are required to become progressively more efficient in three more "phases" reaching a 30% reduction between 2025 and 2030 for applicable ship types.

The rest of this chapter is organized as follows: Sect. 2 provides an overview of the EEDI regulations. Section 3 outlines the details of the EEDI calculation formula. Section 4 describes the EEDI survey and verification process. Section 5 describes the minimum propulsion power requirements under EEDI Regulation 21. Section 6 discusses weaknesses of EEDI. Finally Section 7 includes suggestions on way forward for improvement of EEDI.

2 Overview of EEDI Regulations: MARPOL Annex VI

The primary changes that the new energy regulations brought to MARPOL Annex VI can be categorized as follows:

- Amendments to existing regulations as a result of energy efficiency
- Introduction of new regulations specifically for energy efficiency

2.1 Amendments to Existing Regulations

A summary of the changes are briefly described here and also shown in Table 3.1.

Regulation 2 Introduction of definitions for "new ship" that are applicable to various Phases of EEDI regulations, "major conversion," "conventional/non-conventional propulsion," and "ship types" for which EEDI regulations apply. Since EEDI only applies to new ships and those ships that undergo major conversions beyond 1 January 2013, the exact definition of the "new ship" and "major conversion" terms were required, see IMO (2014a). Additionally, terms such as "Attained EEDI" and "Required EEDI" were defined.

Regulation 5 Requirements were specified for surveys including an initial survey for newly built ships, a full or partial survey in case of a major conversion of existing ships, a survey for a Ship Energy Efficiency Management Plan (SEEMP) to verify

Table 3.1 Existing regulations/amended regulations shown in red

Resolution MEPC.176(58)	Resolutions MEPC.203(62) & MEPC251 (66), …..
Chapter I Reg. 1 Application Reg. 2 Definitions Reg. 3 Exceptions and Exemptions Reg. 4 Equivalents	Chapter I Reg. 1 Application Reg. 2 Definitions Reg. 3 Exceptions and Exemptions Reg. 4 Equivalents
Chapter II Reg. 5 Surveys Reg. 6 Issue or endorsement of a Certificate Reg. 7 Issue of a Certificate by another Party Reg. 8 Form of Certificate Reg. 9 Duration and Validity of Certificate Reg. 10 Port State Control on Operational Requirements Reg. 11 Detection of Violations and Enforcements	Chapter II Reg. 5 Surveys Reg. 6 Issue or endorsement of a Certificate Reg. 7 Issue of a Certificate by another Party Reg. 8 Form of Certificate Reg. 9 Duration and Validity of Certificate Reg. 10 Port State Control on Operational Requirements Reg. 11 Detection of Violations and Enforcements

Source: IMO (2015e)

its existence on board ship, etc. Regulation 5 states that EEDI survey and verification shall be carried out according to relevant IMO guidelines.

Regulations 7 and 8 The changes to these regulations deal with energy efficiency certification. For ships subject to EEDI regulations, an International Energy Efficiency (IEE) Certificate was made mandatory. The responsibility of the Flag Administration was also emphasized:

> An International Energy Efficiency Certificate for the ship shall be issued after a survey in accordance with the provisions of regulation 5.4 to any ship of 400 gross tonnage and above, before that ship may engage in voyages to ports or offshore terminals under the jurisdiction of other Parties.
>
> The certificate shall be issued or endorsed either by the Administration or any organization duly authorized by it. In every case, the Administration assumes full responsibility for the certificate. IMO (2011a)

Regulation 9 The validity aspects of the IEE Certificate were defined. The IEE Certificate has been determined to be valid for the life of the ship unless otherwise invalidated by a major conversion or change of flag or ship withdrawal from service.

The IEE Certificate shall be valid throughout the life of the ship subject to the provisions of paragraph below:

> An IEE issued under this Annex shall cease to be valid in any of the following cases if the ship is withdrawn from service or if a new certificate is issued following major conversion of the ship; or upon transfer of the ship to the flag of another State...... IMO (2011a)

Regulation 10 This regulation specifies how compliance with the EEDI requirements is verified by Port State Control Authorities and defines the extent of the inspection scheme. At present stage, as described in MEPC Resolution 203(62), a Port State Control (PSC) inspection would be limited to verifying that a valid IEE Certificate exists on board the vessel files.

2.2 *Introduction of New Regulations: Chapter 4*

The introduction of EEDI regulations came following a series of discussions at the IMO MEPC sessions. The committee in July 2011 at its 62nd session reached a consensus to add a new Chapter 4 to MARPOL Annex VI, covering the new requirements exclusively. The consensus though was not general as a group of member states primarily consisting of developing countries were strongly opposed to the agreement.

Table 3.2 shows an outline of the newly introduced regulations.

A short description of the main aspects of these new regulations is provided below.

Regulation 19 Regulation 19 specifies the domain of application of the energy efficiency regulations. Chapter 4 of MARPOL Annex VI applies to all ships of 400 gross tonnage (GT) and above that are engaged in international voyages. It gives

Table 3.2 Newly introduced regulations

Resolution MEPC.176(58)	Resolution MEPC.203(62)
Chapter III Reg. 12 Ozone Depleting Substances Reg. 13 Nitrogen Oxides(NOx) Reg. 14 Sulphur Oxides(SOx) and Particular Matter Reg. 15 Volatile Organic Compounds (VOCs) Reg. 16 Shipboard Incineration Reg. 17 Reception Facilities Reg. 18 Fuel Oil Availability and Quality	Chapter III Reg. 12 Ozone Depleting Substances Reg. 13 Nitrogen Oxides(NOx) Reg. 14 Sulphur Oxides(SOx) and Particular Matter Reg. 15 Volatile Organic Compounds(VOCs) Reg. 16 Shipboard Incineration Reg. 17 Reception Facilities Reg. 18 Fuel Oil Availability and Quality
	Chapter IV **Reg. 19 Application** **Reg. 20 Attained EEDI** **Reg. 21 Required EEDI** **Reg. 22 SEEMP** **Reg. 23 Promotion of technical co-operation and transfer of technology relating to the improvement of energy efficiency of ships**
Appendix I ~VI	Appendix I ~VI **Appendix VIII Form of International Energy Efficiency(IEE) Certificate**

Source: IMO (2015e)

limited power to Administrations to waive the requirements for EEDI for a new ship contracted before 1 January 2017 up to a delivery date of 1 July 2019, subject to informing the IMO and other Parties to MARPOL Annex VI of this decision.

The "waiver" clause came about due to significant discussions at MEPC, stressing that some ships may not be able to comply with IMO requirements while considered as good design ships. According to IMO sources, there has been no need for Administrations to use this option.

Regulation 20 This regulation deals with the Attained EEDI and specifies the need for its calculation and verification. Attained EEDI is the actual EEDI of a ship as calculated using EEDI formula. According to Regulation 20:

- Attained EEDI must be calculated for each new ship, each new ship that undergoes a major conversion, or existing ships that undergo so many changes that according to the Administration's judgment are considered as a new ship.
- The Attained EEDI is only applicable to a large number of ship types but not all ships. For example, fishing vessels are not required to have an Attained EEDI.
- The Attained EEDI must be calculated taking into account relevant IMO guidelines.
- The Attained EEDI must be accompanied by an "EEDI Technical File" that contains the information necessary for the calculation of the Attained EEDI and that shows the process of calculation.

- The Attained EEDI must be verified, based on the EEDI Technical File, either by the Administration or by any organization duly authorized by it (see Section B.3 on details of verification).

The following ship types are currently required to comply with the Attained EEDI regulation:

1. Bulk carrier
2. Gas carrier
3. Tanker
4. Containership
5. General cargo ship
6. Refrigerated cargo ship
7. Combination carrier
8. Passenger ship
9. Ro-Ro cargo ship (vehicle carrier)
10. Ro-Ro cargo ship
11. Ro-Ro passenger ship
12. LNG carrier
13. Cruise passenger ship

The definitions of the 13 ship types are described in Regulation 2 of MARPOL ANNEX VI and presented cumulatively in Table 3.3 below:

Of these ship types, EEDI is only applicable to ships with conventional propulsion, i.e., engines that are either direct drive or geared. However, EEDI would not apply to ships not propelled by mechanical means, including floating production storage and offloading assets (FPSO), floating storage units (FSU), and drilling rigs, regardless of their propulsion. Cruise ships however are subject to EEDI regulations when fitted with nonconventional propulsion (such as diesel-electric propulsion, turbine propulsion, or hybrid propulsion systems). Liquefied natural gas (LNG) carriers need to comply when fitted with either conventional or nonconventional propulsion.

Some vessel types are not defined in the regulations. If these types do not fall under 1 of the 13 mandatory vessel types, then it is not mandatory for them to comply with Regulation 20 or Regulation 21.o.

EEDI regulations do not apply to cargo ships with ice-breaking capability but do apply to ice-strengthened ships.

Regulation 21 Regulation 21 provides the requirement and guidelines for calculating the Required EEDI and verifying that a vessel's Attained EEDI is less than the Required EEDI. The Required EEDI is the regulatory limit for EEDI, and its calculation is dependent on a reference line value and a reduction factor.

The basic concepts included in this regulation are:

Reference line A baseline EEDI for each ship type, representing reference EEDI as a function of ship size (DWT). The reference line is a regression, i.e., a mathematical distribution of data representing the average efficiency for ships built between years 1999 and 2009. Reference lines have been developed for each individual ship type

Table 3.3 Definition of each type of ship defined in Regulation 2 of MARPOL ANNEX VI, Chap. 4

Reg.	Ship type	Definition
2.25	Bulk carrier	A ship which is intended primarily to carry dry cargo in bulk, including such types as ore carriers as defined in SOLAS Chap. XII, Regulation 1 but excluding combination carriers
2.26	Gas carrier	A cargo ship, other than an LNG carrier as defined in paragraph 38 of this regulation, constructed or adapted and used for the carriage in bulk of any liquefied gas
2.27	Tanker	An oil tanker as defined m MARPOL Annex I, Regulation 1 or a chemical tanker or an NLS tanker as defined in MARPOL Annex II, Regulation 1
2.28	Container ship	A ship designed exclusively for the carriage of containers in holds and on deck
2.29	General cargo ship	A ship with a multi-deck or single deck hull designed primarily for the carriage of general cargo This definition excludes specialized dry cargo ships, which are not included in the calculation of reference lines for general cargo ships, namely, livestock carrier, barge carrier, heavy load carrier, yacht carrier, and nuclear fuel carrier
2.30	Refrigerated cargo carrier	A ship designed exclusively for the carriage of refrigerated cargoes in holds
2.31	Combination carrier	A ship designed to load 100% deadweight with both liquid and dry cargo in bulk
2.32	Passenger ship	A ship which carries more than 12 passengers
2.33	Ro-ro cargo ship (vehicle carrier)	A multi-deck roll-on-roll-off cargo ship designed for the carriage of empty cars and trucks
2.34	Ro-ro cargo ship	A ship designed for the carriage of roll-on-roll-off cargo transportation units
2.35	Ro-ro passenger ship	A passenger ship with roll-on-roll-off cargo spaces
2.38	LNG carrier	A cargo ship constructed or adapted and used for the carriage in bulk of liquefied natural gas (LNG)
2.39	Cruise passenger ship	A passenger ship not having a cargo deck, designed exclusively for commercial transportation of passengers in overnight accommodations on a sea voyage

Source: ClassNK (2015)

and relate the EEDI value to the vessel's size (deadweight, DWT or gross tonnage, GT). Details of how reference lines are developed including sources of data, data quality checks, number of ships selected and year of build, ship sizes, etc. are fully described in the relevant IMO guidelines, IMO (2013a) and IMO (2013c). Example reference lines developed by the IMO for four indicative vessel types are shown in Fig. 3.1.

The regression equations for each ship type are embodied in Regulation 21 in the form of a formula:

$$\text{Reference EEDI} = a \times b^{-c}$$

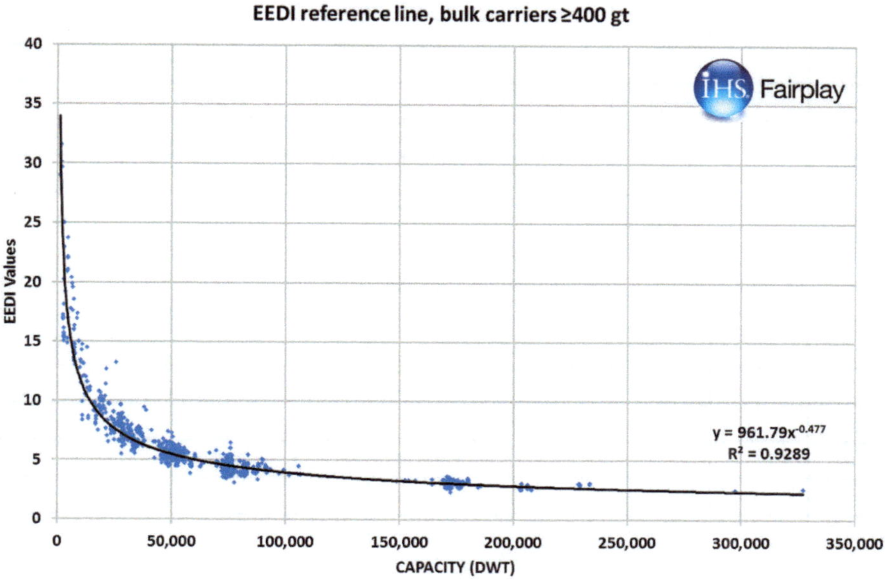

Fig. 3.1 EEDI reference line for bulk carrier developed by the IMO (IMO 2013a, b, c, d, e). (Source: IMO 2015e)

Table 3.4 Parameters for determination of reference line values for the different ship types

Ship type		Reference line
Bulk carrier		$961.79 \times DWT^{-0.477}$
Gas carrier		$1120.00 \times DWT^{-0.456}$
Tanker		$1218.80 \times DWT^{0.488}$
Container ship		$174.22 \times DWT^{-0.201}$
General cargo ship		$107.48 \times DWT^{-0.216}$
Refrigerated cargo carrier		$227.01 \times DWT^{-0.244}$
Combination carrier		$1219.00 \times DWT^{-0.488}$
Ro-ro cargo ship (vehicle carrier)	$DWT/GT < 0.3$	$((DWT/GT)^{-0.7} \times 780.36) \times DWT^{-0.471}$
	$DWT/GT \geq 0.3$	$1812.63 \times DWT^{-0.471}$
Ro-ro cargo ship		$1405.15 \times DWT^{-0.498}$
Ro-ro passenger ship		$752.16 \times DWT^{-0.381}$
LNG carrier		$2253.7 \times DWT^{-0.474}$
Cruise passenger ship having nonconventional propulsion		$170.84 \times GT^{-0.214}$

Source: ClassNK (2015)

Parameters *a*, *b*, and *c* for some of the ship types are given in Table 3.4.

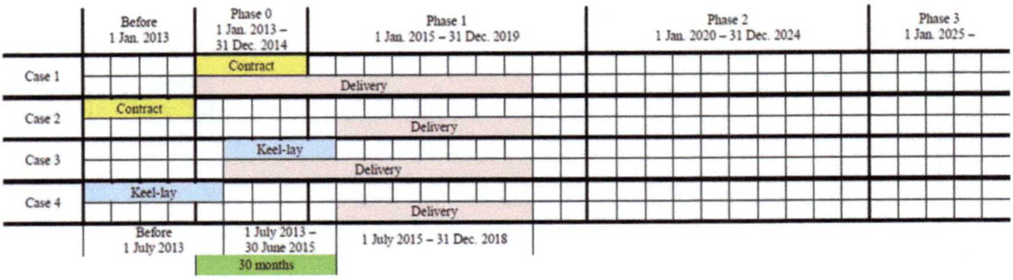

Fig. 3.2 Required EEDI of Phase 0. (Source: ClassNK 2015)

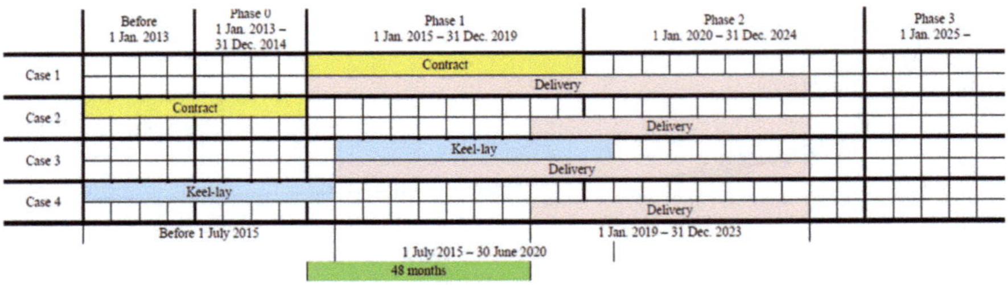

Fig. 3.3 Required EEDI of Phase 1. (Source: ClassNK 2015)

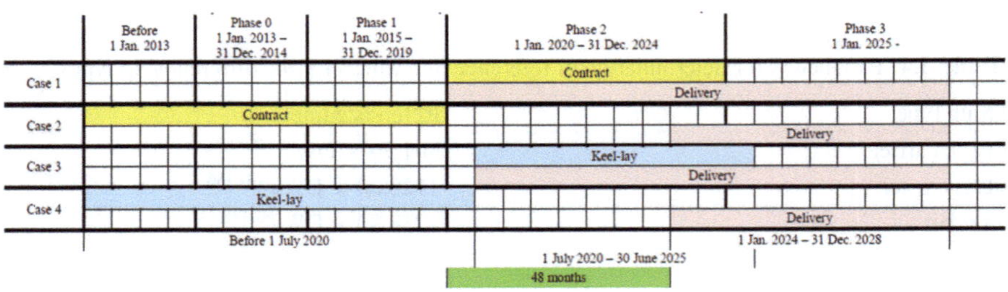

Fig. 3.4 Required EEDI of Phase 2. (Source: ClassNK 2015)

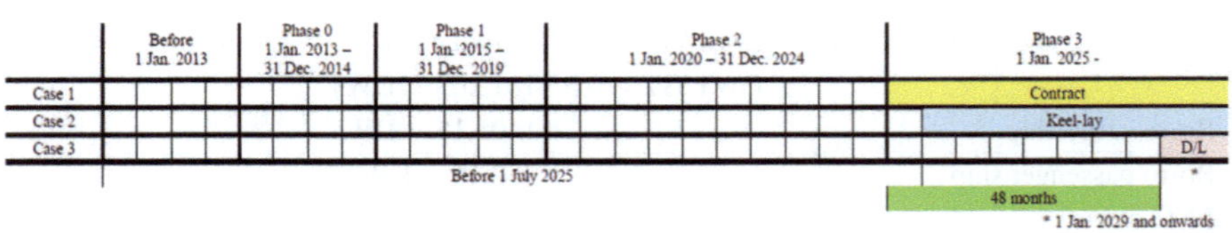

Fig. 3.5 Required EEDI of Phase 3. (Source: ClassNK 2015)

Implementation Phases Required EEDI will be implemented in phases. Currently, it is in Phase 1 that runs from the year 2015 to 2019. Phase 2 will run from the year 2020 to 2024 and Phase 3 starts from the year 2025 onward. Below, the general implementation dates for each of the phases are described and shown in Figs. 3.2, 3.3, 3.4, and 3.5.

Phase 0 (2013–2014) The Required EEDI of Phase 0 is applied to the following new ship:

1. For which the building contract is placed in Phase 0 and the delivery is before 1 January 2019
2. The building contract of which is placed before Phase 0, the delivery is on or after 1 July 2015 and before 1 January 2019, or in the absence of a building contract
3. The keel of which is laid or which is at a similar stage of construction on or after 1 July 2013 and before 1 July 2015 and the delivery is before 1 January 2019
4. The keel of which is laid or which is at a similar stage of construction before 1 July 2013 and the delivery is on or after 1 July 2015 and before 1 January 2019

Phase 1 (2015–2019) The Required EEDI of Phase 1 is applied to the following new ship:

1. For which the building contract is placed in Phase 1 and the delivery is before 1 January 2024
2. The building contract of which is placed before Phase 1, the delivery is on or after 1 January 2019 and before 1 January 2024, or in the absence of a building contract
3. The keel of which is laid or which is at a similar stage of construction on or after 1 July 2015 and before 1 July 2020 and the delivery is before 1 January 2024
4. The keel of which is laid or which is at a similar stage of construction before 1 July 2015 and the delivery is on or after 1 January 2019 and before 1 January 2024

Phase 2 (2020–2024) The Required EEDI of Phase 2 is applied to the following new ship:

1. For which the building contract is placed in Phase 2 and the delivery is before 1 January 2029
2. The building contract of which is placed before Phase 2, the delivery is on or after 1 January 2024 and before 1 January 2029, or in the absence of a building contract
3. The keel of which is laid or which is at a similar stage of construction on or after 1 July 2020 and before 1 July 2025 and the delivery is before 1 January 2029
4. The keel of which is laid or which is at a similar stage of construction before 1 July 2020 and the delivery is on or after 1 January 2024 and before 1 January 2029

Phase 3 (2025+) The Required EEDI of Phase 3 is applied to the following new ship:

1. For which the building contract is placed on or after 1 January 2025.
2. In the absence of a building contract, the keel of which is laid or which is at a similar stage of construction on or after 1 July 2025.
3. The delivery of which is on or after 1 January 2029.

Table 3.5 Implementation phases for bulk carrier, gas carrier, tanker, container ship, general cargo ship, refrigerated cargo carrier, and combination carrier

Delivery \ Contract	Before 1 Jan. 2013	1 Jan. 2013 – 31 Dec. 2014	1 Jan. 2015 – 31 Dec. 2019	1 Jan. 2020 – 31 Dec. 2024	1 Jan. 2025 -
Before 1 July 2015	n/a	Phase 0	Phase 1		
1 July 2015 - 31 Dec. 2018	Phase 0	Phase 0	Phase 1		
1 Jan. 2019 - 31 Dec. 2023	Phase 1	Phase 1	Phase 1	Phase 2	
1 Jan. 2024 - 31 Dec. 2028	Phase 2	Phase 2	Phase 2	Phase 2	Phase 3
1 Jan. 2029 -	Phase 3	Phase 3	Phase 3	Phase 3	Phase 3

Source: ClassNK (2015)

Table 3.6 Implementation phases for Ro-Ro cargo ship (vehicle), Ro-Ro cargo ship, Ro-Ro passenger ship, LNG carrier, and Cruise passenger ship

Delivery \ Contract	Before 1 Jan. 2013	1 Jan. 2013 – 31 Aug. 2015	1 Sep. 2015 – 31 Dec. 2019	1 Jan. 2020 – 31 Dec. 2024	1 Jan. 2025 -
Before 1 July 2015	n/a	n/a			
1 July 2015 - 31 Aug. 2019	n/a	n/a	Phase 1		
1 Sep. 2019 - 31 Dec. 2023	Phase 1	Phase 1	Phase 1	Phase 2	
1 Jan. 2024 - 31 Dec. 2028	Phase 2	Phase 2	Phase 2	Phase 2	Phase 3
1 Jan. 2029 -	Phase 3	Phase 3	Phase 3	Phase 3	Phase 3

Source: ClassNK (2015)

Summary Tables 3.5 and 3.6 also provide the implementation phases *per ship type* by combination of contract and delivery dates.

Reduction Factor This is a phase in percentage value *X* for EEDI reduction relative to the reference line. Reduction factors *X* are dependent on the vessel's type, deadweight, contract and delivery dates and use a structured approach to tighten EEDI regulations over time.

Table 3.7 Reduction factors (in percentage) for the EEDI relative to the EEDI reference line

Ship type	Size	Phase 0 1 Jan 2013–31 Dec 2014	Phase 1 1 Jan 2015–31 Dec 2019	Phase 2 1 Jan 2020–31 Dec 2024	Phase 3 1 Jan 2025 and onward
Bulk carrier	20,000 DWT and above	0	10	20	30
	10,000–20,000 DWT	n/a	0–10[a]	0–20[a]	0–30[a]
Gas carrier	10,000 DWT and above	0	10	20	30
	2000–10,000 DWT	n/a	0–10[a]	0–20[a]	0–30[a]
Tanker	20,000 DWT and above	0	10	20	30
	4000–20,000 DWT	n/a	0–10[a]	0–20[a]	0–30[a]
Container ship	15,000 DWT and above	0	10	20	30
	10,000–15,000 DWT	n/a	0–10[a]	0–20[a]	0–30[a]
General Cargo ships	15,000 DWT and above	0	10	15	30
	3000–15,000 DWT	n/a	0–10[a]	0–15[a]	0–30[a]
Refrigerated cargo carrier	5000 DWT and above	0	10	15	30
	3000–5000 DWT	n/a	0–10[a]	0–15[a]	0–30[a]
Combination carrier	20,000 DWT and above	0	10	20	30
	4000–20,000 DWT	n/a	0–10[a]	0–20[a]	0–30[a]
LNG carrier[c]	10,000 DWT and above	n/a	10[b]	20	30
Ro-ro cargo ship (vehicle carrier)[c]	10,000 DWT and above	n/a	5[b]	15	30
Ro-ro cargo ship[c]	2000 DWT and above	n/a	5[b]	20	30
	1000–2000 DWT	n/a	0–5[a, b]	0–20[a]	0–30[a]
Ro-ro passenger ship[c]	1000 DWT and above	n/a	5[b]	20	30
	250–1000 DWT	n/a	0–5[a, b]	0–20[a]	0–30[a]
Cruise passenger ship[c] having nonconventional propulsion	85,000 GT and above	n/a	5[b]	20	30
	25,000–85,000 GT	n/a	0–5[a, b]	0–20[a]	0–30[a]

Source: ClassNk (2015)
n/a means that no Required EEDI applies
[a]Where a range is given, the lower value is for the lower deadweight segments. The reduction factor increases linearly as the deadweight increases
[b]Phase 1 commences for those ships contracted on 1 September 2015
[c]Reduction factor applies to those ships delivered on or after 1 September 2019, as defined in paragraph 43 of Regulation 2

Reduction factor values have been decided by the IMO and documented in Regulation 21 as shown in Table 3.7.

Figure 3.6 shows a graphic demonstration of the relation between implementation phases and reduction factors.

Cut-Off Levels Smaller size vessels are excluded from having a Required EEDI under certain technical justifications. The size limits are referred to as cut-off levels

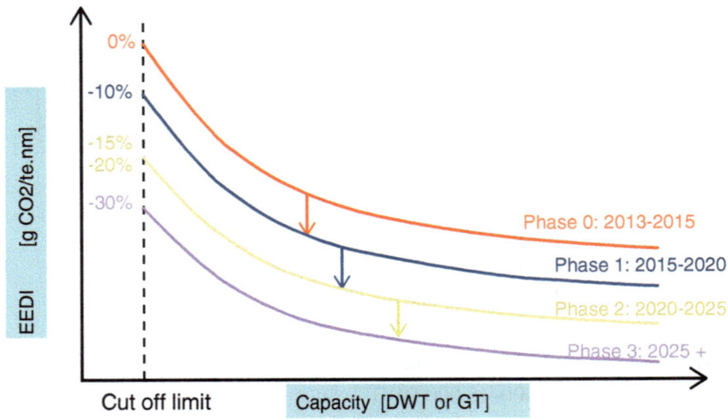

Fig. 3.6 Concept of Required EEDI, reduction factor, cut-off limits, and EEDI phases. (Source: IMO 2015e)

and specified in the regulatory text per vessel type. Cut-off levels are shown in above Table 3.7.

Required EEDI Calculation Formula Using the concepts described above, the following equations show the way Required EEDI is calculated for a ship. As mentioned earlier, for each ship a "Reference EEDI" is calculated using the below equation:

$$\text{Reference EEDI} = a \times b^{-c}$$

where

b	ship capacity
a and c	constants agreed for each ship type and included in the regulation
Reference EEDI	reference value for EEDI

The next step is to establish the reduction factor (X) for the ship. This is dependent on year of ship built and is specified within the regulation (see Table 3.7). Having established the Reference EEDI and X, the Required EEDI is calculated from the following equation:

$$\text{Required EEDI} = \left(1 - \frac{X}{100}\right) \times \text{Reference line value} = \left(1 - \frac{X}{100}\right) \times a(\text{Capacity})^{-c}$$

The Required EEDI applies only to ships defined in column 1 and the ship sizes specified in column 2 of Table 3.7. For these ships, Regulation 21 states that the Attained EEDI must always be less than or equal to Required EEDI:

$$\text{Attained EEDI} \leq \text{Required EEDI} \quad (3)$$

where Attained EEDI: The actual EEDI of the ship, as calculated by the shipyard and verified by a recognized organization (RO)

Note: Regulation 21 does not apply to passenger ships even though vessels falling into this ship type definition are required to have an Attained EEDI calculated and verified subject to Regulation 20.

Regulation 21 additionally stipulates the following:

- If the design of a ship allows it to fall into more than one of the above ship type definitions, the Required EEDI for the ship shall be the most stringent (the lowest) Required EEDI.
- For each ship to which this regulation applies, the installed propulsion power shall not be less than the propulsion power needed to maintain the maneuverability of the ship under adverse conditions as defined in the guidelines to be developed by the organization. The related interim guidelines are introduced in Sect. 6.
- The reference lines and the reduction factors are subject to change. The IMO built two mandatory periods into the regulations when the MEPC would review the status of the currently available technologies and, if necessary, amend the reference lines and reduction factors. The first period was at the beginning of Phase 1, around January 2015, and the second period is midpoint to Phase 2.

Most Recent Developments The IMO MEPC at its 70th session agreed to retain the current reduction rates, time periods, and EEDI reference line parameters in the Phase 2 requirements for ship types other than Ro-Ro cargo and Ro-Ro passenger ships.

For Ro-Ro cargo and Ro-Ro passenger ships, the IMO MEPC adopted amendments concerning the new parameters from Phase 2 that increase the reference line by 20% and introduce a DWT threshold value for larger Ro-Ro cargo ships of 17,000 DWT and Ro-Ro passenger ships of 10,000 DWT, IMO (2018).

A thorough review of EEDI Phase 3 requirements, their early implementation, and of the possibility of establishing a Phase 4 is currently underway. The IMO MEPC has agreed that the review should be finalized in time for adoption of the necessary amendments to MARPOL Annex VI with a view to early implementation of Phase 3 and, if agreed, introduction of Phase 4 as soon as possible.

Regulation 22 The Ship Energy Efficiency Management Plan (SEEMP) is an operational measure that establishes a mechanism to improve the energy efficiency of a ship in a cost-effective manner through the following key steps: planning, implementation, monitoring, self-evaluation, and improvement.

The SEEMP provides an approach for shipping companies to manage ship and fleet efficiency performance over time using, for example, the Energy Efficiency Operational Indicator (EEOI) as a monitoring tool.

The EEOI can be enhanced by applying best practices for fuel-efficient operations as well as deploying latest technological devices for existing vessels. The introduction of initiatives, such as slow steaming, weather routing, antifouling, and trim optimization, can lead to a reduction in fuel consumption for existing vessels as well as contribute to an improvement of ship life cycle environmental performance.

Regulation 22 of MARPOL Annex VI requires that as of 1 January 2013, each ship that is subject to energy regulations shall keep on board a ship-specific Ship Energy Efficiency Management Plan (SEEMP). This may form part of the ship's safety management system (SMS). The SEEMP shall be developed taking into account guidelines adopted by the organization, IMO (2011a).

There are two parts to a SEEMP. Part I provides a possible approach for monitoring ship and fleet efficiency performance over time and some options to be considered when seeking to optimize the performance of the ship. Part II of SEEMP provides the ship-specific methodologies to collect, aggregate, and report ship data with regard to annual fuel oil consumption, distance traveled, hours underway, and other data required by Regulation 22A of MARPOL Annex VI.

Amendments to MARPOL Annex VI under Regulation 22 entered into force on 1 March 2018 to introduce the IMO Data Collection System (DCS) for fuel oil consumption of ships. Beginning January 1, 2019, vessels of 5000 GT and above are required to have a documented plan in place in view of monitoring CO2 emissions.

IMO DCS regulations require companies to update their existing Ship Energy Efficiency Management Plan (SEEMP) to document the methodology that will be used to collect the required data and the processes that will be used to report the data to the ship's Administration for verification, IMO (2016b, 2017).

In summary:

1. Each ship more than 400 GT that is involved in international voyages should have a SEEMP on board.
2. There is no specific reference to a need for review and verification of a SEEMP's content. However, its existence on board must be verified.
3. Currently, the IMO has issued technical guidelines in the form of a basic framework for SEEMP development and implementation. Ship owners and operators should use the IMO guidelines as a basis to develop a vessel's SEEMP, but it is up to them to further identify the appropriate energy KPIs that will stimulate future efficient operational practices.
4. ISO 50001 for Energy Management Systems, which is considered one step beyond SEEMP, is also available to the industry helping companies to improve their energy performance, maximize energy efficiency, and reduce fuel consumption. ISO 50001 requires that energy baselines are established and changes in energy performance are measured against them.

Regulation 23 This regulation was developed at the request of developing countries following a significant debate at IMO MEPC on role of various countries on GHG reduction efforts as well as the technological and financial difficulties that developing countries may face as a result of energy efficiency regulations. This regulation is entitled "Promotion of technical cooperation and transfer of technology relating to the improvement of energy efficiency of ships." It stipulates that:

- Administrations shall, in co-operation with the Organization1 and other international bodies, promote and provide, as appropriate, support directly or through the Organization to States, especially developing States that request technical assistance.
- The Administration of a Party2 shall co-operate actively with other Parties, subject to its national laws, regulations and policies, to promote the development and transfer of technology and exchange of information to States which request technical assistance, particularly developing States, in respect of the implementation of measures to fulfill the requirements of chapter 4 of this annex, in particular Regulations 19.4–19.6. IMO (2011a)

In support of the implementation of the above regulation, IMO MEPC approved a new guideline, IMO (2013d). This document provides a framework for the promotion and facilitation of capacity building, technical cooperation, and technology transfer to support the developing countries in the implementation of the EEDI and the SEEMP. As part of this, the Ad Hoc Expert Working Group on Facilitation of Transfer of Technology for Ships (AHEWG-TT) was set up, and IMO supported relevant meetings and work items. Additionally, IMO has carried out a significant amount of capacity building activities and implemented relevant project in this area.

3 EEDI Calculation

3.1 *The EEDI Calculation Formula*

The Attained EEDI provides a specific figure for an individual ship, expressed in grams of carbon dioxide (CO2) per ship's capacity mile (the smaller the EEDI, the more energy efficient the ship design) and is calculated by a formula based on the technical design parameters for a given ship. A simplified form of the EEDI formula is shown below:

$$\text{EEDI} = \frac{\text{Engine power} \times \text{SFC} \times \text{CF}}{\text{DWT} \times \text{speed}}$$

All terms of the EEDI formula are described in detail in Table 3.8.

At first glance a ship's EEDI appears to be a strong incentive to improve the design efficiency of new ships as an indication of a cost/benefit ratio to society in the form of CO2 emissions. To adapt the formula to a comprehensive calculation method that represents the diverse ship types, propulsion system configurations, fuel systems, and potential energy efficiency technologies, the formula was expanded to its current form:

$$\frac{\left(\prod_{j=1}^{n} f_j\right)\left(\sum_{i=1}^{n\text{ME}} P_{\text{ME}(i)} \cdot C_{\text{FME}(i)} \cdot \text{SFC}_{\text{ME}(i)}\right) + \left(P_{\text{AE}} \cdot C_{\text{FAE}} \cdot \text{SFC}_{\text{AE}}\right) + \left(\left(\prod_{j=1}^{n} f_j \cdot \sum_{i=1}^{n\text{PTI}} P_{\text{PTI}(i)} - \sum_{i=1}^{n\text{eff}} f_{\text{eff}(i)} \cdot P_{\text{AEeff}(i)}\right) C_{\text{FAE}} \cdot \text{SFC}_{\text{AE}}\right) - \left(\sum_{i=1}^{n\text{eff}} f_{\text{eff}(i)} \cdot P_{\text{eff}(i)} \cdot C_{\text{FME}} \cdot \text{SFC}_{\text{ME}}\right)}{f_i \cdot f_c \cdot f_l \cdot \text{Capacity} \cdot f_w \cdot V_{\text{ref}}}$$

Table 3.8 Terms of the EEDI formula

V_{ref}	V_{ref} is the ship's speed measured in knots, in deep water for EEDI loading condition using $\sum_{i+1}^{nME} P_{ME(i)} + \sum P_{PTI(i),Shaft}$ as propulsion shaft power (generally 75% MCR)
	$\sum P_{PTI(i),Shaft} = \sum (0.75 \cdot P_{SM,\max(i)} \cdot \eta_{PTI(i)})_*$
	$P_{SM,\max(i)}$: rated power consumption of each shaft motor measured in kW (if installed)
	$\eta_{PTI(i)}$: efficiency of each shaft motor
	When power to the propulsor is limited by verified technical means, 75% (*) of the limited propulsion power is used to determine V_{ref}
	(*) For steam turbine propulsion systems, 0.75 to be replaced by 0.83
	V_{ref} is subject to the following conditions:
	Deepwater operation
	Calm weather including no wind and waves
	Loading condition corresponding to the capacity
	Total shaft propulsion power at corresponding value of PME
Capacity	*Capacity* for EEDI loading condition is measured in MT and shall be:
	DWT at maximum summer load draft as certified in the vessel's stability booklet approved by the Administration for bulk carriers, tankers, gas carriers, LNG carriers, Ro-Ro cargo ships (vehicle carriers), Ro-Ro cargo ships, Ro-Ro passenger ships, general cargo ships, refrigerated cargo carrier, and combination carriers
	70% DWT for containerships. Draft at 70% DWT may account for a specific trim provided that speed/power curves have been established by dedicated model tests at 70% deadweight and same trim
	Gross tonnage (GT) for passenger ships and cruise passenger ships
$P_{ME(i)}$	$P_{ME(i)}$ is 75% of the engines maximum continuous rating (MCR) for each main engine (*i*), measured in kW
	For LNG carriers with diesel-electric propulsion, $P_{ME(i)}$ is calculated as:
	$P_{ME(i)} = 0.83 \times MPP_{Motor}/\eta_i$ where,
	$MPP_{Motor(i)}$ is the rated output of motor per certified document
	$\eta_i = 91.3\%$
	$\eta_i = \eta_{gen} \cdot \eta_{transf} \cdot \eta_{cov} \cdot \eta_{motor}$ (weighted average)
	For LNG carriers with steam turbine propulsion, $P_{ME(i)}$ is to be taken:
	$P_{MEi} = 0.83 \times MCR_{Steam\ Turbine}$
P_{PTO}	In case shaft generator(s) are installed, $P_{PTO(i)}$ is 75% of the rated electrical output power measured in kW of each shaft generator: $P_{PTO(i)}$: $0.75 \times MCR_{PTO(i)}$*
	For calculation of the effect of the shaft generators, two options are available:
	Option 1:
	$\sum_{i+1}^{nME} P_{ME(i)} = 0.75 \times \left(\sum MCR_{ME(i)} - \sum P_{PTO(i)}\right)$
	Maximum allowable deduction for calculation $0.75 \times \sum P_{PTO(i)} \leq P_{AE}$
	Option 2:
	$\sum P_{ME(i)} = 75\%$ of limited power
	Applicable *only* if installed main engine power is limited by verified technical means
	*In case shaft generators are fitted to steam turbine, 0.75 to be replaced by 0.83

(continued)

Table 3.8 (continued)

P_{PTI}	If shaft motors are installed, $P_{PTI(i)}$ is 75% of the rated power consumption of each shaft motor, measured in kW, divided by the weighted average efficiency of the generators:
	$\sum P_{PTI(i)} = \frac{\sum(0.75 \cdot P_{SM,max(i)})}{\eta_{\overline{Gen}}}(*)$
	$\eta_{\overline{Gen}}$ = weighted average efficiency of generator(s)
	$P_{SM,\,max\,(i)}$: rated power consumption of each shaft motor measured in kW
	*In case shaft motors are fitted to steam turbine, 0.75 to be replaced by 0.83
P_{AE}	P_{AE} is the required auxiliary engine power to supply normal maximum sea load and includes necessary power for propulsion machinery/systems and accommodation
	For ships with total propulsion power of 10,000 kW or above:
	$P_{AE(\sum MCR_{ME(i)} \geq 10{,}000\ kW)} = \left[0.025 \times \left(\sum_{i=1}^{nME} MCR_{ME(i)} + \frac{\sum_{i=1}^{nPTI} P_{PTI(i)}}{0.75}\right)\right] + 250$
	For ships with total propulsion power below 10,000 kW:
	$P_{AE(\sum MCR_{ME(i)} < 10{,}000\ kW)} = \left[0.05 \times \left(\sum_{i=1}^{nME} MCR_{ME(i)} + \frac{\sum_{i=1}^{nPTI} P_{PTI(i)}}{0.75}\right)\right]$
	P_{AE} calculations have specific rules for LNG carriers with re-liquefaction plant or compressors to supply boil of gas BOG to the engines refer to IMO (2014c)
	For cases where calculated P_{AE} is significantly different from actual P_{AE}, the ship Electric Power Table (EPT) should be used to estimate P_{AE}
C_F	C_F is the nondimensional conversion factor between fuel consumption and CO_2 emission. The value of C_F corresponds to the fuel used when determining the SFC listed in the NOx Technical File. C_F shall be determined separately for main engine(s) C_{FME} and auxiliary engine(s) C_{FAE}
SFC	Certified-specific fuel consumption, g/kWh, for main engine(s) SFC_{ME} and auxiliary engine(s) SFC_{AE} obtained from NOx Technical File
	SFC for steam turbine installations should be calculated by manufacturer and verified by ABS
	For those engines with power output below 130 kW, which do not have a test report included in a NOx Technical File, the SFC specified by the manufacturer and endorsed by a competent authority should be used
	For LNG-driven engines for which SFC is measured in kJ/kWh, the SFC value is to be converted to g/kWh using the standard lower calorific value of the LNG (48,000 kJ/kg), referring to the 2006 IPCC Guidelines
f_w	Weather factor, f_w, accounts for a decrease in speed in representative sea conditions of wave height, wave frequency, and wind speed and determined as follows
	$f_w = 1.0$ for the Attained EEDI calculated under Regulations 20 and 21 of MARPOL Annex VI
	$f_w \neq 1.0$ is applicable only to vessels that consistently operate in rough weather on their trade
	Attained EEDI calculated based on $f_w \neq 1.0$ is to be referred to as "Attained $EEDI_{weather}$"
	f_w can be determined using either of the two methods:
	Ship-specific simulation of performance in representative sea conditions following IMO (2012e)
	Standard f_w table/curves, expressed as a function of capacity, for bulk carriers, tankers, and containers provided in IMO (2012e)
	f_w and Attained $EEDI_{weather}$ are to be listed in the EEDI Technical File if calculated

(continued)

Table 3.8 (continued)

$f_{eff(i)}$	$f_{eff(i)}$ is the availability factor for innovative energy efficiency technology
	$f_{eff(i)} = 1.0$ for waste heat recovery systems. Other technologies may have $f_{eff(i)}$ factors less than 1.0 as their output may only be available intermittently. Refer to IMO (2013e)
f_j	f_j is the power correction factor and is applicable to:
	Ice-classed ships
	Shuttle tankers with propulsion redundancy (80,000–160,000 DWT)
	Ro-Ro ships, all types
	General cargo ships
	For detailed information on how f_j is assigned to each of the above categories, refer to IMO (2014c)
f_i	f_i is the capacity correction factor applicable as:
	f_{iVSE} for ship-specific voluntary structural enhancements
	f_{iCSR} for ships built in accordance with the Common Structural Rules (CSR) and assigned the class notation CSR
	f_{iICE} for ice-strengthened ships
	$f_i = 1.0$ for all other ship types
	For detailed information on how f_i is assigned to each of the above categories, refer to IMO (2014c)
f_C	f_c is the cubic capacity correction factor. It is applicable to chemical carriers, gas carriers which carry LNG, with direct diesel propulsion systems, and Ro-Ro passenger ships having a DWT/GT ratio of less than 0.25 where DWT is the capacity and GT is in accordance with tonnage measurement conventions
	f_c should not be applied to LNG carriers that fall into the ship definition of Regulation 2.38 of MARPOL Annex VI
	$f_c = 1.0$ for all other ship types
f_l	*f_l is the crane and cargo gear correction factor for general cargo ships*
	f_l compensates for a loss of deadweight of the ship due to cranes and cargo gear
	$f_l = f_{cranes} \cdot f_{sideloader} \cdot f_{roro}$
	$f_l = 1.0$ for all other ship types

Figure 3.7 explains how each of the terms included in the EEDI formula affects the vessel's Attained EEDI.

The items that primarily influence EEDI are:

- Installed main engine power and energy needed for propulsion; this is represented by the first term in the numerator of the formula.
- Auxiliary power requirements of the ship; this is represented by the second term in the nominator.
- Innovative electrical technologies on board such as electricity from waste heat recovery or solar power. These are represented by the third term in the nominator.
- Innovative mechanical technologies that provide power for ship propulsion such as wind power (sails, kites, etc.). This is the last term in the nominator.
- Ship capacity and ship speed are represented in the denominator. Their product represents the value of transport work.

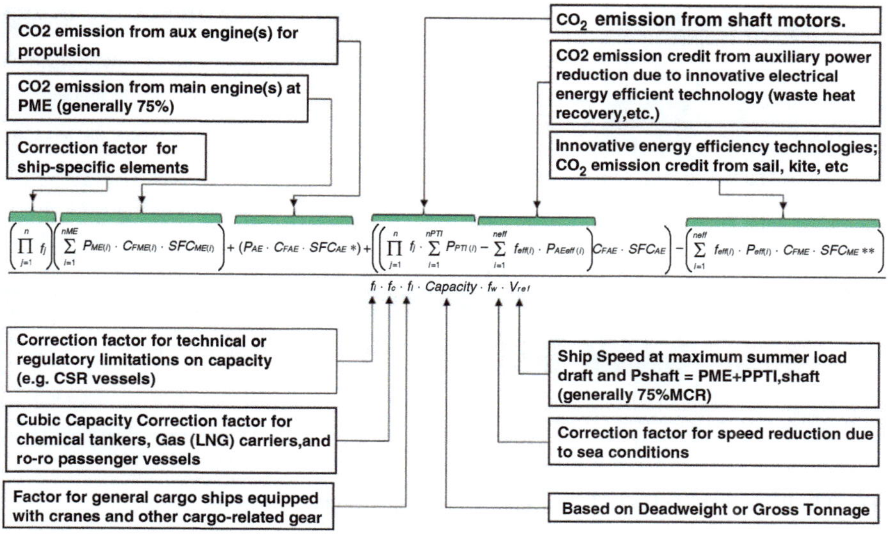

Fig. 3.7 The EEDI calculation formula input parameters

For the majority of ships for which EEDI data have been reported to the IMO and made publically available, several of the parameters in this formula are taken as 0 or 1. More specifically:

1. Correction factor $f_j = 1$ as it represents ship-specific design elements of ice class vessels, Ro-Ro ships, general cargo, or shuttle tankers with propulsion redundancy.
2. Availability factor $f_{\text{eff}} = 0$ as reported innovative technologies are currently limited to numbered waste heat recovery system installations for electrical power generation. It should be pointed out here that the effect of more common energy-saving devices (e.g., pre-swirl stators, rudder bulbs) cannot be separated from the overall performance of the vessel and is accounted for in the EEDI reference speed (V_{ref}) during model tests and speed trials.
3. Correction factor $f_l = 1$ as this is only applicable to general cargo vessels.
4. Correction factor $f_w = 1$ as the weather factor f_w demonstrates the reduction of ship speed in representative sea conditions of wave height, wave frequency, and wind speed (e.g., Beaufort Scale 6). IMO Guidelines on f_w calculation are for now only interim.

Therefore the EEDI formula simplifies to the equation shown below:

$$\frac{\left(\sum_{i=1}^{nME} P_{ME(i)} \cdot C_{FME(i)} \cdot SFC_{ME(i)}\right) + (P_{AE} \cdot C_{FAE} \cdot SFC_{AE})}{f_i \cdot f_c \cdot \text{Capacity} \cdot V_{\text{ref}}}$$

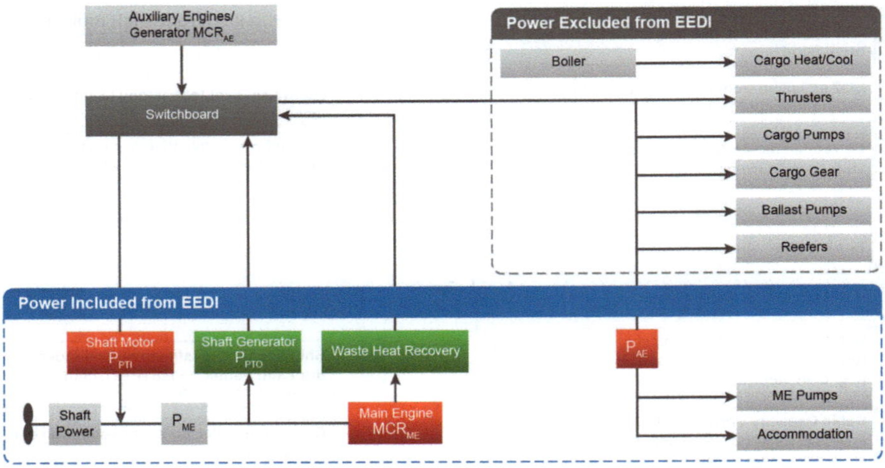

Fig. 3.8 Power included in EEDI calculation – example graph

Figure 3.8 gives a simplified outline of the vessel's power plant in order to demonstrate which machinery components are taken into account in the EEDI calculation.

As a general rule:

- All the cargo-related energy uses on-board are outside the scope of the EEDI calculations (not included in the formula).
- Auxiliary boilers are also excluded from the formula; assuming that under normal sea-going conditions, boilers will not be operating.

Therefore, electricity needed for cargo pumps, cargo handling equipment, ship thrusters, etc. is out of scope of EEDI calculations.

3.2 *Terms in the EEDI Formula*

Table 3.8 gives a cumulative summary description of all the terms used in the EEDI calculation and how these shall be applied according to IMO guidelines, IMO (2014c).

3.3 *EEDI Technical File*

For verification, implementation, and enforcement purposes by Flag Administrations and Port States, all the relevant terms used in the EEDI calculation and their values are required to be recorded in the "EEDI Technical File" along with the calculation methodology applied and then submitted to the verifiers (normally recognized organization on behalf of flag state) that will carry out the certification on behalf of Flag Administration. The "EEDI Technical File" needs to be kept on board and forms a supplement to International Energy Efficiency Certificate (see Sect. 5).

The IMO in its EEDI survey and verification guidelines IMO (2014b) has provided a sample "EEDI Technical File." A similar example is also given in the Procedural Requirement 38 of the International Association of Classification Societies (2016) and attached Industry Guidelines, IMO (2015d). The examples identified are non-exhaustive but provide comprehensive guidance on the use of all data necessary for verification purposes including all the terms defined in Table 3.8 that need to be recorded in the EEDI Technical File.

4 EEDI Survey and Verification

EEDI verification is conducted on behalf of the vessel's Flag Administration by recognized organizations (ROs) according to "2014 Guidelines on survey and certification of the Energy Efficiency Design Index (EEDI)," IMO (2014b). For vessel's equipped with innovative energy efficiency technologies, guidance is provided in the "2013 Guidance on treatment of innovative energy efficiency technologies for calculation and verification of the Attained EEDI," IMO (2013e).

EEDI Verification is performed at two separate stages:

- Preliminary stage
- Final stage

Verification at the preliminary stage is done during the ship's initial design and pre-construction period. Final verification is carried out after construction following the vessel's sea trials and prior to delivery. Relevant ship design data, tank test data, and speed trial data will be subject to scrutiny and verification by ROs. The aforementioned IMO guidelines on EEDI verification are developed to ensure consistency of verification, although some important issues such as certain constraints applicable to the execution and witnessing of tank (model) tests, speed-power scaling methods, as well as standardized approaches used for sea trial correction hold room for further review and improvement.

Figure 3.9 shows the overall process diagram for EEDI verification.

4.1 Preliminary Verification

For the preliminary verification at the design stage, the following should be submitted to the verifier:

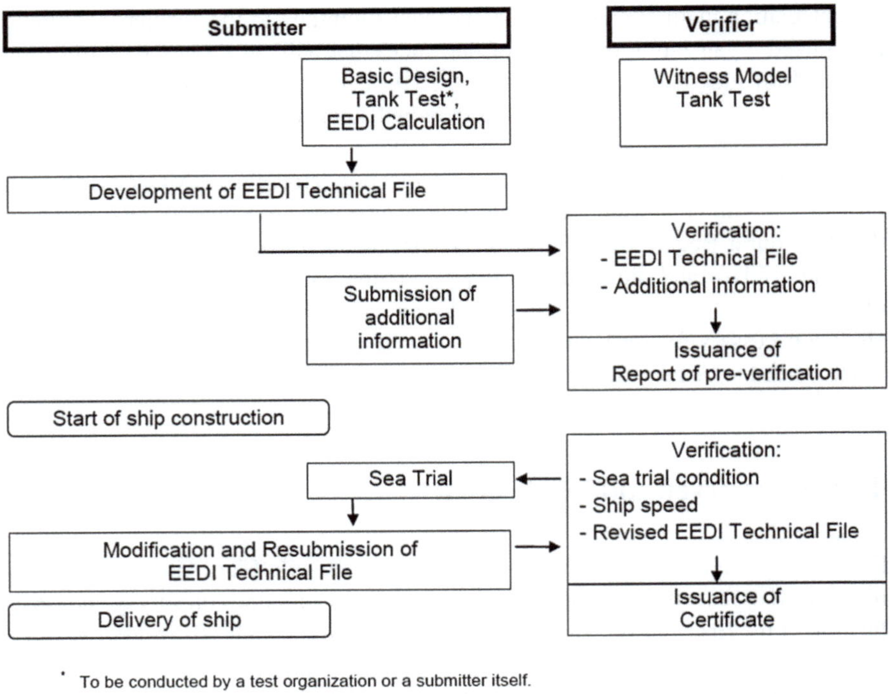

Fig. 3.9 The EEDI verification process (Source: IMO 2014b)

- An application for an initial survey.
- Preliminary "EEDI Technical File" containing the necessary information.
- Relevant background documents and information.

The EEDI Technical File should be developed by the submitter (ship designer or shipyard) and must include of all the data required.

Additional background documents and information necessary for the verifier include but are not limited to:

- Model Test Report complete with towing tank test results and full-scale tabulated power/speed predictions for below two (2) loading conditions:

   I. EEDI loading condition is based on maximum summer load line draft as certified in the approved Stability Booklet and applies for different vessel types as follows:

      Capacity is 100% DWT for bulk carriers, tankers, gas carriers, LNG carriers, Ro-Ro cargo ships (vehicle carriers), Ro-Ro cargo ships, Ro-Ro passenger ships, general cargo ships, refrigerated cargo carrier, and combination carriers.

      Capacity is 70% DWT for containerships.

      Capacity is gross tonnage for passenger ships and cruise passenger ships.

II. Intended sea trial condition (vessel loading condition during sea trials if different from EEDI loading condition, which is required for final verification of EEDI)

- Description of the tank test facility including test equipment and calibrations.
- Lines of the model and the actual ship for the verification of the similarity of model and actual ship.
- Lightweight of the ship and displacement table for the verification of the deadweight. This may require submission of available ship stability data for verification purposes.
- Calculation process of the ship reference speed.
- Reasons for exempting a tank test, if applicable.
- Copy of the NOx Technical File and documented summary of the SFC correction for each type of engine with copy of engines' (Engine International Air Pollution Prevention) EIAPP certificate.
- Electric Power Table (if P_{AE} is significantly different from the value computed using the formula defined in the IMO Calculation Guidelines)
- Other specific data for specific ships: For example for ships using gas as primary fuel, the verifier may request data on gas fuel and liquid fuel tank arrangement and capacities for CF calculation purposes.

The most important element of preliminary verification is the ship's model tank test. According to the IMO guidelines IMO (2014b):

> The speed power curve used for the preliminary verification at the design stage should be based on reliable results of tank test. A tank test for an individual ship may be omitted based on technical justifications such as availability of the results of tank tests for ships of the same type. In addition, omission of tank tests is acceptable for a ship for which sea trials will be carried under the "EEDI Condition"5, upon agreement of the ship-owner and shipbuilder and with approval of the verifier. For ensuring the quality of tank tests, the International Towing Tank Conference (ITTC) quality system should be taken into account. Model tank test should be witnessed by the verifier.

4.2 *Final Verification*

At the final EEDI verification stage, the submitter shall prepare a dedicated sea trial plan in accordance with the International Organization for Standardization ISO 15016:2015 guidelines. The sea trial plan will be the guiding document during the execution of the ship's commissioning trials. Adherence to the process ensures that the ship's final speed-power curve and EEDI reference speed, $V_{ref,}$ are determined accurately; this is an essential step of the final EEDI verification.

Afterward, all relevant parameters of the EEDI calculation will be revisited and verified. Aspects that need to be considered for sea trail are elaborated further here using the IMO guidelines IMO (2014b).

4.3 Calculation and Verification of Innovative Technologies

The verification of innovative energy efficiency technologies is an involved process and is fully documented in the guidelines, IMO (2013e). This is an interim guidance document and will evolve over time as experience is gained as a result of future use of these technologies.

The evaluation of the benefit of innovative technologies on EEDI is to be carried out in conjunction with the hull form and propulsion system with which it is intended to be used. Results of model tests or sea trials of the innovative technology in conjunction with different hull forms or propulsion systems may or may not be applicable.

4.4 Categorization of Technologies

Innovative energy efficiency technologies are allocated to category (A), (B), and (C), depending on their characteristics and the way they influence the EEDI formula. Furthermore, innovative energy efficiency technologies of categories (B) and (C) are categorized to two subcategories (categories (B-1) and (B-2) and (C-1) and (C-2), respectively).

- Category (A): Technologies that directly influence and shift the ship speed-power curve, which results in the change of combination of propulsion power (*PP)* and V_{ref}. For example, such technologies at constant V_{ref} can lead to a reduction of *PP*; or for a constant *PP*, they could lead to an increased V_{ref}. All technologies that directly impact the ship hydrodynamics could have such impacts.
- Category (B): Technologies that reduce the *PP* at a V_{ref} but do not generate electricity. The saved energy is counted as P_{eff}.
  - Category (B-1): Technologies which can be used at all times during the operation (e.g., hull air lubrication); thus the availability factor (f_{eff}) should be treated as 1.00.
  - Category (B-2): Technologies which can be used at their full output only under limited conditions and periods (e.g., wind power). The setting of availability factor (f_{eff}) should be less than 1.00.
- Category (C): Technologies that generate electricity. The saved energy is counted as PAEeff.
  - Category (C-1): Technologies which can be used at all times during the operation (e.g., waste heat recovery); thus the availability factor (f_{eff}) should be treated as 1.00.
  - Category (C-2): Technologies which can be used at their full output only under limited condition (e.g., solar power). The setting of availability factor (f_{eff}) should be less than 1.00.

Table 3.9 Number of double runs for EEDI trials based on current correction method

Current correction	Power setting	Lead vessel	Sister vessel
Mean of means	Below EEDI	2	1
	Around EEDI	2	1
	Above EEDI	2	1
Iterative method	Below EEDI	1	1
	Around EEDI	2	1
	Above EEDI	1	1

4.5 Sea Trials: Observation

In order to ensure accurate EEDI calculation, sea trial conditions should be set close to the "EEDI Condition," if possible. As mentioned earlier, the vessel's sea trial plan should be submitted to the verifier for approval and confirmation that the conditions and processes described follow the ISO 15016:2015 guidelines. EEDI trial requirements include but are not limited to the following:

Ship's actual Displacement measured prior commencement of speed power trials shall be less than 2% of required displacement as derived from dedicated model tests.

The power settings and number of double runs for EEDI speed-power trials are based on the current correction method to be applied and whether the vessel trialed is a lead ship or sister ship:

- Power settings should be distributed within the range from 65%MCR to 100%MCR.
- Number of double runs for Lead and Sister vessels depends on agreed current correction method (Table 3.9).

Each double run shall be conducted heading into and following the dominant wave direction over the same ground area. Duration of each speed run shall be at least 10 min at steady-state ship state.

Speed-power trials should be conducted soon after launching and/or with the hull and propeller clean.

Trial location and heading of forward/return runs shall be consistent for all double runs of the progressive speed trial. Changes in heading (e.g., reversal of forward/return run direction) are not recommended. Recorded parameters may provide inaccuracies in speed trial analysis results (ISO 15016:2015).

The speed-power trials shall be conducted in a location free of hindrance by small boats and commercial traffic where the environmental conditions are expected to be constant with limited wind, waves, and current.

The test procedure should include, as a minimum, descriptions of all necessary items to be measured and corresponding measurement methods. The verifier should attend the sea trial and confirm the following parameters shown in Tables 3.10 and 3.11 are measured and recorded as accurately as possible.

Table 3.10 Parameters measured and recorded prior to speed trials

Measure	Device	Unit
Water density	Salinity sensor, conductivity density Temperature (CDT) sensor	kg/m^3
Water temperature	Thermometer, CDT sensor	°C
Air temperature	Thermometer	°C
Air pressure	Barometer	hPa, mb
Torsion meter zero setting	Torsion meter with calibrated torque sensor or strain gauges	kNm
Trial area	Geographical position (Lat-Long) by DGPS	dddd-mm
Vertical position of anemometer	General arrangement plan of the ship	m
Drafts	Physical observation and/or calibrated draft gauges	m

Table 3.11 Parameters measured and recorded during speed trials

Measure	Device	Unit
Ship track	DGPS	Lat../Long, deg
Speed over ground	DGPS	Knots
Shaft torque	Torsion meter with strain gauges or torque sensor	kNm
Shaft power	Calculated from torque and RPM	kW
Shaft RPM	Pickup, optical sensor, ship revs counter	RPM
Propeller pitch	Bridge replicator	Deg or m
Time	GPS time, stopwatch	s
Water depth	Ship echo sounder and nautical charts	m
Ship heading	Gyro compass or DGPS	deg
Relative wind	Anemometer	m/s, deg
Bow acceleration (STAWAVE-1)	Acceleration meter	m/s^2
Wave height, period and direction	Radar scanner, wave buoy (minimum of three observers)	m, deg
Drafts	Observation, draft gauges	m

4.6 *Speed Trial Analysis*

The main output of the speed trial will be the actual measured ship speed-power curve and its corrected/extrapolated equivalent for the EEDI Condition. A large number of vessels are trialed at ballast condition, for example, bulk carriers and containerships. The speed-power curve representing the actual performance of the vessel at the trial loading condition is derived by analysis calculations that involve the application of a number of corrections related mainly to the prevailing weather and sea state during the course of trials. Once the analysis of the corrected speed-power curve for the trial loading condition is determined, a conversion is done to EEDI loading condition.

The speed trial analysis shall follow the requirements described in ITTC (2017) and ISO (2015).

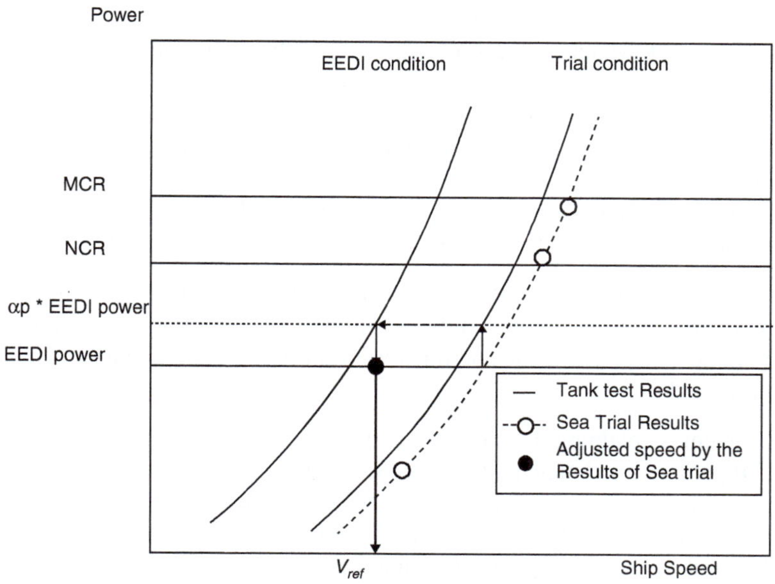

Fig. 3.10 Example scheme of conversion from trial condition to EEDI condition at EEDI power. (Source: IMO 2014b)

The speed adjustment and correction from ballast condition to EEDI condition plays an important role in an accurate estimation of EEDI. An example of a simplified method of the speed adjustment is given in Fig. 3.10 as is included in IMO EEDI survey and verification guidelines, IMO (2014b).

The EEDI reference speed, V_{ref}, is obtained from the results of the sea trials at trial condition using the speed-power curves predicted by the tank tests. The tank tests are also carried out at both drafts: trial condition corresponding to that of the speed-power trials and EEDI condition. For trial conditions the power ratio α_P between model test prediction and sea trial result is calculated for constant ship speed. Ship speed from model test prediction for EEDI condition at EEDI power multiplied with αP is V_{ref}.

The verifier is required to ensure that the sea trial analysis and conversion to EEDI loading condition are done accurately. The collected shipboard data along with a detailed analysis including intermediate results and providing the vessel's EEDI reference speed V_{ref} should be submitted to the verifier.

4.7 Verification of the Attained EEDI for Major Conversions

"Major Conversion" means a conversion of a ship:

- Which substantially alters the dimensions, carrying capacity, or engine power of the ship.

- Which changes the type of the ship.
- The intent of which in the opinion of the Administration is substantially to prolong the life of the ship.
- Which otherwise so alters the ship that, if it were a new ship, it would become subject to relevant provisions of the present Convention not applicable to it as an existing ship.
- Which substantially alters the energy efficiency of the ship and includes any modifications that could cause the ship to exceed the applicable required EEDI.

In case of a major conversion, the owner or shipyard should submit to a verifier an application for an additional survey with the EEDI Technical File duly revised based on the conversion made and other relevant background documents including but not limited to:

- Documents explaining details of the conversion
- EEDI parameters changed after the conversion
- Reasons for other changes made in the EEDI Technical File
- Calculated value of the Attained EEDI, with the calculation summary for each value of the calculation parameters and the calculation process

4.8 EEDI Verification: Scope of Activities

The scope of verification activities may be summarized separately for the preliminary and final stages in the lists below:

Preliminary stage:

- Review the EEDI Technical File, check that all the input parameters are documented and justified, and check that the possible omission of a tank test has been properly justified.
- Check that the ITTC procedures and quality system are implemented by the organization conducting the ship model tank tests. The verifier would audit the quality management system of the towing tank if previous experience is insufficiently demonstrated.
- Witness the tank tests according to a test plan initially agreed between the submitter and the verifier.
- Check that the work done by the tank test organization is consistent with the ITTC recommendations. In particular, the verifier will check that the power speed curves at full scale are determined in a consistent way between test condition and EEDI loading conditions.
- Issue a preliminary verification report inclusive, possibly in the form of a preliminary statement of compliance.

Final stage:

- Review the sea trial plan to check that the test procedure complies with the requirements of the IMO guidelines. It should be noted that the IMO guidelines have endorsed the use of the ISO 15016:2015 standard for all ships trialed after September 2015.
- Survey the vessel to ascertain the ship principle and machinery characteristics conform with those in the EEDI Technical File.
- Attend the sea trial and record the main parameters to be used for the final calculation of the EEDI as discussed before.
- Review the sea trial report provided by the submitter and check that the measured power and speed have been corrected according to the ISO 15016:2015 standard.
- Perform independent speed trial analysis to verify reference ship speed V_{ref}, and confirm the conversion of the speed-power curve to the EEDI loading condition.
- Verify revised EEDI calculation inputs and results.
- Confirm that the vessel's Attained EEDI is less than the required regulatory limit.
- Review the revised EEDI Technical File, if applicable.
- Complete relevant parts of the Record of Construction and endorse.

4.9 International Energy Efficiency (IEE) Certificate and Its Supplements

Following the final EEDI verification, an IEE Certificate is issued, and a Record of Construction for Energy Efficiency will be attached to the certificate. The IEE Certificate has no expiry date, since it will be valid throughout the life of the ship, except in cases where the certificate is rewritten or reissued.

The following two documents are considered as supplements to the IEE Certificate:

- EEDI Technical file
- SEEMP

As specifically stated in MARPOL Annex VI Chapter 4, Port State inspections shall be limited to verifying, when appropriate, that there is a valid IEE Certificate on board, in accordance with Article 5 of the MARPOL Convention.

5 Interim Guidelines for Determining Minimum Propulsion Power (MPP) to Maintain the Maneuverability of Ships in Adverse Conditions

One of the most effective ways of reducing a ship's EEDI is by reducing the ship's design speed by selection of a smaller main engine or main propulsion motor.

Within IMO a debate took place on how far speed reduction could be used for EEDI reduction. As a result, it was decided that there is a need to limit the use of this method of EEDI reduction so that it does not lead to unsafe and underpowered ships that may lose maneuvering capability under adverse weather condition. To ensure safe maneuvering in adverse conditions, a requirement was introduced within the EEDI regulations (Regulation 21.5, Chapter 4 of MARPOL Annex VI):

> For each ship to which this regulation applies, the installed propulsion power shall not be less than the propulsion power needed to maintain the maneuverability of the ship under adverse conditions as defined in the guidelines to be developed by the Organization.

IACS was tasked to develop guidelines for determining minimum propulsion power to enable safe maneuvering. The studies conducted by the IACS working groups served as a basis for the "2013 Interim Guidelines," IMO (2013b), which were further updated in 2015 by IMO (2015a, b).

The IMO guidelines define a methodology for estimating the minimum propulsion power for each ship for safe maneuvering, thus ensuring that choice of the main propulsion engines/motors satisfies these minimum requirements.

The guidelines currently apply to:

- Tankers
- Bulk carriers
- Combination carriers

Investigation showed that the above ship types are most critical with respect to the sufficiency of power for maneuverability in adverse conditions. Views have been expressed by IMO member states that further consideration for other ship types should be done at a later stage.

The applicability of the guidelines from a capacity perspective is currently limited to ships of 20,000 DWT and above. The main reason behind this restriction is that a systematic evaluation of the required standard environmental conditions for ships with deadweight less than 20,000 DWT has not been completed yet. Ongoing studies in the IMO are addressing the issue for these ships, and a solid proposal is envisaged for the future.

The current methodologies for estimating the minimum power are based on two assessment levels or methods that are briefly described.

Assessment Level 1: Minimum Power Lines Assessment A simple approach that involves calculation of the minimum power from a specific line as a function of ship DWT, based on engine power data from already built ships. For this purpose, the verifier should check if the ship has an installed power not less than the minimum power defined by the line represented by the following equation:

$$\text{Minimum Power Line Value [MCR, kW]} = a \times (\text{DWT}) + b$$

where a and b are constants and vary with ship type and given in the IMO guidelines. As can be seen, this is a very simple approach.

It should be noted here that the maximum summer load condition (corresponding to the EEDI condition) has been identified as the "most severe" when estimating required propulsion power in adverse conditions.

Heavy ballast loading condition has been also examined, but the required propulsion power under heavy ballast is typically less than that under full-load conditions.

Furthermore, the normal ballast condition is generally not critical because ship masters generally change from the normal ballast condition to the heavy ballast condition based on weather forecast IMO (2015b).

Assessment Level 2: Simplified Assessment This is a more mathematically involved method of assessment. The assessment procedure consists of two steps:

- Step 1: Definition of the required advance speed in head wind and waves, ensuring course-keeping in all wave and wind directions.
- Step 2: Assessment whether the installed power is sufficient to achieve the above required advance speed.

The Level 2 assessment requires the determination of added resistance of waves by model tests in regular waves; empirical formulae are also referenced albeit not directly specified. To address this challenge, in-depth research was initiated by the EU research project SHOPERA (Energy Efficient Safe SHip OPERAtion) and Japan's JASNAOE research project.

As mentioned above, at IMO MEPC 68, the two assessment levels of the 2013 Interim Guidelines were thoroughly reviewed. It was agreed that the alternative approaches introduced, using inputs from ongoing research projects, could warrant further consideration. The strengthening of existing Level 1 assessment criteria was agreed as a tentative measure and adopted by the IMO MEPC, IMO (2015b).

More specifically, the technical justifications and appropriateness of the formulas embodied in the Level 2 assessment were examined by the Committee in order to confirm whether the current approach correctly evaluates maneuverability and adverse weather conditions and ensures safety.

Because the Level 2 Assessment has not been finalized to date, Level 1 was revised (strengthened) by Resolution IMO (2015b). Figure 3.11 is a sample graph showing old and new Level 1 for bulk carriers.

Discussions in the IMO for Level 2 are currently ongoing. The research project conclusions were examined by the member states at MEPC 71 but considered not mature enough to revise the interim guidelines for calculation of minimum propulsion power.

The IMO MEPC at its 72nd session agreed to extend the 2013 Interim Guidelines to EEDI Phase 2 and requested government states and participating bodies to continue discussions on the matter in an effort to further develop the revision to the guidelines in the upcoming sessions.

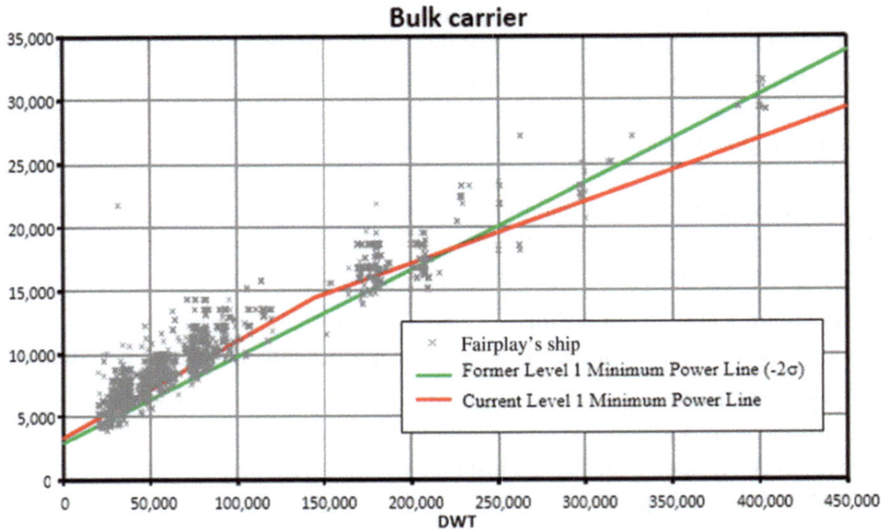

Fig. 3.11 Comparison of former and current Level 1 minimum power lines. (Source: Author's Private Archive)

6 Weaknesses of EEDI

The intent of EEDI is of course to push ship designers and shipyards to design more energy-efficient ships. It has been said that the most effective energy-saving devices are well-designed hull lines, creating the least possible hydrodynamic resistance and a good propulsion coefficient.

EEDI is thus a design index attempting to capture this philosophy, while at the same time giving a measure of how much CO2 is produced under some standard conditions per transport capacity. In simpler terms, EEDI is a measure of the penalty that society pays to enjoy the benefit of goods transportation. Obviously therefore, society wishes a smallest possible index.

At the same time, and in spite of the above intention, EEDI exhibits some weaknesses, which are described below.

6.1 It Is Easy to Comply with the Required EEDI Simply by Reducing the Design Speed, Without Reducing Ship's Resistance or Increasing Its Efficiency

Every ship must comply with the regulation:

$$\text{Attained (actual ship's) EEDI} \leq \text{Required EEDI}$$

"Attained EEDI" grows as a function of speed to the square power (V^2) or even more (and for fast ships V^3 or higher), while "Required EEDI" is a fixed number (from the baselines depending on ship's deadweight). The implication is obvious: Reduce the ship speed (power), and you can reach the Required EEDI.

This weakness was realized early on during the development of the formula and baselines at IMO, and there were calls from few member states (e.g., Greece) to correct it. A solution was proposed by Greece (IMO 2011b) which would make it harder to comply just by reducing speed and thus force designers to refine the ship's hull lines, use better propellers, etc. The proposal was to include speed in the "Required EEDI" so that both sides of the above inequality drop as speed drops. However, the proposal was not accepted; for a more detailed discussion, see Psaraftis (2018).

6.2 Compliance with EEDI Requirements, by Reducing Speed, Leads to Safety Concerns (Possible Underpowering)

Previous IMO Work While fast ships (e.g., containerships) have plenty of room to reduce their design speed safely, slow-speed ships (tankers and bulk carriers) do not. Reducing speed is the direct result of reducing installed power (to lower the Attained EEDI). Early on, concerns were expressed by ship operators that such ships may not have sufficient power to maneuver in adverse weather, leading IMO to examine the issue and publish guidelines on minimum propulsion power (MPP), as described in Sect. 5 of this chapter.

We will simply reiterate here that despite many years of examination and two large projects (SHOPERA and JASNAOE), the MPP Level 2 assessment has not been finalized, while there has been ongoing debate of what constitutes "adverse weather" (Beaufort 7, 8, 9 or 10?). The results from the projects suggest that when high Beaufort numbers are applied, the required power is unrealistically high (much higher than pre-EEDI ships), which is not in line with actual experience (typical pre-EEDI ships have not shown serious adverse weather performance concerns in Beaufort 9 or 10). Thus the project partners proposed Beaufort 7 or 8 as "adverse," with ship operators claiming that this is a relatively mild weather condition. Further adjustment of Level 2 assessment is required to produce results in line with experience.

For the time being, shipyards and operators rely on the MPP Level 1 assessment, which simply is a straight-line regression at the lower ends of pre-EEDI installed powers for various ship sizes.

Further IMO Work After setting the IMO GHG reduction targets in April of 2018 at MEPC 72, there are already calls for various measures in order to achieve the set reduction targets. Among those measures considered "ripe" for fast application is to further strengthen the EEDI requirements. The proposals include to bring the application date of Phase 3 forward and to introduce a more stringent Phase 4 for certain ship types. For slow-speed ships, this might exacerbate the safety concerns, especially for the larger DWT segments (see further below). Even with the current requirements, large-size tankers and bulk carriers cannot easily be made to comply with Phase 3, and if their power drops to the point of compliance, they will have issues even in Beaufort 8. It is recognized that IMO must finalize the minimum power requirements before enacting more stringent EEDI requirements for slow-speed ships.

Proposals have been submitted by certain Flag Administrations (IMO 2015c) to install a proper (safe) size engine, according to this minimum required power, and use a torque limiter at the propeller shaft so the ship operates normally at the EEDI required reduced power. In case of bad weather, the chief engineer can hit the limiter's by-pass button to have all the power available to him. This proposal has been disputed by IMO member states on grounds that it constitutes a dual NOx certification.

Greece has submitted several times, IMO (2011c, 2015c) that the problem is being looked at from a wrong perspective. Instead of setting a minimum power requirement, a minimum required speed (at sea trials) should be set. Speed is a better performance measure than installed power. This way, both safety and better efficiency of future designs could be achieved since, among others, full bodied ships will require large, thirstier engines to achieve the minimum speed than well-designed (slimmer) ships. With a minimum power requirement, there might be no further incentive for the designer to improve a given hull, since in any case he must install the required power. Of course, a minimum power (instead of speed) requirement does not guarantee that this power will be sufficient on a poorly designed ship.

6.3 The Required EEDI Baselines (or Reference Lines) Were Oversimplified

It was decided early on to use one regression line for each ship type and for all ship sizes in the category. As can be seen from Fig. 3.1 of this Chapter, however, smaller-size ships (data points) weigh much more on the regression than larger ships, simply due to the fewer number of larger ships. The result is that, whereas the line is appropriate for the smaller ships, it might penalize the larger ships. This is more profound for tankers and bulkers.

Once the reference lines were set, it became evident that most existing large ships such as very large crude carriers (VLCC), capesizes, etc. fall about 10% above the baseline, i.e., these ships had to drop their EEDI 10% more than other ships, in all EEDI phases going forward. This is part of the reason they are not expected to easily comply with Phase 3 (along with the related power safety issues). This was acknowledged by IMO at the time; however, drawing power lines for each different ship size would be very time-consuming. It was suggested then that a special adjustment "factor" would apply for these ships to be set during a future reviewed period. To date such a factor has not been discussed. However, given that shipyards and owners are now facing the issue, it is expected to be discussed again in view of the calls for strengthening EEDI further.

6.4 "Attained EEDI Weather" Provides a Truer Picture of Efficiency

EEDI is quite a theoretical index, being a snapshot of ship's performance at a rarely used draft (maximum) and in ideal sea conditions (no wind and no waves). As a result, ships with similar EEDI's may have very different performance in real sea conditions. The industry has fresh memories of very full bow ship designs (to increase displacement/deadweight), which at sea trials performed well but which, in real sea conditions of Beaufort 3 or 4, exhibited reduced speed capability and increased fuel consumption compared to similar designs or EEDIs with a more slender bow.

A truer picture of a ship's actual performance could be reflected in the EEDI if the weather coefficient (f_w) was actually used (currently the f_w in the EEDI formula is taken as 1.0), where $f_w = Vw/V_{\text{ref}}$. The weather coefficient f_w is a measure of the drop in ship speed at 75% MCR in weather conditions of Beaufort 6. A typical range for f_w for slow-speed ships (bulk carriers and tankers) is 0.80–0.95, which in itself is an indication of the extreme variation in design efficiency that is not captured in the EEDI (obviously a ship losing less speed – i.e., with $f_w = 0.95$ – is the more efficient.) Each ship design has its own f_w which can be determined experimentally by model tests. IMO (2012e) provides guidelines for the calculation of f_w as well as typical values. Experimental values included in said IMO circular show that, for same deadweight ships, the f_w can vary widely. For example, for 300,000 dwt tankers, f_w ranges from 0.83 to 0.94. Obviously the 0.94 design is far more efficient than the 0.83, dropping only 6% in speed from Beaufort 0 to Beaufort 6, versus 17% for the less efficient design. Since this speed drop is at the fixed power of 75% MCR, it is a direct measure of the efficiency of the ship's hull lines (especially bow shape) (Fig. 3.12).

Including actual f_w in the EEDI formula, resulting in "Attained EEDI weather," would provide a more realistic picture of the ship's efficiency in real operating conditions. At present, both ships in our example may have identical EEDI's, but their actual performance will be very different. Furthermore, the "less efficient" ship, due to smaller engine, may even have a better "Attained EEDI" than the more efficient ship but, in reality, may emit much more CO_2 when operating at the same speed. It has been observed in the past that, due to the large speed drop, such poor designs tend to increase their operating horsepower output to partially recover the loss and thus achieve a more "competitive" speed, thus operating at 90% MCR or more, which increases the engine's specific fuel oil consumption exponentially.

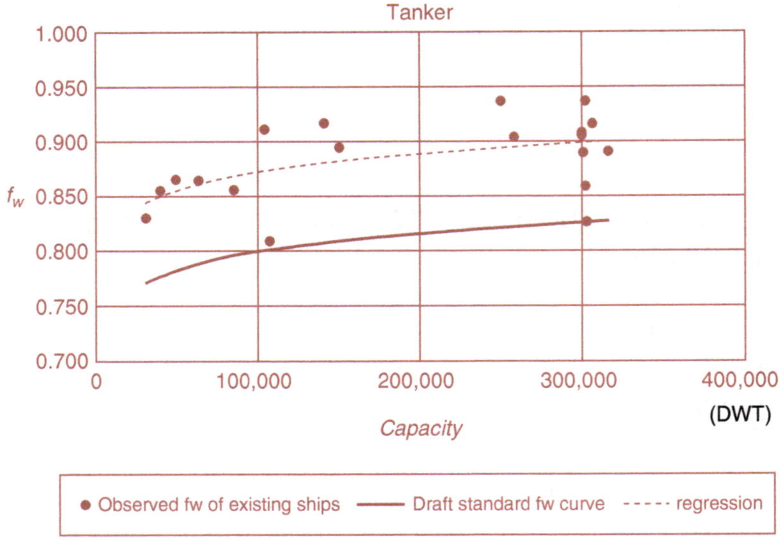

Fig. 3.12 Standard f_w curve for tanker. (Source: IMO 2012e)

Obviously then, f_w and "Attained EEDI weather" are crucial pieces of information of a ship's real efficiency.

Related distortion: Ships with larger engines, having lower actual (in real operating conditions) fuel consumption at same speed and deadweight, will have higher (worse) EEDI than otherwise identical standard ships.

Some owners install larger engines on a shipyard's standard design (e.g., one extra cylinder) in order to be able to operate at the optimum specific fuel oil consumption (SFOC) point of 70–75% MCR in real weather conditions of Beaufort 4–5, instead of 75% MCR at the calm conditions of EEDI. Thus they are able to achieve V_{ref} (or the design speed[1]) in real weather conditions, whereas a typical EEDI ship would need to operate at much higher MCR than 75% to achieve the same speed. The larger-engine ship has a higher (worse) EEDI, typically by about 15%, yet for the same speed and draft, it might save typically 7–7.5% in fuel consumption. This is a direct contradiction of the EEDI premise and could easily be alleviated by using "EEDI weather."

[1]Practically V_{ref} is approximately equal to a typical shipbuilding contract's "design speed." V_{ref} is speed at maximum deadweight (70% DWT for container vessels), at 75% MCR but with no sea margin. Design speed is speed at the reduced design draft, typically at 85% MCR but with 15% sea margin.

6.5 Operational Indices (EEOI, EVDI, etc.) Can Be Meaningless

For real CO_2 reductions, ship performance in real operating conditions should be evaluated. Several "Operational Efficiency Indices" have been devised, but unfortunately none has proven effective in capturing a ship's true operating efficiency. This is because of (a) the inaccuracy of data in the databases used for some indices and mostly (b) the unpredictable and unavoidable effect of bad weather (slower speed – high fuel consumption) and (c) penalization of ballast voyages (consuming fuel without carrying cargo).

The most commonly referred to "operational" index is the Energy Efficiency Operational Indicator (EEOI), with a formula very similar to that of EEDI, but instead of DWT in the denominator, actual amount of cargo carried is used. Also, total CO_2 emitted at the actual voyage speed (not V_{ref} at 75% MCR) is estimated.

The official position on EEOI of the Baltic and International Maritime Council (BIMCO), the largest shipping association with members controlling 65% of the world's tonnage, is as follows: "Operational efficiency indices, such as the IMO Energy Efficiency Operational Indicator (EEOI), are overly simplistic or even misleading on an individual ship basis and therefore irrelevant, and should not be considered for regulatory purposes. Also, such indices could be wrongly perceived as valid selection criteria when assessing the efficiency of a ship prior to chartering."[2]

As BIMCO correctly advises, there are several problems with EEOI, rendering it an unreliable indicator of efficiency. First, the effect of bad weather, where a ship increases fuel consumption to keep a certain speed, penalizes the EEOI value. Secondly, a voyage with less than maximum cargo, and more so zero cargo (on ballast), heavily worsens the EEOI value. Some owners add the fuel consumption from the preceding ballast voyage to the subsequent laden voyage and in this way account for the total fuel consumption to carry a certain amount of cargo per voyage.

Penalizing the value of EEOI for a ship in ballast (empty of cargo) proceeding to a port to load its cargo implies that a ship in ballast condition is not producing a benefit to society. In analogy, this is similar to assuming that an empty ambulance rushing to an accident scene to pick up the injured also does not produce a benefit to society.

Apart from these issues, the EEOI is not really connected to the efforts of the ship operator to operate his ship as efficiently as possible. Bad weather and ballast voyages are mostly out of the control of the operator, and their effect on EEOI may be much larger than any best practices applied by the ship owner (e.g., course optimization, frequent hull and propeller cleaning, etc.) Figure 3.13 shows that even for ships on dedicated (identical) voyages for years, a plot of EEOI rolling average

[2]Statement on BIMCO's web site (https://www.bimco.org/about-us-and-our-members/bimco-statements/04-greenhouse-gases-ghg-emissions, accessed 2 July 2018).

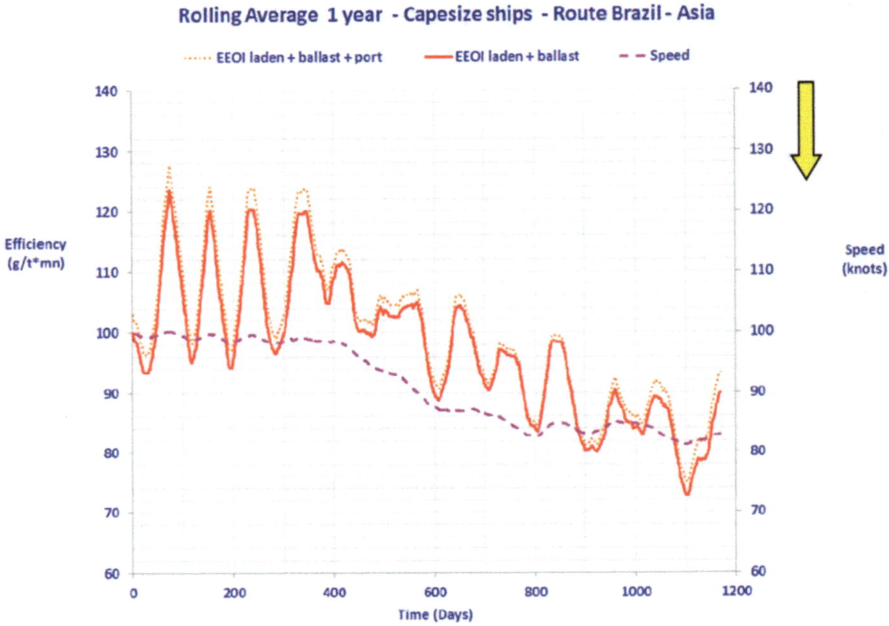

Fig. 3.13 Rolling average EEOI for capesize ships. (Source: IMO 2016a)

(each daily plot being the average value of the previous 365 daily EEOIs) shows no convergence. A most efficient ship with the most prudent operator may have a good EEOI one year and a bad EEOI the next, thanks to the whims of nature and the markets. Several companies, initially using EEOI as an efficiency indicator in their SEEMPs, have since abandoned it for other more targeted individual KPIs, suitable to the specific ship and its trade.

Several other proposed "operational" efficiency indices are more or less variations of EEOI. These include the Annual Efficiency Ratio (AER) proposed by Japan, the Fuel Oil Reduction Strategy (FORS) proposed by Germany, and others. These suffer from similar randomness and non-convergence issues, while several general ship energy efficiency indices used by some ports (for instance, the Clean Shipping Index (CSI), the Environmental Ship Index (ESI), etc.) use the EEOI in evaluating the portion of the ship's CO_2 footprint. Nevertheless, the desire for such indices stems from a desire to give a simple operational "efficiency" rating to each ship and compare it to an average value (e.g., EEOI baselines). That could be appropriate provided the ship index (e.g., EEOI) and the baselines (collection of various EEOIs from similar ship types) had a meaningful connection with ships' actual energy efficiency. So far, no simple index has been found to truly represent real operational efficiency.

Yet another "operational" index used for CO_2 efficiency evaluation is the EVDI (Existing Vessel Design Index), originated by the shipping rating service "Right-Ship." It has the same formula as the simplified EEDI formula, using data for each ship as they appear in public and proprietary databases. Such data have shown to be fraught with errors and inaccuracies, without independent verification of the accuracy of the data. In IMO (2013a, b, c) it was reported that even identical sister ships that were built by the same yard in the same period as part of a series program have EEDIs varying between 8% and 10%, the sole reason being different entries for the design speed recorded in the fleet databases. Such inaccuracies in the data also translate into EVDI, whose scope is broader than EEDI's. Of course, there is no standard % MCR for the reported speed, nor can there be any verification of the supplied data for EVDI. RightShip advises that any ship operator may provide to them corrected or more accurate data for their ship. Practically, however, ships with initially favorable published EVDIs, which may have been calculated from such inaccurate data, would have little incentive to provide "correct" data to RightShip. Lastly, a ship may have a very good EVDI, calculated with accurate or inaccurate data, but may be imprudently operated and still be registered with a favorable energy efficiency rating. Even if the vessel is prudently operated, it is clear that a better EEDI (EVDI) does not necessarily mean an efficient ship in actual operation (see above).

7 Way Ahead: Can EEDI Be Improved?

Despite the weaknesses identified above, we cannot and should not dismiss the usefulness of EEDI, if only for its intention of trying to push designers to design more efficient ships. To achieve CO_2 reductions in actual operation, EEDI must be linked more closely with the operation of ships in real weather conditions. Some studies suggest that the effectiveness of EEDI in improving efficiency so far might be quite small, on the order of 3% according to (Smith et al. 2016). While this may be because EEDI is a theoretical design index, it does increase awareness of energy efficiency. However, when optimizing a ship for maximum overall fuel efficiency, this cannot be achieved simply by minimizing the EEDI. As described in Chap. 2 of this book, to minimize fuel consumption (and hence CO_2 emissions), as part of a proper design process, the hull lines and propulsion system should be evaluated for a set of realistic draft and speed conditions, which are representative for the operational profile of the ship. The EEDI should thus be considered as a design constraint rather than an optimization objective.

What matters also is how the ship performs in real seas of, e.g., Beaufort 4–5, and some prudent owners require model tests in such simulated conditions (seakeeping tests). More and more shipyards care and design their ships for more realistic conditions, simply because owners want to know their future ships' actual projected performance and fuel consumption, in various loaded conditions and various speeds. As stated in the previous section, "EEDI weather" would be a big step forward albeit it still relates to the one condition of maximum draft and speed of 75% MCR. For a complete ship energy evaluation, a matrix of data should be used providing the fuel consumption and CO_2 emissions for various speeds and drafts of a typical operational profile.

It is a small wonder then that exactly that has been used for decades by charterers to choose the most efficient ship from those available to them for hire, i.e., they use a matrix that, as a minimum, includes the charter party owner-guaranteed speeds and fuel consumption figures in laden and ballast conditions, at full speed and slow speed, up to a weather condition of Beaufort 4, without paying much attention to index ratings (EEDI, EVDI, etc.) for the ships under evaluation, which are considered mostly unreliable.

To be more representative of operational conditions, the EEDI could be transformed with the use of f_W, however, at various drafts and speeds, toward a more meaningful index reflective of the real ship efficiency, not only as theoretically designed but also in actual operation.

Based on the above, it is preferable in our opinion that instead of trying to devise elusive "operational" indices which attempt to rate how efficiently an operator operates his ship, the ship should be designed to be most efficient at actual operating conditions. This is done in all other industries (e.g., we don't rate how efficiently a driver drives his car but how the car performs by design at predetermined conditions and cycles). Thus, a next step for EEDI should be to connect it to the expected operational conditions instead of searching for elusive operators' "operational" indices.

Disclaimer The views and opinions expressed in this chapter are those of the authors and do not necessarily reflect the position or views of American Bureau of Shipping or organizations that the authors belong to.

References

ClassNK. (2015). TEC-1048 survey and certification for EEDI and SEEMP required by the Amendments to ANNEX VI of MARPOL 73/7827.

IACS. (2016). Procedural requirement, procedure for calculation and verification of the Energy Efficiency Design Index (EEDI), PR No. 38.

IMO. (2011a). Resolution MEPC.203(62) "Amendments to the Annex of the Protocol of 1997 to amend the International Convention for the Prevention of Pollution from Ships, 1973, as modified by the Protocol of 1978".

IMO. (2011b). MEPC 62/5/6, Submitted by Greece, "Further prospects for EEDI improvement".

IMO. (2011c). MEPC 63/4/15, Submitted by Greece, "Minimum design speed for bulk carriers and oil tankers".

IMO. (2012a). Resolution MEPC.212(63): "2012 Guidelines on the Method of Calculation of the Attained EEDI for new ships".

IMO. (2012b). Resolution MEPC.213(63), "2012 Guidelines for the development of a ship energy efficiency management plan (SEEMP)".

IMO. (2012c). Resolution MEPC.214(63): "2012 Guidelines on Survey and Certification of the EEDI", IMO MEPC.

IMO. (2012d). Resolution MEPC.215(63): "Guidelines for Calculation of Reference Lines for Use With the Energy Efficiency Design Index (EEDI)".

IMO. (2012e). MEPC.1/Circ.796 "Interim Guidelines for the Calculation of the Coefficient f_w*for* Decrease in Ship Speed in a Representative Sea Condition for Trial Use".

IMO. (2013a). Resolution MEPC.231(65): "2013 Guidelines for calculation of reference lines for use with the energy efficiency design index (EEDI)".

IMO. (2013b). Resolution MEPC.232(65): "2013 Interim Guidelines for determining minimum propulsion power to maintain the maneuverability".

IMO. (2013c). Resolution MEPC.233(65): "2013 Guidelines for calculation of reference lines for use with the Energy Efficiency Design Index (EEDI) for cruise passenger ships having non-conventional propulsion".

IMO. (2013d). Resolution MEPC.229(65), "Promotion of Technical Co-Operation and Transfer of Technology Relating to the Improvement of Energy Efficiency of Ships".

IMO. (2013e). MEPC.1/Circ.815: "2013 Guidance on treatment of innovative energy efficiency technologies for calculation and verification of the attained EEDI for ships in adverse conditions".

IMO. (2014a). MEPC.1/Circ.795.rev1 "Unified Interpretations to MARPOL Annex VI".

IMO. (2014b). Resolution MEPC.254(67): "2014 Guidelines on Survey and Certification of EEDI (one amendments made in MEPC 68)".

IMO. (2014c). Resolution MEPC.245(66): "2014 Guidelines on the method of calculation of the Attained Energy Efficiency Design Index (EEDI) for new ships, adopted 4 April 2014".

IMO. (2015a). Determining minimum propulsion power to maintain the manoeuvrability of ships in adverse conditions. (Resolution MEPC.232(65), as amended by Resolution MEPC.255(67)), adopted on 15 May 2015.

IMO. (2015b). MEPC.1/Circ.850/Rev.1 "Amendments to the 2013 Interim Guidelines for Determining Minimum Propulsion Power to Maintain the Manoeuvrability of Ships in Adverse Conditions".

IMO. (2015c). MEPC 68/3/28, Submitted by Greece, "Proposed revision of the 2013 Interim Guidelines for determining minimum propulsion power to maintain the manoeuvrability of ships in adverse conditions".

IMO. (2015d). MEPC 68/INF.30 2015 industry guidelines on calculation and verification of the Energy Efficiency Design Index (EEDI).

IMO. (2015e). TTT course on energy efficient ship operation – M2 ship energy efficiency regulations and related guidelines.

IMO. (2016a). MEPC 69/INF.28: "Further technical and operational measures for enhancing the energy efficiency of international shipping, submitted by Brazil".

IMO. (2016b). Resolution MEPC.282(70): "Guidelines for the Development of a Ship Energy Efficiency Management Plan (SEEMP)".

IMO. (2017). Resolution MEPC.292(71): "Guidelines For Administration Verification of Ship Fuel Oil Consumption Data".

IMO. (2018). Resolution 301(72): "Amendments to MARPOL Annex VI (ECAs and required EEDI for Ro-Ro cargo ships and Ro-Ro passenger ships).

ISO. (2015). 15016: 2015, ships and marine technology – Guidelines for the assessment of speed and power performance by analysis of speed trial data.

ITTC. (2017). Recommended procedures and guidelines, speed and power trials, 7.5-04-01-01.2.

Psaraftis, H. N. (2018). Decarbonization of maritime transport: to be or not to be? *Maritime Economics and Logistics*. https://doi.org/10.1057/s41278-018-0098-8.

Smith et al. (2016). O_2 Emissions from International Shipping - Possible reduction targets and their associated pathways.

Chapter 4
ICT for Sustainable Shipping

Kay Fjørtoft and Svein Peder Berge

Abstract When Titanic hit an iceberg off the coast of Newfoundland in 1912, information about drifting icebergs had not reached the ship officers and navigators, and it took a long time before nearby vessels received a request for assistance. Maxim Gorkiy, which sailed into an ice belt southwest of Svalbard in 1989, experienced a similar lack of information. The hull was damaged, and passengers, crew, and ships were rescued due to extremely good weather conditions and courageous on-scene commanders. For both accidents, had the navigators on board received information in time, they would have been able to choose another and safer route, and the accidents could have been avoided.

From the Titanic days up to now, the ICT maturity has grown rapidly. We are also heading for digital transformation in shipping, that we do not know the consequences of, but we know that shipping sector will be changed, and the ICT will be one of the most important driving factors for sustainability. In parallel with the development, we must ensure that the human interactions will be taken care of. Therefore, the introduction of new technology should include the "human in the loop," the user aspects, and must have focus on the integration between Man, Technology, and Organization (MTO).

In this chapter we will describe some of the central ICT solutions used for sustainable shipping and the way they are operated and give examples on existing and future trends that influence sustainability where the ICT's role in the process is elaborated.

Abbreviations

4C	Commitment, competence, continuous learning, collaboration
AI	Artificial intelligence
AIS	Automatic identification systems

K. Fjørtoft (✉) · S. P. Berge
SINTEF Ocean AS, Trondheim, Norway
e-mail: Kay.Fjortoft@sintef.no; Svein.Berge@sintef.no

H. N. Psaraftis (ed.), *Sustainable Shipping*,
https://doi.org/10.1007/978-3-030-04330-8_4

AR	Augmented reality
BP	British petroleum
CFD	Computational fluid dynamics
DNC	Digital Nautical Charts
DSC	Digital selective calling
ECDIS	Electronic Chart Display and Information System
ENC	Electronic Navigational Charts
FAL	The convention on facilitation of international maritime traffic
GMDSS	Global Maritime Distress and Safety System
GT	Gross tonnage
IAMS	Integrated alarm and monitoring systems
IAS	Integrated automation system
ICT	Information and communication technologies
IMO	International Maritime Organization
IO	Integrated operations
IOT	Internet of Things
IPL	Integrated planning and logistics
IT	Information technology
ITS	Intelligent transport systems
MIMS	Maritime Information Management System
MSW	Maritime Single Window
MTO	Man, Technology, and Organization
OT	Operational technology
PCS	Port community system
RPM	Revolution per minute
S-AIS	Satellite AIS
SW	Single window
TOS	Terminal Operating System
TSW	Trade Single Window
VR	Virtual reality
VTMIS	Vessel Traffic Management and Information Services
VTS	Vessel traffic services

1 Introduction

ICT (information and communication technologies) do not lead to direct environmental benefits, but their smart use can definitely do so. This can be done by increasing the efficiency of the maritime supply chain, improving safety, improving the load factors, and so on. This chapter highlights the background for sustainable shipping, looks into efficiency within maritime transport, describes possibilities for improved safety, and reviews relevant ICT systems in shipping and considers their impact on improving environmental performance. The chapter will reflect on future maritime outlook within digitalization in the maritime sector.

Sustainability can be divided into different categories, such as sustainable vessel design and production, sustainable maritime supply chain, sustainable operations, sustainable production, and technologies for sustainable shipping as examples.

A vessel's life time should cover the whole ship's lifecycle from ship design, operation, and the process of decommissioning the vessel at the end. Sustainability is about designing the vessel with the required capabilities to operate within the planned operational criteria's and with the focus on a solution where the capital costs are in harmony with the operational costs. Sustainable transport is to consider the whole logistic value chain and to understand the collaboration within the chain. That means the vessel should be integrated with the land side, where the focus should be to see the transport demand in the context of offered vessel service. Sustainable shipping will have a view on the vessel itself, where energy consumption and possibilities of smart shipping is considered, with a review from an environmental perspective.

In this chapter we will describe some of the central ICT systems used for sustainable shipping and the way they are operated and give examples on existing and future trends that influence sustainability where the ICT's role in the process is elaborated. This chapter is organized in six sections. Section 1 is a short introduction, background, trends, and drivers for ICT in shipping. Section 2 focuses on how we can use ICT as tools for sustainable design, production, operation, and maintenance throughout the ship's lifecycle. Section 3 covers sustainability in the supporting maritime supply chain. Section 4 is covering the key enabling technologies adapted from the Industry 4.0 applied to the shipping domain, denoted Shipping 4.0. Section 5 presents some major outlooks for ICT in shipping and finally Sect. 6 is a short summary.

1.1 Background

For centuries, sea transport has been a major facilitator of trades between nations, regions, and continents. More recently, together with trade liberalization, telecommunication, and international standardization, it has been a key enabler of globalization (Hoffmann and Kumar 2002). Over the past 40 years, maritime transport has increased by 250%, following the same growth rate as global gross domestic product (GDP) and growing more rapidly than global energy consumption (170%) and global population (90%) as illustrated in Fig. 4.1.

Maritime fuel use, and CO_2 emissions, has increased by 150%, over the same four decades. In annual figures, maritime transport has increased by 3%, fuel consumption by 2%, which is 1% less, which is the average annual energy efficiency improvement from 1970 to 2012. This improvement of one percentage point per year in CO_2 productivity per ton-mile has been forthcoming without large-scale implementation of new technologies or alternative fuels. Lindstad (2013, 2016) notes that improvements since 1970 (speed, size, slenderness) may be viewed as capital substituting for energy – larger, new and slower vessels require capital

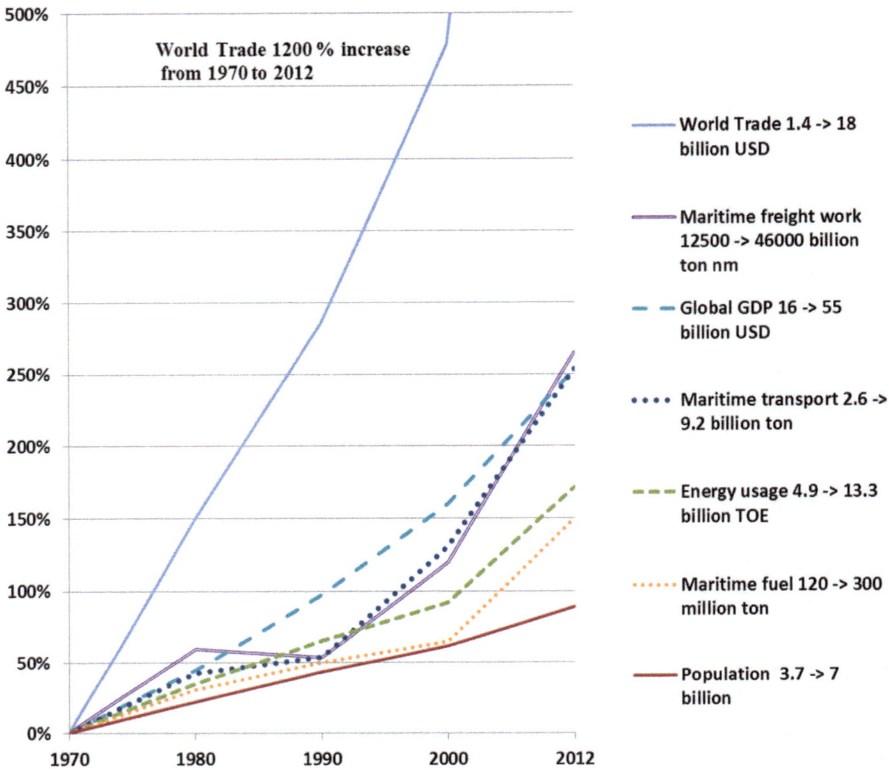

Fig. 4.1 World trade, maritime transport, and other indicators, 1970–2012. (Source Lindstad 2013, 2016)

but spend less fuel and emit less greenhouse gases per ton-mile freight services produced. The same can be said for new technologies, propulsion, and alternative fuels in the future; they can be delivered by investments in research, technological developments, accelerated fleet renovation, ports, and canals.

Environmental sustainability within the shipping sector is of great importance. A very high percentage of the world transport in tonnage is going maritime. The world's maritime fleet increased from 907 million deadweight tons (DWT) in 2005 to 1.69 billion DWT in 2014, equivalent to average growth of 7.2% per year. Deadweight tonnage is the weight measure of a vessel's carrying capacity and includes cargo, fuel, and stores.

The shipping industry has agreed that the CO_2 emissions in 2050 should be reduced by at least 50%, compared with today's shipping emission from the vessels (IMO 2018). In 2100 it should be zero emission according to Norwegian Shipowners' Association (NSA 2018). These ambitions are following the Paris agreement, that was signed 12th of December 2015, where one of the aims was in a long-term goal of keeping the increase in global average temperature to well below 2 °C above pre-industrial levels. In the IMO/MPEC 72, April 2018, this was followed up by the mentioned CO_2 cuts. DNV-GL have stated that the CO_2 reduction from shipping requires a combination of reduction measures in logistics, alternative fuels, technical and operational measures, and offset (DNV-GL 2017).

Maritime transport is global and is traditionally regulated by the International Maritime Organization, in collaboration with the flag states. That also means the technology to be used on board the vessels should be developed with the aim of operating also outside a nation's borders, which implies a global perspective. As regards environmental emissions from the sector, there are many different effects that can be measured, such as local, global, within a region, sector wise, etc. that includes the different vulnerabilities. Examples could be that emissions from a vessel in port in a city such as Bergen in Norway are more dangerous than in another Norwegian town as Trondheim, due to the topographic elements in Bergen where the air change is slower as compared to Trondheim. Another example is that emissions in the Arctic have a higher risk level for the environment as compared to the areas further south, due to Arctic vulnerabilities. The maritime ambition for operation in the high north should be to have no dangerous emissions, low footprint, or an environmental shipping footprint that is better or at least not higher than today.

Sustainability is to perform operations in harmony with the environment. The ICT technology should assist in the process of reducing harmful emissions, to aim for smart operations, to share information and to have an integrated logistics chain, where the operations are safe and cost-efficient. ICT and digitalization are drivers for a sustainable development of maritime shipping.

2 Sustainable Vessel Design, Production, Operation, and Maintenance

The maritime industry is somehow different from other transport industries. Each ship is unique in both design and operation as compared to the automotive industry with serial production lines. The process from design, production to operation must be time effective and profitable to maintain a competitive edge.

The shipbuilding industry is always looking for new opportunities in new types of designs. One possible shift is to focus on environmentally friendly solutions. One characteristic is that the relationship between investment in materials and operations needs to be done more in line with what the industry can pay for.

Technologically, the ship systems must be integrated in a completely different way today than the practice was before. They must be adapted to the intended operation to minimize complexity, price, and security that will require a completely different degree of multidisciplinary design and operation. Environmental technology has a global market, so the customer base is high. The challenge is to get good business models that provide a good economy to realize new environmentally friendly technology.

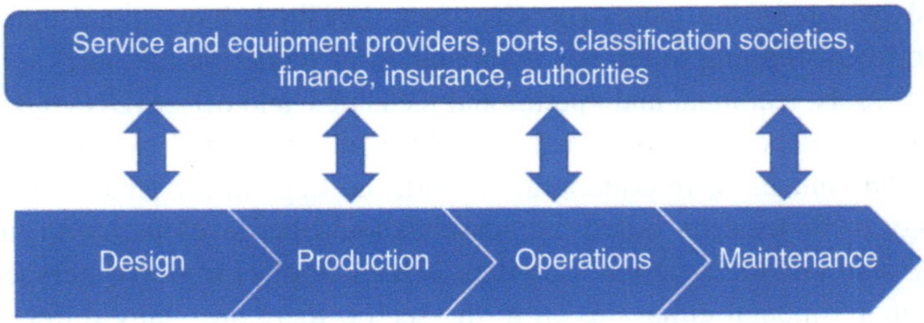

Fig. 4.2 Value chain. (Source: SINTEF Ocean)

The maritime industry consists of a comprehensive value chain from the design of ships and ships systems for operation, and operation of ships in efficient logistics systems for goods and passenger transport in a global market. The lifecycle therefore includes the entire value creation chain as illustrated in Fig. 4.2.

The advantage of a sustainable approach is not primarily savings in energy consumption on an individual trip and for a single part of the value chain but must cover the total transport and operational value chain.

Sustainable vessel design will include different stakeholder groups, from service and equipment providers, the shipping companies, user of a service, the classification and financial bodies, as well as governmental stakeholders as examples. They have different interests in different parts of a value chain, but at the end the interaction between them gives the rise of a good maritime solution and design that are developed to suite different requirements. As example of this statement, it is likely to understand that a vessel design is done out of a harmonization of the requirements, i.e., a vessel should be built to operate in the environment that fulfills the demands. If a vessel has a depth or height that is too large to sail in to a port to serve a customer, something has gone wrong in the design process. The same counts for a design that is tailored for a much higher goods volume than required from a market perspective, which will lead to a more expensive service to the customers because the total capital cost of building the vessel is higher than necessarily.

The interaction between stakeholder groups means that a good business model is crucial for success. We see the need for completely new types of ships that are more adapted to their operating field than what characterizes today's practice. In particular, it can be economically possible to scale down the ship's size to get more flexibility in the transport chains, where environmentally friendly propulsion and smart technology for loading and unloading operations combine for increased effects. It is important to understand the green shift also from a safety perspective. The introduction of new energy forms or new technologies, such as hydrogen or battery energy sources, must focus on personnel, material, and environmental safety. New applications of technology in a maritime atmosphere put new demands on preparedness. It is important to obtain a correct and comprehensive picture of potential accident cases, possible consequences based on the type of accident, and take the necessary measures to increase safety for both humans and equipment. The ICT's role in the preparedness picture is crucial.

2.1 Sustainable Vessel Design

In ship design it will be especially important to ensure that the design and production of ships and equipment is as cost-effective as possible. In addition, ships and equipment must be designed to operate in the most cost-effective, environment-friendly, and safe manner possible. The maritime green shift will not come as the result of a single technology or a simple new operational practice. It will come as a result of an integrated interaction between technology and operational practices.

Interaction also means that the maritime industry should use environmentally friendly energy such as LNG, battery, or hydrogen. To succeed fully with the introduction of, i.e., LNG, it is necessary that the infrastructure expands in parallel with the introduction of new technology on board the vessels. Should it use LNG, we need LNG stations at the ports. Should ship engines be based on pure battery as main energy source we need to development a charging infrastructure on land. Vessels arriving a port on a short stay will require quick and flexible connection and a short recharging period of the batteries. This will require a different infrastructure than is usual for traditional land power plants for larger ships. Large-capacity ships will also require the development of the local distribution network on land, which in some areas is too weak to tolerate quick charging of batteries with power levels in the megawatt (MW) class as examples.

The digital transformation will be essential to achieving these aims. Simulation, virtual prototyping and virtual testing will enable the testing of complete systems early in the design phase. Using data from actual operations and data on sea conditions will make it possible to test in a virtual environment how ships and equipment will perform under realistic conditions (Maritime21 2016). Digital interfaces must be developed and standardized, components and equipment must be developed and tested, and data must be stored and shared, where the stakeholders must cooperate to realize ambitions. Digitalization will facilitate innovation in products and processes that will enhance companies' productivity.

Examples of common used ICT tools and methods for sustainable vessel design include:

1. *Simulation-based design and virtual prototyping*. Designing new ships and ship systems is a multidisciplinary exercise involving many stakeholders working concurrently on the same concept. Concurrent engineering is getting the right people together at the right time to identify and resolve design problems. Typical constraints in the design phase is to make trade-offs in assembly, availability, operability, performance, quality, risk, and maritime safety. The design phase is an iteratively process that requires ICT tools like numerical calculations, simulations, and virtual environment to validate different concepts in an early phase. Key technologies like augmented reality (AR) and virtual reality (VR) are used to bring the components of the digital worlds into a person's perception of the real world.
2. *Numerical tools and model testing.* Optimizing ship concepts with regard to satisfactory hydrodynamic performance and energy consumptions is done

with different tools like computational fluid dynamics (CFD) and other codes available for ship designers. There will always be a trade-off between different tools and required accuracy in the different design stages. Early ship concept analysis using numerical tools will play an important role to find the best initial concepts. When the design is chosen, model tests in hydrodynamic laboratories can be used to validate the concept in model scale and then scale the results to full scale. Numerical optimization methods and use of existing operational data from full-scale measurements can further be used to improve the design.

2.2 Sustainable Production

Furthermore, automation and robotization will improve the efficiency of the production phase. Design data can be directly used to generate engineering data and for automatic production preparation. Digital production systems can optimize the production sequence and improve the efficiency of operational planning.

Compared to the automotive industry, production of ships is more small series or even single production compared to mass production of cars. This requires highly cost-efficient production methods and higher degree of automation in the production line. To obtain this we need to digitalize the components, e.g., digital twins, generate documentations automatically for production, using intelligent robotics, develop standards for components and procedures, and finally have flexible system for planning and production. Digitalization of the whole maritime value chain, effective production process, and introduction of new materials will be needed to obtain sustainable production. Motivated from the German Industry 4.0 strategy, the Shipping 4.0 strategy has been introduced for the maritime industry. Shipyards, equipment providers, and subcontractors need to exchange data securely using digital platform and reference architectures based on new standards covering data management, data and service architecture, software architecture, and security architecture (IDSA 2018).

2.3 Sustainable Operation

Vessel operations must be as cost-effective, safe, and environment-friendly as possible, and digitalization will play a key role in achieving this. Data from the same digital value chain can be used to simulate and plan complex operations as well as to train personnel with the help of simulators. Data from actual operations can be compiled and later used for improving simulation models.

Each ship will have its own digital infrastructure. By using the established Internet of Things (IoT) technology, ship sensors and ship system are able to connect and exchange data on-board. Data are aggregated and processed to provide decision support for operators and for maintenance. More intelligence (higher level of automation) is done locally using embedded computing system. With the help of communications technology, the ships can be monitored and remote-controlled from land to enhance efficiency and safety and support to the vessel and crew when needed.

Creating this digital infrastructure will require developing necessary technology and components along with the digital interface to tie these together. It will also be important to give adequate consideration to the human element when developing technology. Equipment and components from different suppliers must be able to "communicate," and the stakeholders involved will have to work together to achieve this. In sea transport, advanced commercial and operative decision support systems for operations can yield a higher degree of efficiency, utilization, reliability, and safety. More complex analysis models will help to enhance understanding of vessel performance under different conditions. Good, inexpensive communications solutions will lead to new operational and business models that can boost the competitiveness of shipping actors as a result of lower costs and improved customer service. Furthermore, predictive systems for decision support and data analysis will be drivers for optimizing and safeguarding technical and commercial operations through more effective prevention and management of undesirable events (e.g., predictive maintenance).

Examples of commonly used ICT tools for sustainable operations include:

1. *Onboard decision support system.* During operation, data measured from different sensors and other external sources (e.g., weather information, route planning, other operational constraints) can be used to give the operator advice how to operate with less use of energy. In fact, an advanced offshore vessel with a hybrid solution (both diesel and electric powered) have several combinator curves mapping desired thrust to revolution per minute (RPM) and pitch commands to the propulsion system. However, each combinator is not necessarily optimal with respect to lowest fuel consumption. Real-time operational data from different sensors and other information can be fused, integrated, and combined with numerical models to provide or suggest more energy-efficient use of the installed power system. Both energy and power management system can in a more intelligent way give decision support during operations in different operational phases to reduce both mechanical and electrical losses in the power chain.
2. *Increased connectivity and data security.* Due to improved connectivity onshore and offshore, it is possible to transfer operational data for online support and monitoring for maintenance purposes. Secure data transfer and ownership of operational data will require digital frameworks where ship owners, operators, and equipment providers can share data based on digital and secured contracts.
3. *Autonomous transport system.* Sustainable shipping will cover the whole transport chain, including ports, authorities, and intelligent transport systems (ITS) which must be a complete integrated transport system that needs to be optimized.

Cost/benefit analysis technology of transport system design, as well as systems for monitoring and reporting, is therefore an essential ICT tool that must be included into an operational context.

2.4 Sustainable Maintenance

Preventive maintenance is a common strategy in shipping industry. The service interval is based on the manufacturer recommendations and classification society requirements. However, the preventive approach is not perfect and cannot guarantee zero failure when shipping. In addition, due to increased pressure on cost, maintenance must reduce the downtime. Condition-based maintenance (CBM) is a different strategy where operational data can be aggregated to detect early signs of failure or to adjust maintenance plans. Increased connectivity between ship and shore offers remote monitoring possibilities.

3 Sustainable Maritime Supply Chain

The maritime industry accounts for about 2.2% of the total global CO_2 man-made greenhouse gas emissions (IMO 2014). Work is ongoing to make this figure lower and as low as possible. The success will come from a combination of the integration between operational and technical issues. Sustainable transport chains count for an integrated chain where several transport modes are working together in a smooth operational context.

There are therefore two important key drivers of knowledge that are important for achieving a difference, where the goal is to achieve sustainability within a transport:

1. *Operational knowledge*: Improvement and reduction of energy consumption in existing transport systems by better planning and smarter operation of a vessel
2. *Technological knowledge*: Transition to alternative forms of energy or development of new technological solutions that increase performance in a more sustainable way than existing solutions

Integrated operations (IO) is a term normally used within the oil and gas sector, which focuses on the integration of people, organizations, work processes, and information technology to make smarter decisions. It is enabled by global access to real-time information, collaborative technology, and integration of multiple expertise across disciplines, organizations, and geographical locations.

The interaction between technology, organization, and people is essential for success. But of course, legislative and regulatory frameworks must be in place. A close interaction and collaboration between authorities, infrastructure managers, cargo owners, and service providers and a holistic approach to the goal of sustainable shipping is desirable.

3.1 Operational Knowledge

In the context of a sustainable transport chain, integrated operations (IO) is considered as a good structure and a way of organizing the collaboration between sea operations, between sea and land, and between people, technology, and organizations. ICT will play an important role in IO, but we should not forget the people and organizations that are operating or controlling the ICT. "It's 80 percent about the people, 15 percent about processes and organizations and 5 percent about technology" says David Latin, Vice President, E&P Technology (BP 2009).

About sustainable transport, the whole value chain must be considered, not only the ship performance but also the utilization of the transport means along the chain. A vessel, truck, or a train must aim to avoid empty transport. An integration between transport demand and transport services, as example between a customer and a vessel operator, is therefore the key for an integrated sustainable shipping transport approach, where the tailoring of the transport means is done based on the cargo to be transported. As regards the transport aspect, in some cases, slow steaming is best practice; in other cases, the speed is decided based on the next leg in the value chain, i.e., the interaction between a vessel and a train, truck, and a deep-sea operator that are carrying the goods to the next destination. To be able to do such planning, it is important to integrate the logistics systems between the actors involved in the transport, which is called integrated planning.

From the oil sector, we see the planning activities and the move of responsibilities by sending more and more duties to land-based centers, such as the planning of a work orders at an offshore installation as a trend. The same trend is expected to come to the shipping sector.

The transfer of the IO principles to the planning domain has led to the development of the concept of integrated planning and logistics (IPL). The concept represents a holistic perspective on planning, emphasizing the interplay between planning horizons, between organizational units, and among cross-organizational partners. It is defined as a holistic, cross-domain planning enabling optimal resource allocation and activity prioritization for safe and efficient operations (Ramstad et al. 2013).

This work has confirmed that information sharing and cooperation across disciplines/organizations represent significant challenges related to planning. It has showed that current work processes in planning do not support integrated planning; neither does the organizational structure nor the performance management systems. Consequently, there is a need for better coordination between activities as well as identification of dependencies across discipline and organizations. Moreover, there is a clear need to develop tools that can help the industry to develop the existing planning processes into a more integrated practice. Furthermore, we have not seen the use of models that contribute significantly to the decision-making in overall planning. We believe this also counts for the maritime sector where planning and logistics should have more attention than today, where not only the sailing from port-to-port is considered but also the integration with land-based services such as the terminal and the next transport leg for the cargo, normally by truck out of the terminal.

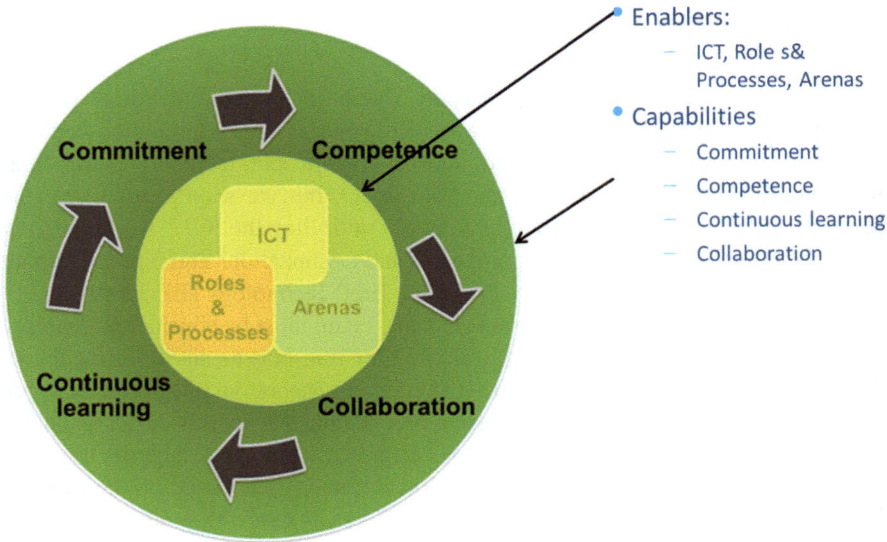

Fig. 4.3 Integrated planning and logistics. (Source: SINTEF Ocean)

The IPL model is used for achieving integrated planning and promoting a proactive planning culture, as illustrated in Fig. 4.3. It introduces key enablers for design and implementation, where ICT, roles and processes, and arenas for plan coordination are included. It includes the basic capabilities and cultivating a 4C culture (commitment, competence, continuous learning, collaboration). The key is to build trust and knowledge between the stakeholders, the planners, and those that are executing a transport. Main efforts of the work carried out have therefore been focused upon answering the following questions:

Processes and roles:

- How can the industry implement a best practice for integrated planning?
- What characterizes the planning domain, which roles and actors are involved, and how can decision-making across roles and actors be facilitated?
- How can KPIs (key performance indicators) support implementation of a good plan, and which indicators are crucial at different planning levels?

Optimization tools:

- How can new conceptual solutions for optimization tools support integrated planning (e.g., optimization of logistics operations, smart shipping, efficient cargo transfer)?
- How can information be shared and used for situational awareness and decision support?
- How can autonomy, automation, and automatic reporting assist in releasing demands to operators, at the same time as error within reporting is reduced?

Arenas for plan coordination:

- What types of arenas are necessary for efficient plan coordination and what are significant requirements for designing and implementation of the arenas?
- How can the industry facilitate collaboration, teamwork, and committed participation on the arenas?
- How can the whole value chain be considered and not only a single transport mode?

Integrated planning is key to cost-effective and secure operation, where it is crucial to have a comprehensive picture of all operational activities related to, e.g., a maritime operation, in order to make better decisions. This requires all parties involved to interact with a common goal to achieve desired results and shared gains. Here it is important that everyone understands the importance of developing good plans and understanding the totality and value of everyone's contributions. For companies, the introduction of integrated planning involves facilitating effective interaction through both organizational and technological means.

The lesson learned from the work within the oil and gas sector can be applied to the sustainable transport chain domain. As previously mentioned, the technology itself cannot make a difference without a close link to the instances operating the technology, the organizations, and the humans.

ICT for sustainable shipping is therefore the interaction between the technology, the collaboration and planning of operations, and the human and organizational influence in the execution stage, including governance.

Changes in business models may be necessary to realize the energy savings potential of shipping. As an example, to let a ship optimize speed, it may also have to adjust port arrival and departure times which requires increased cooperation between charterers, ship operators, and ports. Speed as well as arrival/departure time optimization is today difficult to implement due to a prescriptive contract regime between the parties of the transport operation. To change this, one may need new business models, including increased transparency.

3.2 *Technological Knowledge*

Technology should be used to optimize the best chain performance, by understanding the operational criteria and by suggesting decision support that should be sent either automatic to a control system or to humans as a filter before the commands should be entered to the technical systems.

While maritime transport is very efficient, there is still up to a 75% CO_2 emission savings potential (IMO 2009). Significant parts of this potential savings depend on more efficient port operations both to reduce idle berth or anchorage times and to optimize voyage execution. Thus, measures that improve efficiency of trade will in general also contribute to the greening of transport operations.

Technical knowledge will generate new technical solutions and products/systems that have not yet fully been taken into use today. This may include new propulsion systems like energy saving devices on a vessel and general improvements with more energy-efficient solutions, or systems used for trade purposes. New alternative energy based on carbon neutral production from wind, hydropower and waves can contribute to reduced greenhouse gas emissions. This might require battery systems and new energy converters on board the vessel.

Other factors that may be important for achieving reduced emissions in maritime transport are new transport patterns and logistics solutions, higher utilization rates on ships (land-to-sea cargo) and increased competitiveness by introducing a tax level that promotes maritime transport in competition with land-based transport such as truck transport. This also counts for an efficient loading process, where the handling technology is modern and cost-efficient, maybe also autonomous.

An interesting field and subject area that will be important for the green shift is the more efficient use of large amounts of data. What can be learned from available data and how to integrate them to provide added value for the development of new technology? The world today is heading toward the digital age. We must exploit this to research on new environmentally friendly and smart solutions. These are systems where cost-effectiveness, security, and commercial values are important. Energy management is an example of how to learn operational practices where, for example, the use of multiple energy sources is done to increase the efficiency and utilization of the energy available. One of the keys is to provide data in the form of continuously monitored by a ship's control center, which is brought back to research to promote new innovative technologies.

Figure 4.4 shows how Maritime Information Systems relates to both regulatory and commercial systems, both cargo details and maritime operation, and both public information (tariffs, safety, fees) and commercial information (logistics, ownership, liability) (Rødseth et al. 2017). Traditionally, Single Window (SW)

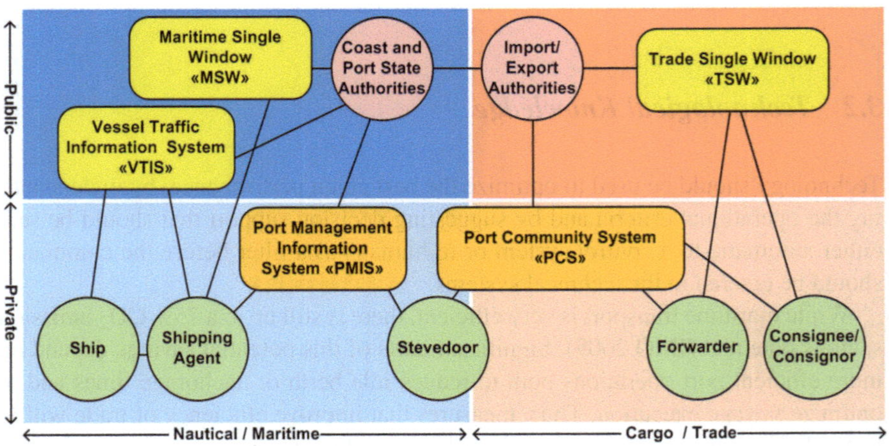

Fig. 4.4 Maritime information systems and the role in a Single Window context. (Source: SINTEF Ocean)

System is defined as "A facility that allows parties involved in trade and transport to lodge standardized information and documents with a single-entry point to fulfill all import, export, and transit-related regulatory requirements. If information is electronic then individual data elements should only be submitted once." The role of a system in this context is to be the "glue" at national level to achieve single entrance of data, coordination among all actors, and added value services both related to trade (cargo handling) and maritime actors and related to regulatory and commercial actors.

The figure divides the port clearance processes into four quadrants dependent on the Single Window (rounded rectangle). It mainly handles the ship or cargo import/export and where the process domain is mainly in the public or private sector. This is a generic figure and in real implementations, one or more of these Single Windows may be integrated or further subdivided. The main Single Window functions are:

- *Maritime Information Management System (MIMS)*: This is a system dedicated to collecting ship movement data for safety and security purposes and which in Europe typically is integrated with SafeSeaNet (2009). The MIMS will typically handle AIS reports, ship reports as mandated in SOLAS, and other data.
- *Maritime Single Window (MSW)*: The Single Window concept, as strongly encouraged or even directed by international and EU policy initiatives today, appears in a number of forms, where it primarily addresses the need for collaborative, efficient electronic transactions between governmental and business trade and transport entities (Rødseth et al. 2011). This is the system that handles ship clearance as defined by the Facilitation of International Maritime Traffic, the FAL convention (IMO 1965). Note that the EU Directive 2010/65/EC stipulates that most functions of the MIMS shall be integrated with the MSW.
- *Trade Single Window (TSW)*: This is one or more Single Window systems dedicated to import and export of cargo. These are typically operated by Customs authorities or other public entities with responsibility for tariffs, contraband, inland security, or related issues.
- *Port community system (PCS)/Terminal Operating System (TOS)*: These are typically commercially operated systems that coordinate logistics operators in the port or in a terminal.

The upper half of the picture is "controlled" by authorities directly or indirectly based on statutory law, while the lower half is controlled by private parties and is governed by commercial contracts or other private agreements. The left half (blue) is related to ships and maritime services, while the right-hand side is related to cargo and trade.

In this picture, "Trade Single Window" is an authority-operated single window that caters for document flows related to import and export clearance of cargo, usually for several transport modes. The operation will be regulated mainly by national legislation although the actual tariffs and documentation requirements are usually based on multilateral agreements. The Maritime Single Window is an authority-operated single window for clearance of ships, including the cargo they carry (whether intended for import, export, or transit) partly regulated through the FAL Convention (IMO 1965).

Vessel Traffic Management and Information Services (VTMIS) is a nautical control system for ship movements in and approaching port where functions are mainly regulated through SOLAS (IMO 1980). An information system can be used to generate value-added services by coordinating regulatory information from VTMIS, for instance, automatic identification system (AIS) information (dark-blue quadrant) and logistics actors (light-blue quadrant). In this context, in Fig. 4.4, Short Sea Navigation fits into the Maritime Single Window part (dark blue), while Customs fits into the Trade Single Window part (dark orange).

Supply chain management systems are systems that are used to control a door-to-door transport, across different modes and transport providers and to integrate transport demands and put the demands into transport. Supply chain systems organize transport data, cargo data, and statuses and provide status reporting along the transport chain, by managing the flow of information and services. The data within a supply chain management system can be used for optimization and for allocation of transport orders. In this context a supply chain management system is important to achieve a sustainable transport chain, by providing tools and solutions for an integrated management, where most relevant applications within the maritime sector are either managed within the management system or integrated with the system. Below is a list of concepts, terms, and systems that have a central place in a maritime transport chain:

Single Window is defined as "A facility that allows parties involved in trade and transport to lodge standardized information and documents with a single-entry point to fulfill all import, export, and transit-related regulatory requirements. If information is electronic then individual data elements should only be submitted once" (UN/CEFACT 2005).

The Single Window concept, as strongly encouraged or even directed by international and EU policy initiatives today, appears in a number of forms, where it primarily addresses the need for collaborative, efficient electronic transactions between governmental and business trade and transport entities (EU 2010). Ship formalities, cargo declarations, and safety and security notifications are all services that should be rationalized and offered in a harmonized manner by a transport and trade Single Window application. Modern process definition and information systems development methods and technologies can significantly support a Single Window application design and implementation process.

The one stop shop business model has been exhaustively researched and applied in the context of e-business and e-government service provision over the last decade (Wimmer 2002 and Fjørtoft et al. 2011). In a similar vein, in the trade, transport, and shipping sector, the "Single Window" (SW) concept was formalized by the United Nations Centre for Trade Facilitation and Electronic Business, to enhance the efficient exchange of information between trade and government agencies (UN/CEFACT 2005).

Port Community Systems The main goal of a PCS is to control the trade-related activities within a port, by managing information, and to perform mandatorily reporting to authorities. A port community system (PCS) is a system that is operated by the ports. It is an information management system that has been implemented normally closely integrated with Terminal Operating Systems (TOS) and public services, such as customs, immigration, and police e-services.

The role of a port community system differs some from country to country and from port to port, due to the trade types within the port. As PCS is widely recognized as a critical instrument in facilitating national and international trade. In Europe, nautical authorities under the auspices of the EU and EMSA have develop national Single Window systems for nautical information that is in turn integrated into the European SafeSeaNet infrastructure. These initiatives do help in providing a more integrated environment for the maritime business actors and the authorities, but as policies and implementations differ among countries and ports, they do also create a relatively ambiguous and complex environment.

e-Freight This is solutions that will encompass legal, organizational, and technical frameworks to enable transport operators, shippers/freight forwarders, customs, and other government administrations to seamlessly exchange information in order to improve the efficiency and quality of freight transport logistics.

In 2016, more than 50% of the global air trade relied on paper-based processes. In the MARNIS project (MARNIS 2009), it was identified more than 40 reports and statements to be sent from a container vessel arriving and departing a port coming from an abroad port. A shipment can generate many paper documents, and many of the processes, such as track and trace, still depend on human intervention. One of the aims will be to automate this reporting as much as possible, to avoid unnecessarily burden on navigational personnel.

The e-freight initiative is a program that aims to build an end-to-end paperless transportation process made possible through the regulatory framework, modern electronic messages, and high quality of data.

e-Customs The EU Commission outlined a course of action for a more robust and unified EU Customs Union by 2020. The Customs 2020 Programme maintains the support for coordination between the customs administrations of EU Member States by providing a platform for the electronic exchange of information and the development of common guidelines and IT systems. In parallel with this initiative, EMSA, supported by DGMove, introduced the eManifest pilot project, which aims at demonstrating how different cargo notifications used for maritime or customs purposes can be consolidated in an eManifest and reported electronically in a harmonized manner to a Maritime Single Window, together with the other reporting information covered by the Reporting Formalities Directive 2010/65/EU.

IMO e-Navigation The International Maritime Organization (IMO) has described e-Navigation as "the harmonized collection, integration, exchange, presentation and analysis of maritime information onboard and ashore by electronic means to enhance berth to berth navigation and related services, for safety and security at sea and protection of the marine environment" (IMO 2014b).

e-Navigation is intended to meet present and future user needs through harmonization of marine navigation systems and support of shore services. It is primarily related to safety management and aids to the nautical operators. e-Navigation is an ongoing initiative by the IMO to implement next generation navigation and safety systems for shipping. e-Navigation will be supporting ship critical information as well as vessel reporting to shore-based sites. Information coming from the AIS-transponders, the ISPS-documentation, as well as more ship-specific information will be of relevance.

ECDIS ECDIS stands for Electronic Chart Display and Information System. It is a geographic information system used for nautical navigation that complies with International Maritime Organization (IMO) regulations as an alternative to paper nautical charts. IMO refers to similar systems not meeting the regulations as Electronic Chart Systems. An ECDIS system displays the information from Electronic Navigational Charts (ENC) or Digital Nautical Charts (DNC) and integrates position information from position, heading, and speed through water reference systems and optionally other navigational sensors. Other sensors which could interface with an ECDIS are radar, Navtex, automatic identification systems (AIS), and depth sounders as examples.

AIS The automatic identification system (AIS) is an automatic tracking system used for collision avoidance on ships and by vessel traffic services (VTS). The tracking system is either based on terrestrial receivers, from satellites, or from the VHF transceivers which are used when vessels are within range of each other's. When satellites are used to detect AIS signatures, the term Satellite-AIS (S-AIS) is used. Information provided by AIS equipment, such as unique identification, position, course, and speed, can be displayed on a screen or an ECDIS. The International Maritime Organization's International Convention for the Safety of Life at Sea requires AIS to be fitted aboard international voyaging ships with 300 or more gross tonnage (GT) and all passenger ships regardless of size.

The automatic identification system (AIS) gives important information on vessels' identity, position, speed, and course. Also, sensors at port, in containers, or in the vessels may give information of interest to the user.

Navigation and Bridge Systems Bridge systems are normally used for a navigation purpose or to control the vessels safety and propulsion system. It is integrated with other control systems but is capable of operating without a direct link. The bridge system will be different between types of vessels. Examples of types of systems can be described as following:

- Wind system: measuring and monitoring wind speed and direction for monitoring and input to control system (e.g., dynamic positioning system)
- Radar: track of other vessels within a radio range
- Electronic sea map: mapping of the chart systems; normally it includes dynamic data such as weather and wind forces

- Echo sounder: gives the depth at the vessels fore and aft positions
- Navtex: gives navigational warnings from messages broadcasted from both international and local channels
- Bridge systems: integrates all navigation systems as well as gives the opportunity to provide a bridge navigation watch alarm
- Speed log: gives a speed through water and distance traveled
- Autopilot: Heading control, Advanced Auto utilizing automatic ground tracking control, Course or Precision Cross Track control when integrated with navigation sensors
- GPS Plotter/GPS: position sensor for Radar, AIS, ECDIS, autopilot, echo sounder, and other navigation and communications equipment
- Compass: provides accurate heading data for autopilot, radar, AIS, Sonar, and plotting systems
- ECDIS: gives the navigators a tool for precise route planning, monitoring, and navigation data management
- Weather facsimile receiver: receives and writes the weather map and satellite pictures

Integrated Automation System (IAS) IAS is used to control and monitor different onboard systems such as engines, propulsion, and thruster and vessel performance. It covers information systems and alarm monitoring systems, power, and energy management systems. Each solution is custom-made to a vessel's operating profile. Ships are increasingly using systems that rely on digitization, integration, and automation, which calls for cyber risk management on board. As technology continues to develop, information technology (IT) and operational technology (OT) onboard ships are being networked together – and more frequently connected to the Internet. This brings the greater risk of unauthorized access or malicious attacks to ships' systems and networks. Risks may also occur from personnel accessing systems on board, for example, by introducing malware via removable media. The safety, environmental, and commercial consequences of not being prepared for a cyber incident may be significant toward ICT for sustainable shipping.

4 Key Enabling Technologies for Sustainable Shipping

"Industry 4.0" is a national strategic initiative from the German government through the Ministry of Education and Research (BMBF) and Ministry for Economic Affairs and Energy (BMWI). It aims to drive digital manufacturing forward by increased digitization and the interconnection of products, value chains, and business models. Industry 4.0 has become a trend word for the ongoing digital transformation in every industry domain.

Industry 4.0 has been established as a collaborative effort, not only at the European level but also in collaborative with international initiatives. It represents the fourth revolution in manufacturing and industry. Industry 4.0 is the current transformation with the key enabling technologies, Robotics-Autonomy, secure data

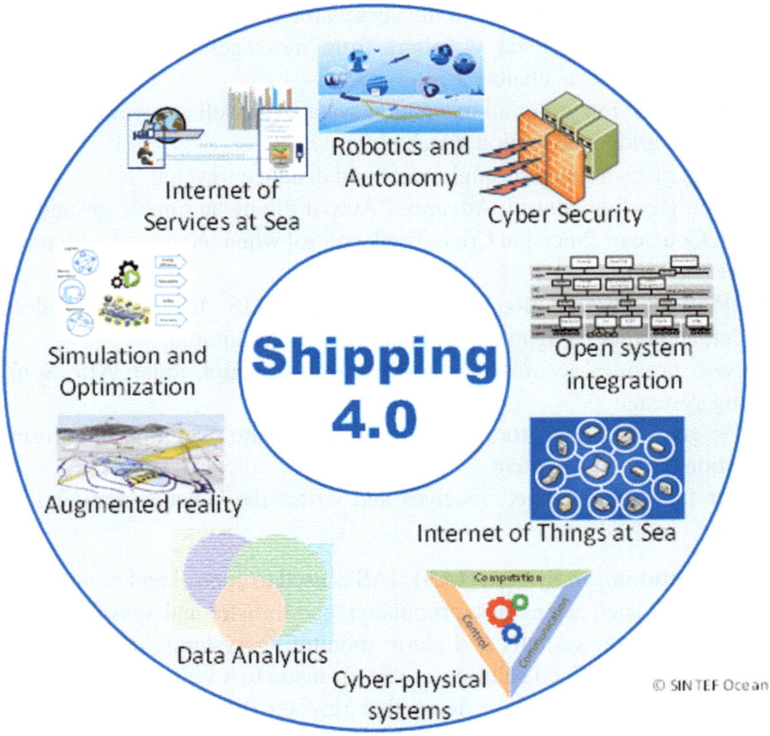

Fig. 4.5 Shipping 4.0: key technologies. (Source: SINTEF Ocean)

exchanges, cloud, cyber-physical systems, robots, Big Data, Artificial Intelligence (AI), Internet of Things (IoT), simulation, and other emerging technologies as shown in Fig. 4.5.

The Industry 4.0 has been adapted to the maritime domain, where Maritime 4.0 or Shipping 4.0 has been introduced. However, there are some key technologies that cannot be directly used from Industry 4.0, due to limited connectivity to ships and lack of both national and international regulations and standardization. Examples of technologies that are adapted to the shipping domain are described in the following parts.

4.1 Key Enabling Technologies Adapted to Shipping Domain

Some of the key technologies will have more impact than others for the maritime domain. Compared to the manufacturing industries, the main challenge is the have sufficient connectivity between ship and shore side to access data from the cloud and other data storage system. In essential the following technologies will be essential and game changers for shipping.

- *Internet of Service at Sea (IoS@Sea)*: Due to increased connectivity at sea, manual paperwork onboard a ship can be digitalized. Digitalization of working process will include automatic transfer of performance reports and other reports required to different stakeholders (ports, authorities, ship owner, ship operators, etc.)
- *Internet of Things at Sea (IoT@Sea)*: Technology based on the Internet of Things (IoT) will allow stakeholders in the value chain to exchange and track operational data in real time. Where problem cannot be solved onboard, notification from ship to shore that the right parts and technician can be ready and waiting when vessel berths. Every single component with an IP-address onboard a ship can in principle send and receive data. Digital twins can be used to store and analyze operational data which will lower the maintenance cost. However, the bandwidth can still be limited to use the cloud for data exchange. Another technology-like edge computing allows the data processing to be done near the source of the data and reduces the communication bandwidth needed between the sensors and central data center.
- *Open system integration at Sea*: Traditionally, each equipment and service providers must integrate their equipment on every ship. Uncoordinated development across the industry and a conservative approval regime result in fragmented solutions with low user-friendliness and relative high cost for integration and classification. An open system integration process will typically include standardized user interaction, open system integration architecture, and test methodology for functional testing (not system or component testing).
- *Robotics and Autonomy at Sea*: Automation and control systems reduce the manual work in shipping and reduce the environmental footprint. Remote control allows decision to be made outside the ship, and work processes will move from ship to shore. The process moving from manual to fully unmanned autonomous ship will naturally involve remote control to ensure sufficient maritime safety. Autonomy means being that a system or a device can sense its environment and take decision without any human input. Level of autonomy in shipping will increase, and more advanced algorithms and more sensors will be integrated. Key technologies are artificial intelligence, prediction, and big data analytics to be used for decision support.
- *Simulation and optimization*: Simulation technologies are an essential technology covering design, manufacturing, training, and operation of ships. Each role or component is modeled and integrated into complete dynamic models including control systems. Virtual prototyping requires simulation infrastructure and standards for data exchange. Virtual prototyping forces experts from different knowledge domains to collaborate on the same digital mode to develop the optimal trade-off solution. Distributed simulation systems also allow different stakeholders to work on the same digital model.

4.2 Communication

There are different needs for communication based on the vessels type, position, and the trade a vessel is operating in (Rødseth et al. 2009). Example, a fishing vessel have other requirements to communication and bandwidth than a cruise vessel. Further, vessels used in the oil and gas sector are more advanced than a container vessel sailing between two destinations and therefore requires higher communication capacity.

- *Minimum requirements:* This is defined by SOLAS and can as a minimum involve radio communication equipment with no digital capacity beyond what is implemented in GMDSS (DSC, NAVTEX, etc.). This class requires voice communication for all but some distress notifications.
- *Efficient reporting*: This would imply that a ship can send and receive mandatory and operation-related reports without problems. A rough estimate is a transmission requirement of below 64 kb per 24 h. Receive requirements are probably lower. This could easily be handled with line speeds of below 9600 bits/s.
- *Efficient operation:* Given that operational processes are more moved to shore, efficient operation would require more (automatic) reporting from the ship, e.g., on machinery condition monitoring, remaining consumables, various special requirements for port calls, and so on. Line speeds of 9600 bits/s will probably provide the required transmission rates although volumes may easily be doubled or more compared to the previous class.
- *Online ship:* This concept covers a ship that can be put online for remote maintenance, system diagnostics, or other purposes. This could include updates of digital weather forecasts, navigational maps, etc. This requires a relatively high total capacity and also higher available bandwidth. A somewhat uninformed guess is on the order of a megabyte per 24 h and minimum 64 kbit/s. This would also be sufficient for online emergency coordination, i.e., exchange of emergency-related status and planning information between ship and shore.
- *Broadband ship:* Passenger ships, research ships, and other ships that require transmissions of large quantities of data also require high-capacity communication links. This may mean rates of 1 megabit/s or more and correspondingly high volumes of data.

One should note that data requirements change as the ship enters or leaves various sea areas. Typically, data transmission requirements will normally increase substantially as the ship near port. This opens up for more cost-effective communication by using a combination of communication mechanisms and channels, which could be a mix between satellite-based services and terrestrial solutions.

One should also note that new services or applications that are coming to market normally are more bandwidth hungry than it was only a few years back in time. We also see the trend that many new applications are integrated with other applications, as example through cloud technology, which requires an infrastructure that is online all the time.

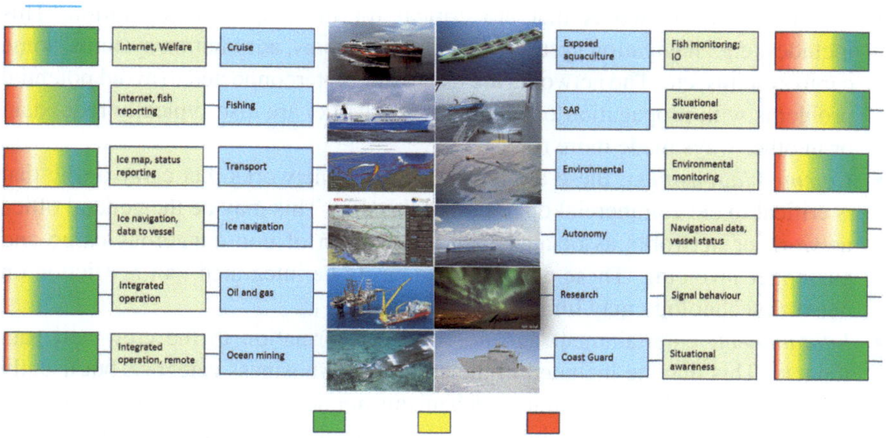

Fig. 4.6 Bandwidth demand per user category. (Source SINTEF Ocean)

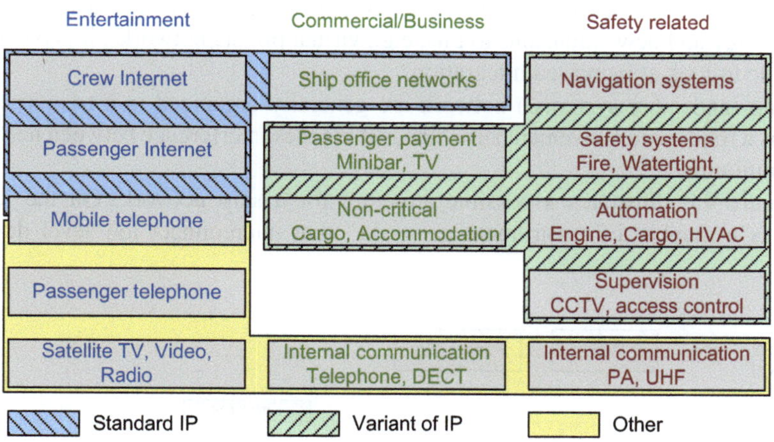

Fig. 4.7 Common ship network types. (Source SINTEF Ocean)

Figure 4.6 is an example on sectorial drivers as regards communication in an Arctic environment. It is based on typical services within each ship category, and it shows the expected demand for bandwidth, where green color represents high demand, yellow moderate, and red low. The picture is used as an illustration and is not the through picture of communication needs based on empiric data analysis, it is based on different results from previous projects and from a "light" analysis of the data traffic from a vessel.

Although this section is mainly concerned by communication between ship and shore, it may also be useful to look at the communication networks used onboard the ship. The networks have been grouped vertically in three main groups, as illustrated in Fig. 4.7:

- *Safety related*: networks that are critical for the ships safe operation. This includes, e.g., navigation systems, fire alarm systems, automation systems, some cargo systems, etc. The networks are usually not interconnected to avoid potential problems with propagation of faults. The networks are also typically redundant or in other ways made more robust.
- *Commercial/business*: these are business critical networks that carry traffic that is critical to the commercial operation of the ship. This may be the general office network or networks associated with supervision of non-dangerous cargo and passenger invoicing. These networks can easily be integrated, but restriction on network technology may limit open integration.
- *Infotainment*: entertainment and noncritical information to the crew and passengers. This may be included in the commercial/business category, but usually invoicing is handled by separate mechanisms not necessarily being dependent on the network itself, i.e., invoicing applications in gateway or management systems.

As noted above, safety and security considerations have so far limited the degree of interconnection between networks. However, with the increasing use of IP technology and new extensions to this, e.g., virtual private networks, it is becoming feasible to look at increased integration.

Ship (and other systems') network are generally organized in a tree structure. This is a robust and convenient structure as it reduces interference between networks to a minimum.

Figure 4.8 illustrates a possible structure for a ship network. On the lowest level is a number of "instrument networks" that interconnect low-level devices,

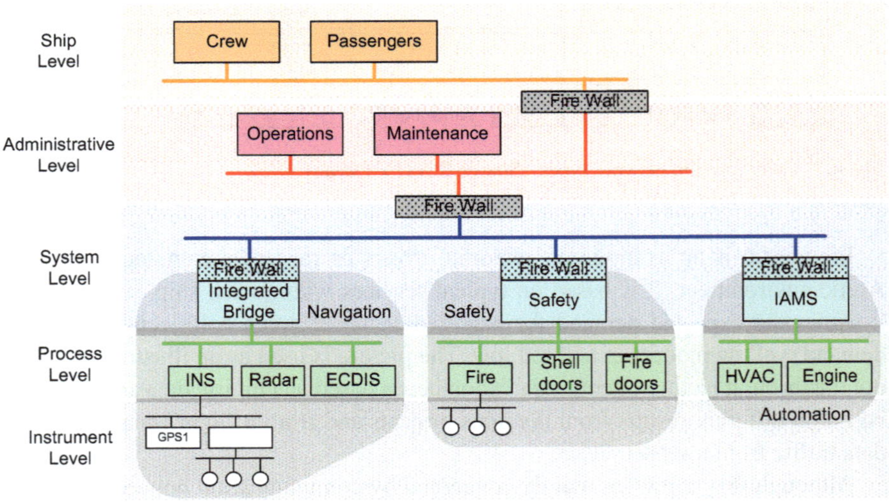

Fig. 4.8 Layered ship network. (Source SINTEF Ocean)

e.g., GPS receivers, heading sensors, and so on. On the level above this, a number of "process networks" interconnect devices and computers related to one process onboard. In this figure, the navigation or bridge system is shown as one process, the safety systems as one and the integrated alarm and monitoring systems (IAMS) as a third. The latter will typically include engine control as well as heat and ventilation control. On the next level is a system-level network interconnecting the different process segments to achieve integrated ship control. As an example, alarms from fire alarm systems must be transferred to the IAMS, and fire alarms may also trigger specific smoke extraction functions in the ventilation system. This layer is mostly implemented as point-to-point links today. On the next level is the technical and operational functions collected on the administrative or ship office network. This may again be connected to a ship level network.

Note the use of "fire walls" between process networks and between higher layers. The purpose of these fire walls is twofold:

- Ensure that faults or malicious actions in one network segment cannot influence others. This includes limiting the number on frequency of external access requests to a safe level. This is critical for the process networks.
- Ensure that data only is made available to authorized users. This is not directly safety related but covers protection of sensitive commercial or personal information.

Note also that fire walls may be embedded in parts of the system or in the communication means used. As an example, a talker-type serial line protocol can transport data from one system to another while ensuring that no problem in the second system propagates back to the first.

Different ships will show different topologies, and one cannot normally expect to find a pure tree structure in all cases. However, from the instrument level to the gateway to the administrative or ship level networks, one will normally employ a very strict tree structure to ensure integrity in the different functions and to ensure that faults in one process or in one network propagate to others.

Today, most of these networks are mostly decoupled to maintain safety and robustness. Some integration takes place, but this is very limited and usually just to connect one system to another through a point-to-point serial link. Some systems also allow remote access through an Ethernet gateway, typically via use of virtual private network (VPN) technology. Thus, the figure more indicates a principle than an actual realization.

Another element of interest as regards communication is called cybersecurity. An increased use of ICT has made a vessel more vulnerable. Cybersecurity comprises technologies, processes, and controls that are designed to protect systems, networks, and data from the unauthorized exploitation of systems, networks, and technologies. The trend today is that vessels are getting more and more digitalized with higher level of automation. Normally, automation systems are interacting with external sources which rises the danger for cyber threats. Therefore, good preventive information security is important. By information security, we mean that information is protected against unauthorized access. Cybersecurity is the protection of data and systems connected to the Internet.

5 Maritime ICT Outlook

The shipping industry is heading for a new game changer. A large percentage of the goods and volume loads are transported at sea, which has been the trademark of shipping. In the coming years, we will see new opportunities emerge for generating renewable energy, increased food production, and harvesting of other natural resources, minerals, and medicines from the sea, according to the CEO Harald Solberg, Norwegian Shipowners' (Association 2018b). He further says that:

- We must reach international accord in the UN maritime organization IMO on an ambitious strategy for reducing CO2 emissions from shipping
- Shipping must seize the opportunities presented by increased digitalization, a development that will impact every aspect of our members' operations
- The industry must contribute to solutions for sustainable development and cultivation of our oceans

One suggestion mentioned from the Norwegian Shipowners' Association according to sustainability is to equipping ships with sensors to collect data and track the health of the oceans and harnessing the power of innovation in the industry to devise technology for removing garbage from the sea. The ships can collect the garbage, but land-based depots to deliver the garbage must be in place. It will be hard to succeed if not thinking the whole value chain, which counts not only for garbage collection but for all shipping trades in general.

In the list below, there are some research questions recommended illuminating for a sustainable shipping approach.

- *A clean ocean*: technology for monitoring and combating pollution in the sea and air. For example, how can knowledge be used for designing emission-free crafts, how to make a business model that supports the green shift, and how to create robust technology for the purpose.
- *Environmentally friendly operations at sea*: How can we utilize the resources we have in an environmentally friendly environment? Keywords include increased utilization, coordinated logistics, good forecasting systems, and sound management.
- *New technology*: How can we obtain sufficient knowledge and data for good operations? Technology is an interaction between structures (boat, fleet, cages, terminals, subsea installations) and technology (communication, sensor, engine, control systems, propulsion technology) and the people who will operate the technology.
- *Climate-friendly operations*: How can we develop the best propulsion technology for the environment? This may mean different suitability of solutions, from

battery operation to hybrid solutions, and use of heavy oil vs MGO (marine gas oil), low-carbon fuel, renewable energy sources, new energy recovery systems, etc.

- *Operational logistics*: How to stimulate integrated solutions where the vessel's schedule and speed are dependent on the load and the terminal the vessels are sailing to? Schedule is based on the shortest journey, the most environmentally friendly journey, where external factors like tide, wind, and weather are playing together with the logistic planning of the journey.
- *Design of transport solutions*: In this, the need to build vessels and constructions is based on the operational areas that the vessel will be operated in. The balance between CAPEX and OPEX should be optimized for each vessel's operational purpose.
- *An integrated picture*: How can the entire life cycle count into the climate calculations, where interaction with all stakeholders in the value chain counts? We believe that many of the solutions need to be operated and created across domains and not as individual results supporting only one country, port, or one vessel. Integration of data sources is essential for new knowledge.
- *Monitoring and control*: It is and will be important to be able to monitor maritime operations to establish best practice that is done in an environmentally friendly manner. The interaction between operation, technology, and people to detect unregulated operations as an example to controlling illegal activities.
- *Secure technology*: The sustainable shift also means fighting accidents and emissions in the best possible way. What will a discharge mean to the surrounding environment for fish and plants, and how can we handle an emission efficiently and environmentally are questions to be answered together with the demand for new technology to battle the threats.
- *Automated solutions*: New technology such as unmanned or low-powered vessels can have a positive climate gain if they are operated properly. Today's technology can be divided into several levels, such as fully autonomous solutions that make all decisions independent of information and commands from the outside world into what is likely to be a future solution in automated solutions where technology takes decisions after they have obtained information from interoperable systems (e.g., from weather sensors). Use of intelligent transport systems (ITS) in the maritime sector will be important.
- *Propulsion systems*: New and improved propulsion systems including hybrid solutions with energy storage on board. Development, testing, and verification of new solutions; focus on low-carbon fuel; renewable sources of energy and better management of the available energy. The main goal is to verify that the technology is climate-friendly and that it gives the winnings it is designed for. We need approved technology for this purpose.
- *Use of large amounts of data*: In an environmental and sustainable perspective, it is important that we build new knowledge from available data that we can trust. Here is much undone, and we need new knowledge in combining data sources that will make the situational understanding (situational awareness) better. Not least, this is important when new environmentally friendly technologies are to be

introduced and developed, for example, test of new ship designs, with available data in laboratories before the building process starts.

- *Efficient ship concepts*: This relay on a good hydrodynamic design, including both hull and propulsion solutions. In general, the operational profile will determine what type of hull propulsion and machinery solution that will best fit the profile to have the minimum fuel consumption over the operating profile. Hull that has low water resistance means less energy needs. As a rule of thumb, additional speed requires twice amount of thrust and third times amount of power. Increased speed speaks against a more sustainable shipping. However, there will always be a trade-off depending on the environmental footprints and business models. Optimization of hull and propulsion systems can be investigating using CFD tools and also model tests in hydrodynamic laboratories.
- *Power systems*: Zero/low-emission power systems are reflecting new power systems. Power system is used mainly for producing power to the propulsions systems but also for other power consumers like hotel loads (air-condition system, cranes, winches, other ship systems, etc.). New trends are going from diesel to more non-carbon energy sources like solar, fuel cells, hydrogen, and battery. In general, sustainable solutions will be combinations with multiple energy sources that depend on different environmental, operational, and safety requirements. Power systems need highly advanced ICT systems to manage energy efficiency.

6 Summary

There are many challenges to be addressed to obtain a sustainable shipping industry. But for sure, it will be impossible to reach ambition goals of no or low emissions without focusing on the ICT technology. The interaction between technology and operation is extremely important for success.

Sustainable shipping is an area with good opportunities for success. The interaction between industry and government is essential. As an example, the focus on battery as energy source for car ferries will be of great importance for the development of expertise and solutions in this field that other types of maritime operations can get valuable knowledge from. Here, the authorities can stimulate new innovations in the form of supporting programs and schemes that help create new technology.

The regulations must in many cases be formulated in parallel with the development of new technology. What will the introduction of autonomous solutions mean to the sector as an example, and how can we develop new regulations for the purpose, and how can it be tested in a controllable environment are keys for success. As an example, the Coastal Administration plays an important role when they are working together both with the service providers, the users, and the regulatorily bodies when new laws and regulations are developed. The role of being close to the maritime industry and to be able to provide good advices and assessments along the way that the industry can achieve inspiration from and acquire new environmentally friendly technology is an important governmental role.

As mentioned, ICT for sustainable shipping contains many elements. It is trade-related, operation-related, and safety-related and will involve not only technology but the operational knowledge within the maritime domain. It is cross-discipline-oriented and has ambitious goals for the future to reduce emission from the sector.

If the Titanic expedition had happened today, it is likely it could have been avoided. New solutions, better communication infrastructure, and a more integrated picture are the case today. The good element from the Titanic accident was that it started a process of generating needs for communication and development of new ICT solutions. It provided the VHF frequency to be used for the maritime sector and highlighted safety as important. But the sector is about to enter into a new epoch where exchange of information in a digital format will be the future. More shore-based control and follow-up and a better governmental control on all maritime operations. Even service providers are preparing themselves for a better control of technology supported to end users. The interaction between the operational and the technological aspect will be much stronger than today. The sustainable element in it will of course be one of the key drivers.

References

BP. (2009). David Latin, vice president E&P technology, British Petroleum, comments to the Digital Energy Journal, 2009.

DNV. (2017). Consequences for NSA members from CO2 regulations, navigating a low-carbon future. Norwegian Shipowners' Association, Report No.: 2017-0205, Rev. 0.

EU. (2010). Directive 2010/65/EU of the European Parliament and of the Council of 20 October 2010 on reporting formalities for ships arriving in and/or departing from ports of the Member States and repealing Directive 2002/6/EC Text with EEA relevance.

Fjørtoft, K. E., et al. (2011). Maritime transport single windows: Issues and prospects. *International Journal on Marine Navigation and Safety of Sea Transportation*. September 2011.

Hoffmann and Kumar. (2002). Chapter 3 globalization: The maritime nexus. In C. Grammenos (Ed.), *Handbook of maritime economics and business* (pp. 35–62). London: Informa, Lloyds List Press.

IDSA. (2018). International Data Space Association. https://www.internationaldataspaces.org/en/.

IMO. (1965). Convention on facilitation of international maritime traffic (FAL). http://www.imo.org.

IMO. (1980). International convention for the safety of life at sea (SOLAS), 1974.

IMO. (2009). Second IMO GHG study 2009: Executive summary and final report.

IMO. (2014). Third IMO GHG study 2014: Executive summary and final report.

IMO. (2014b). Strategy for the development and implementation of e-navigation, MSC 85/26/Add.1.

IMO. (2018). Marine Environment Protection Committee (MEPC), 72nd session. http://www.imo.org/en/MediaCentre/IMOMediaAccreditation/Pages/MEPC72.aspx.

Lindstad, H. (2013). Strategies and measures for reducing maritime CO2 emissions, Doctoral thesis PhD. Norwegian University of Science and Technology – Department of Marine Technology. ISBN 978-82-461-4513-6.

Lindstad, H., et al. (2016). Lindstad, Haakon Elizabeth; Eskeland, Gunnar, Environmental regulations in shipping: Policies leaning towards globalization of scrubbers deserve scrutiny. *Transportation Research Part D: Transport and Environment, 47*, 67–76 2016.

Maritime21. (2016). An integrated maritime strategy for research, development and innovation publication. A maritime strategy report for the Norwegian Ministry of Food and Fisheries, coordinated by the Research Council of Norway.

MARNIS. (2009). Maritime navigation and information services. Sixth framework programme priority [1.6.2], Sustainable Surface Transport, Final Report, Deliverable code: D-MT-15.

NSA. (2018). Norwegian Shipowners' Association. Report: Think Ocean, Maritime Outlook Report 2018. http://www.rederi.no/.

Ramstad, L. S., et al. (2013). MARINTEK, implementing integrated planning–organizational enablers and capabilities. http://www.iocenter.no/system/files/sites/default/files/IO2/Papers/Final%20IGI%20Chapter%20IPL.pdf.

Rødseth, Ø. J., Kvamstad, B., Tjora, Å., Drezet, F. (2009). Ship-shore communication requirements. Flagship Deliverable D-D1.3, 31st December 2009. EU Project contract number TIP5-CT-2006-031406. http://www.mits-forum.org/communication.html.

Rødseth, Ø. J., et al. (2011). Ornulf Jan RODSETH, Kay Endre Fjørtoft, Maria A. LAMBROU. Web technologies for maritime Single Windows. Proceedings of MTEC2011.

Rodseth, O. J., & Kapidani N. (2017). A taxonomy for Single Window environments in seaports (pp. 26–28). Proc. of the MTEC2017 (2017).

SafeSeaNet. (2009). Directive 2009/17/EU of the European Parliament and of the Council of 23 April 2009 amending Directive 2002/59/EC establishing a Community vessel traffic monitoring and information system.

UN/CEFACT. (2005). The United Nations Centre for Trade Facilitation and Electronic Business. ISBN 92-1-116924-0. Recommendation and Guidelines on establishing a Single Window, to enhance the efficient exchange of information between trade and government. Recommendation No. 33.

Wimmer, M. A. (2002). Integrated service modeling for online one-stop government. *Electronic Markets, 12*(3), 1–8.

Chapter 5
Oil Pollution: Sustainable Ships and Shipping

Nikolaos P. Ventikos, Konstantinos Louzis, and Panagiotis Sotiralis

Abstract The aim of this chapter is to highlight the most significant attributes of oil pollution in the context of the sustainable shipping. The chapter presents the current legislative framework for the environmental protection against oil pollution and depicts the utility of the implementation of risk control options (RCOs). Furthermore, the measures of containment of the oil pollution cost are illustrated along with the incorporation of the Environmental Risk Evaluation Criteria in IMO's Formal Safety Assessment. Finally, the chapter discusses feasible ways of achieving a sustainable future without undermining the environmental integrity.

Abbreviations

ALARP	As Low As Reasonably Practicable
AMVER	Automated Mutual Assistance Vessel Rescue System
CAF	Cost of Averting a Fatality
CATS	Cost of Averting a Tonne of Spilt oil
COW	Crude oil washing
CV	Contingent Valuation
DWT	Deadweight
ECDIS	Electronic Chart Display and Information System
ENC	Electronic Navigation Chart
EU	European Union
FSA	Formal Safety Assessment
GT	Gross tonnage
HEA	Habitat Equivalent Analysis
ICAF	Implied Cost of Averting a Fatality
IMO	International Maritime Organization
IOPCF	International Oil Pollution Compensation Funds

N. P. Ventikos (✉) · K. Louzis · P. Sotiralis
National Technical University of Athens (NTUA), Athens, Greece
e-mail: niven@deslab.ntua.gr

H. N. Psaraftis (ed.), *Sustainable Shipping*,
https://doi.org/10.1007/978-3-030-04330-8_5

ITOPF	International Tanker Owners Pollution Federation Limited
MARPOL	International Convention for the Prevention of Pollution from Ships
MEPC	Marine Environment Protection Committee
NOAA	National Oceanic and Atmospheric Administration
OILPOL	International Convention for the Prevention of Pollution of the Sea by oil
OPA	Oil Pollution Act
OPRC	International Convention on Oil Pollution Preparedness, Response and Co-operation
OSIR	Oil Spill Intelligence Report
P&I	Protection and indemnity
PDF	Probability density function
RCOs	Risk control options
SOLAS	International Convention for the Safety of Life at Sea
UK	United Kingdom
UN	United Nations
USA	United States of America
VHL	Value of Human Life

1 Introduction

Sustainable development has been defined by the United Nations (UN) as "*development that meets the needs of the present without compromising the ability of future generations to meet their own needs*" (Brundtland 1987). Although the concept of sustainability for transportation systems has not been standardized, it is mostly perceived through the impacts on the environment, the social well-being, and the economy (Jeon and Amekudzi 2005).

The maritime transportation system that carries crude oil and its refined products is an essential part of modern society, as these commodities are driving many activities, such as transportation, defense, technology, industry, commerce, research, development, etc. This undeniable widespread use of petroleum also leads to accidental and intentional discharges (Burgherr 2007). Maritime accidents, oil well blowouts, and pipeline ruptures are some illustrative causes of oil pollution. It is a fact that tanker vessels attract the most attention when it comes to oil pollution, compared to other ship types, due to the amount of the spilt oil and the resulting environmental consequences. Major oil spill accidents in the past have demonstrated the large potential for environmental damage from the maritime transport of oil, a fact that is incompatible to the goals of sustainability. As will be illustrated in a subsequent section of this chapter, environmental damage comprises only one aspect of the cost of oil spills. Another aspect is socio-economic losses, which have a direct impact on social well-being, both in terms of lost income and the inability to conduct recreational activities. Finally, there is the cost of clean-up operations that have an impact on the broader maritime economy as the costs the polluter is required to pay

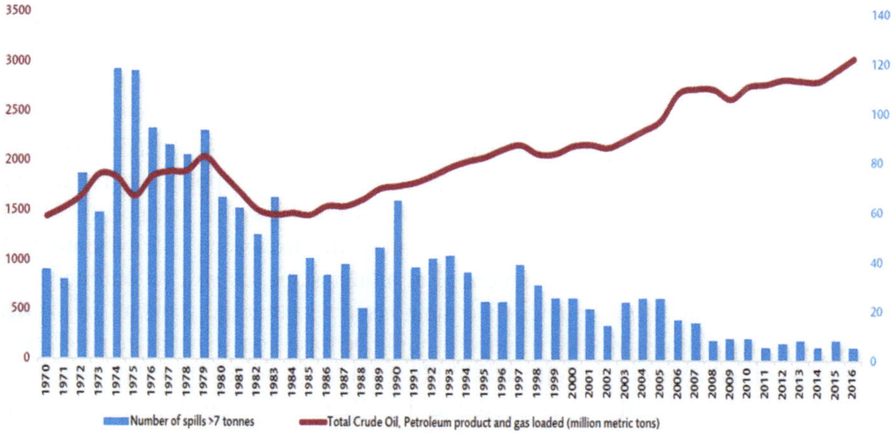

Fig. 5.1 The development of the number of tanker oil spills contrasted to the volume of global oil trade (incl. crude and petroleum oil and natural gas). (Source: ITOPF 2018)

in claims may be extremely high and therefore profitability is indirectly affected. Therefore, the prevention of maritime accidents that lead to marine oil pollution promotes sustainable development of the industry in all its dimensions.

During the past decades, there has been great improvement in this sector of the maritime industry. According to ITOPF (2018), the magnitude of oil spills from tanker ships has been steadily declining since the 1970s despite a constant increase in the volume of global oil trade (Fig. 5.1). A major contributing factor has been the development of a strict international regulatory framework that has imposed several design (e.g. double hulls for tanker ships, etc.) and operational risk control options (RCOs), such as the Electronic Chart Display and Information System (ECDIS). The mission of these RCOs has been to prevent oil pollution and/or mitigate the resulting environmental consequences of maritime accidents. The first two sections of this chapter provide a brief description of the most significant international and regional conventions and the most significant RCOs that are stipulated in the international regulations.

Although the focus of the International Maritime Organization (IMO) has shifted during the last decades from marine oil pollution toward minimizing air pollution and the transfer of invasive species through water ballast discharge, the risk of oil pollution from ships needs to be further reduced because even small oil spills in sensitive environments (e.g. the Arctic region) may be unacceptable from the viewpoint of sustainable shipping. Last but not least, despite a series of implemented measures (relating to the structure of the ship and its operation) by the IMO resulting in identifiable improvements, oil spills, especially from tanker vessels still pose a great environmental threat because many navigational routes intersect with the "Large Marine Ecosystems", which are sea areas near coastal waters with primary productivity higher than the open sea waters (Afenyo et al. 2016). The fundamental concept of sustainable development refers to the simultaneous fulfillment of two goals that are usually considered as mutually exclusive: the preservation and protection of the marine environment and economic development (Olawuyi 2012).

Following this rationale, it is imperative to achieve a balance between environmental protection and the economic goals of the maritime industry.

The balance between economy and environmental protection largely depends on the ability of the IMO to propose economically sound RCOs that are effective in reducing the costs of oil pollution. The Formal Safety Assessment (FSA) process of the IMO is indeed an instrument that is intended to support the regulatory process and the development of cost-effective RCOs. However, because its initial focus was on evaluating the risk the human safety, the discussion on Environmental Risk Evaluation Criteria was commenced at a later stage, as will be outlined in a subsequent section of this chapter. The main issue was that the standard by which RCO cost-effectiveness would be judged should be based on an accurate prediction of the true total cost of oil spills. Cost-effectiveness of an RCO in the context of environmental oil pollution is perceived as the difference between the oil spill cost that is averted and the cost implemented. This chapter also provides a survey of the international literature that relates to modeling oil spill costs, as these studies fuelled the discussion at the IMO level.

An industry that transports about 90% of goods globally and is therefore a significant part of the global supply chain will inevitably play an important part in shaping a sustainable future economy. However, this may not be achieved if the industry experiences maritime accidents that are devastating to the environment. In fact, whether maritime transport will continue to be viable in a sustainable future may depend on the sustainability of maritime transport of oil. Therefore, the question of how this sector of the industry may become sustainable is as current as ever and is briefly referenced in the concluding section of this chapter.

2 Regulatory Framework

A series of international and regional conventions led to the adoption of specific measures for the enhancement of the structural integrity of tanker vessels and the improvement of their operational effectiveness. These measures were intended to reduce the amount of oil spilt in the event of a maritime accident. This section illustrates the existing international regulatory framework concerning marine oil pollution as well as their evolution throughout the last two centuries, by briefly describing each convention. In addition, the most important regional conventions in the European Union (EU) on the prevention of marine oil pollution are also presented (Fig. 5.2).

2.1 The Evolution of the Marine Pollution International Law

Public concern with respect to oil pollution was aroused shortly after World War I. The necessity to adopt specific measures to deal with oil pollution in an efficient manner was expressed initially by the USA and subsequently by the League of

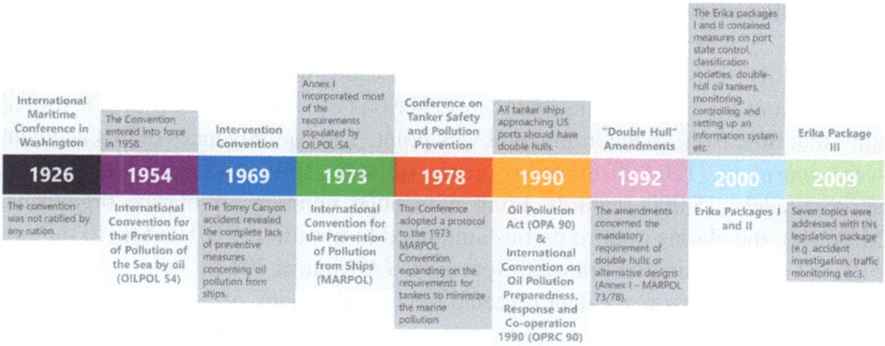

Fig. 5.2 Timeline of the most significant international and EU regulatory framework concerning marine oil pollution

Nations – the organization now known as the United Nations (UN). Prior to this time, only a handful of states had tried to enforce pollution legislation beyond their national waters (Gold 1998). The first international convention on the prevention of marine oil pollution was an International Maritime Conference that was held in 1926 in Washington, USA. Although the technical aspects and legal implications of oil pollution were addressed during the conference, the convention was not ratified by any other nation.

2.1.1 OILPOL 54

The negative impact of oil pollution was evident even after the World War II due to a large number of damaged and sunken vessels. Along with the increasing demand for oil due to the growing world economy during the 1950s, there was a renewed concern over the problem of marine oil pollution. The International Convention for the Prevention of Pollution of the Sea by oil (OILPOL 54) was the first international regulatory instrument that aimed to deal with oil pollution of the marine environment from shipping. The conference for this convention was held in London, where the involved parties agreed to take specific preventive measures, and the Convention entered into force in 1958. OILPOL 54 prohibited the discharge of oily wastes that resulted from washing the oil cargo tanks with water (a common practice up to that point) within a certain distance from land and in sensitive environmental areas. Furthermore, special records with respect to the oil discharges should be registered in an oil record book, which should be regularly inspected. The flag state was responsible for the implementation and enforcement of this convention. In 1962, the geographical limits were extended following an amendment adopted by the IMO. In 1965, the IMO established the Subcommittee on Oil Pollution to deal with oil pollution issues.

2.1.2 The Intervention Convention 1969

On March 18, 1967, the supertanker SS Torrey Canyon ran aground near the southwest coast of the UK resulting in the most serious maritime accident ever recorded until that time from an environmental viewpoint. The consequences of this accident were enormous; 120,000 tonnes of crude oil were spilt leading to the death of approximately 15,000 seabirds along with huge numbers of marine organisms. Moreover, the claims to third parties amounted to 6 million GBP and to 40 million FRF in Great Britain and France, respectively (Anyanova 2012).

The Torrey Canyon accident revealed the complete lack of preventive measures concerning oil pollution from ships and exposed the deficiencies in the existing system for providing compensation following accidents at sea (IMO 2018a). This accident led to a series of actions toward a stricter international policy on oil pollution. In 1969, the OILPOL 54 Convention was amended by incorporating a more efficient oil industry procedure, the so-called load on top. This process focused on oil savings and the reduction of pollution by separating the oil and water from the washings of the tank cleaning. Furthermore, a Legal Committee was established by the IMO to improve the legal framework of liability and compensation for oil spill damage. In addition, the Sub-Committee on Oil Pollution was renamed to Sub-Committee on Marine Pollution, which eventually became the Marine Environment Protection Committee (MEPC). The MEPC expanded the scope of its work by handling all the aspects of marine pollution. Finally, the IMO adopted the International Convention Relating to Intervention on the High Seas in Cases of Oil Pollution Casualties 1969 (INTERVENTION 1969). This international convention stipulated that a coastal state could take measures if a ship accident posed a pollution threat to its coastline (Özçayir 2004). Finally, the rapid growth of the size of the oil tankers due to the increasing demand for oil and other chemical substances resulted in the perception that the OILPOL 54 Convention was inadequate despite the various improvements.

2.1.3 MARPOL 73/78

The dissatisfaction of the international community regarding the effectiveness of the OILPOL 54 Convention led in1973 to the adoption of the International Convention for the Prevention of Pollution from Ships (MARPOL). Annex I incorporated most of the requirements stipulated by OILPOL 54 and its amendments, whereas the other annexes were dedicated to polluting substances apart from oil carried in packaged form, sewage, and garbage.

The slow process of the ratification of MARPOL along with a series of tanker accidents (e.g. the stranding of MV Argo Merchant, etc.) in 1976–1977 strengthened the need for stricter legislation for dealing with accidental and operational marine oil pollution. In February 1978, after an initiative by the USA, the IMO organized a conference – the Conference on Tanker Safety and Pollution Prevention – that led to the adoption of a Protocol to the 1973 MARPOL Convention (IMO 2018a).

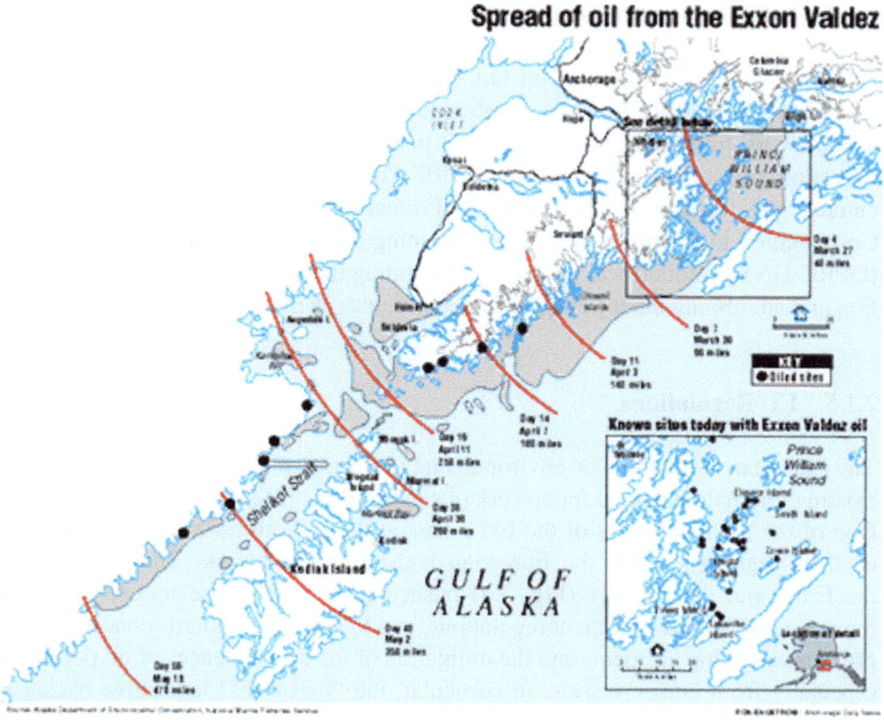

Fig. 5.3 Areas impacted by Exxon Valdez oil spill. (Source: Butricks 2014)

The accident of the oil tanker Exxon Valdez was the motivator for the most significant alteration of the Convention (Annex I) since the adoption of the 1978 Protocol. On March 24, 1989, the Exxon Valdez struck Prince William Sound's Bligh Reef and spilt about 40,000 tons of crude oil (Fig. 5.3). The oil spill is considered the second largest oil spill in the USA, in terms of quantity. As a result, the USA decided to enact the first oil pollution legislation, the Oil Pollution Act of 1990 (OPA 90). OPA 90 dictated that all tanker ships approaching US ports should have double hulls. Furthermore, the USA requested that the IMO would implement double hulls as a global mandatory requirement for tankers. After extensive discussions with the involved members of the IMO, the MEPC agreed to make mandatory the requirement of double hulls or alternative designs that ensure a specific level of protection against marine oil pollution. The relevant amendments were adopted in March 1992 and entered into force in July 1993. Finally, the accidents of the Erika and the Prestige in 1999 and in 2002, respectively, accelerated the substitution of single-hull with double-hull tanker vessels.

2.1.4 Pollution Preparedness and Response

The International Convention on Oil Pollution Preparedness, Response and Co-operation 1990 (OPRC 90) was an international maritime convention that provided a framework for dealing with marine oil pollution incidents nationally and in co-operation with other countries. The OPRC Convention was adopted in 1990 and entered into force in 1995. In 2000, a Protocol on Preparedness, Response and Co-operation to Pollution Incidents concerning hazardous and noxious substances (OPRC-HNS Protocol) was adopted by extending the initial content of the Convention and addressing other polluting factors, such as chemicals (IMO 2018b).

2.1.5 EU Regulations

The European regulations on environmental protection play a supportive role to the existing international legal framework (e.g. MARPOL 73/78) overseen by the IMO. One of the main concerns of the EU relies on the smooth transfer and adaptation of IMO regulations into the European legislative framework. The accidents of the Erika and the Prestige (Fig. 5.4) pushed the EU to revise its existing legal regime, to adopt more efficient regulations, and to set new standards concerning the prevention of ship accidents and the mitigation of the consequences of oil pollution, especially from tanker vessels. In particular, the "Erika" EU legislative packages adopt maritime laws aiming at improving safety and minimizing the environmental impact of ship accidents. The Erika I package contained measures on port state control (Directive 2001/106/EC), classification societies (Directive 2001/105/EC), and double-hull oil tankers (Regulation 417/2002). The Erika II package consisted of measures on monitoring, controlling, and setting up an information system (Directive 2002/59/EC), a fund to compensate victims of oil pollution (Proposal 2000/0326(COD)) and the establishment of the European Maritime Safety Agency (EMSA – Regulation 1406/2002). The focus of the Erika III package was preventing

Fig. 5.4 The Erika and the Prestige maritime accidents. (Source: Officer of the Watch 2013)

accidents at sea and improving the existing legislative system for dealing with the consequences of ship accidents. As a result, the following seven topics were addressed:

- The quality of flag states (Directive 2009/21/EC)
- Classification societies (Directive 2009/15/EC)
- Port state control (Directive 2009/16/EC)
- Traffic monitoring (Directive 2009/17/EC)
- Accident investigation (Directive 2009/18/EC)
- Liability of carriers (Athens Convention) (Regulation (EC) No. 392/2009)
- Insurance (Directive 2009/20/EC)

3 Risk Control Options to Prevent Oil Pollution from Ships

RCOs can generally be classified into two main categories: active safety measures and passive safety measures. Active safety measures focus on reducing the frequency or probability of occurrence of an accident (e.g. collision, grounding, etc.). Passive safety measures are design and technical solutions that remain inactive until called into action. They become active during the accident and aim to mitigate the magnitude of the consequences. Collisions and groundings usually result in larger oil spills compared to other types of accidents, such as fires/explosions and contacts (Yip et al. 2011). Hence, the RCOs have been mainly designed either to prevent oil pollution or to minimize the environmental impact of these two prevailing types of maritime accidents concerning oil pollution.

3.1 Active Safety Measures

Most of the following RCOs are included in the International Convention for the Safety of Life at Sea (SOLAS). The SOLAS Convention is globally accepted as the most significant international convention for ensuring maritime safety and, indirectly, for preventing oil pollution (Ornitz and Champ 2002). The mandatory installation of communication (Chap. 4) and navigational systems (Chap. 5) onboard contribute to preventing maritime accidents and the potential oil pollution.

3.1.1 Crude Oil Washing (COW)

A technical solution for reducing the quantity of oil spilt is the mandatory installation of the COW system, which according to Regulation 33 of MARPOL (Annex 1) must be installed and requires every new tanker over 20,000 deadweight (DWT) tonnes. The COW system is a system for cleaning oil tanks using crude

Fig. 5.5 Installation of the FOR system in a double-hull vessel. (Source: Sanguri 2017)

oil instead of water, which aims to make the cleaning process more effective and prevent operational discharges of oily wastes.

3.1.2 Oily Water Separator (OWS)

Regulation 15 of Annex I (MARPOL) stipulates that any ship larger than 400 gross tonnage (GT) may discharge the oily mixtures after being processed through an OWS system (Fig. 5.5). The technical requirements of the OWS are presented thoroughly in Regulation 15, whereas the operational requirements (e.g. distance from the nearest land, discharge rate, total quantity of oil discharged, etc.) of the discharges of oily mixtures are included in Regulation 34.

3.1.3 Fast Oil Recovery (FOR) System

Apart from the mandatory technical and operational solutions that are required under the SOLAS and MARPOL 73/78 Conventions, there are also other proposed solutions that even though their installation is not obligatory, they are highly effective toward the prevention of oil pollution. An indicative example is the FOR system, which consists of emergency connectors and enables fast and easy access to the fuel and cargo tanks aiming at the retrieval of oil or other hazardous substances from fully submerged shipwrecks (Fig. 5.6). From these access points, salvors may lighter the contents of the tank regardless of the inclination of the ship structure. The FOR system can be installed to both new and existing ships and is compatible with

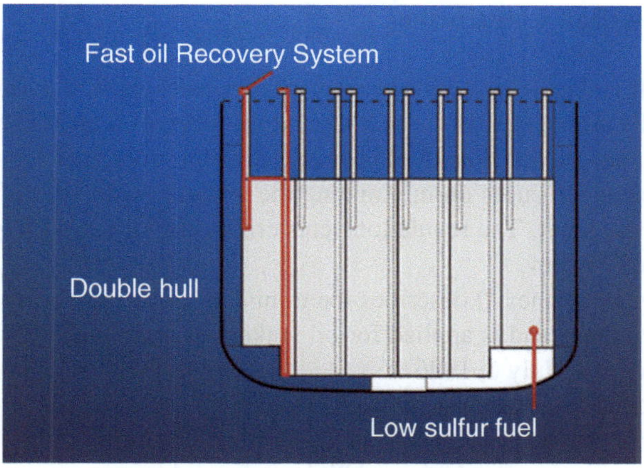

Fig. 5.6 Installation of the FOR system in a double-hull vessel. (Source: CMA-CGM 2017)

all ship types (i.e. containerships, general cargo vessels, bulk carriers, chemical and crude oil tankers, RoPax, and cruise ships). The system requires no maintenance by the crew members due to the simplicity of its design.

3.2 Passive Safety Measures

The following RCOs have been integrated into MARPOL regulations, where the design requirements regarding subdivision and stability that tankers should meet to provide adequate protection against accidental oil outflow are analytically described (IMO 2018c).

3.2.1 Segregated Ballast Tanks

A significant regulation of MARPOL 73/78 is Regulation 18 (Annex I) which requires segregated ballast tanks on new tankers over 20,000 DWT tonnes. These tanks ensure that ballast water (required for the stability of the ship) will be separated from the cargo oil and the fuel oil systems of the ship to avoid ballast water contamination from cargo oil. As a result, the potential discharge of oily ballast water mixtures into the marine environment is prevented. Furthermore, according to Paragraphs 12–15 of Regulation 18, these tanks should be located in specific areas (protective location) where the possibility of impact due to collision or grounding is higher. Following this structural approach, the amount of oil spilt will be significantly less.

3.2.2 Double-Hull Construction

A double-hull tanker is one with two layers of watertight hull surface. The inner and outer hulls are located at the bottom and the sides of a tanker ship. The space between the inner and the outer hull is dedicated to the storage of water ballast. This enhanced structure focuses on minimizing the amount of oil spilt after a collision or grounding accident. The regulations concerning the structural requirements are presented briefly below.

Regulation 19 (Annex I) describes the minimum dimensions of wing tanks and double bottom tanks and is applied for oil tankers of more than 5000 DWT tonnes delivered on or after July 6, 1996.

Wing tanks or spaces are given by (5.1):

$$w = \min\left\{0.5 + \frac{\mathrm{DWT}}{20,000};\ 2.0\ \mathrm{m}\right\} > 1.0\ \mathrm{m} \tag{5.1}$$

Double-bottom tanks or spaces are given by 5.2):

$$h = \min\left\{\frac{B}{15};\ 2.0\ \mathrm{m}\right\} > 1.0\ \mathrm{m} \tag{5.2}$$

Regulation 20 (Annex I) contains the double-hull and double bottom requirements for oil tankers delivered before July 6, 1996.

Regulation 23 (Accidental oil outflow performance – Annex I) is applicable to oil tankers delivered on or after Jan 1, 2010, and sets the limits for the mean oil outflow parameter ($\mathrm{O_M}$), along with the procedure for its calculation. The mean oil outflow parameter is calculated independently for side damage and bottom damage in the event of stranding or collision and then combined in a nondimensionalized form by applying formula (5.3):

$$O_{\mathrm{M}} = \frac{(0.4\ O_{\mathrm{MS}} + 0.6\ O_{\mathrm{MB}})}{C} \tag{5.3}$$

In (5.3) the parameters are:

O_{MS} = mean outflow for side damage, in m^3
O_{MB} = mean outflow for bottom damage, in m^3
C = total volume of cargo oil, in m^3, at 98% tank filling

The double-hull design has proven to be very effective, compared to single-hull, in reducing marine oil pollution from tanker ship accidents. Yip et al. (2011) have found that oil spill size has been reduced by 62% for tanker ships and 20% for tank barges.

4 Estimating the Total Cost of Oil Pollution

The total cost of an oil spill is a complex quantity with components that are affected by several interdependent factors. As a result, and because each oil spill case is essentially unique, the problem of determining the total cost is a non-trivial one. The total cost of an oil spill is used for quantifying the damages that result from such an event. It is also a crucial parameter for evaluating alternative ship designs and policy measures for reducing oil pollution (Kontovas et al. 2011), in terms of the benefits against the cost of implementing an RCO. There are various studies in the international literature with statistical models for estimating the costs associated with an oil spill as a function of oil spill volume. However, large variability in the available total oil spill cost data and the fact that the corresponding databases have several limitations reduce the reliability of the derived statistical regression models.

4.1 Components of Oil Spill Cost

According to Liu and Wirtz (2006), the total cost of oil spills may be classified into the following five categories: environmental damages, socio-economic losses, removal, research and other cost categories. Kontovas and Psaraftis (2008) have reclassified these categories into the following three groups: clean-up (incl. removal, research, and other costs), socio-economic losses, and environmental costs. It should be noted that other costs may include the costs that are incurred by the loss of cargo and/or the vessel itself, repairs, etc. The losses that are related to an oil spill accident include both direct and indirect costs.

Direct costs are tangible quantities that include lost income from the potential market value of the oil lost at sea, lost income for the affected parties (e.g. from potential future reduction in economic activities), property damage, the cost of the response and clean-up operations, as well as claims paid for compensation to third parties such as the fishing and the tourism industry that are the major contributors to the local economy of a coastal area. It should be noted that the number of claims has been increasing under pressure from environmental groups and public opinion (Hendrickx 2007) indicating the low public tolerance of large-scale environmental damage. The total cost of eligible compensation claims to third parties depends on how much the oil spill has affected revenues from economic activities and the expenses paid for the response operations and restoration of the environment. The latter may fall upon the local government (EMSA 2012), which must bear the burden of restoration, especially in case of large-scale environmental damage.

Indirect costs are mostly intangible quantities relating to the magnitude of the environmental damage and the corresponding societal effects. These include the cost of lost recreational activities, which relates to the environmental damage caused by the oil pollution and the restoration of the environment (Helton and Penn 1999).

4.2 Factors That Influence the Cost of Oil Spills

The total cost of oil spills is related to the conditions that led to the accident that resulted in the spill, as well as to the consequences from the said accident and ultimately to the adopted mitigation strategies. The factors that shape the total cost of oil spills are multiple and are not independent of each other, a feature that introduces considerable complexity in the effort to determine the true cost of an oil spill. The international literature contains several studies that identify the factors (see Table 5.1) that influence the cost of oil spills (Etkin 1999a; White and Molloy 2003; Ventikos et al. 2009; Yamada 2009; Ventikos and Sotiropoulos 2014). There are various types of accidents that may lead to an oil spill, such as foundering, grounding, collision, and fire/explosion. According to several studies (Yamada 2009; Kontovas et al. 2010; EMSA 2012), the most expensive oil spills are related to foundering and grounding accidents.

Because some of the most expensive oil spills are linked to relatively small ships, it seems that the size of the ship does not affect the cost of the oil spill. This is a reasonable conclusion if one considers that while the potential risk of oil pollution from a larger ship is greater because of the larger quantities of transported oil involved, it is rather improbable that the entire cargo would be lost in case of an accident.

White and Molloy (2003) conclude that the most important factors that determine the cost of oil spills are the type of oil, the location of the oil spill, and the characteristics of the affected area, the quality of the contingency plan, and the effectiveness of the actual response operations. The following is a brief overview of how location and spill size affect the total cost of an oil spill.

Table 5.1 Major factors affecting the cost of oil spills

Oil characteristics	Geography factors	Accident particulars	Response and remediation factors
Size of the oil spill	Location of the accident (near or far from the shore)	Type of accident (e.g. grounding, collision, sinking, discharge of oil)	Response and remediation strategies applied
Type of oil	Characteristics of the affected area (e.g. physical, biological, economic, etc.)	The leakage rate of oil into the sea	The effectiveness of response and remediation operations
			Detection speed
			Weather conditions
	Regulatory regime and the degree of enforcement		Season of the year

4.2.1 Location

The location factor consists of two main aspects, namely, distance from the coast and the specific geographical region where the oil spill has occurred. When an oil spill occurs near the coast or near ports, then there is an increased chance that the response operation will not begin fast enough to prevent the oil from reaching the shore and therefore the potential environmental damage increases (White and Molloy 2003). The implication is that the cost of the clean-up operation will be much greater, as it is generally more difficult to deal with an oil spill onshore rather than at sea (Etkin 1999b, 2001b). The geographical region where the oil spill occurs relates to factors such as the degree of regulatory enforcement, the perception of environmental conservation, capability and effectiveness of equipment used in response, and remediation operations (Etkin 1999b, 2000). Another parameter that varies with the geographical region is how reliant the local economy is on fishing, fish farming, and tourism (e.g. restaurants, hotels, etc.), activities that are directly affected by oil spills in coastal areas and may experience long-term effects. Fishing may be affected in the event of an oil spill as ports may suspend their operations to prevent further spreading, fishing boats may need to halt their activity for cleaning, fish populations may migrate, and the demand in the fish market may decrease for fear of contamination. Fish farming is often affected by oil spills as they are very sensitive to environmental contaminants. The magnitude of the effects on tourism depends on the time of year when the oil spill occurred, on the stance of the mass media and their criticism of the response and restoration operations, and the rate of the remediation activities.

4.2.2 Oil Spill Size

The size of an oil spill is a significant factor that determines the total cost. Larger oil spills are generally more expensive compared to small spills, as more effort needs to be implemented for removal and remediation. Dunford and Freeman (2001) also noted that higher costs are generally linked to larger spills that are close to the shore and endangered species are affected. However, there have been cases of expensive small oil spills, which complicate the relationship between total cost and oil spill size. In a series of studies, Etkin (1999b, 2000) found that the unit cost of response and remediation per tonne of oil is, in fact, greater for small spills, due to various fixed costs for response resources (i.e. spill mitigation crews and equipment) that are required by law to be mobilized. Although there is a clear relationship between oil spill size and the ensuing total cost, comparing spill costs only on the basis of the volume of the oil spilt may be misleading (White and Molloy 2003).

4.3 Oil Spill Cost Modeling

The task of statistical modeling of oil spill costs is a non-trivial one, as the complex relationships between contributing factors introduce large variabilities in the available data. Helton and Penn (1999) noted that the variation in total costs (including the compensation to third parties) may be attributed to the widely different environmental impacts of oil spills that depend on factors such as location and environmental conditions. From a statistical viewpoint, the implication of large variability is that it is difficult to isolate the effect of a single contributing factor and use that as an explanatory variable. However, there is a consensus as to the possibility of employing spill size (in terms of volume) as the determining factor in oil spill cost modeling.

Kontovas and Psaraftis (2008) have noted that there are at least four different methods that may be employed for calculating the total cost of an oil spill:

1. Estimating the individual cost components (i.e. clean-up, socio-economic losses, and environmental damages).
2. Estimating the clean-up costs via modeling and applying an assumed equivalence ratio for the other two categories.
3. Estimating the total cost directly through modeling.
4. Estimating the total cost directly by assuming that it may be approximated by the compensation that was eventually paid to third parties. The main source of such information is the annual reports of the International Oil Pollution Compensation Funds (IOPCF).

A series of studies have investigated the relationship between spill size and oil spill costs by employing a statistical approach. Different approaches to oil spill cost modeling include models that estimate clean-up costs, socio-economic losses, and environmental damages separately and models that estimate the total cost directly. The following is a brief survey of the available oil spill cost models from the international literature, adapted from the works of Psarros et al. (2011), Kontovas et al. (2011), Kontovas et al. (2012), and Prendergast and Gschwend (2014).

4.3.1 Models That Estimate Clean-Up Costs

The following approaches estimate oil spill clean-up cost as a statistical average (weighted or unweighted) or as a linear function of oil spill size. Harper et al. (1995) have estimated average values for clean-up costs (shoreline and offshore), which vary due to a dependence on spill size and the remoteness of the location. Sirkar et al. (1997) calculated an average value equal to $92,138/t of oil spilt (USD 2009 value). In the work of Etkin (2000), clean-up costs were calculated as a function of the geographical region, the distance to the shore, the size of the oil spill, the degree of shoreline oiling, and the techniques used in the operation. The model was based on worldwide historical cost data, derived from the Oil Spill Intelligence Report

(OSIR) International Oil Spill Database and validated with multiple hypothetical scenarios in a subsequent study (Etkin 2001a). The developed model is linear and expressed with the simplified Eq. (5.4):

$$C_i = C_n t_i s_i m_i \tag{5.4}$$

In (5.4), i is the scenario that is examined, C_i is the response cost per unit spilt, C_n is the average response cost per unit spilt for country (geographical region) n, t_i is a cost modifier for considering different oil types, s_i is a cost modifier for considering different spill sizes, and m_i is a cost modifier for the employed clean-up technique. The cost modifiers represent the percentage difference from the overall median spill cost.

Etkin found that the clean-up costs varied widely with the geographical region, even by multiple orders of magnitude. This was attributed to factors such as spiller liability, clean-up standards, labor costs, and the scarcity of data for some regions. Vanem et al. (2008b) use the estimates from Etkin (2000) to calculate a worldwide weighted average for oil spill costs. This was accomplished by weighting the estimated clean-up costs per region with the percentage distribution of oil tanker traffic from the Automated Mutual Assistance Vessel Rescue System (AMVER). The calculated weighted average was equal to $16,000/t of oil spilt. Shahriari and Frost (2008) developed a regression model for clean-up costs based on the oil spill size, the oil density, and the level of preparedness as estimated by the International Tanker Owners Pollution Federation Limited (ITOPF) on how well different countries respond to oil spill incidents. A limitation of this study is that it used a methodologically unorthodox approach where two different models were derived and proposed to be used alternately depending on the interval of the total cost estimation. This model resulted in a minimum value for clean-up cost equal to $8127/t of oil spilt (USD 2009 value). Ventikos et al. (2009) calculated the oil spill response cost in Greece, based on a dedicated database that was specifically constructed for the study. The parameters that were considered for the estimation included the type of oil, the quantity of oil, and the impact to the shoreline. The resulting figure amounted to 25,000 euro/t of oil spilt.

4.3.2 Estimating Socio-economic Losses

According to Liu and Wirtz (2006), socio-economic losses include lost income and property damage. Property damage may be estimated by considering the cost of repairing and/or cleaning various economic resources such as facilities and vessels. According to Kontovas and Psaraftis (2008), lost income includes the value of lost oil as well as income lost during the recovery period from activities such as fishing and tourist facilities (e.g. hotels, restaurants, etc.) in the affected area. The authors noted that even though some of the socio-economic losses are straightforward to quantify, this is a very difficult subject and one that will likely never become an exact science.

4.3.3 Estimating Environmental Damages

Environmental damages are especially difficult to quantify as they involve mostly intangible (i.e. non-market) quantities. However, several approaches have been developed that attempt to estimate the economic value of non-market impacts. The estimates are calculated either by indirectly linking environmental resources to market goods or by estimating a value for the willingness to pay by assuming a hypothetical market where the public is asked to pay for these resources. A brief reference will be made below to indicative examples of approaches that have been used for estimating the environmental damages from an oil spill incident.

Contingent Valuation (CV) is a non-market valuation method that is widely used by environmental economists to value environmental goods and services, where the calculated values are contingent upon a specific scenario (Kontovas and Psaraftis 2008). According to Carson et al. (2003), this method is based on a survey approach "*designed to create the missing market for public goods by determining what people would be willing to pay (WTP) for specified changes in the quantity or quality of such goods or, more rarely, what they would be willing to accept (WTA) in compensation for well-specified degradations in the provision of these goods*". The CV methodology has been used to estimate the impacts of various oil spills from tankers such as the Exxon Valdez (Carson et al. 2003) and the Prestige (Loureiro et al. 2007).

Another commonly employed method, especially by the National Oceanic and Atmospheric Administration (NOAA), is the Habitat Equivalent Analysis (HEA). According to Dunford et al. (2004), the objective of the HEA method is to estimate the ecological value of lost resource services by estimating how many resources, or habitats, would have to be provided for compensation without assigning specific monetary values. In a critical review of this methodology, Desvousges et al. (2018) note that HEA continues to be used as a negotiation tool in oil spill cases and is rather successful in cases where the oil has affected a limited geographical area. In the same review, the authors have identified the complex theoretical framework and multiple economic and ecological assumptions as the main limitation of the HEA method.

4.3.4 Models That Estimate the Total Cost

The following approaches consider the total oil spill cost as a non-linear function of oil spill size based on statistical regression analyses. Liu and Wirtz (2009) have employed a simulation-based approach to estimate total costs and identify relationships between key parameters. This approach involved constructing a series of scenarios by using oil spill fate modeling techniques and subsequently calculating the total costs for each scenario based on a spill cost model that was developed. The cost model included the following parameters: a time-dependent recovery function, a monetary value per unit resource and year, a yearly revenue for the economic sectors of interest, and a price of using a response facility per hour. Figure 5.7

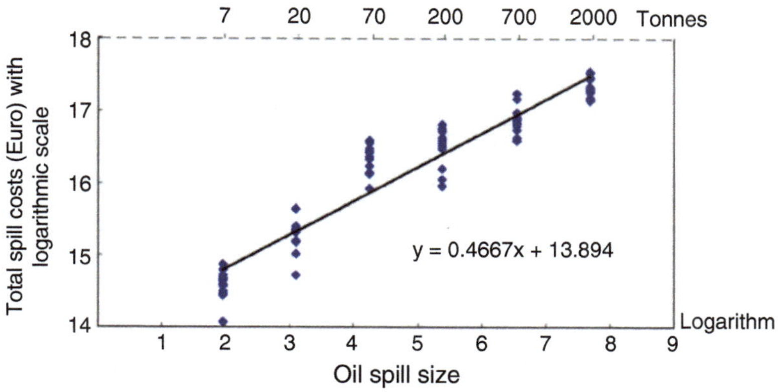

Fig. 5.7 Total spill costs (in Euros) as a function of oil spill size (log-log scale). (Source: Liu and Wirtz 2009)

shows the derived log-linear relationship between total spill costs and oil spill size ($R^2 = 0.9036$).

Yamada (2009) estimated the total cost as a function of spill size through a regression analysis, based on IOPCF data (101 spill cases during the period 1979–2005), excluding spills less than 1 tonne. It should be noted that this study was the basis for the submission to the MEPC from Japan and the basis for the decision of the MEPC to recommend a volume-based approach for determining Environmental Risk Evaluation Criteria in the context of the FSA (see the section on Environmental Risk Evaluation Criteria). The non-linear regression formula (5.5) was derived from the analysis, where W is the weight of the spilt oil:

$$\text{Total cost} = 35,951\,W^{0.68} \quad (5.5)$$

Psarros et al. (2011) performed a statistical regression analysis on a combined dataset consisting of IOPCF data and the data from the accident database developed in the context of the EC-funded research project SAFECO II. The relationship that was derived from the total unit cost and the oil spill size is expressed by formula (5.6) in USD/tonne:

$$\text{Total unit cost} = \frac{61,155}{W^{0.3528}} \quad (5.6)$$

The data points and the results from the log-log regression analysis for the total cost and the unit total cost are shown in Fig. 5.8.

Kontovas et al. (2010) conducted a non-linear regression analysis of total cost as a function of oil spill volume based on IOPCF data. At a later point, the authors updated their analysis (Kontovas et al. 2012), based on a consolidated oil spill

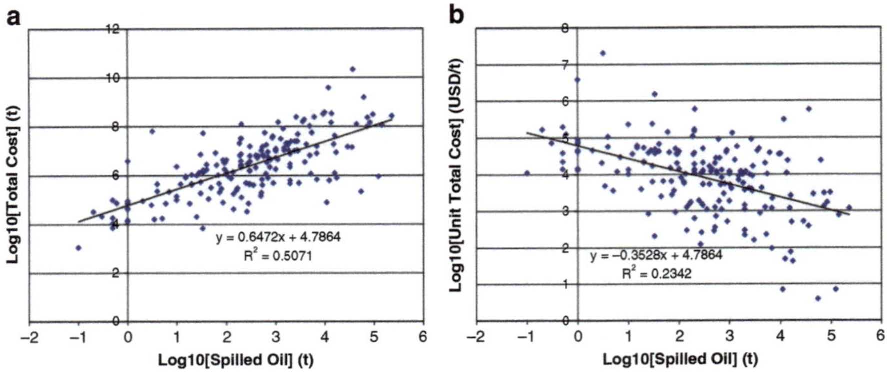

Fig. 5.8 Total costs and unit total cost as a function of oil spill size (log-log scale). (Source: Psarros et al. 2011)

database that integrated IOPCF, USA, and Norway data (see section on the history of the discussion at the IMO on Environmental Risk Evaluation Criteria), which resulted in the following non-linear regression formula (5.7):

$$\text{Total cost (2009 USD)} = 68,779V^{0.593} \tag{5.7}$$

The authors also conducted various regression analyses on subsets of the data that excluded small spills ($V < 0.1$ tonne) and spills before 1990 and concluded that the original formula they had derived overestimated the true total cost of oil spills. Figure 5.9 shows the comparison among the various regression formulas.

4.4 Limitations of the IOPCF Dataset

IOPCF compensation data is the most commonly used dataset in studies that attempt to correlate the total cost of an oil spill with spill volume. As is the case with any database, the IOPCF dataset has some limitations that relate to the geographical coverage of recorded cases, the type of costs, and the type of oil spill cases included. It should be noted that statistical analyses on biased datasets make generalization an ambiguous process that may very well result in misleading conclusions. The following is a brief overview of the limitations found in the IOPCF data.

Geographical Coverage The IOPCF currently has 115 member states of the 1992 Fund and 31 member states of the Supplementary Fund, but the USA remains an Observer State of the 1992 Fund (IOPCF 2017). As a result, major oil spill cases such as the Exxon Valdez are not included in the analysis. Furthermore, the dataset has a bias toward small, expensive claims caused by mishandling of oil supply that

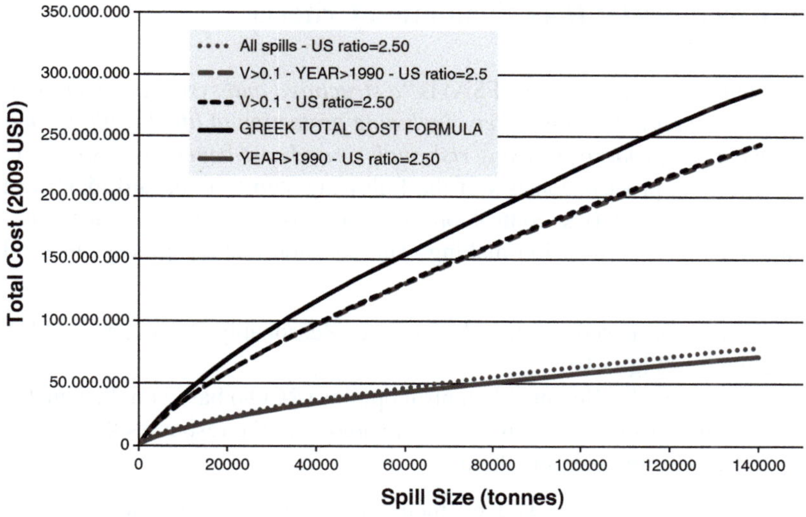

Fig. 5.9 Comparison of various regressions of the total cost (2009 USD) as a function of oil spill size. (Source: Kontovas et al. 2012)

have occurred in Japan, which is the major contributor of the IOPCF. Therefore, extrapolation to large spills based on the IOPCF data should be treated with caution.

Type of Costs The costs reported to the public from the IOPCF are not "real" oil spill costs, but the amount that was ultimately agreed to be paid to claimants for compensation. Therefore, Kontovas et al. (2011) have raised the question whether these figures may be used either as a reasonable approximation or a realistic "surrogate" for actual oil spill costs. These costs are difficult to assess as the amount of compensation depends heavily upon political and indemnity expenditure, as well as the public perception that is reflected by the media coverage of a specific oil spill case (Psarros et al. 2011). Furthermore, the IOPCF data have an upper bound, since admissible claims cannot be paid in full especially in the case of large oil spills. The amount of compensation paid is limited by the 1992 Civil Liability Convention (CLC) and the 1992 Fund. The implication of these limitations is a discrepancy among damage estimates from economic valuation methodologies, claims for compensation, and the amount of compensation eventually paid to claimants (Thébaud et al. 2005). Another major issue with the IOPCF dataset is that it does not contain environmental damage costs, because admissible claims are restricted to quantifiable costs for reasonable measures to reinstate the contaminated environment (IOPCF 2017). In fact, Helton and Penn (1999) have found that fewer than 1% of admissible claims even included Natural Resource Damage (NRD).

Type of Oil Spill Cases The IOPCF only provides compensation for damages resulting from spills of persistent oil from tankers. Therefore, operational spills from other ship types are not included in the dataset.

5 Environmental Risk Evaluation Criteria

The Formal Safety Assessment (FSA) is "*a structured and systematic methodology, aimed at enhancing maritime safety, including protection of life, health, the marine environment and property, by using risk analysis and cost-benefit assessment*" (IMO 2015). The FSA process consists of the following steps: (1) hazard identification, (2) risk analysis, (3) risk control options, (4) cost-benefit assessment, and (5) recommendations for decision-making. The decision-making process in the FSA is driven by:

- Risk evaluation based on the "As Low As Reasonably Practicable" (ALARP) principle (see Fig. 5.10)
- The evaluation of different risk control options (RCOs) based on the benefit they provide, in terms of risk reduction, compared with the resources that need to be committed to implementing them

Therefore, RCOs are ranked by their "practicability" expressed in monetary terms. The aim is to reduce the quantified risk to an acceptably low level by applying the technical and/or operational solutions that achieve the most risk reduction with the expenditure of reasonable amounts of resources (i.e. the RCOs that are most cost-effective). At present, there are no universally accepted risk evaluation criteria, either for individual risk or societal risk (i.e. the boundaries of the ALARP region) as the FSA Guidelines provide only recommendations on this matter.

The FSAs that have been submitted to the IMO have quantified the consequences to human life (i.e. fatalities and injuries) and used the Cost of Averting a Fatality (CAF) approach that is proposed in the FSA guidelines for evaluating the cost-effectiveness of different RCOs. The CAF index is expressed in gross and net terms (GCAF and NCAF, respectively) and is defined by formulas (5.8) and (5.9):

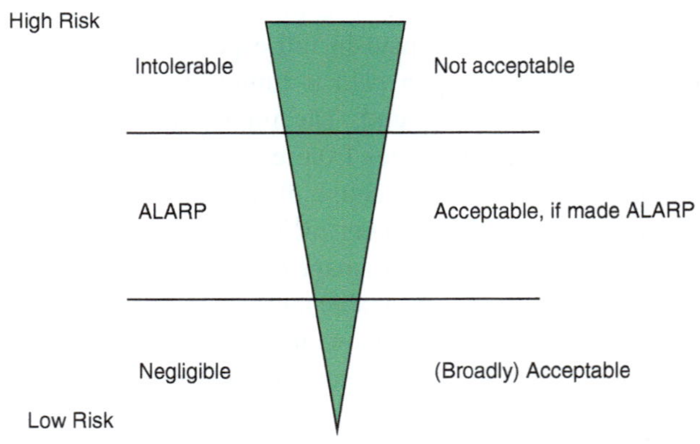

Fig. 5.10 The ALARP Principle for risk evaluation. (Source: Revised FSA Guidelines IMO 2015)

$$\mathrm{GCAF} = \frac{\Delta C}{\Delta R} \tag{5.8}$$

$$\mathrm{NCAF} = \frac{\Delta C - \Delta B}{\Delta R} \tag{5.9}$$

In (5.8) and (5.9), ΔC is the implementation cost of the RCO, ΔB is the potential economic benefit resulting from the implementation of the RCO, and ΔR is the expected risk reduction (in terms of the number of fatalities) after the implementation of the RCO. To determine cost-effectiveness in absolute terms, a limiting value is needed that will render a proposed RCO as acceptable or unacceptable. The commonly used limit for safety issues is ICAF = USD $3m$, where ICAF stands for Implied Cost of Averting a Fatality. Following this rationale, an RCO is considered cost-effective and a suitable candidate for recommendation if the following condition applies: GCAF *and* NCAF < USD $3m$.

Initially, the focus of the FSA as a risk management tool was, primarily, the safety of human life at sea and, secondarily, the safety of the ship itself (property). The accident consequences that had been quantified involved only possible fatalities (and indirectly injuries, based on the risk equivalence concept), while environmental considerations were discussed at the IMO at a secondary stage. Finally, these discussions led to amendments in the FSA Guidelines that incorporate basic recommendations on the use of Environmental Risk Evaluation Criteria in the cost-benefit step of the process. It should be noted that, at present, Environmental Risk Evaluation Criteria have been proposed only in relation to oil pollution without considering other environmental impacts from shipping such as emissions, ballast water, fouling, etc.

The major issue that is needed to be resolved to reach a consensus on this matter was the appropriate way to express the total cost of an oil spill as a function of oil spill size (see the section on Estimating the total cost of oil pollution for an overview of the relevant international literature). An additional consideration and an important difference between global risk evaluation criteria for safety and the risk of oil pollution from shipping is that while the risk of the loss of human life is independent of the geographical region where the loss occurred, this is not the case for oil pollution. The risk of a tonne of oil spilt in an ecologically sensitive region is obviously far greater than in other less sensitive regions.

5.1 History of the Discussion at the IMO

The discussion on Environmental Risk Evaluation Criteria at the IMO commenced with the deliberations of the MEPC 55 in 2006, where a major topic was how to define risk evaluation criteria for accidental oil releases to the marine environment (IMO 2006a). The motivation for this discussion was the definition of the Cost of Averting a Tonne of Spilt Oil (CATS) criterion in a report of the EU project SAFE-DOR (Skjong et al. 2005), as an environmental equivalent to the CAF criterion. At

MEPC 56 in 2007, the delegation from Greece (IMO 2007a) highlighted several issues that relate to the suitability of the CATS criterion for environmental risk evaluation. The issues concerned the specific threshold value of USD 60,000/tonne that was proposed for the CATS criterion – the proposed value was updated at a later point to USD 80,000/tonne by Psarros et al. (2011) – as well as the broader issue of whether any single, constant monetary value per tonne of spilt oil is appropriate for evaluating environmental risk from accidental oil spills. The concern was that any constant value per tonne would not be a representative statistical metric due to the large variability of average total costs. Indicatively, Etkin (2000) calculated the average clean-up costs per country, given in 1999 USD/tonne, for various spill cases that range from 6.09 in Mozambique up to 25,614 in the USA and an extreme value of 76,589 in Malaysia. The same variability is evident in average clean-up costs as a function of spill volume, as the approximately 40,000-tonne oil spill from the Exxon Valdez resulted in USD 107,000/tonne (USD 2007 value), while the 85,000-tonne oil spill from the Braer amount to only USD 6/tonne.

In a critical review of the state of the art on estimating the disutility cost of oil spills, Kontovas and Psaraftis (2008) identified limitations of the approach that was used to derive the CATS criterion. These are related to the estimation of the global average for total oil spill costs and the specific values that were assumed for the assurance factor.

Estimation of Total Oil Spill Costs The implicit assumption behind the estimation of the global weighted average for clean-up costs is that regions with greater oil traffic density are more potentially at risk to experience oil spills and therefore increased clean-up costs. However, oil spill costs are a complex function of interdependent factors such as different environmental conditions, either physical or regulatory, and different technologies (ships, traffic control schemes, etc.). Considering that most of the oil spilt in the marine environment comes from a small number of accidents, oil traffic may not be positively correlated with either oil spill volume or clean-up costs. According to Devanney and Stewart (1974), finding an appropriate exposure variable for the distribution of the number and volume of oil spills is not a trivial matter, reflecting on the poor statistical correlation between oil spill volume and regional traffic. In addition, it is not clear why the authors decided against a direct computation of the global average clean-up cost that would be derived by dividing the global clean-up cost by the global sum of tonnes of oil spilt, which would result in fairly different results and, according to Kontovas and Psaraftis (2008), most probably in lower costs. Regarding the other components of the total cost (i.e. environmental damages and socio-economic losses), the ratio of 1.5 relative to clean-up costs also seems to be poorly supported.

Assurance Factor The assurance parameter is intended to reflect the society's or the maritime policy-makers' willingness to pay to avert oil pollution from shipping. The authors assume that it would be rational for this parameter to range from 1 to 3. This assumption was based on a reverse engineering approach that assumes that previous legislative action to prevent pollution had been correct. Even though the authors state that the exact values are indicative and should be assessed by the

FSA analyst, the concept of an upper bound does not seem rational, as society or maritime policy-makers could very well decide to spend even larger amounts to prevent an oil spill. In addition, the concept of a lower bound equal to 1 is also debatable, as society could, in theory, decide to pay for mitigation instead of prevention. Furthermore, another fact that complicates the calculation of the exact value of the assurance factor is that the stakeholders who are typically burdened with clean-up and environmental costs are different than those who will pay for preventive measures.

At MEPC 57 in 2007, it was agreed that a Correspondence Group (CG) would be established under the coordination of Greece (the chair of the CG was assigned to Prof. H. Psaraftis) with the aim to review the draft Environmental Risk Acceptance Criteria in the FSA. The CG acknowledged that, apart from spill size, there are also other significant factors that determine the severity of an oil spill accident such as oil type, location, weather conditions, and the characteristics of the shoreline. However, it was decided that the volume of oil spilt would be used as the determining variable for estimating relevant costs (IMO 2007b, 2008a, 2009). The deliberations of this CG lasted 4 years, and the results were discussed at MEPC 60 in 2010. The conclusion was that oil spill cost estimates should be based on a non-linear function of spill volume. Greece proposed a specific function for further analysis (IMO 2010), which is detailed in the work by Kontovas et al. (2010). The regression analysis was based on IOPCF data and was updated at a later publication (Kontovas et al. 2012). This function was preferred over the proposals submitted to the IMO by Japan (Yamada 2009) and Norway (Psarros et al. 2011) as it produced a higher cost for the same spill volume, and therefore it was considered a more conservative approximation of the actual total cost. In 2011 a research group under the initiative of Germanischer Lloyd with members from Japan, the USA, Greece, and Germany performed a different set of regression analyses to be submitted at MEPC 62. The new regressions were derived from a consolidated oil spill database that included updated IOPCF data, as well as data from the USA and Norway. The results (see Table 5.2) were communicated to the IMO by several joint submissions to MEPC 62 by Germany, Japan, and the USA and an independent submission by Greece (IMO 2011a, b, c, d, e, g).

Compared to the original analysis by Greece (that was solely based on IOPCF data), the results of the new regressions produce lower total costs even though the consolidated dataset contained expensive spills that had occurred in the USA and Norway. Following the deliberations of a working group (IMO 2011f), MEPC 62

Table 5.2 Non-linear total oil spill cost functions, based on the consolidated oil spill database

Oil spill category (V in tonnes)	Total oil spill cost (2009 USD)	Reference
All spills	$67{,}275V^{0.5893}$	MEPC 62/INF.24
Spills larger than 0.1 tonnes	$42{,}301V^{0.7233}$	MEPC 62/18

Source: IMO (2015)

endorsed the consolidated dataset and the non-linear functions that were derived from the regression analyses. The decisions made at MEPC 62 were that:

- Other functions could be used in an FSA if they are supported by the data.
- The consolidated oil spill database would be made available publicly.
- The main recommendations that resulted from the discussion on this topic would be included in a future amendment to the FSA guidelines.

5.2 An Alternative Approach

Psaraftis (2008) proposed an alternative approach to integrate Environmental Risk Evaluation Criteria for evaluating RCOs that deal with oil pollution from shipping in the cost-benefit stage of the FSA process. In this framework, the differential of the expected annual total cost of oil spills before and after the implementation of a proposed RCO is compared with the expected cost of the RCO, which may be expressed by formula (5.10):

$$\Delta K < \Delta E\,(\mathrm{TOT}) = E\,(\mathrm{TOT}) - E_{\mathrm{RCO}}\,(\mathrm{TOT}) \tag{5.10}$$

In (5.10), ΔK is the total cost for implementing the RCO, E(TOT) is the expected annual total cost of all spills globally before the implementation of the proposed RCO, and E_{RCO}(TOT) is the corresponding expected annual cost after the implementation. The term "expected" is used in its probabilistic sense as the framework is based on the concept that the oil spill generation process is determined by several independent random processes. The random processes represent factors such as the time or the region the spill occurred, the volume and type of oil spilt, or the prevailing weather conditions. The effect of an RCO for reducing the total oil spill cost may either be on the oil spill frequency or the probability density function (PDF) of the oil spill volume distribution.

Psaraftis breaks down the total cost into the following two components: total damage cost and total clean-up cost. Damage cost includes the economic consequences (i.e. shipowner, cargo owner, fisheries, tourism, and other affected industries) and quantifiable environmental damages. Clean-up cost includes the cost of the response and remediation operations at sea and/or at the shore, which depends on the response level and the employed strategy. Generally, these cost categories are assumed to be non-linear functions of the volume of the oil spilt that may be derived from appropriate data (e.g. ITOPF, IOPCF, etc.).

After calculating the annually expected cost differential, an RCO may be considered cost-effective if the following applies: $\Delta K < \Delta E$(TOT). The author also notes that the RCO with the greatest positive difference $\{\Delta K - \Delta E(\mathrm{TOT})\}$ is preferable among alternative RCOs that are cost-effective. Essentially, the differential ΔE(TOT) reflects the expected benefit in monetary terms from the implementation of an RCO. Therefore, the difference of this framework to the CATS criterion (and in fact any criterion based on ratios) is that RCOs are compared based on their net benefit (expressed by differential costs) rather than on their cost-effectiveness (expressed by a ratio). It should be noted that the main problem with ratios is that they ignore scale, which is not irrelevant when thinking in monetary terms. However, the two approaches are not unrelated, as Psaraftis demonstrated that his framework may be considered as a generalization of the CATS approach even though it is based on different calculations. The successful implementation of this approach depends on the availability of appropriate datasets, which has been identified by Psarros et al. (2011) as a limitation that would make the approach impractical. They also argued that the approach might be difficult to communicate to a wide audience.

The approach may be readily integrated with the existing CAF risk evaluation criteria for safety to consider RCOs that aim to reduce both environmental risk from pollution and the risk of fatalities. Psaraftis recommends formula (5.11):

$$\Delta K < \Delta E\,(\mathrm{TOT}) + \mathrm{VHL} \cdot \Delta R \tag{5.11}$$

In (5.11), VHL is the value of human life (currently estimated by the IMO at USD 3 million per person), and ΔR is the expected reduction of fatalities from the implementation of the proposed RCO. The rationale for evaluating alternative RCOs and the concept of using differentials rather than ratios remain the same. Finally, the approach may easily be extended to include environmental consequences other than oil pollution (e.g. emissions, ballast water, fouling, etc.).

5.3 FSA Guidelines Status

The latest revision of the FSA Guidelines (IMO 2015) recommends the use of the updated regression formulas that were derived from the consolidated oil spill database. Figure 5.11 shows all the data points that are available in the consolidated database, classified according to their original source (i.e. IOPCF, US, and Norway data). Considering the results of the MEPC 62, the IMO decided not to recommend a constant value as an environmental risk evaluation criterion, such as CATS and its proposed value of at least USD 60,000/tonne.

However, even though the CATS criterion was not officially adopted by the MEPC, it has been employed in the following instances: a study for evaluating ECDIS and Electronic Navigation Chart (ENC) for navigational safety (Vanem et al. 2007, 2008a), which contributed toward establishing the IMO requirement for ECDIS; an FSA on crude oil tankers submitted to the IMO by Denmark (IMO 2008b); and structural reliability studies for hull girder safety (IMO 2006b; Hørte et al. 2007).

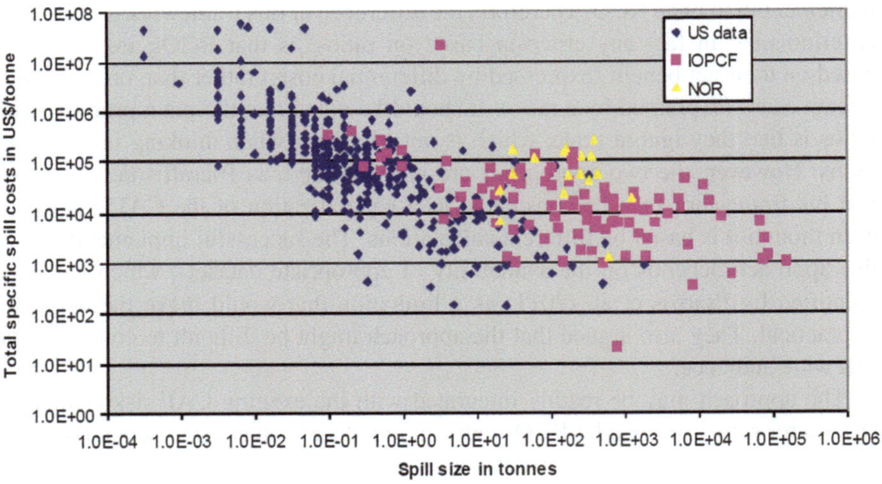

Fig. 5.11 Scatter plot of the data points in the consolidated oil spill database in 2009 USD. (Source: Revised FSA Guidelines IMO 2015)

The guidelines note that the FSA analyst may use a different regression formula based on new, properly documented, oil spill data for determining the total oil spill cost as a function of the volume of oil spilt. The FSA analyst is also free to conduct a regression analysis for determining this function by covering a percentile different than 50%, provided that this choice is well documented. Based on the specific regression formula that will be used, the societal oil spill costs may be estimated by formula (5.12):

$$SC(V) = F_{\text{assurance}} \cdot F_{\text{uncertainty}} \cdot f(V) \tag{5.12}$$

In (5.12), $F_{\text{assurance}}$ is the assurance factor that reflects the society's willingness to pay to avert accidents, $F_{\text{uncertainty}}$ is an uncertainty factor for considering the uncertainties in the available oil spill cost data, and $f(V)$ is the volume-dependent total cost function. The guidelines do not provide specific values for either the assurance factor or the uncertainty factor, but stipulate that if values different than 1.0 are used, then a cost-effectiveness analysis with $F_{\text{assurance}} = 1.0$ and $F_{\text{uncertainty}} = 1.0$ should also be included as a reference point.

Regarding the issue of environmental risk evaluation criteria, the latest FSA Guidelines stipulate that the FSA analyst should conduct a cost-benefit and cost-effectiveness evaluation of the proposed RCOs. For RCOs that affect only oil spills, an RCO is deemed cost-effective if the following condition (5.13) applies:

$$\Delta C < \Delta SC = \text{Expected benefit of the RCO} \tag{5.13}$$

The expected benefit is calculated as the difference between the expected societal cost without and with the implementation of the RCO. It should be noted that the FSA guidelines recommend formula (5.14) for RCOs that affect safety and the environment:

$$\text{NCAF} = \frac{\Delta C - \Delta \text{SC}}{\Delta \text{PLL}} \tag{5.14}$$

In (5.14), ΔC and ΔSC are defined as in the previous formula, and ΔPLL is the expected reduction of fatalities due to the implementation of the RCO.

5.3.1 Open Issues

In an updated review of the FSA, Psaraftis (2012) identified several limitations of the recommended approach in the IMO guidelines that relate to the latitude that the FSA analyst is given in conducting an environmental FSA. This directly impacts the methodological soundness of the FSA, as different solutions would be provided by different teams working on the same problems. Consequently, the results of different FSA studies may be rendered useless in a regulatory setting because the followed approach and the underlying assumptions may be significantly different. The following is a brief description of these open issues.

Assurance Factor The exact value for this factor is left to the FSA analyst, provided that proper documentation is included, as there has not been an agreement at the IMO level. However, some delegations have suggested that this value should only be determined by policy-makers. Even though some delegations support a value well over 1.0, Psaraftis argues that this rationale is not supported by evidence. In addition, the author questions the fact while an assurance factor has not been included in safety-related risk evaluation criteria, this concept should be employed for environmental criteria.

Uncertainty Factor The concept of the uncertainty factor is new and is intended to deal with the inherent uncertainties in the available oil spill cost data. For example, a 1.5 uncertainty factor implies that the real costs are 50% higher compared to the available recorded costs. However, according to Psaraftis, the exact value for such a quantity may not be computed with a reasonable degree of confidence and may even be less than 1.0 because spill claims are typically inflated.

Regression Analysis at a Level Different from 50% This concept is also new and provides the FSA analyst with the possibility to conduct a methodologically non-typical regression by choosing a regression line over the 50% level, contingent on proper justification. In the context of oil spill costs, such regressions would result in higher values and ultimately different results concerning the evaluation of proposed RCOs based on their cost-effectiveness.

Use of Different Cost Functions, Based on Different Data The possibility to use different oil spill cost data leaves the process open to manipulation, while it does not preclude the use of a constant value environmental risk evaluation criterion.

6 Sustainable Maritime Transport of Oil

Although the terms "sustainability" and "sustainable development" are often used interchangeably, a distinction should be made between the two concepts. According to Litman (2011), sustainability is "*a condition in which economic, social and environmental factors are optimized, taking into account indirect and long-term impacts*". Sustainable development, on the other hand, may be considered as the process by which we may achieve the goals of sustainability. These goals outline certain characteristics that a system should have to be sustainable that relate to how resources are used and to the extent of environmental damage that results from the functioning of the system. A sustainable system should follow the following long-term rules (Ornitz and Champ 2002):

- Consumption of renewable resources should not exceed their production.
- Consumption of non-renewable resources should not exceed the production of a renewable substitute, which, according to the authors, in the case of oil production could be achieved by investing part of the income in developing alternative energy sources.
- Emissions and pollution to the environment should not exceed the natural ability of the ecosystem to recirculate, absorb, or render harmless.

Jeon (2007) presented the goals in each dimension of the concept of sustainability (environmental sustainability, sociocultural sustainability, and economic sustainability) that a transport system should fulfill to be considered sustainable (Fig. 5.12). Chatzinikolaou and Ventikos (2011) identified that most of the existing definitions of sustainable transport systems answer the question of "what" but not the question of "how" such a system may become sustainable. The authors attempted to provide an operational definition of maritime transport that would conceptually integrate the principles of sustainable development (i.e. equal opportunities for future generations and continued development) and provide a roadmap for measuring sustainability. The definition they provided was that *maritime transport is sustainable when it has the capability to maintain non-declining and efficient accessibility in time*. The term "efficient accessibility" in the context of a sustainable maritime transport system would mean less mobility with greater accessibility.

In recognition of the importance of sustainability for the future of shipping, the IMO defined the concept of a Sustainable Maritime Transportation System as the *safe, secure, efficient and reliable transport of goods across the world, while minimising pollution, maximising energy efficiency and ensuring resource conservation* (IMO 2013). The concept focuses on the following key areas with specific goals and actions:

Fig. 5.12 Essential elements of a sustainable transportation system. (Source: Jeon 2007)

- Safety culture and environmental stewardship
- Education and training in maritime professions and support for seafarers
- Energy efficiency and ship-port interfaces
- Energy supply for ships
- Maritime traffic support and advisory systems
- Maritime security
- Technical co-operation
- New technology and innovation
- Finance, liability and insurance mechanisms
- Ocean governance

Andersson et al. (2016) argue that if shipping is to become sustainable and fulfill the vision of zero harmful emissions, several challenges that relate to the broad categories of environmental awareness, regulations, and enforcement, and technical solutions should be addressed.

The maritime transport of oil is essential for an oil-based society, and, according to Ornitz and Champ (2002), the future viability of the industry depends on whether this activity will become sustainable. The authors argue that good environmental stewardship will become a reality if costly accidents are prevented and if environmental impact after an oil spill is minimized by activating optimized response strategies. The key issues for achieving proactiveness are for the industry to adopt a safety culture, as opposed to the current culture of compliance and the culture of "cutting corners" for cost-effectiveness (i.e. a reactive approach), and to provide quality training to qualified mariners. The development of a safety culture implies that the industry will outgrow simple regulatory compliance into self-regulation and self-improvement. In addition, the authors identify the following policy considerations regarding the optimization of response strategies:

- Planning long-term contingency/vessel response
- Optimizing the use of technology and science
- Considering clean-up and restoration operations as a unified process

The proactive approach will facilitate the minimization of the social and environmental impacts that result from oil pollution incidents and accidents, which are two of the three pillars of sustainability. However, considering the economic pillar of sustainability, the industry should aim to minimize the total cost of oil spills by adopting cost-effective RCOs based on scientifically sound environmental risk evaluation criteria. Cost-effectiveness is an issue that cannot be overlooked, as sustainability is also linked with reasonable profitability, and therefore for RCOs to be successfully implemented, the benefits must outweigh the relevant costs. Finally, the concept of "efficient accessibility" implies that the maritime transport of oil will become sustainable by maintaining a non-declining level of efficiency both in terms of economic and environmental performance.

References

Afenyo, M., Veitch, B., & Khan, F. (2016). A state-of-the-art review of fate and transport of oil spills in open and ice-covered water. *Ocean Engineering, 119*, 233–248.

Andersson, K. et al. (2016). Shipping and the environment. In *Shipping and the environment: Improving environmental performance in marine transportation* (pp. 3–27).

Anyanova, E. (2012). Oil pollution and international marine environmental law. In *Sustainable development – Authoritative and leading edge content for environmental management* (pp. 29–54).

Brundtland, G. H. (1987). *Our common future: Report of the world commission on environment and development, United Nations Commission.*

Burgherr, P. (2007). In-depth analysis of accidental oil spills from tankers in the context of global spill trends from all sources. *Journal of Hazardous Materials, 140*(1–2), 245–256.

Butricks, T. (2014). *Progress since the Exxon Valdez oil spill?, Cornell University, ROPER Center for Public Opinion Research.* Available at: https://ropercenter.cornell.edu/progress-since-the-exxon-valdez-oil-spill/

Carson, R. T., et al. (2003). Contingent valuation and lost passive use: Damages from the Exxon Valdez oil spill. *Environmental and Resource Economics, 25*, 257–286.

Chatzinikolaou, S. D., & Ventikos, N. P. (2011). Sustainable maritime transport: An operational definition. In *Sustainable maritime transportation and exploitation of sea resources (IMAM 2011)* (pp. 931–939).

CMA-CGM. (2017). *Preserving oceans and their biodiversity*. Available at: https://www.cma-cgm.com/the-group/corporate-social-responsibility/environment/sea-biodiversity-protection

Desvousges, W. H., et al. (2018). Habitat and resource equivalency analysis: A critical assessment. *Ecological Economics, 143*, 74–89.

Devanney, J. W., & Stewart, J. R. (1974). A bayesian approach to oil spill statistics. In *New England Section Meeting of SNAME*. p. 85.

Dunford, R. W., & Freeman, M. L. (2001). A statistical model for estimating natural resource damages from oil spills. In *International Oil Spill Conference* (pp. 225–229).

Dunford, R. W., Ginn, T. C., & Desvousges, W. H. (2004). The use of habitat equivalency analysis in natural resource damage assessments. *Ecological Economics, 48*(1), 49–70.

EMSA. (2012). *EU states claims management guidelines: Claims arising due to maritime pollution incidents.*

Etkin, D. S. (1999a). Estimating cleanup costs for oil spills. *International Oil Spill Conference Proceedings, 1999*(1), 35–39.

Etkin, D. S. (1999b). Estimating cleanup costs for oil spills. *International Oil Spill Conference Proceedings, 1999*(1), 35–39.

Etkin, D. S. (2000). Worldwide analysis of marine oil spill cleanup cost factors, *Arctic and Marine Oil spill Program Technical Seminar*, (June), p. 14.

Etkin, D. S. (2001a). Comparative methodologies for estimating on-water response costs for marine oil spills. *International Oil Spill Conference Proceedings, 2001*, 1281–1289.

Etkin, D. S. (2001b) Methodologies for estimating shoreline cleanup costs. In *Proceedings of 24th Arctic and Marine Oil spill Program Technical Seminar*, pp. 647–670.

Gold, E. (1998). *Handbook on marine pollution* (2nd ed.). Arendal: Assuranceforeningen Gard.

Harper, J., Godon, A., & Allen, A. A. (1995). Costs associated with the cleanup of marine oil spills. In *International Oil Spill Conference*, 27–30.

Helton, D., & Penn, T. (1999). Putting response and natural resource damage costs in perspective. In *1999 International Oil Spill Conference*, 1–22.

Hendrickx, R. (2007). Maritime oil pollution: An empirical analysis. In *Shifts in compensation for environmental damage* (pp. 243–260).

Hørte, T. et al. (2007). Probabilistic methods applied to structural design and rule development. In *RINA conference on Developments in Classification and International Regulations*. London, UK.

IMO. (2006a). *MEPC 55/18, Formal Safety Assessment, Outcome of MSC 81, Note by the Secretariat.*

IMO. (2006b). *MSC 81/INF.6, Goal-based new ship construction standards-linkage between FSA and GBS, Submitted by the International Association of Classification Societies (IACS).*

IMO (2007a). *MEPC 56/18/1, Formal Safety Assessment, Environmental risk evaluation criteria, Submitted by Greece.*

IMO. (2007b). *MEPC 57/17, Formal Safety Assessment, Report of the Correspondence Group on Environmental Risk Evaluation Criteria, Submitted by Greece.*

IMO. (2008a). *MEPC 58/17, Formal Safety Assessment, Report of the correspondence group on environmental risk evaluation criteria, Submitted by Greece, Co-ordinator of the Correspondence Group.*

IMO. (2008b). *MEPC 58/INF.2, Formal Safety Assessment-FSA-crude oil tankers, Submitted by Denmark.*

IMO. (2009). *MEPC 59/17, Formal Safety Assessment, Report of the Correspondence Group on Environmental Risk Evaluation Criteria, Submitted by Greece on behalf of the Coordinator of the Correspondence Group.*

IMO. (2010). *MEPC 60/17/2, Formal Safety Assessment, Comments on the Correspondence Group report on Environmental Risk Evaluation Criteria, Submitted by Japan.*

IMO. (2011a). *MEPC 62/18, Combining environmental and safety criteria and selection of a severity matrix, Submitted by Greece.*

IMO. (2011b). *MEPC 62/18/1, Further experience with oil spill databases and upd ate of non-linear oil spill cost functions, Submitted by Greece.*

IMO. (2011c). *MEPC 62/18/2, Consideration on the Environmental Risk Evaluation Criteria, Submitted by Japan.*

IMO. (2011d). *MEPC 62/18/3, Consolidated dataset on oil spills and further progress made with regard to the development of a CATS value, Submitted by Germany, Japan and United States.*

IMO. (2011e). *MEPC 62/18/4, Consideration on the Cost of Averting a Tonne of Oil Spilt (CATS) threshold Function, Submitted by Germany, Japan and United States.*

IMO. (2011f). *MEPC 62/24, Report of the Marine Environment Protection Committee on its sixty-second session.*

IMO. (2011g). *MEPC 62/INF.24, Consolidated dataset on oil spills, Submitted by Germany, Japan and United States*.

IMO. (2013) *World Maritime Day-A concept of a sustainable maritime transportation system*.

IMO. (2015). *Revised Guidelines for Formal Safety Assessment (FSA) for use in the IMO rule-making process*. London, UK.

IMO. (2018a). *Marine environment-pollution prevention-oil pollution-background*. Available at: http://www.imo.org/en/OurWork/Environment/PollutionPrevention/OilPollution/Pages/Background.aspx.

IMO. (2018b). *Pollution preparedness and response*. Available at: http://www.imo.org/en/OurWork/Environment/PollutionResponse/Pages/Default.aspx.

IMO. (2018c). *Tanker safety – preventing accidental pollution*. Available at: http://www.imo.org/en/OurWork/Safety/Regulations/Pages/OilTankers.aspx.

IOPCF. (2017). *International oil pollution compensation funds: Annual Report 2017*.

ITOPF. (2018). Oil tanker spill statistics 2017, *The international tanker owners pollution federation limited*, (January), p. 16p.

Jeon, C. M. (2007). *Incorporating sustainability into transportation planning*. Georgia Institute of Technology, Atlanta.

Jeon, C. M., & Amekudzi, A. (2005). Addressing sustainability in transportation systems: Definitions, indicators, and metrics. *Journal of Infrastructure Systems, 11*(1), 31–50.

Kontovas, C. A., & Psaraftis, H. N. (2008). Marine environment risk assessment: A survey on the disutility cost of oil spills. *2nd International Symposium on Ship Operations*.

Kontovas, C. A., Psaraftis, H. N., & Ventikos, N. P. (2010). An empirical analysis of IOPCF oil spill cost data. *Marine Pollution Bulletin, 60*(9), 1455–1466.

Kontovas, C. A., Ventikos, N. P., & Psaraftis, H. N. (2011). Estimating the consequence costs of oil spills from tankers. In *SNAME 2011 Annual Meeting* (pp. 1–13).

Kontovas, C. A., Ventikos, N. P., & Psaraftis, H. N. (2012). An updated analysis of IOPCF oil spill data: Estimation of the disutility cost of tanker oil spills. *Sustainable maritime transportation and exploitation of sea resources – Proceedings of the 14th International Congress of the International Maritime Association of the Mediterranean, IMAM 2011*, 2.

Litman, T. (2011). Sustainability and livability-summary of definitions, goals, objectives and performance indicators, *Victoria Transport Policy Institute* (pp. 1–5).

Liu, X., & Wirtz, K. W. (2006). Total oil spill costs and compensations. *Maritime Policy & Management, 33*(1), 49–60.

Liu, X., & Wirtz, K. W. (2009). The economy of oil spills: Direct and indirect costs as a function of spill size. *Journal of Hazardous Materials, 171*(1–3), 471–477.

Loureiro, M. L., Loomis, J. B., & Vazquez, M. X. (2007). *Estimating the non-market environmental damages caused by the prestige oil spill*.

Officer of the Watch. (2013). *Potential costs of an offshore accident*. Available at: https://officerofthewatch.com/2013/07/16/potential-costs-of-an-offshore-accident/.

Olawuyi, D. S. (2012). Legal and sustainable development impacts of major oil spills. *Consilience: The Journal of Sustainable Development, 9*(May), 1–15.

Ornitz, B. E., & Champ, M. A. (2002). *Oil spills first principles, oil spills first principles*.

Özçayir, O. (2004). *Port state control* (2nd ed.). London: LLP.

Prendergast, D. P., & Gschwend, P. M. (2014). Assessing the performance and cost of oil spill remediation technologies. *Journal of Cleaner Production, 78*, 233–242.

Psaraftis, H. N. (2008). Environmental risk evaluation criteria. *WMU Journal of Maritime Affairs, 7*(2), 409–427.

Psaraftis, H. N. (2012). Formal safety assessment: An updated review. *Journal of Marine Science and Technology (Japan), 17*(3), 390–402.

Psarros, G., Skjong, R., & Vanem, E. (2011). Risk acceptance criterion for tanker oil spill risk reduction measures. *Marine Pollution Bulletin, 62*(1), 116–127.

Sanguri, M. (2017). *20 important factors for oily water separator operation on ships, Marine Insight*. Available at: https://www.marineinsight.com/tech/ows/20-important-factors-for-oily-water-separator-operation-on-ships/.

Shahriari, M., & Frost, A. (2008). Oil spill cleanup cost estimation-developing a mathematical model for marine environment. *Process Safety and Environmental Protection, 86*(3), 189–197.
Sirkar, J., et al. (1997). A framework for assessing the environmental performance of tankers in accidental groundings and collisions. *SNAME Transactions, 105*, 253–295.
Skjong, R., Vanem, E., & Endresen, Ø. (2005). *Risk evaluation criteria. SAFEDOR report D 4.2.*
Thébaud, O. et al. (2005). The cost of oil pollution at sea: an analysis of the process of damage valuation and compensation following oil spills. *Economic, social and environmental effects of the Prestige Oil Spill de Compostella, Santiago* (pp. 187–219).
Vanem, E. et al. (2007). Cost-effectiveness of preventing grounding with ECDIS. In *4th International conference on Collision and Grounding of Ships, ICCGS 2007*. Hamburg.
Vanem, E., et al. (2008a). Electronic chart display and information systems–navigational safety in maritime transportation. *European Journal of Navigation, 6*, 28–36.
Vanem, E., Endresen, Ø., & Skjong, R. (2008b). Cost-effectiveness criteria for marine oil spill preventive measures. *Reliability Engineering and System Safety, 93*(9), 1354–1368.
Ventikos, N. P., & Sotiropoulos, F. S. (2014). Disutility analysis of oil spills: Graphs and trends. *Marine Pollution Bulletin, 81*(1), 116–123.
Ventikos, N. P., Chatzinikolaou, S. D., & Zagoraios, G. (2009) The cost of oil spill response in Greece: Analysis & Results. In *13th congress of Intl. Maritime Assoc. of Mediterranean IMAM 2009* (pp. 12–15). Istanbul.
White, I. C., & Molloy, F. C. (2003). Factors that determine the cost of oil spills. *IOSC, 2003*, 1–15.
Yamada, Y. (2009). The cost of oil spills from tankers in relation to weight of spilt oil. *Marine Technology, 46*(4), 10.
Yip, T. L., Talley, W. K., & Jin, D. (2011). The effectiveness of double hulls in reducing vessel-accident oil spillage. *Marine Pollution Bulletin, 62*(11), 2427–2432.

Chapter 6
Ship Recycling

Nikos Mikelis

Abstract This chapter addresses the recycling of ships, otherwise known as dismantling, ship breaking, scrapping, and demolition. The size and age profile of the world fleet, the conditions that lead to ending the operating life of a ship, and the countries where the recycling of ships is concentrated are first examined. This is followed by an analysis of the economic drivers of ship recycling, which have resulted in the industry being dominated by five countries and also analyzes steelmaking as the main driving force for ship recycling. We then discuss the sale and purchase market for end-of-life ships, explain the roles of brokers and cash buyers, and provide a simplified inventory of the components that are recycled out of a ship. We outline the efforts to implement existing international legislation to ship recycling, and the development of the Hong Kong Convention, and provide a critical analysis of the development of regional legislation by the European Union. We finally discuss the combination of voluntary and legislative mechanisms that will secure the global implementation of minimum standards for safe and environmentally sound ship recycling.

Abbreviations

AFS	International Convention on the Control of Harmful Anti-fouling Systems on Ships
COP	Conference of the Parties
CSR	Corporate social responsibility
DASR	Document of Authorization to conduct Ship Recycling
EAF	Electric Arc Furnace
EC	European Commission
EU SRR	European Union Ship Recycling Regulation

N. Mikelis (✉)
Mikelis Consulting, Chiddingly, East Sussex, UK
e-mail: nikos@mikelis.com

H. N. Psaraftis (ed.), *Sustainable Shipping*,
https://doi.org/10.1007/978-3-030-04330-8_6

EU	European Union
GT	Gross tons
HBCDD	Brominated flame retardant
HKC	Hong Kong International Convention for the safe and environmentally sound recycling of ships, 2009
ICIHM	International Certificate on Inventory of Hazardous Materials
IGO	Inter-governmental Organization
IHM	Inventory of Hazardous Materials
ILO	International Labour Organization, or International Labour Office
IMO	International Maritime Organization
IRRC	International Ready for Recycling Certificate
LDT	Light displacement tonnage
MARPOL	The International Convention for the Prevention of Pollution from Ships
MEP	Member of European Parliament
MEPC	Marine Environment Protection Committee
MoA	Memorandum of Agreement
NGO	Non-governmental organization
OBC	Oxygen blown converter
OECD	Organisation for Economic Co-operation and Development
PFOS	Perfluorooctane sulfonic acid and its derivatives
SOC	Statement of Compliance
SOLAS	International Convention for the Safety of Life at Sea
SRFP	Ship Recycling Facility Plan
SRP	Ship Recycling Plan

1 Introduction

1.1 The World Fleet and Ship Recycling

At the end of December of 2017, the world fleet of ships in service of 100 gross tons (GT) and above comprised of 115,761 ships totaling 1,291,046,701 GT (IHS Maritime & Trade, World Fleet Statistics 2017, Table 20). The majority of ships in the world fleet are small vessels, mostly trading in domestic waters. In fact, 73% of the ships in the world fleet (84,708 ships) were less than 5000 GT. If on the other hand, we focus on the fleet above the size limit used in relevant international regulations, namely, 500 GT, at the end of 2017 there were 62,503 ships of 1,277,729,875 GT. Looking at the aging of the world fleet over 500 GT, we see that 21,817 ships of 135,325,025 GT were over 20 years old. In other words, 34.9% of the ships, but only 10.6% of the tonnage of the fleet over 500 GT was over 20 years old, reflecting a skewed size-age distribution, with smaller ships having a much larger average age than the large, ocean going ships. An additional examination of the 2017 data shows that 55.0% of ships between 500 GT and 5000 GT were 20 years old or older, whereas only 7.9% of ships larger than 20,000 GT were of that age.

In general, while small ships used in domestic or regional trading tend to have longer lives, larger ships tend to be sent for recycling at around 25 years of age (or a few years later when demand for ships is high, or earlier, in periods of low demand). Ships retain significant economic value at the end of their life, as their steel, nonferrous metals, and machinery and equipment are sold for recycling, reconditioning, and reuse. This value can often represent 10% or more of the newbuilding value, such a figure varying with the price trends in the newbuilding market and separately in the recycling market. In general ships reach their end of life when their secondhand sale value for further trading drops below their recycling value. This may happen due to a ship's deteriorating condition with increasing age that may necessitate uneconomic repairs, due to the demand for ships being lower than the available supply, due to specific regulatory requirements (as is the case with the forthcoming requirements for the retrofit of expensive ballast water treatment systems or the "phaseout" requirements for single-hulled tankers in the early 2000s), or very occasionally due to the introduction of innovative technology (transition from steam to diesel) or abrupt changes in trading patterns (as was the case with the recent widening of the Panama Canal which devalued Panamax-sized ships).

It may just be relevant to make the point here that, for as long as end-of-life ships have economic value, there is no alternative to recycling them. If on the other hand the liabilities of end-of-life ships were to grow and become larger than residual value, we would then most probably witness abandonment of ships or deliberate scuttling on a large scale.

1.2 Countries that Recycle Ships

Figure 6.1 depicts the annual tonnage (GT) of recycled ships analyzed by country of recycling, while Fig. 6.2 expresses the same data in terms of the countries' market share. The two figures, together with Table 6.1, underline five important facts of the ship recycling industry.

Firstly, the ship recycling business is seen to be particularly cyclic, providing the recycler with no guarantees of future employment and no guarantees of a smooth depreciation of investment. This is caused by the cyclic nature of the shipping industry's supply and demand imbalance and, importantly, by the fluctuations in the price of steel internationally. The second key fact of the industry is that, for the last 20 years, ship recycling yards in the five leading ship recycling countries (Bangladesh, China, India, Pakistan, and Turkey) have been recycling 97–98% of all the tonnage that is recycled in the world. Table 6.1, detailing the worldwide ship recycling activities in 2017, illustrates this point.

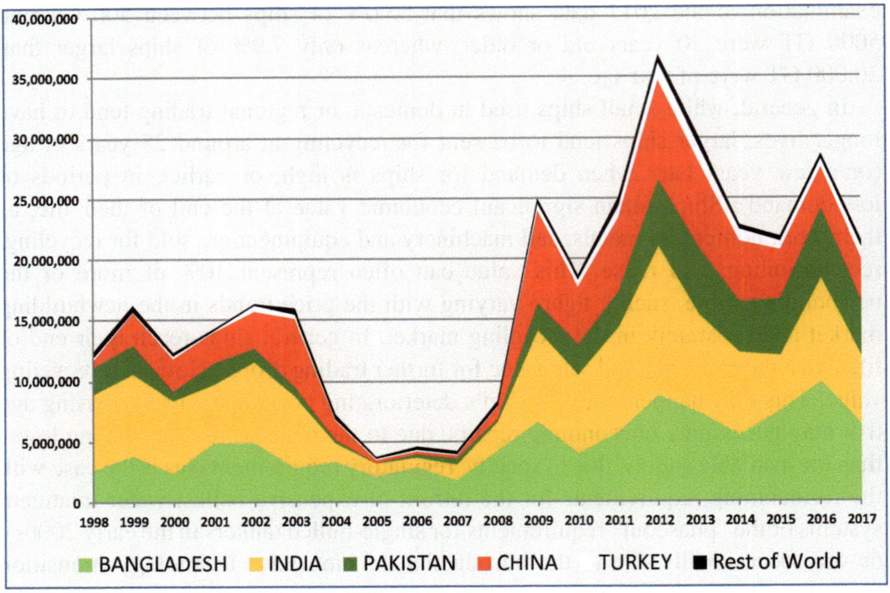

Fig. 6.1 Annual tonnage (GT) of recycled ships analyzed by country of recycling. (Adapted from IHS Global Ltd, World Casualty Statistics 2017, Table 7C; also, back issues of the same publication from 1998 to 2016)

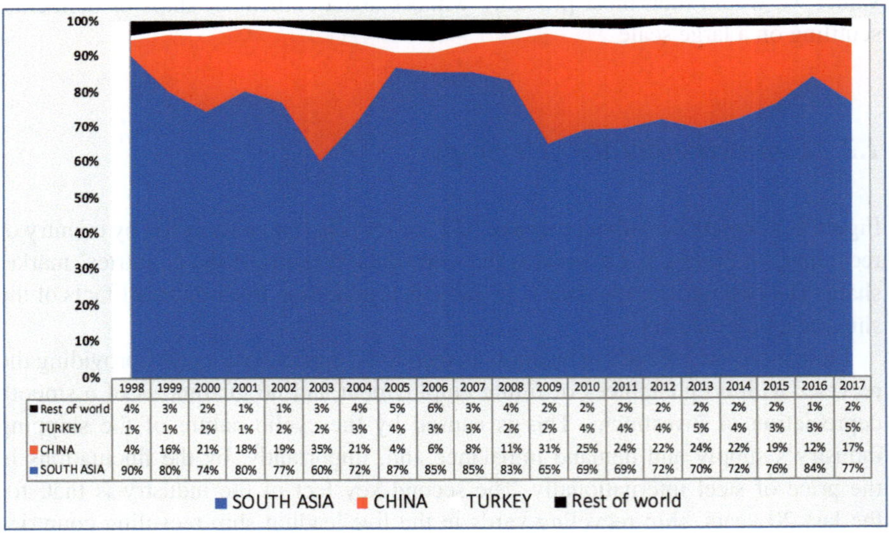

	1998	1999	2000	2001	2002	2003	2004	2005	2006	2007	2008	2009	2010	2011	2012	2013	2014	2015	2016	2017
Rest of world	4%	3%	2%	1%	1%	3%	4%	5%	6%	3%	4%	2%	2%	2%	2%	2%	2%	2%	1%	2%
TURKEY	1%	1%	2%	1%	2%	2%	3%	4%	3%	3%	2%	2%	4%	4%	4%	5%	4%	3%	3%	5%
CHINA	4%	16%	21%	18%	19%	35%	21%	4%	6%	8%	11%	31%	25%	24%	22%	24%	22%	19%	12%	17%
SOUTH ASIA	90%	80%	74%	80%	77%	60%	72%	87%	85%	85%	83%	65%	69%	69%	72%	70%	72%	76%	84%	77%

SOUTH ASIA CHINA TURKEY Rest of world

Fig. 6.2 Market share of ship recycling (expressed in % of GT). (Adapted from IHS Global Ltd, World Casualty Statistics 2017, Table 7C; also, back issues of the same publication from 1998 to 2016)

Table 6.1 Worldwide ship recycling activity in 2017

	Recycling country	No.	GT	Average GT	Average age	% of world tonnage	Cumulative % of tonnage	Cumulative % of ships
1	Bangladesh	197	6,361,485	32,292	26	29.7%	29.7%	19.3%
2	India	232	5,755,526	24,808	28	26.9%	56.6%	42.1%
3	Pakistan	119	4,302,798	36,158	26	20.1%	76.6°%	53.7%
4	China, People's Republic Of	167	3,573,932	21,401	25	16.7%	93.3°%	70.1%
5	Turkey	134	971,278	7,248	37	4.5%	97.9%	83.2%
6	Unknown	47	222,760	4,740	39	1.0%	98.9%	87.8%
7	United States of America	4	58,041	14,510	48	0.3%	99.2%	88.2%
8	Denmark	25	29,870	1,195	42	0.1%	99.3%	90.7%
9	Cuba	4	23,545	5,886	26	0.1%	99.4%	91.1%
10	Indonesia	13	22,436	1,726	30	0.1%	99.5%	92.4%
11	Canada	6	22,024	3,671	58	0.1%	99.6%	92.9%
12	Belgium	11	18,358	1,669	42	0.1%	99.7%	94.0%
13	Netherlands	6	10,711	1,785	42	0.0%	99.8%	94.6%
14	Latvia	3	10,666	3,555	29	0.0%	99.8%	94.9%
15	Japan	16	7,540	471	36	0.0%	99.8%	96.5%
16	Ecuador	3	7,173	2,391	37	0.0%	99.9%	96.8%
17	Spain	4	4,472	1,118	35	0.0%	99.9%	97.2%

(continued)

Table 6.1 (continued)

	Recycling country	No.	GT	Average GT	Average age	% of world tonnage	Cumulative % of tonnage	Cumulative % of ships
18	Korea, South	5	4,233	847	31	0.0%	99.9%	97.6%
19	Philippines	2	3,307	1,654	34	0.0%	99.9%	97.8%
20	Russia	2	3,120	1,560	40	0.0%	99.9%	98.0%
21	Ukraine	1	2,516	2,516	44	0.0%	100.0%	98.1%
22	Nigeria	3	2,040	680	45	0.0%	100.0%	98.4%
23	Norway	5	1,997	399	50	0.0%	100.0%	98.9%
24	United Kingdom	3	1,937	646	38	0.0%	100.0%	99.2%
25	Azerbaijan	2	1,295	648	35	0.0%	100.0%	99.4%
26	Vietnam	4	1,160	290	24	0.0%	100.0%	99.8%
27	Finland	1	710	710	57	0.0%	100.0%	99.9%
28	Croatia	1	130	130	40	0.0%	100.0%	100.0%
	Grand total	**1,020**	**21,425,060**					
Of which	Five leading countries	849	20,965,019	24,694		97.9%		83.2%
	South Asia (B+I+P)	548	16,419,809	29,963		76.6%		53.7%
	European Union	54	76,854	1,423		0.4%		5.3%

Adapted from IHS Global Ltd, World Casualty Statistics (2017), Table 7C

The third notable fact is the fluctuation of the volumes recycled in China, who for relatively long periods has recycled 25–30% of the world's tonnage and then for other periods has almost withdrawn from the market. Furthermore, in the Spring of 2018, China's Ministry of Ecology and Environment announced that from the end of 2018, the import of ships for recycling will be banned, thus at a stroke of a pen diminishing the market of China's ship recycling industry to just domestic tonnage (Lloyd's List 2018). The fourth striking feature of the industry, clearly depicted in both figures, is the dominant position of the three South Asian countries, namely, Bangladesh, India, and Pakistan, who, for the last 20 years, have been recycling more than two thirds of the world's recycled tonnage (their combined market share in 2016 was 84% and in 2017 77%). They achieve this dominance by being able to pay the most competitive prices for buying end-of-life ships. The fifth fact is the very limited relevance the ship recycling industry of the European Union has for the international shipping industry, as is illustrated by the 2017 data of Table 6.1 showing eight EU States having recycled small ships and boats totaling just 0.4% of the total recycled tonnage.

Figures 6.3, 6.4, 6.5, 6.6, and 6.7 depict recycling yards in the leading five recycling countries.

The rest of this chapter is organized as follows: Sect. 2 examines the economic drivers of ship recycling, which have resulted in the industry being dominated by five countries and also analyzes steelmaking as the main driving force for ship recycling. Section 3 discusses the sale and purchase market for end-of-life ships, explains the roles of brokers and cash buyers, and provides a simplified inventory

Fig. 6.3 View of recycling yard in Bangladesh

Fig. 6.4 View of recycling yard in China

Fig. 6.5 View of recycling yard in India

Fig. 6.6 View of recycling yards in Pakistan

Fig. 6.7 View of recycling yards in Turkey

of the components that are recycled out of a ship. Section 4 discusses the efforts to implement existing international legislation to ship recycling and the development of Hong Kong Convention, this being a new but not yet in force international Convention that was developed specifically for ship recycling. Section 5 provides a critical analysis of the development of regional legislation by the European Union and Sect. 6 discusses the combination of voluntary and legislative mechanisms that will secure the global implementation of minimum standards for safe and environmentally sound ship recycling.

2 The Economic Drivers of Ship Recycling

2.1 The Dominance of South Asia in Ship Recycling

Compared to China and Turkey, the three South Asian countries (India, Pakistan, and Bangladesh) are currently less developed and poorer. As poverty is usually linked to lower safety, social welfare, and environmental standards, it is often claimed that the market dominance of the South Asian recycling countries is owed to their lower labor costs and lower compliance costs. This however is only one part of South Asia's competitive advantage.

The next section examines the sale and purchase process for end-of-life ships and provides an illustrative breakdown of an Indian recycler's income from selling the materials and equipment of a recycled ship. Although the data that is provided would not be applicable to a specialized ship (such as a ship with stainless steel tanks) nor would it apply to China or Turkey, the data nevertheless points to the importance of ferrous scrap to the recycler, representing more than 80% of the ship's value. The same data also points to the additional income ship recyclers in South Asia derive from selling equipment, machinery, furniture, stores, parts, etc., in the impressive secondhand markets that exist in Alang, Chittagong, and Gadani. This does not only provide additional competitiveness to South Asia' recyclers but is also a paradigm of a more environmentally friendly utilization of resources.

South Asia's ship recycling industry has a further advantage, helping it dominate the international ship recycling market. In South Asia there are large numbers of rerolling mills making steel products, such as reinforcing bars for the construction industry, by heating and reshaping semifinished steel products, such as billets or plates from recycled ships. The rerolled steel does not reach its melting point and, compared to making new steel, the process requires lower temperatures. Because the chemical composition of rerolled steel is not controlled, the quality of the products is not considered to be equal to new steel. Nevertheless, for appropriate applications rerolled steel products offer good economic alternatives. Furthermore, as the chemical composition and therefore quality of all structural steel that goes into shipbuilding is certificated by Classification Societies, steel plate from ship recycling competes with billets as the raw material for South Asia's rerolling mills. Consequently, South Asia's recyclers have the advantage of commanding better prices for flat rerollable steel compared to scrap steel destined for melting.

2.2 *Steelmaking as the Driver for Ship Recycling*

There are two main processes in modern steelmaking: (i) melting of steel scrap in electric arc furnaces (EAF), which in 2017 accounted for 28.0% of the world's production of new steel, and (ii) smelting of iron ore in oxygen blown converters (OBC), accounting for 71.5% of the steel production. The EAF is the more environmentally friendly of the two methods as the OBC requires more energy input, it requires the burning of coking coal, and also it produces more wastes. As however the price of iron ore has dropped in the last few years, the economics have somewhat shifted in favor of the OBC, as can be seen from the drop in EAF's world market share in Fig. 6.8 (see curve for "world average").

In 2017 a total of 1690 million tonnes of new steel were made worldwide using 600 million tonnes of steel scrap (note: this is more than the 28.0% share of production by EAF, as some steel scrap is also needed when smelting iron ore).

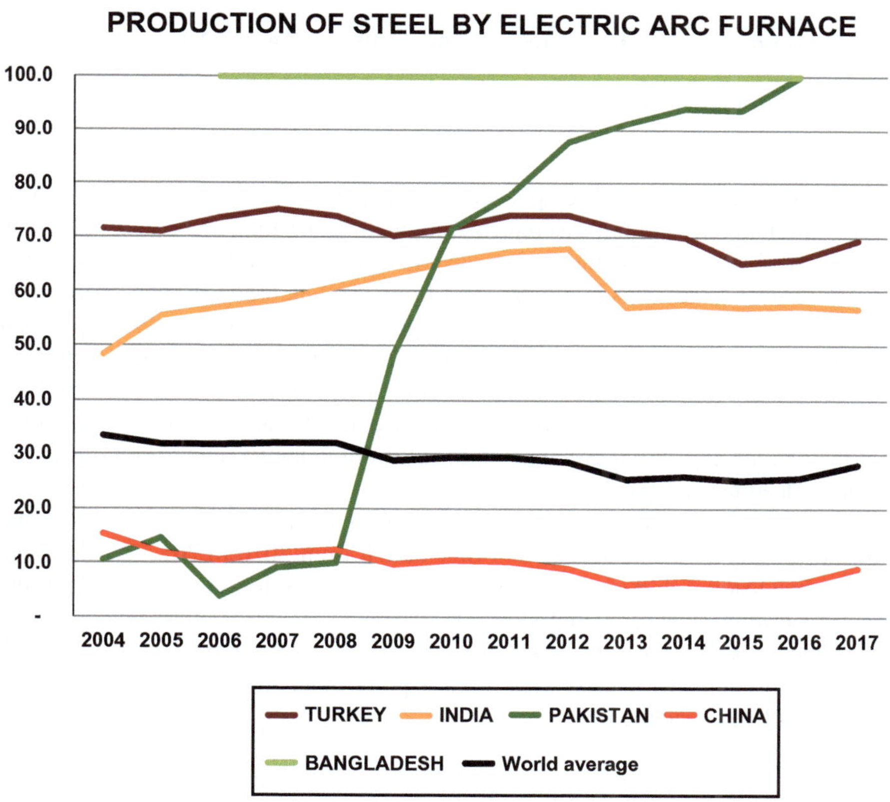

Fig. 6.8 Production of crude steel by EAF in the ship recycling countries. (Sources: For 2004–2016 from: Steel Statistical Yearbook 2017, Table 8; and from earlier issues. Preliminary data for 2017 from: World Steel in Figures 2018, page 10)

Table 6.2 World's leading steel producers together with Pakistan and Bangladesh's production (in million tonnes)

		2010	2011	2012	2013	2014	2015	2016	2017
	World total	1432.8	1538.0	1560.1	1650.4	1669.9	1620.0	1627.0	1690.0
1	**China**	**638.7**	**702.0**	**731.0**	**822.0**	**822.3**	**803.8**	**807.6**	**831.7**
2	Japan	109.6	107.6	107.2	110.6	110.6	105.1	104.8	104.7
3	**India**	**69.0**	**73.5**	**77.3**	**81.3**	**87.3**	**89.0**	**95.5**	**101.4**
4	USA	80.5	86.4	88.7	86.9	88.2	78.8	78.5	81.6
5	Russia	66.9	68.5	70.2	69.0	71.5	70.9	70.4	71.3
6	Korea Rep.	58.9	68.5	69.1	66.1	71.5	69.7	68.6	71.0
7	Germany	43.8	44.3	42.7	42.6	42.9	42.7	42.1	43.3
8	**Turkey**	**29.1**	**34.1**	**35.9**	**34.7**	**34.0**	**31.5**	**33.2**	**37.5**
9	Brazil	32.9	35.2	34.5	34.2	33.9	33.3	31.3	34.4
10	Italy	25.8	28.7	27.3	24.1	23.7	22.0	23.4	24.1
11	Taiwan	19.8	20.2	20.7	22.3	23.1	21.4	21.8	22.4
12	Ukraine	33.4	35.3	33.0	32.8	27.2	23.0	24.2	22.7
...	...								
...	**Pakistan**	**1.4**	**1.6**	**1.6**	**1.8**	**2.4**	**2.9**	**3.6**	**N/A**
	...	...	...	...	...	...	...	...	
..	**Bangladesh**	**0.05**	**0.1**	**0.1**	**0.1**	**0.1**	**0.1**	**0.1**	**N/A**

There are three sources of scrap steel for steelmaking: (1) "own arisings," which arise internally in steel mills as rejects from melting, casting, and rolling; (2) "new steel scrap" which is generated when steel is fabricated into finished products; and (3) "old steel scrap" which is scrap steel from obsolete products (including ships) sold to steel plants for remelting. This category forms around 40–44% of the total steel scrap used in steelmaking (World Steel Recycling in Figures 2012–2016).

Table 6.2[1] shows the total world production of crude steel from 2010 to 2017, in the 12 largest steel-producing countries and also in Pakistan and Bangladesh. Three of the five leading ship recycling countries feature in the top eight positions of the world's leading steel producers.

Table 6.3 shows the world's leading importers of steel scrap together with the quantities imported by Pakistan and Bangladesh from 2010 to 2017. Again, three of the five leading ship recycling countries feature in the top seven positions of the world's leading steel scrap importers.

Table 6.4 shows the world's leading exporters of steel scrap from 2010 to 2017 and also the top two net exporters (i.e., exports minus imports), these being the European Union and the USA.

[1]Source of data for Tables 6.2, 6.3, and 6.4: World Steel Recycling in Figures 2013–2017 and earlier issues; data on Pakistan and Bangladesh from Tables 6.1 and 54 of Steel Statistical Yearbook 2017).

Table 6.3 World's leading ferrous scrap importers together with Pakistan and Bangladesh's imports (in million tonnes)

		2010	2011	2012	2013	2014	2015	2016	2017
1	**Turkey**	**19.20**	**21.45**	**22.42**	**19.73**	**19.07**	**16.25**	**17.72**	**20.98**
2	Korea Rep.	8.09	8.63	10.13	9.26	8.00	5.76	5.85	6.17
3	**India**	**4.64**	**6.18**	**8.18**	**5.64**	**5.70**	**6.71**	**6.38**	**5.36**
4	USA	3.77	4.00	3.71	3.88	4.22	3.51	3.86	4.64
5	Taiwan	5.36	5.33	4.96	4.45	4.27	3.37	3.16	2.92
6	EU-28	3.65	3.71	3.20	3.19	3.14	2.85	2.74	3.14
7	**China**	**5.85**	**6.77**	**4.97**	**4.47**	**2.56**	**2.33**	**2.16**	**2.33**
	...								
	Pakistan	**0.79**	**0.96**	**0.92**	**0.87**	**1.34**	**2.12**	**2.39**	N/A
	...								
	Bangladesh	**0.24**	**0.32**	**0.20**	**0.28**	**0.46**	**0.95**	**2.01**	N/A
	...								

Table 6.4 World's leading ferrous scrap exporters and the two top NET exporters (in million tonnes)

		2010	2011	2012	2013	2014	2015	2016	2017
1	EU-28	19.03	18.81	19.58	16.81	16.95	13.74	17.77	20.05
2	USA	20.56	24.37	21.40	18.50	15.34	12.98	12.82	15.02
3	Japan	6.47	5.44	8.59	8.13	7.34	7.84	8.70	8.22
4	Russia	2.39	4.04	4.35	4.52	5.77	5.65	5.52	5.19
5	Canada	5.15	4.83	4.25	4.52	4.51	3.42	3.63	4.41
6	China	–	–	–	–	–	–	–	2.23
1	EU-28	15.38	15.10	16.38	13.62	13.81	10.89	15.03	16.91
2	USA	16.79	20.37	17.69	14.61	11.13	9.47	8.96	10.38

As pointed in the previous section, for the last 20 years, the ship recycling yards in Bangladesh, China, India, Pakistan, and Turkey have dominated the industry having recycled 97–98% of the worldwide recycled tonnage. Figure 6.9 shows the light displacement tonnage (LDT) recycled in each of the five countries. Published data on LDT are generally not available, so the author has collected annually the data presented here from the ship recyclers' associations of each of the five countries.

The data shown in Tables 6.2, 6.3, and 6.4 and in Fig. 6.8 go some way to help explain the success of the five recycling countries, all of which are importers of scrap steel:

China is by far the largest steelmaking country, currently producing 49% of the world's steel. As seen in Fig. 6.8, China relies heavily in OBC for its steel production, with the EAF's share having dropped from 15.3% in 2004 down to 5.9% in 2015 and then up to 9.0% in 2017, while in the same period its steel production more than tripled from 272 to 831.7 million tonnes. China's imports of steel scrap have been reducing from a record 13.7 million tonnes in

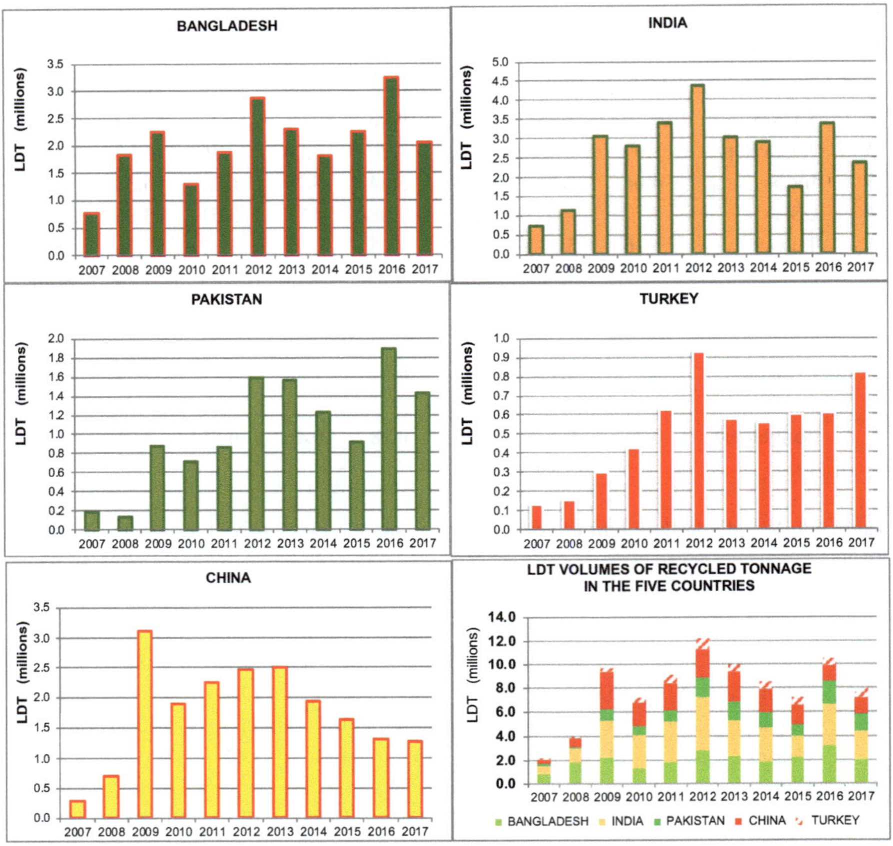

Fig. 6.9 LDT of recycled tonnage in the five-ship recycling countries

2009 (Steel Statistical Yearbook 2017) down to a net import of 0.1 million in 2017 (see Tables 6.3 and 6.4). Figure 6.2 illustrates the fortunes and problems of China's ship recycling industry: In 1998 and in the period 2004–2008, China's ship recycling market share was small, while in 2009 China was the leading ship recycling country with 31% market share. From 2010 the Chinese share declined to the fourth position in the last 2 years. Also, since 2013 Chinese ship recyclers have imported very few ships and have had to rely on a "scrappage" subsidy that the government offered to Chinese-flagged ships from 2013. The subsidy which was very generous (around US$395/LDT) was maintained until 2018. In the next few years, China is expected to increase the EAF's share of its steelmaking in order to combat its severe industrial pollution. This will result in increased imports of scrap steel and could have propelled again China to a leading position in ship recycling. Nevertheless, the recent ministerial announcement that China will ban the import of ships for recycling from the end of 2018 (Lloyd's List

2018) has placed a big question mark on the future of China's well-developed and large-capacity ship recycling industry.

India in the last 3 years has overtaken the steelmaking industry of the USA and is heading to become the world's second largest steelmaker. The majority of its steel is being made with the EAF process making India a major importer of scrap steel. Some of the imported scrap steel comes from its ship recycling industry, which has been the world's largest in terms of LDT in 7 of the last 10 years. As discussed in the next section, 90% of a ship's LDT is steel, which can be subdivided into melting steel (30% of LDT) and rerollable steel (60% of LDT). In India, as also in Bangladesh and Pakistan, ship recyclers separate flat plates, lengths of girders, beams, and angle bars from smaller irregular pieces of metal. The smaller pieces become melting scrap, while the larger items attract higher prices as they can either be used directly in construction, or road building, or can be heated and rerolled into bars and rods in rerolling mills. It therefore follows that a fraction (possibly of the order of 30%) of the annually recycled LDT serves the country's needs for new steelmaking, whereas the majority (around 60%) is rerolled.

Turkey has substantial steelmaking industry, currently being the eighth largest in the world. Its steelmaking relies to a great extent on the EAF process and is thus characterized by high demand for scrap steel. Turkey has been and continues to be the world's largest importer of scrap steel. Its ship recycling industry is the smallest of the major five-ship recycling countries but recycles more tonnage than the rest of the world put together (excepting of course the four major ship recycling countries). There is little rerolling of ship's plates in Turkey, and therefore the main outlet of ship recycling is new steelmaking.

Pakistan produces relatively little steel, although its production has increased in the last few years, mostly through the addition of EAFs. Consequently Pakistan's ship recycling industry has been growing fast, providing scrap steel for rerolling and for melting while enjoying additional economic benefits from secondhand markets for machinery, equipment, spare parts, etc., as do India and Bangladesh.

Bangladesh currently produces very little new steel, all based on EAF. Its ship recycling provides scrap steel for the rerolling market, which is very active due to the urbanization of this very densely populated country. Consequently, Bangladesh's ship recycling industry has been the world's largest in terms of LDT in 2 of the last 10 years, including 2015.

Other countries have not recycled any significant quantities of tonnage in the last 20 years. Of course, it is uneconomic for a small ship or for a damaged ship to sail thousands of miles to reach one of the main ship recycling centers, and for this reason ship recycling facilities also exist in many countries, even some that have no need for ferrous scrap (see Table 6.1). Ship recycling in such cases can be seen as a service for disposing boats and ships, rather than an industry driven by the economy of steelmaking. Although the five main ship recycling countries dominate the industry, it is certainly possible that, in the proximate future, another country might join the major league. Such a country would most probably be a developing country with low labor costs and will be an importer of scrap steel for its steelmaking industry. Vietnam or the Philippines could be such a new entrant.

On the other hand, environmental activists and some European politicians have been promoting in recent times the development of facilities for the recycling of large ships in Europe, claiming that this will provide best practice ship recycling services to international shipping and, in so doing, create much needed jobs and economic prosperity. The reality however is that Europe is the world's largest net exporter of scrap steel, as seen in Table 6.4. The vast majority of the ferrous scrap exports from the European Union go to Turkey, with some quantities also being exported to Egypt, Pakistan, and India (World Steel Recycling in Figures 2013–2017). It makes no sense whatsoever to recycle large ships in Europe to produce scrap that will have to compete with the large quantities of other European ferrous scrap in order to be sold and transported to countries most of which already recycle ships.

3 Sale and Purchase of End-of-Life Ships

3.1 Selling of Ships for Recycling

Almost all recycling sale and purchase transactions are quoted in US$ per lightship (long) ton. The long ton is an imperial measurement unit equating to 2240 pounds (lb) or 1.016 tonnes. Lightship (or light displacement tonnage or LDT) is defined as the extreme displacement of an unloaded ship, with or without the bunkers and lubricants of the main and auxiliary engines, the hydraulic oil contained in hydraulic systems, and the water needed to fill the ship's boilers up to working level. Lightship excludes crew, passengers, stores, fuel, ballast, potable water, paints, cargo, liquids, and constants in the system and all other items not affixed to the vessel. Lightship is relevant for ship recycling transactions because it provides the basis for estimates of the weight of the ship's steel and approximate quantities of various other commercially valuable materials belonging to the vessel that can be obtained from the ship's recycling.

Often the question arises as to what the lightship content of different ship types and sizes is and what is the relation between lightship, gross tonnage, and deadweight. These quantities are related empirically, and therefore these questions can be answered using tabulated data, such as those shown in Table 6.5, which were obtained by interviewing an experienced ship broker (Mikelis 2007).

To put ship recycling prices in perspective, consider the sale of a middle-sized ocean-going ship, say a Panamax tanker of around 10,000 LDT (GMS Weekly 2006). Figure 6.10 provides historic price data for each of the five main recycling centers (data compiled from GMS Weekly, published during 2006–2018). The graph shows that the three South Asian countries compete with each other very closely on price, whereas the prices offered by China and Turkey tend to be separated further

Table 6.5 Approximate estimates of lightship content of different ship types

Ship type	Cargo carrying capacity (DWT tonnes)	Gross tonnage (GT)	Lightship (LDT)
Tanker VLCC	300,000	159,000	35,000
Tanker Suezmax	150,000	80,000	22,000
Tanker Aframax	80–120,000	45–67,000	15–18,000
Tanker Panamax	70,000	40,000	10–13,000
Tanker Handysize	35,000	22,000	7000
Capesize bulk carrier	150–170,000	78–86,000	20–21,000
Panamax bulk carrier	70,000	40,000	10–12,000
Handysize bulk carrier	35,000	22,000	7000

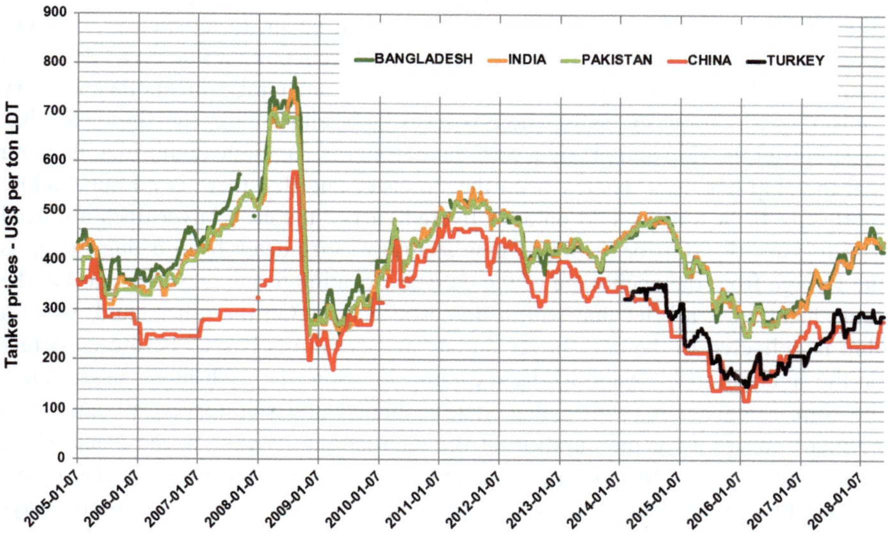

Fig. 6.10 Weekly tanker recycling prices by country of recycling. (Source: GMS Weekly, published during 2006–2018)

away on the pricing spectrum, be it lower or (rarely) higher (as has been the case with China). As an example, in May 2018, a 10,000 LDT Panamax tanker could be sold to South Asian recyclers for around US$4,400,000 but only for US$2,800,000 to Turkey or China.

A shipowner who is contemplating the sale of a ship for recycling will normally contact a ship broker that specializes in ship recycling. The broker would then market the ship to different "cash buyers" (i.e., companies that specialize in trading end-of-life ships). The broker will represent and advise his client (the seller) during the sale negotiations that lead to the drawing of the sale contract, known as the MoA, or Memorandum of Agreement. On completion of the sale, the broker receives a commission for his services from the seller, which usually is an agreed percentage of the value of the contract (the industry standard being 1%). It is important to underline that at no time does the broker own the ship.

Nearly all merchant ships are sold for recycling via cash buyers, who purchase ships for cash (as opposed to by letter of credit) and then sell them (usually at a profit) to the recycler, who normally pays the cash buyer with a bank letter of credit. Unlike a broker, the cash buyer takes legal ownership of the vessel (albeit for a limited time). Cash buyers are an integral part of the industry because they provide indispensable services to the shipowner, namely, expertise in a specialized and a difficult market, reduction to the shipowner's risk, payment in cash of a sizeable advance on signing of the MoA, and of the balance on delivery (as opposed to payment by letter of credit).

Completion of the sale occurs with the payment of the balance of the purchase price to the shipowner and execution of the Protocol of Delivery and Acceptance (the "PoDA") between the shipowner and the cash buyer. In most cases, delivery takes place at the anchorage of the recycling yard (reflecting a sale on a "delivered basis") or, less frequently, at an agreed port or anchorage in another country (reflecting a sale on an "as-is where-is basis"). In either case, the shipowner has to deregister the ship and obtain a certificate from the flag State authorities showing that the ship has been deleted from their register and that there is no outstanding mortgage. If the ship is sold on a "delivered basis," the cash buyer does not need to reregister the ship or to obtain new statutory certificates from a flag State, as the voyage from the anchorage of the delivery location to the recycling yard at the same location is a brief one within domestic waters. On the other hand, when a ship is delivered to the cash buyer on an "as-is where-is basis," before departing on the international voyage to the recycling location, the cash buyer has to crew the ship, reregister it (with a flag State), obtain valid statutory certificates, and normally insure it for the duration of the international voyage to the place of the final delivery. A number of open registers facilitate such short-term registrations, and for this reason, statistics of ship recycling by country of registration always show a disproportionate number of recycled ships for these flags compared to their fleet of ships in service.

3.2 *Purchasing of Ships for Recycling*

On a delivered deal, the recycler will normally take delivery of the ship from the cash buyer at the anchorage; however, the cash buyer will have terms in the MoA with the shipowner that require the shipowner's crew to move the ship from the anchorage to the recycling yard. The ship recycler normally pays for the ship with a bank letter of credit. In addition to the purchase price, the recycler will incur financial costs, insurance costs (related to the yard and his recycling labor force), (import) taxes and duties related to the vessel, yard rental costs, investment costs (yard equipment, etc.), costs of consumables and utilities (oxygen, LPG, diesel, electricity), and labor costs. For the purpose of illustration, we can approximate these costs to around 15–

20% of the purchase price of the ship (note: this is a crude simplification, as the purchase price can vary by large amounts, as already seen in Fig. 6.10).

An approximate breakdown of a ship's LDT is as follows: 5% of LDT is assumed to be waste and losses due to corrosion and aging over time. Another 5% of the LDT is made up of equipment, machinery (excluding the main engine), cables, shafting, fittings, spares, lubricants, and nonferrous metals. The remaining 90% is steel, which can be subdivided into melting steel (30% of LDT and which includes the ship's main engine) and rerollable steel (60% of LDT). In South Asia, flat rerollable steel attracts a higher price than scrap steel (by around 10%), as plate can be utilized in rerolling mills for shaping it into long or flat steel products, without having to go through the more costly process of making new steel in a mill.

Data obtained from a recycler in India provides an illustrative analysis of income from the sale of different components of a ship. It should be stressed that the following figures are changeable as the prices of steel and of nonferrous metals are volatile: steel 82%, nonferrous metals and cables 10%, electrical panel and various machineries 1.4%, motors and winches 1.3%, shafting 1.3%, generators 1.2%, spare parts and lubricants 1%, compressors 0.8%, and other items 1%.

The recycler usually obtains from his bank a letter of credit in US dollars for a period of 180 days, although in some cases it can be for a longer period. For an average-sized ship of 10,000 LDT, it might take 100–120 days to complete the recycling work. From around the 40th day from the commencement of work and until completion, the recycler sells the ship's metals, machinery, equipment, and other materials.

The recycler has to contend two key volatilities: (a) domestic steel plate prices and (b) domestic currency exchange rate with the US dollar. The Indian recycler of our example borrows US dollars, then starts earning rupees from around the 40th day, and finally has to buy US dollars with rupees in order to pay back his loan on or before the 180th day. If in that time the price of steel in India has moved up, the recycler will receive extra income, as was the case in the last 2 months and as depicted in Fig. 6.11. Had the recycler based his budget in the beginning of April on a steel price of 29,000 rupees per tonne, 40 days later his income would be noticeably higher. However, as can be seen from Fig. 6.11, the converse situation is equally likely.

The second source of volatility and risk to the recycler is the exchange rate between his currency and the currency of his loan. Figure 6.12 shows the actual fluctuations of rupee against US dollars from December 2017 to end of May 2018. As the recycler cannot predict the movements of the exchange rate, he has to face and factor the risk of incurring higher costs (or the bonus of cheaper US dollars).

Ship recycling is an informal industry wedged between two powerful players, the steel industry and the shipping industry. When demand for shipping is healthy and charterers are paying well for the hire of ships, the volume of tonnage offered for recycling decreases. Ship recyclers may increase their offers to tempt more tonnage to come out of trading, but the prevailing price of steel forms a natural ceiling on how much recyclers can afford to pay. When recyclers cannot attract sufficient

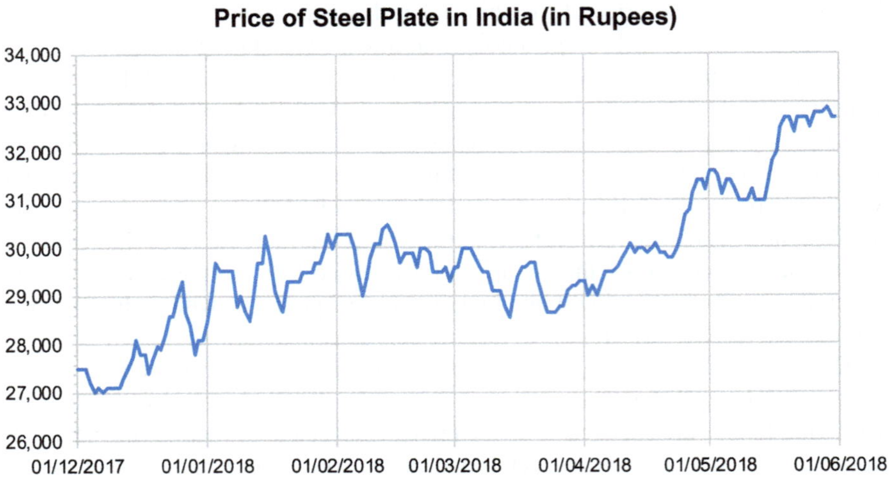

Fig. 6.11 Price of 12 mm steel plate in India. (Data compiled by GMS)

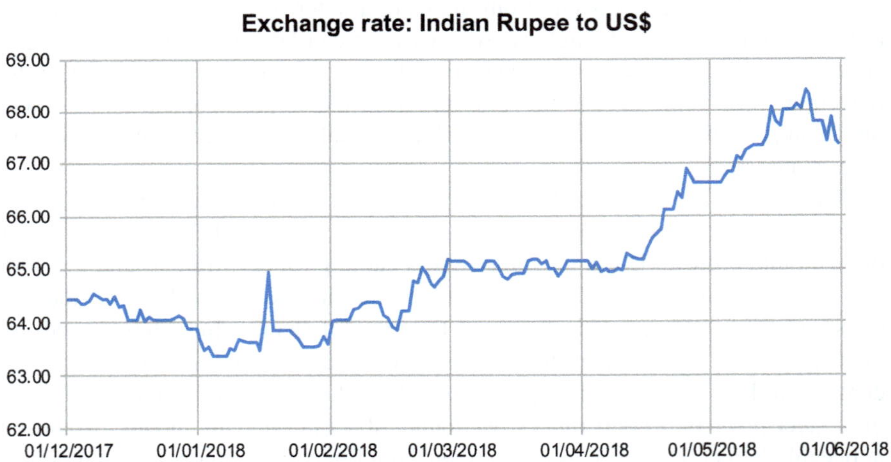

Fig. 6.12 Exchange rate fluctuations – Indian rupee to US$. (Data compiled by GMS)

tonnage at affordable prices, a number of recycling facilities face temporary (or permanent, depending on the severity) closures.

Conversely, when the shipping markets are depressed, more tonnage is offered for recycling and, consequently, recyclers can reduce the prices they pay for ships. If at such times steel prices happen to be high, ship recycling becomes more profitable, attracting more recycling capacity through the reactivation of closed yards.

4 Hong Kong Convention

In the 1990s international attention focused on the poor working conditions in ship recycling yards following reports of recurring accidents with fatalities and also of degradation of the environment through persistent pollution. At that time it was not uncommon for some yards to clean cargo holds and tanks of beached ships by drilling holes on the side shell and then pumping seawater into the cargo spaces. Often, the removal of a ship's propeller led to spilling of the tail shaft hydraulic oil onto the beach. Figure 6.13 depicts a scene from a major ship recycling yard in Bangladesh, as recently as 2008.

During the 1990s environmental activists led by Greenpeace International campaigned to bring ship recycling into the public attention. The environmental NGOs also took their campaign to the meetings of the Basel Convention that was established under the United Nations Environment Programme and which is the forum for Ministries of Environment of Member States of the UN. Around the same time, the government of Norway also led some first discussions at the International Maritime Organization (IMO) on the need to address in the future the recycling of ships with an international regulatory instrument.

Fig. 6.13 Temporary storage of waste oil in a recycling yard in Bangladesh, 2008

4.1 The Basel Convention and Its Implications

The Basel Convention (or, to give it its full title, "The Basel Convention on the Control of Transboundary Movements of Hazardous Wastes and their Disposal") was adopted in March 1989 and entered into force in May 1992. To date it has been ratified by 186 countries (i.e., most of the world with the important exception of the USA). The Basel Convention provides controls for the international movement of hazardous wastes and for their environmentally sound management. These controls are implemented through the establishment of a chain of communications, aimed to reach consent for the shipment, between the authorities of the country exporting the hazardous wastes with the authorities of the importing country and with the involvement of the authorities of any transit State. The consent is based on the understanding that the hazardous waste in question will be treated in an environmentally sound manner in the importing country. In most countries the implementing authorities of the Basel Convention are Ministries of Environment.

Toward the end of the 1990s, the subject of ship recycling[2] first entered the agendas of the Conference of the Parties (COPs) of the Basel Convention. In December 1999 COP 5, in its decision V/28, instructed its technical working group to develop guidelines in collaboration with IMO for the environmentally sound management of the dismantling of ships. It also instructed its technical working group, together with its legal working group "to discuss the legal aspects under the Basel Convention relating to the issue of the full and partial dismantling of ships", this in effect being a formal request to assess whether Basel Convention could be implemented to regulate ship recycling. COP 6 in December 2002 adopted the Basel Convention's *Technical Guidelines for the Environmentally Sound Management of the Full and Partial Dismantling of Ships* and also extended the mandate of the working group on the legal aspects of ship dismantling for another intercessional period. It should be pointed out here that guidelines are voluntary standards that do not have the mandatory role of international conventions. Incidentally, the growing international interest in ship recycling also resulted in the publication by the International Labour Office (ILO)[3] of a further set of voluntary guidelines in 2004 on *Safety and Health in Shipbreaking Guidelines for Asian Countries and Turkey*.

COP 7 of the Basel Convention, in its decision VII/26 in October 2004, reached an ambivalent compromise position when addressing the question on whether Basel Convention can regulate the movement of end-of-life ships. The decision said: "Noting that a ship may become waste as defined in article 2 of the Basel Convention and at the same time it may be defined as a ship under other international rules." Importantly, the same decision VII/26 also "Invites the International Maritime Orga-

[2]Note: While IMO uses the term "ship recycling," Basel Convention refers to "ship dismantling," whereas ILO uses "ship breaking." In the shipping industry, the term "ship scrapping" still persists.

[3]The International Labour Office is the Secretariat of the International Labour Organization (ILO), which is a specialized agency of the UN for setting labor standards, developing policies, and devising programmes promoting decent work for all women and men.

nization to continue to consider the establishment in its regulations of mandatory requirements, including a reporting system for ships destined for dismantling, that ensure an equivalent level of control as established under the Basel Convention and to continue work aimed at the establishment of mandatory requirements to ensure the environmentally sound management of ship dismantling, which might include pre-decontamination within its scope." IMO responded positively to this invitation by developing a Convention specific to ship recycling, namely, the "Hong Kong International Convention for the Safe and Environmentally Sound Recycling of Ships, 2009," also known as *the Hong Kong Convention* (hereafter also the *HKC*).

It is important to realize that whereas the Basel Convention may have been most successful in fighting against illegal exports of hazardous wastes to countries that are unable to process and dispose of them in an environmentally sound manner, on the other hand, the Convention is unsuitable for defining minimum standards for the recycling of ships. The author's critical view is that the attempt by the international community in the early 2000s to establish and enforce Basel Convention as the international regulatory regime for the recycling of ships was an avoidable mistake, encouraged by the persistent lobbying of environmental activists and aided by the fact that the Convention was already in force and therefore could be implemented without delay. Nevertheless, it is a fact that Basel Convention does not contain any requirements that are relevant to ships and to ship recycling facilities nor does it concern itself with issues on workers' safety. The only relevant requirement of Basel Convention to ship recycling is its generic requirement that the wastes should be managed in an environmentally sound manner. Furthermore, the mechanism for achieving the Convention's "prior informed consent" relies on the establishment of communications between the exporting and importing countries, which, when applied to end-of-life ships, means in practice the authorities of the State from where the ship departed for its last voyage and the authorities of the recycling State. This is because the Convention is not cognizant of the concept of flag State that is central to the United Nations Convention on the Law of the Sea and to all maritime Conventions and therefore leaves no option but to consider the State from where the ship departed for its last voyage as being the exporting State.

Implementing the Basel Convention to control the movements of end-of-life ships creates a number of problems: (a) it takes inordinate amount of time to arrange for the necessary communications between exporting, importing and any transit States, communications which in any case have no effect whatsoever in improving the standards under which ships are recycled; (b) the managers of the ship will most often have no connection with the country that is deemed to be the State of export; (c) a number of countries are unwilling to recognize that Basel Convention should regulate the recycling of end-of-life ships, thus making the communications between the managers of the ship, the State of export, and the other involved States even more cumbersome; and (d) the decision to recycle a ship may not be taken, or may not be finalized, or may not be admitted until after the ship has departed from the port and is in international waters, in which case there is no exporting State to lead the inter-State communications envisaged by Basel Convention. In practice the above problems make the Basel Convention unenforceable to ship recycling. The difficulty in applying the provisions of the Basel Convention to ship recycling and the circumvention of the Convention's controls by ships destined for recycling are acknowledged in the website of the Basel Convention.[4]

Notwithstanding the above, in October 2010 following intense lobbying by environmental activists, COP 10 of the Basel Convention failed to reach conclusive consensus that Hong Kong Convention can replace Basel Convention for the recycling of ships. Instead its decision BC-10/17[5] maintains all options open:

1. Notes that while some parties believe that the Hong Kong International Convention for the Safe and Environmentally Sound Recycling of Ships provides an equivalent level of control and enforcement to that established under the Basel Convention on the Control of Transboundary Movements of Hazardous Wastes and Their Disposal, some parties do not believe this to be the case.
2. Encourages parties to ratify the Hong Kong Convention to enable its early entry into force.
3. Acknowledges that the Basel Convention should continue to assist countries to apply the Basel Convention as it relates to ships.

It is certainly hoped and expected that following the future entry into force of Hong Kong Convention, the Parties to Basel Convention will come to a formal decision that the recycling of ships shall fall under the scope of the former Convention.

4.2 The Ban Amendment and the European Waste Shipment Regulation

In an effort to strengthen protection to developing countries, COP 2 of the Basel Convention adopted in March 1994 its "Ban Amendment" banning the export of hazardous wastes from OECD to non-OECD countries. However, the Ban Amendment is not yet in force internationally. It will enter into force between member states that have ratified it 90 days after it has been ratified by at least three-fourths (66) of the 87 countries that were Parties to the Convention at the time the Amendment was adopted. By May 2018 the Amendment had received 93 ratifications, 63 of which from States who were Parties at the time of its adoption. The Amendment is therefore expected to enter into force relatively soon.

The Ban Amendment however has already been enforced unilaterally in the European Union, through the European Waste Shipment Regulation, which imple-

[4]See section: *Overview on Ship Dismantling* www.basel.int.

[5]http://ec.europa.eu/environment/waste/ships/pdf/COP_10%20Decision_10_17.pdf.

ments the Basel Convention and the Ban Amendment in European Union law. The European Union had implemented the Basel Convention into European law from as early as February 1993. In 2006 the Union replaced its earlier regulation by the *Waste Shipment Regulation (EC) No. 1013/2006*, which additionally implemented unilaterally the Ban Amendment, forbidding the export of hazardous wastes from member states of the European Union to any developing (i.e., non-OECD) countries.

When EU countries and the European Commission tried to enforce the Waste Shipment Regulation to end-of-life ships, they faced many difficulties and much evasion, as was seen with ships such as the *Otapan*, the *Sea Beirut*, the *Sandrien*, the *Margaret Hill*, the *Tor Anglia*, etc. This was primarily because, in enforcing the Ban Amendment, the European Waste Shipment Regulation deems illegal the recycling in Bangladesh, China, India, or Pakistan of any ship that has started its last voyage from a European Union port (i.e., exporting EU State, regardless of the flag the ship flies). The simple reality is that these four non-OECD countries consistently recycle around 95% of the world's tonnage. In fact, a study by the European Commission in 2011 reported that (at least) 91% of ships under the scope of the regulation had ignored or circumvented its requirements. This led the European Commission in 2012 to propose the development of a new European Regulation on Ship Recycling that is discussed in the next section.

For reasons that appear to have nothing to do with improving standards in the ship recycling industry, some environmental NGOs are to this day continuing to lobby for the enforcement of the Basel Convention to regulate end-of-life ships. They are particularly active in Brussels where they have managed to attain undeserved influence in the European Commission and the European Parliament.

4.3 The Mechanisms and Spirit of Hong Kong Convention

In December 2003 the 23rd session of IMO's Assembly adopted with its Resolution A.962 (23) the *IMO Guidelines on Ship Recycling*. It would have been clear by that time, however, that what was needed was international regulation rather than another set of voluntary guidelines. Therefore, 2 years later, and following the invitation to IMO by COP 7 of the Basel Convention, the 24th session of IMO's Assembly with Resolution A.981(24) agreed in December 2005 that IMO would develop a "new legally binding instrument on ship recycling that would provide regulations for:

1. The design, construction, operation, and preparation of ships so as to facilitate safe and environmentally sound recycling, without compromising the safety and operational efficiency of ships;
2. The operation of ship recycling facilities in a safe and environmentally sound manner; and
3. The establishment of an appropriate enforcement mechanism for ship recycling (certification/reporting requirements)"

Following concentrated work for over 3.5 years, IMO's Marine Environment Protection Committee (MEPC) completed the draft text of the new international Convention, which was submitted to a Diplomatic Conference that was convened in Hong Kong and China from 11th to 15th of May 2009. The Diplomatic Conference was attended by representatives of 63 member states, two associate members, representatives from the Secretariats of the Basel Convention and of ILO, and observers from 1 IGO and 8 NGOs. The Conference unanimously adopted the final text of "Hong Kong International Convention for the Safe and Environmentally Sound Recycling of Ships, 2009", also known as *the Hong Kong Convention* (or *the HKC*).

The main part of the Convention contains 21 articles that establish the Convention's main legal mechanisms. This is followed by the Annex to the Convention that contains 25 regulations divided in 4 chapters: (1) General (regulations 1–3), (2) Requirements for ships (regulations 4–14), (3) Requirements for ship recycling facilities (regulations 15–23), and (4) Reporting requirements (regulations 24–25). Lastly, HKC has seven appendices, with lists of hazardous materials, standard formats for certificates, etc. The text of the Convention also makes reference to six guidelines that were developed by IMO's MEPC in the years following the adoption of the Convention. Although guidelines are nonmandatory texts, they are considered indispensable in providing clarifications, interpretations, and uniform and effective implementation and enforcement of the relevant requirements of the Convention. It is worth noting that when the text of an international Convention requires that a certain set of guidelines are to "be taken into account," frequently this is implemented and enforced by administrations and their recognized organizations as if the guidelines are mandatory requirements. MEPC 68, in May 2015, completed the development of the sixth set of the guidelines for the Hong Kong Convention[6] listed in Table 6.6.

Table 6.6 Guidelines adopted by IMO for HKC

Development of the guidelines of the Hong Kong Convention	
Guidelines	Adoption
2015 guidelines for the development of the inventory of hazardous materials (inventory guidelines)	Revised guidelines adopted by resolution MEPC.269(68)
2011 guidelines for the development of the ship recycling plan (SRP guidelines)	Guidelines adopted by resolution MEPC.196(62)
2012 guidelines for safe and environmentally sound ship recycling (facility guidelines)	Guidelines adopted by resolution MEPC.210(63)
2012 guidelines for the authorization of ship recycling facilities (authorization guidelines)	Guidelines adopted by resolution MEPC.211(63)
2012 guidelines for the survey and certification of ships under the Hong Kong Convention	Guidelines adopted by resolution MEPC.222(64)
2012 guidelines for the inspection of ships under the Hong Kong Convention	Guidelines adopted by resolution MEPC.223(64)

[6]For the texts see: http://www.imo.org/OurWork/Environment/ShipRecycling/Pages/Default.aspx.

The key elements of the mechanisms underlying HKC are the following:

- The Convention applies to all ships,[7] except (a) ships below 500 GT, (b) government-owned noncommercial service ships, and (c) ships operated throughout their lives exclusively in waters of the State whose flag the ship is flying.
- Inventory of hazardous materials (IHM): The Convention requires that ships will be provided with an IHM detailing the locations and approximate quantities of hazardous materials listed in the Convention's Appendices 1 and 2. Note that the materials listed on Appendix 1 are already controlled by other international Conventions, such as SOLAS, AFS, Montreal Protocol, etc. For *new ships*, i.e., those built after the entry into force of the Convention, it is required that (a) materials listed in Appendix 1 must not be used and (b) any materials listed in Appendix 2 and used on the ship must be shown in the IHM. For *existing ships*, i.e., those built before HKC's entry into force, it is required that (a) any preexisting materials listed in Appendix 1 must be shown in the ship's IHM, while the same materials must not be used on the ship subsequent to the Convention's entry into force; and (b) the inclusion in the ship's IHM of any Appendix 2 materials used on the ship is encouraged but not mandated.
- International Certificate on Inventory of Hazardous Materials (ICIHM): Once HKC is in force, ships will be issued the ICIHM, after an initial or renewal survey, by their flag State, or the delegated classification society. The purpose of this certificate, whose validity will be for 5 years, is to ensure that the IHM continues to correctly reflect the hazardous materials that are on the ship.
- Ship Recycling Facility Plan (SRFP): Recycling yards located in countries that are Parties to HKC will document in their SRFP the yard's systems and processes for ensuring safety and environmental protection.
- Document of Authorization to conduct Ship Recycling (DASR): This will be issued by the competent authorities in recycling States Parties to HKC to each authorized yard within their jurisdiction. The DASR will list any limitations that are imposed to the yard, such as size or type of ship and quantities of any specific hazardous materials that the yard may not be qualified to accept. This certificate will be valid for up to 5 years.
- Ship Recycling Plan (SRP): Recycling yards in countries that are Parties to HKC, prior to commencing the recycling of a ship, will have to produce a plan based on the specific ship's IHM and other particulars. The SRP will detail how the yard will dispose of the ship's hazardous materials and what precautions will be taken against unsafe situations. The SRP will normally be approved by the competent authority of the recycling State.
- International Ready for Recycling Certificate (IRRC): Prior to the commencement of the recycling of a ship, the IRRC will be issued by the ship's flag State

[7]In HKC "ship" is defined as "a vessel of any type whatsoever operating or having operated in the marine environment and includes submersibles, floating craft, floating platforms, self-elevating platforms, floating storage units (FSUs), and floating production storage and offloading units (FPSOs), including a vessel stripped of equipment or being towed."

or by its delegated classification society, following a final survey. The survey and certificate will confirm the validity of the IHM and the suitability of the selected recycling yard (on the basis of the IHM, DASR, and SRP).
- Other elements in HKC include notification by the recycling yard to its competent authority of the commencement of recycling, notification to competent authority and flag State of completion of recycling, port State control by Party coastal States, ability of a ship flying the flag of a non-Party State to be recycled in a yard at a Party State as long as that ship meets the requirements for IHM, and inability of a ship flying the flag of a Party State to be recycled in a yard at a non-Party State.

4.4 Implications of Hong Kong Convention

Leaders from the ship recycling industry on occasions have reflected and complained that whereas the underlying requirements of HKC for recycling yards are onerous (in terms of improvements in systems, procedures, training, equipment, and infrastructure), on the other hand, the requirements on ships are very light (the costs for the provision of an IHM and for the associated surveys and certification are relatively small). Nevertheless, as discussed in more detail in Sect. 6.2, the equitability between the ship recycler and the shipowner in HKC lies in the fact that a ship flying the flag of a Party to the Convention will have to be recycled in a Party (HKC) yard, and therefore the costs of compliance to the Convention would pass this way back to the shipowner (unless of course the ship changes flag to a non-Party flag and is recycled in a non-Party yard – note: in the future, avoidance of HKC obligations through reflagging will cease to be possible after all five main recycling countries become Parties to HKC).

Hong Kong Convention has been opposed and is frequently criticized by organized civil society activists for not banning the beaching method of ship recycling. This persistent lobbying by NGO activists has led to beaching being widely associated with poor ship recycling standards and vice versa. Whereas during the development of HKC there were repeated proposals to ban beaching, the developers of the Convention realized that banning beaching through the Convention would not be viable as presently around three-quarters of the world's recycling capacity utilizes this method. Had HKC banned beaching, eventually this would have led to two distinct regimes in the world, one in accordance with the standards of the Hong Kong Convention and the second one being the unregulated (by international standards) recycling yards of the countries that continue to employ beaching. As it will always be legal for ships to be sold and to change flag (and thus avoid any flag State requirements to implement the Hong Kong Convention), shipowners would therefore have the choice under which regime to recycle their ships. By keeping the South Asian countries that employ beaching outside the influence of HKC, IMO and its Convention would in effect have turned their backs to the parts of the industry that were in most need for the improvements that are envisaged by HKC.

Furthermore, and most importantly, it would not be possible for HKC to enter into force without the recycling capacity of at least one of the three South Asian ship recycling countries.

The Hong Kong Convention addresses the systematic prevention, reduction, minimization, and, where practicable, elimination of risks to human health and safety and to the environment through mandatory requirements on worker safety and training, the protection of human health and the environment, emergency preparedness and response, and systems for monitoring, reporting, and recordkeeping. This way the Hong Kong Convention has a truly realistic chance of being ratified by all recycling countries, including the three South Asian countries, and thus of providing a single international standard for the recycling of all ships.

4.5 Entry into Force of Hong Kong Convention

The Convention will enter into force 24 months after the date on which 15 States, representing 40% of world merchant shipping by gross tonnage, have either acceded to it or have ratified it.[8] Also, the combined maximum annual ship recycling volume of those States must be no less than 3% of their combined merchant shipping tonnage.[9] As the size of the world fleet changes every year, so do the second and third of the three conditions. Table 6.7 shows the growth of the world fleet in the last 10 years.

Table 6.7 Growth of world fleet (2008–2017)

	2008	**2009**	**2010**	**2011**	**2012**
World fleet	830,704,412	882,634,804	957,982,304	1,043,081,509	1,081,204,742
40% of world Fleet	332,281,765	353,053,922	383,192,922	417,232,604	432,481,897
3% of 40%	9,968,453	10,591,618	11,495,788	12,516,978	12,974,457
	2013	**2014**	**2015**	**2016**	**2017**
World fleet	1,122,649,460	1,166,847,462	1,211,223,165	1,248,583,186	1,291,046,701
40%of world fleet	449,059,784	466,738,985	484,489,266	499,433,274	516,418,680
3% of 40%	13,471,794	14,002,170	14,534,678	14,982,998	15,492,560

Source: IHS Maritime & Trade, World Fleet Statistics (2017), and earlier years

[8] A country wishing to become a contracting Party to an international Convention can do this by accession to the Convention or by a two-stage process that involves first signing the intent to become Party and then ratifying its signature.

[9] For more information on the calculation of the recycling capacity for meeting the entry-into-force conditions of HKC, refer to *resolution* MEPC.178(59) and to document MEPC 67/INF.2/Rev.1.

During 2018[10] the requirements for entry into force of HKC are that it has to be ratified (or acceded) by at least:

1. Fifteen States
2. Whose fleets amount to at least 516,418,680 GT (i.e., 40% of the 2017 world fleet of 1,291,046,701 GT)
3. Whose recycling facilities' combined maximum annual ship recycling volume is at least 15,492,560 GT (i.e., 3% of the tonnage condition of 516,418,680 GT)

To date, six countries have ratified or acceded to Hong Kong Convention (Norway, France, Belgium, Republic of Congo, Panama, and Denmark), while Turkey has obtained parliamentary approval for ratification. Together, these seven countries currently control 20.8% of the world's fleet. A number of maritime countries are making progress toward accession or ratification, and it is expected that it will not be too difficult to fulfill the first two conditions for entry-into-force, especially if shipowners feel the need to urge some of the open registries to accede.

The third condition in 2018 requires a "combined maximum annual ship recycling volume" of at least 15,492,560 GT. The 2017 capacities of the five-ship recycling countries and of the rest of the world were as follows:

India	12,210,082 GT
Bangladesh	9,888,137 GT
China	8,167,710 GT
Pakistan	5,703,133 GT
Turkey	1,540,800 GT
Rest of the world	624,848 GT

This data shows that Turkey plus India (or Turkey plus China) do not meet the required 15.49 m GT. The key to HKC's entry into force is therefore accession by two of the four large recycling nations (ideally India and China as the hazardous waste management infrastructure and many of the yards of both countries are already well developed, in line with the requirements of HKC).

It is a reasonable expectation that within the next 4–7 years, HKC will enter into force. In the meantime, meaningful progress has been taking place as the main ship recycling countries are working toward implementing tighter safety and pollution prevention requirements. The Turkish administration has implemented most of the requirements of HKC into its rules. In India the Ministry of Shipping introduced the Shipbreaking Code 2013, which replicated the full requirements of HKC to its ship recycling industry. Furthermore, the Indian Minister of Transport, Mr. Nitin Gadkari, told IMO's 30th Assembly in November 2017: "I am confident that we will ratify the Hong Kong Convention in the not-too-distant future". In Bangladesh the

[10]In 2019 the criteria will change according to the then published figure of the total GT of the world fleet as of end of December 2018 (to be published in April or May 2019).

Ministry of Industries has been working together with the recycling industry, with IMO and the Secretariat of the Basel Convention under a Norwegian-funded project for the creation of hazardous waste management facilities and for the development of training courses for ship recycling workers and managers. China, on the other hand, while it had implemented stricter requirements for yards authorized to import foreign ships for recycling, in 2018 announced a ban to the import of ships for recycling from the end of the year (Lloyd's List 2018). Lastly, the ship recycling industry in Pakistan does not appear to have embarked yet on the necessary work to improve safety and environmental standards. Nevertheless, following the appalling explosion on a tanker which killed 28 workers on November 1, 2016, at a Gadani yard, promises have been made by the Pakistani government and also by the ship recycling industry of significant improvements in the near future.

Recyclers from South Asia have expressed their concern that once their countries are Parties to HKC, then any powerful group of opponents to beaching may introduce an amendment to HKC to ban beaching and to close down the ship recycling industries in South Asian countries after the Convention has entered into force (note: it is not possible to amend a Convention before it enters into force). However, this cannot happen as provisions in HKC's Articles 18 (*Amendments*) and 19 (*Denunciation*) afford protection to all Parties by ensuring that the introduction of amendments will have to be done in a spirit of compromise and cooperation and by making it impossible to force an amendment to any Parties that do not agree to it. For example, Article 18.4 provides that: "Any Party that has declined to accept an amendment to the Annex shall be treated as a non-Party only for the purpose of application of that amendment", and furthermore, Article 19.1 provides that: "This Convention may be denounced by any Party at any time after the expiry of two years from the date on which this Convention enters into force for that Party." Conversely, a country that is not a Party to the Convention at a time an amendment is accepted will not be in a position to enjoy the protection described above, as Article 17.4 (*Entry into force*) provides that: "After the date on which an amendment to this Convention is deemed to have been accepted under Article 18, any instrument of ratification, acceptance, approval or accession deposited shall apply to the Convention, as amended." In other words, for a country to be in a position to control any amendments proposed after the Convention's entry into force, it should be a Party to the Convention at the time the amendments are discussed and negotiated.

In the longer term, the expectation is for the establishment of HKC as the single global standard. This is a realistic and feasible target to achieve in the next 7–10 years, simply by the accession of all five-ship recycling countries to HKC. Afterward, all flag States will have no option but to also accede to the Convention, while all shipowners will also have no option but to recycle their ships in line with HKC.

5 The European Union Ship Recycling Regulation

5.1 *The Mechanisms and Spirit of the New EU Regulation*

The progress that has taken place with the voluntary implementation of Hong Kong Convention and the, admittedly slow, uptake of the Convention by IMO member states should probably have provided sufficient comfort to activists and to the authorities within the European Union. This was not the case. Early in 2012, the European Commission having publicly recognized[11] that the enforcement of its own Waste Shipment Regulation to the recycling of ships was not working, it embarked on the development of new legislation for the recycling of European-flagged ships.

In accordance with the political system of the European Union, the Commission is the body responsible for initiating new legislation. On March 23, 2012, the Commission published its *Proposal for a Regulation of the European Parliament and of the Council on ship recycling.*[12] The document provided the Commission's version of the proposed Regulation and also its *Explanatory Memorandum.* The following three extracts from the Explanatory Memorandum convey the Commission's thinking at that time:

> A significant recycling capacity exists outside the OECD in China, India, Pakistan and Bangladesh. It is expected that facilities located in the OECD, in China as well as some facilities located in India will be able to comply with the requirements of the Hong Kong Convention by 2015.
>
> The objective of the Ship Recycling Regulation is to reduce significantly the negative impacts linked to the recycling of EU-flagged ships, especially in South Asia without creating unnecessary economic burdens. The proposed Regulation brings into force an early implementation of the requirements of the Hong Kong Convention, therefore hastening its entry into force globally.
>
> While it is difficult to expect the current 'beaching' facilities to be able to meet these requirements, it is possible that upgraded facilities might be able to fulfil these criteria in the future.

The Draft Article on "Requirements for ship recycling facilities" in the text of the proposed Regulation virtually reproduced the text of HKC, making no attempt to ban the beaching method.

The draft Regulation then went through the formal European process of negotiations between a Working Group of the European Council (officials from the ministries of the 28 member states) and the Environment Committee of the European Parliament, which in this instance was led by an MEP of the Green Party (*the rapporteur*). Unfortunately, during these negotiations the subject became unduly politicized through the persistent efforts of the rapporteur to enact a ban on beaching. In the end, after a long process of meetings and discussions, the three

[11] https://www.youtube.com/watch?v=nxk_c0Abhos.

[12] http://ec.europa.eu/transparency/regdoc/rep/1/2012/EN/1-2012-118-EN-F1-1.Pdf.

Parties ("the trilogue" between Council, Parliament, and Commission) agreed a compromise text on June 27, 2013, which removed all mentions of banning of beaching or of exclusion of South Asia's yards. It was published on December 10, 2013 in the Official Journal of the European Union, and on the December 30, 2013 the new "European Regulation on Ship Recycling (EU) No 1257/2013" (or simply here, "the EU SRR") entered into force.

While the European Council's representatives succeeded in rejecting the Parliament's preferred text which was banning beaching and also managed to preserve the full set of mechanisms of HKC, on the other hand, in order to accommodate political sensitivities, the final negotiations introduced some imprecise and ambiguous terms in the text, such as a requirement that ship recycling facilities shall *operate from built structures*[13]. Naturally, the persons involved in these negotiations did not attempt to define the meaning of "built structures" but left its interpretation to be given at a future time by the Commission. Furthermore, elsewhere the Regulation requires that, in order to be approved, a ship recycling facility (a) shall "demonstrate control of any leakage, in particular in intertidal zones[14]" and (b) shall ensure "the handling of hazardous materials and waste generated during the ship recycling process only on impermeable floors with effective drainage systems.[15]" Whereas the new European Regulation is very similar to HKC, the above two requirements as well as the requirement for facilities to operate from "built structures" are not from HKC. While these requirements may look noncontroversial and not unreasonable, it now appears that they might be used to justify a ban to beaching, as will be discussed in Sect. 5.2.

Notwithstanding the uncertainty introduced by the ambiguous terms that were invented in the final negotiations, the European Regulation replicates well the standards and the mechanisms of HKC and even requires the implementation of the guidelines that were developed by IMO for HKC. It is relevant to quote from the preamble to the Regulation: "This Regulation is aimed at facilitating early ratification of the Hong Kong Convention both within the Union and in third countries by applying proportionate controls to ships and ship recycling facilities on the basis of that Convention."

There are two noteworthy areas where the European Regulation differs from HKC: (a) in the way yards are authorized and (b) in defining two additional hazardous materials that need to be controlled in EU-flagged ships.

For yards located in EU countries, the Regulation requires each member state to enforce the requirements of the Regulation and to authorize the operation of yards in its jurisdiction. On the other hand, as the Regulation is not an international Convention, the EU does not have the power to enforce its requirements on yards that are located outside the EU nor can it expect the administrations of non-EU countries to authorize yards within their jurisdiction in line with the European

[13] Article 13.1 (c).

[14] Article 13.1 (f).

[15] Article 13.1 (g)(i).

Regulation. For this reason, yards located outside the European Union wishing to be included in "the European List of approved facilities" are required to apply to the European Commission, providing evidence of their compliance with the requirements of the Regulation, together with certification by an "independent verifier" who has inspected the site. Additionally, yards must accept the possibility of being subject to site inspection by the Commission or its agents.

The second difference between HKC and the EU SRR is that the latter includes two additional hazardous materials, which will need to be controlled on EU-flagged ships. The first of these materials is included in Annex I[16] of the EU SRR and is already banned in European Union law (perfluorooctane sulfonic acid and its derivatives, or PFOS, the main application on board ships being in some firefighting foams). The second material is included in Annex II of EU SRR (brominated flame retardant, or HBCDD, the main application on board ships being in expanded polystyrene used for cryogenic insulation, such as for liquefied gas tanks but also for refrigerator areas). A relevant footnote in Annex I to the EU SRR states that the control on PFOS "is not applicable to ships flying the flag of a third country," while the HBCDD, being a material of Annex II, strictly speaking need only be included in IHMs of newbuildings, plus in any retrofits involving changes to structure and equipment of existing ships. As the EU SRR treats all ships flying the flag of a third country as existing ships, regardless of their date of built (see EU SRR's Article 12(1), referring to Art. 5(2)), it follows that the inclusion of information on either of these materials will not be required on IHMs of non-EU-flagged ships, unless the HBCDD has been installed during a retrofit. Conversely, IHMs compiled for EU-flagged ships after the date of application of the EU SRR will fully satisfy the requirements of HKC. Note that good descriptions of the properties and typical uses of PFOS and of HBCDD can be found in a Norwegian submission[17] to IMO, which proposed in 2008 their inclusion as controlled hazardous materials under Appendices 1 and 2, respectively, of the HKC. The Norwegian submission was rejected at that time by IMO.

On December 30, 2013, the European Union brought into force the "European Regulation on Ship Recycling (EU) No 1257/2013." The provisions of the Regulation did not take effect immediately, but instead the Regulation specifies a schedule of application, whereby the first version of the European list of approved yards would be published not later than December 31, 2016. Thereafter, EU-flagged ships will have to have an Inventory of Hazardous Materials, be surveyed, be certificated, and be recycled in accordance with the new Regulation, from the earlier of the following two dates (termed as "the date of application"): (a) 6 months after the European List of approved yards reaches a combined capacity of 2.5 million LDT or (b) the end of December 2018. From the date of application, European-flagged

[16]Note that with regard to new and to existing ships, Annexes I and II of EU SRR have the same functionality as Appendices 1 and 2 of HKC.

[17]See IMO document MEPC 57/3/19: http://merchantmarine.financelaw.fju.edu.tw/data/IMO/MEPC/57/.

ships will be excluded from the scope of the European Waste Shipment Regulation, whereas non-European-flagged ships departing from European Union ports and destined for recycling will continue to be subject to the Waste Shipment Regulation, which forbids their export to developing countries. Furthermore, all ships visiting European Union ports, regardless of their flag, will be required from December 2020, to be provided with inventories of hazardous materials (IHMs).

It is unclear why the European Commission delayed until the middle of 2016 its invitation of applications from ship recycling facilities located outside the European Union. Applications were received from the USA (2 yards of 72,868 LDT), China (4 yards of 1,767,215 LDT), Turkey (7 yards of 450,903 LDT), and India (initially 5 yards of 323,497 LDT and subsequently a further 4 yards) of a combined maximum annual capacity of 2.6 million LDT. Due to further delays, by the beginning of 2018, none of the non-EU applicant yards had been inspected or approved by the Commission. In the meantime, the Regulation's requirement for the publication of the first European List by or before December 2016 was satisfied, as the Commission published at the end of 2016 its first list, which included 18 yards in 10 EU member states with maximum annual recycling capacity of 303,065 LDT. On 4th of May 2018 the Commission updated its first List with its Implementing Decision (EU) 2018/684[18] thereby increasing the total number of EU-based approved yards to 21 facilities of 329,917 LDT maximum annual recycling capacity (of which 86,815 LDT correspond to three UK yards that will lose their approval on Brexit in March 2019). At the time of writing, the Commission had not approved any non-EU-based yards, and it is therefore a fair guess that the date of application of the Regulation will be the end of December 2018, and not earlier, as it is improbable that the Commission will approve yards of 2.5 million LDT capacity by the middle of 2018.

5.2 *Implications of the EU Regulation*

For the last 20 years, environmental activists have been campaigning against unsafe and polluting practices in ship recycling. Initially Greenpeace was the lead NGO in this campaign. In 2005 the Brussels-based *NGO Shipbreaking Platform* ("the Platform") was set up to coordinate the activities of 19 environmental, human rights, and labor rights organizations interested in ship recycling. It is indisputable that the activists have made a great contribution to the development of awareness among the public, the regulators, and the shipping industry. Without their relentless demands, it is conceivable that the HKC may not have even been developed. On the other hand, the activists, and more specifically the Platform, have shown a total lack of knowledge and interest to learn how the shipping and the ship recycling industries

[18] See the European Commission's official site for the European List of approved facilities: https://eur-lex.europa.eu/legal-content/EN/TXT/PDF/?uri=CELEX:32018D0684&from=EN.

work. The result is that their campaigns are, more often than not, impractical or unworkable. Over the years, the Platform has campaigned tirelessly for the enforcement of the Basel Convention, of the Ban Amendment and of the European Regulation on Waste Shipment to ship recycling, regardless of the numerous cases that have demonstrated these regimes as impractical and unenforceable to ships. Furthermore, since its inception the Platform has campaigned for the banning of beaching and for stopping the recycling of ships in South Asia. The Platform has opposed HKC, primarily because the Convention does not ban beaching. To start with, they strenuously opposed the development of the European Regulation on Ship Recycling, as they saw this as an admission by the European Union of the failure of the Basel Convention and the related European Regulation on Waste Shipment. However, when the Green Party assumed the leadership of the Environment Committee of the European Parliament for the development of the new regulation and as the Green Party adopted all the policies and arguments of the Platform, the NGO appeared to change its view on the usefulness of the new regulation as a vehicle for banning beaching. Since the adoption of the new regulation in 2013, the Platform has been pressing the European Commission to interpret the new Regulation as banning beaching.

In parallel to the Platform's lobbying in Brussels, in the summer of 2013, after the final version of the text of the Regulation had been agreed between the Council and the Parliament, there was a highly irregular intervention by an adviser of the Green Party in the European Parliament who managed, unnoticed, to make some small changes to the agreed text in some of the EU languages, including English. The changed text in essence requires that "the handling of hazardous materials, and of waste" must be done on impermeable floors, as opposed to the agreed text, which referred to "the handling of hazardous materials and waste" (i.e., the word hazardous applying to both, materials and wastes). As in European regulatory language an "end-of-life ship" is considered "waste," but not necessarily "hazardous waste," it follows that the changed text can be interpreted as requiring that nothing from the ship must touch the beach, not any clean steel blocks, and not even a table and chairs! This point of detail obviously has been invented as an impassable obstacle to beaching. It took EU member states some time before noticing this small but potentially crucial change. When it was also realized how the changed text could be interpreted, the European Council proposed to the Parliament and the Commission a Corrigendum (i.e., a formal correction) to reinstate the agreed text. It is understood that a faction of the Parliament strenuously opposed this and consequently the issue remains unresolved to this day.

In accordance with the European system, the European Commission is the body that enforces and interprets European law, although it should be noted that the final arbiter to interpret the Regulation would be the European Court. The Commission so far has given mixed messages as to what it intends to do about the beaching method of recycling. If the Commission interprets the new European Regulation in line with the rational of Hong Kong Convention and does not invent reasons to ban beaching, then EU SRR will result in motivating and in providing commercial benefits to those yards that have invested in improvements. In doing so, the EU will be encouraging the accession to HKC of the countries in South Asia. If on the other hand the Commission chooses to interpret the Regulation as providing a ban on beaching (and in so doing use the surreptitiously modified English version of the text, ignoring the originally agreed text which luckily survives in some of the European Union's languages), it will block European-flagged ships from using much of the world's recycling market. In that case, it is very likely that many, if not most, of these ships will change flag and go for recycling to South Asia, thus electing to ignore Brussels. This scenario has become even more likely following the Chinese government's ban to the import of ships for recycling beyond the end of 2018 (Lloyd's List 2018). But even if the European ships do not reflag but choose to comply with the Regulation, what a hollow victory the Commission will have scored! By preventing the progressive minded European market from using HKC compliant yards in South Asia, the EU Regulation would torpedo the progress that has taken place so far. Without the demand for responsible recycling that is currently filling the HKC compliant yards in South Asia, one of the major driving forces for change would be removed. All the European Commission will have achieved is to abandon the majority of the world's ship recycling workers and infrastructure to the realm of the noncompliant, which is not in line with the HKC's goal of raising standards at all yards across the world.

6 Enabling Mechanisms for the Improvement of Standards in the Ship Recycling Industry

6.1 The Responsibility of Shipowners

It is often said that shipowners are, or should be, responsible for the standards under which their ships are recycled. However, as was discussed in Sect. 3, the disposal of an end-of-life ship involves the transfer of ownership, first from the shipowner (the ship-owning company to be more precise) to the cash buyer and then from the cash buyer to the recycler. At the instant ownership is transferred, the old owner of the ship ceases to have the benefits and the responsibilities of owning the ship, regardless of whether the ship is sold for recycling or for further operation. It is therefore not realistic to expect shipowners to be legally responsible for what happens after their ship is sold, unless the seller is violating some specific law by the way he is selling his ship. It is thought that this will be the case under the new European Union Ship Recycling Regulation, which forbids European-flagged ships from being recycled in yards that are not included in the European List of approved yards. However, even in this case, as shipowners do not sell their ships directly to yards, there may be serious doubt whether a European-flagged ship sold through a cash buyer and recycled outside the scope of the European Regulation could be breaking the law. This is because as the moment ownership is transferred to the cash buyer, the ship has to be deregistered from its (European) registry and a subsequent sale to a non-approved yard would logically be taking place outside the scope of the

European Regulation. Even if European courts were to decide in the future that a seller of a European-flagged ship can be held responsible for the actions of a new owner, i.e., the cash buyer, the shipowner will still have the escape route of changing the flag of his ship prior to selling it for recycling so that the ship does not fall under the scope of the European Regulation.

The above considerations would suggest that, at least presently, no legal responsibility can be attributed to shipowners for the way they recycle their ships. Nevertheless, NGO activists operating under the umbrella organization of "NGO Shipbreaking Platform" have been pursuing "name and shame" campaigns against individual shipping companies who have recycled ships in countries whose standards are judged by the Platform as being low. Aside from the claim of the Platform that its activists are qualified and even capable to judge the standards of individual yards (normally done by the Platform on the basis of whether a yard is located in South Asia or not), the activists ignore whether legal responsibility applies or not and instead imply a moral responsibility for the shipowner. But as we know, the world of commerce does not work like that. In the main, shipowners do not decide where to build their ships, or where to repair them, or what cargoes to carry on the basis of subjective judgments of self-appointed watchdogs on safety and environmental conditions in building yards, or in repair yards, or in the mines where ores are extracted. Nevertheless, separate from legal and from moral responsibility, there is another kind of responsibility that is gaining popularity nowadays. This is the corporate social responsibility (CSR), which in fact is a business approach that relies on voluntarily taking responsibility for a company's effects on the environment and on social well-being. CSR applies to efforts that go beyond what may be required by regulation. A main benefit of incorporating CSR in a company's policies and procedures is a marketing advantage among the company's clients and the wider public. This could explain why most of the shipping companies that have embraced CSR are either publicly listed companies or companies whom, or whose clients, are directly exposed to the vagaries of public opinion.

The above discussion leads us to the conclusion that CSR is the one kind of responsibility that can motivate a shipowner to consider and select the most appropriate recycling yard for his end-of-life ship. And this is not just theory but is what has been happening in the market in the last 4 years, with a small number of well-known and influential shipping companies having managed to create a two-tier market between normal recycling and responsible (or "green") recycling. Interestingly, at least one of these companies have interpreted their CSR policy on ship recycling as meaning that their ships will not be recycled by the beaching method. The rest of CSR companies have expended considerable effort and resources to select and supervise beaching yards that have improved their infrastructure, procedures, and training of their workforce. The one company that has vowed to stay away from beaching has done so apparently by pressure applied by the NGO Platform to the local government who are shareholders of the company. If the ideal of corporate social responsibility is to strive to improve the well-being of people and the environment in the wake of the corporation, then the choice to totally abstain from recycling your ships in the very places that need your motivating influence appears to be just a poor cop-out.

6.2 The Role of Regulations

Issues of safety, prevention of pollution, and even social justice cannot be left to the industry at large in the hope that good sense will prevail nor to the small sector of the industry that may voluntarily adopt high standards. This is the reason why virtually all aspects of ship safety and pollution prevention are regulated by IMO's international conventions. Most safety issues are regulated by SOLAS (the International Convention for the Safety of Life at Sea 1974), which at the end of 2017 had 163 Contracting Parties representing 99.14% of the GT of the world fleet. Safety is also the subject of COLREG (Convention on the International Regulations for Preventing Collisions at Sea 1972), which in December 2017 had 157 Contracting Parties representing 99.13% of the GT of the world fleet. Prevention of overloading of ships is regulated by the LL Convention (the International Convention on Load Lines 1966), which at the end of 2017 had 161 Contracting Parties representing 99.13% of the GT of the world fleet. Prevention of pollution is regulated by MARPOL (the International Convention for the Prevention of Pollution from Ships 1973), which had 155 Contracting Parties at the end of 2017, representing 99.14% of the GT of the world fleet (IMO 2017). These and other Conventions are enforced by the flag States of the ships when these are Parties to the Conventions and are also policed by the port State control (PSC) officials in ports where the ships load or discharge cargo. The PSC system is enshrined in all of the above Conventions, as is also what is known as the "no more favorable treatment" that allows port States that are Parties to a Convention to demand that a ship flying the flag of a non-Party satisfies the requirements of the subject Convention. For example, a ship that flies the flag of a non-Party to MARPOL will still be expected to satisfy the requirements of MARPOL when it sails in a port of a State that is MARPOL Party. In other words, the ship receives no more favorable treatment by flying the flag of a non-Party. The combination of high percentage of the world fleet being Parties to a Convention, the policing at ports by PSC, and the provision of no more favorable treatment mean that safety and pollution prevention issues are implemented globally and to 100% of ships, and therefore a shipowner cannot gain a commercial advantage over his competitors by reducing costs through noncompliance.

The majority of IMO's Conventions apply exclusively to ships. Hong Kong Convention is one of a handful of IMO Conventions whose scope extends beyond the ship, in also regulating safety and environmental protection on land-based facilities. Incidentally, this simple fact makes it more complicated and time consuming for a government to ratify or accede to HKC, as the concurrence of a number of ministries becomes necessary. The process will usually involve the Ministry of Transport (or shipping) being the IMO focal point and the responsible ministry for shipping matters; the Ministry of Labor being responsible for issues relating to the health and safety of workers in the recycling facilities; the Ministry of Environment being responsible for the treatment, storage, and disposal of hazardous materials; plus other ministries or government departments providing input on matters such as law, customs and excise, testing for explosive conditions, etc.

This dual nature of the regulatory regime for ship recycling (i.e., the need to have jurisdiction on ships and also on land facilities) creates profound difficulties for the strict enforcement of the Convention (or for that matter for the strict enforcement of the regional European Union Ship Recycling Regulation). It is instructive to explain these difficulties here, as this will provide the reader with a much clearer understanding of ship recycling.

It has been said that Hong Kong Convention imposes high costs to ship recycling facilities, while it requires very little expenditure from the shipowner. At first, this may appear to be a valid claim, as the owner of a recycling facility would invariably have to spend considerable resources on infrastructural improvements, on training of the workforce, and on developing and implementing working procedures which would inevitably extend the time it takes to recycle each ship, thereby increasing ship recycling costs. On the other hand, the shipowner only needs to procure an Inventory of Hazardous Materials for his ship, perform some surveys, and obtain certification, all of which do not amount to a significant cost. This seemingly unequal distribution of costs appears even more unfair when considering that the great majority of ship recycling facilities are located in less developed countries, while the push for improvements of ship recycling standards has come from the most developed countries in the northern hemisphere. However, the justification for what may appear to be an unfair allocation of investment costs for compliance with the Convention is relatively simple. In the first place, it is only natural that the owner of the recycling facility and the shipowner will each invest on their properties. But more importantly, if the shipowner has no option but to send his ship for recycling only to facilities that fulfill the standards of the Convention, then the market forces of supply and demand will adjust the purchase price of ships to cover the cost of the investments made by the owner of the recycling facility. Simply expressed, the shipping industry will have no option but to pay all the costs for the improvements demanded by Hong Kong Convention. Furthermore, in the longer term, the shipping industry will also recoup its costs of compliance with the Convention from its clients, the world's consumers.

The above rationalization relies on one simple but vital assumption, namely, that owners of ships flying the flags of Parties to the Convention will send their ships for recycling only to Hong Kong Convention compliant yards. As discussed in Sect. 1.2, presently five countries recycle around 98% of the tonnage recycled worldwide. It follows that if all five recycling countries are Parties to the Convention when this enters into force, then shipowners will truly have no option but to recycle their ships in compliant yards. If, on the other hand, one or more of these five countries are not Parties to the Convention when this enters into force, their recycling yards at that time would be operating at a lower-cost basis compared to compliant yards in Party countries and would therefore be in a position to pay higher prices for purchasing ships. This is the situation that gives rise to the profound difficulties mentioned above for enforcing a strict regulatory regime to ship recycling.

As Hong Kong Convention make provisions for port State control in its Article 8 and for "no more favorable treatment" in its Article 3.4, it should follow that the requirements of the Convention that apply to ships in service (i.e., provision of IHM, survey, certification, and restrictions on installation of hazardous materials) will be implemented and enforced on all ships, including those that fly the flag of non-Parties by virtue of the provision for "no more favorable treatment". However, while one or more of the main ship recycling countries remain non-Parties, the uniform and strict enforcement of the Convention's provisions for the recycling of ships flying the flag of Party States cannot be guaranteed. Whereas a ship will be able to demonstrate to PSC inspections throughout its operating life that it fulfills the requirements of the Convention, on the other hand, at the time the ship is sent for recycling, it will be possible (and certainly not illegal) for the shipowner to take advantage of any better prices that may be offered by non-Convention yards, either by selling the ship to a cash buyer on an as-is-where-is basis or by reregistering the ship to a non-Party flag. The cost of changing flag for an average-sized ship is of the order of US$1 per LDT, which is quite insignificant if a non-Convention yard pays, say US$30 to US$50 per LDT, more than a Convention yard.

The above discussion must not be taken as suggesting that yards in non-Convention countries would be having a clear-cut marketing advantage, as it is quite probable that Convention yards would profit from having unhindered access to end-of-life ships of compliant shipowners. Nevertheless, the conclusion and the plain truth is that unlike most Conventions that regulate the shipping industry, the dual nature of Hong Kong Convention will allow shipowners to avoid their obligations for as long as there are ship recycling countries that are not Parties to the Convention. Conversely, when all five main ship recycling countries are Parties to the Convention, then its requirements will become the universal standard for all ships and all recycling yards. This is one additional reason why IMO, during the development of the Convention, turned down proposals to ban beaching.

The dynamics discussed in this subsection also apply to the enforcement of the European Regulation on Ship Recycling. If the European Commission approves South Asian facilities for the recycling of European-flagged ships, it will motivate the progressive uptake of the Regulation's standards, which are almost the same as those of Hong Kong Convention. If on the other hand the Commission ignores the improvements that are taking place in South Asia, then the intelligent reader should be able to predict easily the outcome of Europe's involvement with ship recycling.

6.3 Steps Toward a Global Regulatory Regime for Ship Recycling

The above discussion should have demonstrated that, for the global establishment and strict enforcement of minimum standards in ship recycling, there is no alternative to a universally implemented international Convention. This is the intended role and the future of Hong Kong Convention, as long as it is not derailed by the efforts of an overzealous Europe. However, as we are approaching the 10-year anniversary since the adoption of the Convention, it is fair to ask what is keeping all the counties who unanimously adopted the Convention in 2009 from acceding to it. Looking at other IMO Conventions, it would appear that long delays between the adoption and the entry into force of Conventions are quite normal. Furthermore, as already discussed, the nature of Hong Kong Convention, combining regulations for ships and for land facilities, increases the complexity of accession. Another contributory factor that could be delaying accessions is a perceived conflict between the second and third conditions for entry into force, requiring that countries who have ratified/acceded to the Convention control (a) no less than 40% of the world's fleet and (b) a proportionate (3%) ship recycling capacity. It is presently understood that this may be holding back some large open registries from acceding because of a fear that too much tonnage under the second condition could make it very difficult to satisfy the third condition. However, on closer examination of the fleet and ship recycling data presented in Sect. 4.5 on entry into force of Hong Kong Convention, it would appear that the risk of too much tonnage jeopardizing the satisfaction of the third condition is far too remote: as the ideal minimum, the third condition would require accession by India and China. The recycling capacities of these two countries correspond (according to the 3% formula of the third condition) to 52.6% of the world fleet in 2018. Furthermore, if the recycling capacity of Turkey, which is due to complete its ratification of the Convention, is added to those of India and China, the capacity of the three countries corresponds to 56.6% of the world fleet, providing ample cushioning against the perceived risk of conflict between the two conditions.

The Diplomatic Conference that adopted Hong Kong Convention in May 2009 also adopted six Conference Resolutions, including one on "Early Implementation of the Technical Standards of the Hong Kong International Convention for the Safe and Environmentally Sound Recycling of Ships, 2009." The Resolution, which has no binding power, in apparent anticipation of the delay in the Convention's entry into force, invites member states of the Organization to consider applying the technical standards contained in the Convention on a voluntary basis to ship recycling facilities under their jurisdiction and also invites the industry to apply the technical standards contained in the Convention to ships and to ship recycling facilities. The government of India adopted its "Shipbreaking Code, 2013," which replicated substantial parts of the technical standards of Hong Kong Convention and applied these to India's recycling industry. At the end of 2017, the government of India commenced pre-legislative consultations with stakeholders on its draft "Safe and Environmentally Sound Recycling of Ships Bill, 2017," to give effect to the provisions of Hong Kong Convention. The government of Bangladesh also commenced the process of aligning its regulations with Hong Kong Convention with the adoption of "The Ship Breaking and Recycling Rules, 2011." On January 23, 2018, the country's Parliament adopted the "Bangladesh Ship Recycling Act 2018," which aims to create the necessary capacity and infrastructure for Bangladesh to meet the requirements of HKC within the next 5 years. In Europe, the "Proposal

Fig. 6.14 Yard in Alang with HKC SOC by ClassNK

for a Regulation of the European Parliament and of the Council on ship recycling," which led to the development of the European Regulation on Ship Recycling, had the stated aim to "bring into force the requirements of Hong Kong Convention, therefore hastening its entry into force globally."

The delays of governments to accede/ratify Hong Kong Convention, concerns over a ban to beaching by the European Regulation, the propaganda of the NGO Platform, and the decline of the Chinese ship recycling market led a number of quality shipping companies, first from Japan and then from Europe, to work closely with selected recycling yards in India who agreed to upgrade their infrastructure, training, and procedures so as to comply with Hong Kong Convention. Initially four recycling yards decided to invest in improvements, on the expectation that they would benefit financially from the custom of quality shipping companies who needed the availability of yards that can recycle ships in compliance to Hong Kong Convention. Following more than 1 year's work, toward the end of 2015, the four yards were awarded Statements of Compliance (SOCs) with Hong Kong Convention by Japan's ClassNK. Figures 6.14 and 6.15 were taken at two of these yards. What followed can be described as a virtuous cycle at work. With growing demand for responsible recycling from shipowners, a two-tier market developed with a price differential between normal recycling and responsible (or green) recycling. The four compliant yards enjoyed demand for their services, which was reflected in profitable

Fig. 6.15 Yard in Alang with HKC SOC by ClassNK

contracts. Profiting from compliance with Hong Kong Convention incentivized numerous other recyclers in Alang to start upgrading and to seek Statements of Compliance for their yards. Whereas in 2015 the majority of the recycling industry in Alang was openly hostile toward Hong Kong Convention, attitudes changed and as of April of 2018, 61 of Alang's 120 recycling yards had obtained Statements of Compliance with HKC from IACS classification societies, while a further seven yards were working toward their certification.

Following the voluntary initiatives taken by the shipping and the ship recycling industries, the European Commission is now in a position to further motivate the virtuous cycle of improved standards for improved rewards, by approving the leading yards in India and, in this way, helping increase the number of ships that are seeking responsible recycling in the traditional recycling centers.

Whereas so far limited progress has taken place in Bangladesh (and even more limited in Pakistan), one of the largest yards in Bangladesh has taken notice of the growing international expectations for improved ship recycling standards and has responded with startling improvements to its infrastructure and working procedures, as can be seen in Fig. 6.16.

Until governments finally bring Hong Kong Convention into force, the shipping industry will need to continue to support and to channel its business to those yards that are investing in improved standards.

Fig. 6.16 PHP Shipbreaking and Recycling Industries Ltd. in Chittagong Bangladesh with HKC SOC by RINa

Acknowledgments The photographs used in this chapter are the author's with the exception of the photographs on Fig. 6.6 (which has been provided by Mrs. Susan Wingfield, Programme Officer of the Secretariat of the Basel, Rotterdam, and Stockholm Conventions), Fig. 6.13 (provided by Dr. Claude Wohrer of the Secrétariat Général de la Mer of the French Prime Minister's Office), and Fig. 6.16 (provided by Mr. Mohammed Zahirul Islam, Managing Director of PHP Shipbreaking and Recycling Industries Ltd). The chapter draws on material from a booklet produced by GMS (Mikelis 2018).

References

GMS Weekly (issues from 2006 to present). Back issues starting from 7th January.

IHS Global Ltd, World Casualty Statistics. (2017). Table 7C. Also back issues of the same publication from 1998 to 2016.

IHS Maritime & Trade, World Fleet Statistics. (2017). Table 20. Also back issues of the same publication from 2007 to 2016.

IMO. (2017). Status of multilateral conventions and instruments in respect of which the International Maritime Organization or its Secretary-general performs depositary or other functions. International Maritime Organization.

Lloyd's List. (2018). Import ban on scrap ships deals a heavy blow to Chinese shipbreakers, 4 May.

Mikelis, N. (2007). A statistical overview of ship recycling, International Symposium on Maritime Safety, Security & Environmental Protection (SSE07). Laboratory for Maritime Transport at the National Technical University of Athens, 20–21 September, Athens.

Mikelis, N. (2018). The recycling of ships. http://www.gmsinc.net
Steel Statistical Yearbook. (2017). Also, earlier issues. World Steel Association.
World Steel in Figures. (2018). World Steel Association.
World Steel Recycling in Figures. (2013–2017). Also, earlier issues. Bureau of International Recycling, Ferrous Division.

Chapter 7
Reducing Sulfur Emissions: Logistical and Environmental Considerations

Thalis P. V. Zis and Harilaos N. Psaraftis

Abstract In recent years the issue of sulfur emissions from maritime transport has seen newfound attention. This chapter presents an overview of the main issues of sulfur emissions and the legislative framework that seeks to reduce the sulfur footprint of the maritime sector. It also analyzes potential modal shifts toward less efficient land-based modes which may happen as a result of sulfur regulations and investigates the related potential economic damage to ship operators. To that effect, this chapter presents findings from a recently finished project at DTU and the developed methodological framework that can be used to estimate such modal shifts, as well as to measure the efficacy of policy and ship operators' measures to reverse such shifts.

Abbreviations

BAF	Bunker adjustment factor
CBA	Cost-benefit analysis
CO	Carbon monoxide
CO_2	Carbon dioxide
EC	European Commission
EEDI	Energy Efficiency Design Index
ETS	Emission trading system
EU	European Union
GHG	Greenhouse gas
HFO	Heavy fuel oil
IMO	International Maritime Organization
LNG	Liquefied natural gas
MDO	Marine diesel oil

T. P. V. Zis (✉) · H. N. Psaraftis
DTU Management Engineering, Technical University of Denmark, Kongens Lyngby, Denmark
e-mail: tzis@dtu.dk; hnpsar@dtu.dk

H. N. Psaraftis (ed.), *Sustainable Shipping*,
https://doi.org/10.1007/978-3-030-04330-8_7

MGO	Marine gas oil
NO_x	Nitrogen oxides
Ro-Pax	Ro-Ro with passengers
Ro-Ro	Roll-on roll-off
SECA	Sulfur Emission Control Area
SO_x	Sulfur oxides
SSS	Short sea shipping
WHO	World Health Organization

1 Introduction

The purpose of this chapter is twofold: (a) present the main general issues associated with the reduction of sulfur emissions in maritime transport and (b) focus on a specific study which has investigated possible modal shifts as a result of sulfur regulations, as well as possible actions to mitigate and reverse such modal shifts.

1.1 Background: What Are SO_x

In order to understand the recent attention of the shipping community on the issue of sulfur emissions, it is important to first present this pollutant species in brief. Sulfur oxides (SO_x) refer to the family of chemical compounds formed from atoms of sulfur and oxygen. SO_x may refer to one of the following: SO (sulfur monoxide), SO_2 (sulfur dioxide), SO_3 (sulfur trioxide), S_2O (disulfur monoxide), S_2O_2 (disulfur dioxide) and lower sulfur oxides (S_nO, S_7O_2 S_6O_2). Most of these SO_x are unstable and rarely encountered in nature. However, SO_x produced by fuel combustion in marine engines predominantly contain SO_2 emissions which is the main subject of this section.

Sulfur dioxide is a chemical compound that consists of two oxygen atoms and a sulfur atom. Its molar mass is equal to 64.066 g per mol. SO_2 is in gaseous form at standard temperature and pressure and exists in Earth's atmosphere in very small concentrations of approximately 1 ppbv (Pickering and Owen 1997). Chemically, SO_2 is produced following the oxidation of sulfur or other materials that contain sulfur according to the following chemical reaction:

$$S + O_2 \rightarrow SO_2$$

Fossil fuels (coal and oil) contain varying amounts of sulfur according to their purpose which typically varies between 0.10% and 5%.

SO_2 has a dual nature as a global and local pollutant. It is considered a hazardous pollutant that can cause nerve stimulation in the lining of the nose and throat and affects people with asthma. Further oxidation of SO_2 occurring in the presence of NO2 (which is also generated during fuel combustion) forms sulfuric acid (H_2SO_4).

This contributes substantially to form acid rain which in turn interferes with the growth of flora and fauna and also affects water-life (Cullinane and Cullinane 2013).

Sulfur dioxide is generally not considered a greenhouse gas (GHG). In fact, it has been argued that SO_2 only has a cooling effect. In the pre-industrial era, climate change is credited to have started by major changes in volcanic activity which release large quantities of SO_2 (Crowley 2000). Major historic volcanic eruptions have led to the cooling of the earth's surface; however, Ward (2009) concludes that this happens only when such volcanic eruptions are sporadic and large. Large eruptions in quick succession (e.g. at least 1 each year for a period of 10 or more years) would impair the oxidizing capacity of the atmosphere, leading to the accumulation of GHGs. Ward therefore shows that SO_2 can indirectly contribute to global warming. The combustion of fossil fuels every 1.7 years is emitting as much SO_2 as one large volcanic eruption (as measured in Greenland). The largest source of SO_2 emissions is fossil fuel combustion at power plants (approximately 73%) and other industrial facilities (20%), while smaller sources include the burning of high sulfur containing fuels by locomotives, non-road equipment and large ships (EPA 2013). The World Health Organization (WHO) recommends that a concentration of 500 $\mu g/m^3$ should not be exceeded over averaging periods of 10 min duration and that the 24-h mean should be less than 20 $\mu g/m^3$. The 10-min period is justified as such short periods of exposure of asthmatics to SO_2 may provoke changes in pulmonary function and respiratory symptoms. Longer-term exposures to SO_2 together with other pollutants contribute to changes in mortality, morbidity or lung function (WHO 2005). In the European Union, legislation suggests that the 1-h mean should not exceed 350 $\mu g/m^3$, whereas the 24-h mean should be less than 125 $\mu g/m^3$. The maximum allowed number of exceedances each year is 24 and 3, respectively, (European Commission 2008) which is less strict than the WHO guidelines. Comparing the two standards, the EU limit is more tolerant on its daily mean value which can be interpreted as a consequence of the better air quality in comparison to countries that rely more on coal and fossil fuels for their energy production.

1.2 Relevant Regulation

In response to these limits, the IMO and the EU have set specific limits on the sulfur content of fuel used near and at ports.

In 1997, the Kyoto Protocol of the UN framework convention on climate change requested that the IMO considered and addressed the issue of ship emissions. In response, the IMO commissioned a study on GHG emissions with a focus of identifying feasible reduction strategies. In 1977 limits for the main air pollutants from exhaust gas from ships were introduced through the MARPOL Annex VI. The second GHG study was commissioned and estimated the global contribution of shipping in CO_2 emissions at 2.7% (Buhaug et al. 2009). The revised MARPOL Annex VI introduced limits on the maximum sulfur content allowed in bunker oil and created designated sulfur emission control areas (ECAs) where tighter limits applied. The first ECA for SO_x emissions was the Baltic Sea which was designated in 1997. However, it was not until 2005 that the ECAs were enforced. The next ECA was the North Sea which was adopted in 2005 and enforced in 2006. In 2010 the North American ECA was designed and enforced in 2011. In 2011 the United States Caribbean Sea ECA was adopted and enforced at the beginning of 2013. Activity in the ECAs would be in effect 1 year after the area was enforced. Figure 7.1 presents the progression of the maximum sulfur content in bunker oil within and outside SECAs.

The regulation dictates that from 2020 onward, the global limit of sulfur content will be 0.50% (outside SECAs), a decision that was confirmed at the 70th session of IMO's Marine Environment Protection Committee (MEPC 70) in October 2016 after a long debate on low-sulfur fuel availability. The further designation of SECAs has been suggested by Cullinane and Bergqvist (2014) due to the social benefits due to the regulation. Regarding NO_x emissions, marine diesel engines on-board ships must meet certain performance standards defined by MARPOL VI. This will also provide a positive impact on the near-port air quality. The US ECAs are also NECAs (Nitrogen Emission Control Areas), whereas for the European ECAs (North and Baltic Sea), the NO_x limits will be enforced from 2021. The specifications for NO_x emissions are presented in Fig. 7.2 for the different tiers of the regulation (based on the ship construction date), and the resulting emission limits are plotted.

Fig. 7.1 Maximum allowed sulfur fuel content (%)

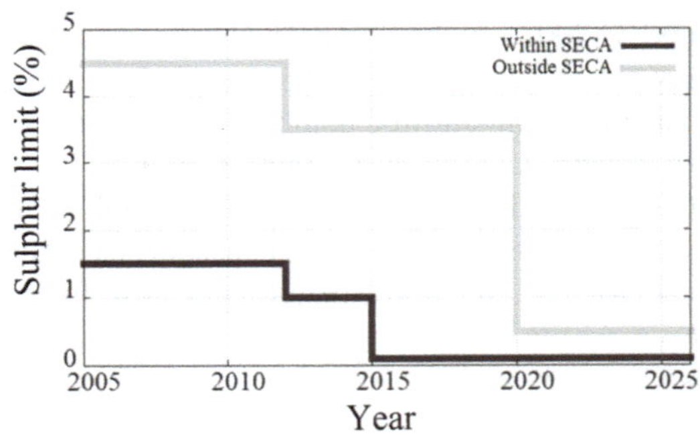

		NOx Limit (g/kWh)		
		Rated Engine Speed (rpm)		
Tier	Year	<130	130-2000	>2000
I	2000	17	$45 \cdot n^{-0.2}$	9.8
II	2011	14.4	$45 \cdot n^{-0.23}$	7.7
III	2016	3.4	$9 \cdot n^{-0.2}$	1.96

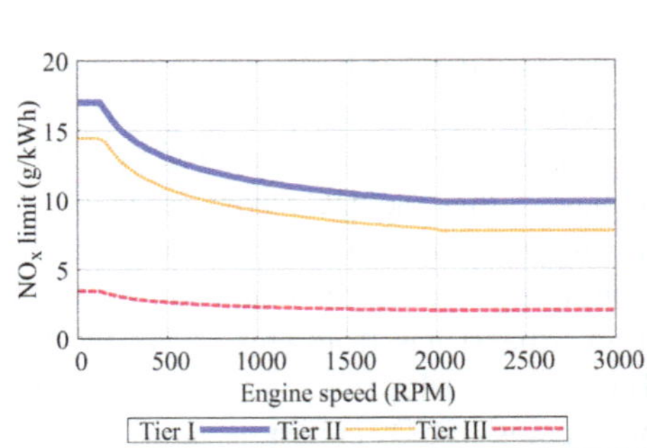

Fig. 7.2 NO_x emission limits (MARPOL Annex VI). (Source: Zis 2015)

Tier II describes the current NO_x limits for engines anywhere in the world, whereas Tier III will apply for ships built on or after the 1st of January 2016 and sailing in NO_x ECAs. For these vessels sailing outside ECAs, the Tier II limits will apply. As the limits in sulfur content will soon reach very low values there, it is envisioned that existing zones will be expanded and new areas introduced.

The introduction of SECAs and their progressively stricter limits have shown to be a considerable success on limiting SO_x emissions of the sector. The most cited figure on the contribution of international shipping on SO_x emissions is the one provided by Eyring et al. (2005) which ranges between 5% and 8%, but that was at a time where no sulfur limits were in place. Zis and Psaraftis (2018a) provide a more recent estimate on sulfur emissions that is extracted from data of the Organization for Cooperation and Development (OECD). For 2015, OECD estimates that transportation (mobile sources) accounted for 3.45% of the total sulfur oxides emissions within the OECD countries, 14% of which was attributed to road transport (OECD 2017). This shows that the recent regulation on limiting SO_x emissions from the maritime sector has seen considerable success, and it can be expected that post-2020, the global contribution will be even lower when the global sulfur cap is enforced.

1.3 Compliance with the Regulation

To comply with the regulation within SECAs, ship operators can either use ultra-low sulfur fuel (such as MGO or MDO) or rely in abatement technologies that result in the same SO_x emissions reduction (e.g. use of scrubber systems).

Low-sulfur MGO is pure distillate oil that contains less than 0.10% sulfur and can be used in conventional marine engines within ECAs and other sulfur-regulated areas (e.g. EU ports). This fuel can be used without major modifications, but one drawback is that it has to be stored at a different tank for vessels that use fuel switching. MGO in general has a lower viscosity than HFO, and as a result additional lubrication must be used to avoid damage in the engine's pumps (MAN 2014). Historically, fuel with lower sulfur content is more expensive than regular bunker oil. Low-sulfur fuel requires additional refining which can also result in additional transportation (from production facility to refinery) with environmental (e.g. increased carbon footprint) and economic implications. The price differential among different fuel types is not constant, and as all fuel prices, it is also characterized by significant volatility.

Scrubber systems are neutralizing sulfur oxides by filtering the exhaust gases through water which results in sulfate containing waste water that is recirculated into the sea. Three main types of scrubber systems are currently used depending on the water use to wash out the sulfur oxides. These are:

- Seawater systems (open loop)
- Freshwater systems (closed loop)
- Hybrid systems

The first type can use seawater for the scrubbing process so long as the alkalinity of the water is sufficient (Henriksson 2007). In other cases (notable examples are the waters in the Baltic Sea and near Alaska), it is necessary to use freshwater systems. Finally, hybrid systems allow the change of water depending on where the vessel is operating. All types of scrubber systems can be installed on both new builds and older vessels (retrofitted). The latter is more costly, and there are additional considerations on the space capacity available to install the system and where necessary the freshwater tanks (DMA 2012). In terms of environmental performance, freshwater scrubber systems are reported to reduce SO_x by up to 97% and PM emissions by an estimated 30–60% when HFO of up to 2.5% sulfur is used (EMSA 2010). However, these emission reductions do not take into account the increase in the overall fuel consumption that is associated with the scrubber systems' energy requirements.

The total capital cost required to install a scrubber system depends on the type (open or closed loop) and size of the installation. Rough estimates include a cost range of between 100 and 200 Euros per kW of installed power on new builds and 200–400 Euros for retrofitting installations. More detailed information from EMSA is given in Table 7.1 below.

Additionally, there are operating and maintenance costs associated with the use of scrubbers. A very important extra cost stems from the increased fuel consumption to cover the energy requirements of the scrubbers. This varies per technology type and is estimated at approximately 1–3% for seawater systems and 0.5–1.5% for freshwater scrubbers. Therefore, no matter which compliance option the affected ship operators choose, their operating costs will increase.

Table 7.1 Capital costs for scrubber systems

Scrubber system	Vessel	Cruise ferry (~40 MW)	Cargo ship (~20 MW)
Seawater system	New build	3 M€	2.1 M€
	Retrofit	3.5 M€	2.4 M€
Freshwater system	New build	2.4 M€	1.9 M€
	Retrofit	3.4 M€	2.4 M€
Hybrid system	New build	3.8 M€	2.6 M€
	Retrofit	4.3 M€	3 M€

Source: EMSA (2010)

1.4 Impacts of Sulfur Regulations on Short Sea Shipping

As low-sulfur fuel (marine gas oil (MGO) or marine diesel oil (MDO)) is substantially more expensive than heavy fuel oil (HFO), there is little or no room for the short sea shipping sector to absorb such additional costs, and thus significant price increases must be expected on their freight rates. Unlike its deep-sea counterpart, in short sea shipping such a freight rate increase may induce shippers to use land-based alternatives (mainly road). A reverse shift of cargo would go against the EU policy to shift traffic from land to sea to reduce congestion and might ultimately (under certain circumstances) increase the overall level of CO_2 emissions. If the shipping price is no longer competitive with road transport, this will likely have one or more of the following ramifications:

- Modal shifts toward land-based options and possible congestion
- Loss of cargo volumes for shipping companies
- Reduced profits or increased losses
- Potentially more CO_2 in the overall supply chain
- Increased cost of the produced goods, making these products uncompetitive as compared with sourcing from other areas, including areas outside the EU and thus among others, additional transport-related emissions
- The loss of business may hinder the shipping routes non-viable and thus candidates for closure. A consequence is that all of the remaining cargoes on such routes will need to find alternative transport routes, most likely road

This chapter will present the issues that SO_x emissions may create on the shipping sector, with a particular focus on the short sea shipping (SSS) sector and more specifically an application on Ro-Ro services in the North and Baltic Sea.

These increased operating costs will partially be passed on to shippers via increased freight rates. This could have more severe consequences for short sea shipping (SSS) operators as these compete heavily with other transportation modes (such as rail and road). As a result, modal shifts could be triggered due to the regulation.

1.4.1 Anticipated Impacts Before the New Limit

In fact, there were numerous media reports during 2013 and 2014 (before the 0.10% limit) that hinted on potential closures of SSS services in the affected areas, as well as possible new rail links to absorb an increased transportation demand arising from previous closures.

Some ship operators started investing in scrubber systems years ahead of the new limit, in response to the new regulation. To assist the early adapters, the European

Commission provided subsidies for capital investment costs, reaching up to 20% of the total system installation costs.[1]

A cost-benefit analysis (CBA) study by Jiang et al. (2014) compared investment in scrubber systems vs the use of low-sulfur fuel to comply with the regulation. They concluded that scrubber systems are more beneficial when installed in new builds than retrofits but also noted that only if the lifespan of the vessel is more than 4 years, scrubbers are worth consideration. These conclusions agree with the overall view in the industry that scrubber systems were the way forward ahead of the new limits. However, these conclusions were drawn based on the high fuel prices at the time and the important price differential between the different fuels. In a more recent study, Zis et al. (2016) argued that with the unexpectedly low fuel prices for both MGO and HFO during 2014–2015, the payback period of an investment in scrubber systems (retrofit) has in some cases more than doubled, reaching 10 years for small vessels operating most of their time within SECA. Therefore in retrospect, the low fuel prices support the argument that investments in scrubber systems were not the best option. This can be further supported by the fact that in 2020, the global cap of sulfur content will be lowered to 0.50% (certainly in European waters regardless of the outcome of the IMO review on postponing the limit to 2025), and thus the fuel price differential will be lower once the limit kicks in.

In 2010 the Institute of Shipping Economics and Logistics (Lemper et al. 2009) estimated that modal shifts due to the new limit after 2015 could reach on average 22% (considering container and Ro-Ro shipping). The study anticipated an increase in sea transport costs for all fuel scenarios (high and low prices); however, even the low fuel price scenarios were actually much higher than the actual fuel prices. In 2013, a study from the North Sea Consultation Group (Odgaard et al. 2013) examined the potential modal shifts following the establishment of NECA and SECA in the North Sea and the Baltic. The study reports an anticipated increase in sea transportation costs ranging between 8% and 16%, reduced to 5–13% when the road haulage is included.

1.4.2 What Actually Happened After the New Limit

The previous section tried to capture the gloom and doom that was the prevailing mood of the industry at the time before the new limit. Certain operators had already shut down routes (Stena line, DFDS) or significantly altered their service schedules (different sailing speeds, lower frequencies, new fleet assignment). There were also fears that manufacturing, mining and forest industries in the area could relocate due to the increased transportation costs. Despite the concerns, most SSS operators saw a very positive year in 2015. In fact, some of the larger Ro-Ro operators reported record revenues over the year. This unexpected turn of events cannot be attributed

[1] Source: http://www.cosbc.ca/index.php/international/item/1748-eu-hands-out-scrubber-subsidies

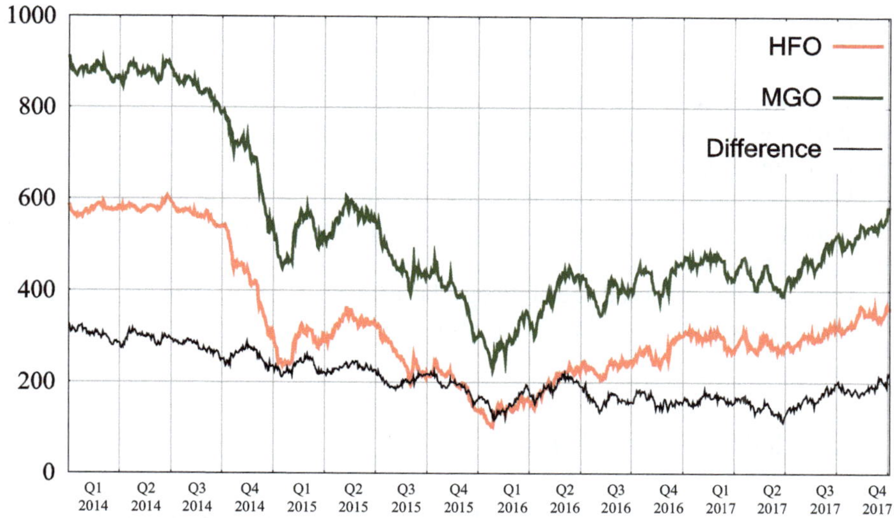

Fig. 7.3 Fuel prices 2014–2017. (Data source: www.bunkerworld.com)

to the lower sulfur limit but rather to the significant and entirely unexpected drop in fuel prices since the mid of 2014. This is shown in Fig. 7.3 that presents the fuel price fluctuation in the period of 2014–2017, for MGO in the green line (0.10% sulfur content), HFO in the salmon line (1% sulfur content) and their price differential in the black trend line.

The price differential between the two types of fuel can be seen to gradually decline in absolute terms during the 1st year of the new limit. One can observe that in 2015, the MGO price was lower than the HFO price in 2014. This means that despite the stricter regulation, the fuel cost was actually lower for ship operators compared to the year before the limit. This would in turn allow ship operators to offer similar (and in some cases lower) freight rates as in 2014 but operate on lower overall costs. This fact may explain the record revenues recorded in 2015. It has to be noted though that fuel prices have started increasing again since 2016, a trend which if continued could have major implications on modal shifts to land-based options if they continue to do so and reach previous higher levels. The repercussions of the global sulfur cap on fuel prices are also an important question for the near future.

1.5 Structure of the Rest of This Chapter

The previous sections presented in very brief the main challenges that SO_x emissions have put on the maritime sector and particularly the SSS sector that was immediately affected by it. The next sections of this chapter present work undertaken in the context of the RoRoSECA project at DTU during the period 2015–2017 and a methodology that can be useful in answering the following questions:

- What are the economic impacts of sulfur legislation?
- What are the environmental impacts, and what trade-offs emerge?
- How can modal shifts be modeled?
- What measures can ship operators take to mitigate and reverse the situation?
- What policy measures can assist in ensuring that the sulfur legislation is not hurting the SSS?
- How can the proposed methodology be applied in other shipping sectors to examine the impacts of the global sulfur cap?

2 Modeling Modal Shifts

This section presents an enhanced modal split model that was developed to estimate modal shifts caused by the introduction of the new sulfur limit. The model was subsequently applied in a set of Ro-Ro routes within the North and Baltic Sea. Section 2 of this chapter is an adaptation of the work of Zis and Psaraftis (2017).

2.1 Logit Models

Modal split models are useful to facilitate the simulation of travel demand between an O-D pair among a set of different transport modes. The underlying theory is based on the assumption that the decision-maker (in this case the shipper) seeks to maximize his utility (or minimize his disutility) by selecting the optimal transport mode. In the context of this chapter, the shipper aims to minimize their perceived generalized cost of transport.

In theory, this would lead to an all or nothing assignment, as a simple enumeration of the total generalized costs could show which option has the minimum. However, in reality each decision-maker will have a different perception of what the lower cost is. In transportation, the majority of modal split models used are falling in the category of logit models, as these are found to fit mode choice behavior quite well (Panagakos et al. 2014). Logit models are special cases of regression models where the dependent variable is discrete. The purpose of logit models is to predict the probability of particular outcomes (mode choice) based on one or more explanatory variables.

Logit models can have a binary structure (where there are only two options available), an N-way structure (three or more options uncorrelated to each other) or a hierarchical (nested) structure where the decision-maker selects first a group of options that are related to each other and subsequently selects one of the options within that group (nest). These three forms are shown schematically in Fig. 7.4.

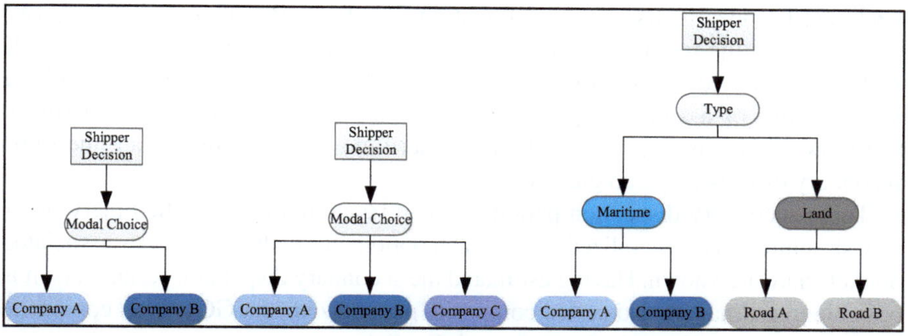

Fig. 7.4 The different types of nested models (binary, N-way, nested)

In this work, the assumption is that a shipper follows a hierarchical decision-making process, where they need to first select whether to make use of a maritime mode or not and subsequently decide which transport suppler (which maritime company) to use or which route if using a land-based mode. The assumption is that the maritime modes are correlated to each other and the land-based modes, respectively. Let GC_i expressed in € per lane meter (€/lm) be the generalized cost of transport via mode *i*. In this case, a first split between the maritime (*M*) and land (*L*) modes is assumed. The probability p_n of choosing a maritime mode *M* is given by Eq. 7.1.

$$p_{\mathrm{M}} = \frac{e^{-\lambda_1^i \cdot \mathrm{GC_M}}}{\sum_{n=M,L} e^{-\lambda_1^i \cdot \mathrm{GC_N}}} \tag{7.1}$$

Where λ_1^i is the dispersion parameter between the two nests. GC_N represents the composite generalized cost for nest *N* and is a function of the generalized cost of all *j* alternatives in nest *N*.

Assuming that the first decision revolves around which type of mode is selected (*M* or *L*), and that the decision is a maritime mode $j \in M$, the shipper must now decide which of the available *j* maritime options to use. The hierarchical structure is then assuming that the conditional probability $P_{j/N}$ of choosing mode *j* when nest *N* is selected:

$$P_{j/M} = \frac{e^{-\lambda_M \cdot \mathrm{GC}_{j/M}}}{\sum_{j \in M} e^{-\lambda_M \cdot GC_{j/M}}} \tag{7.2}$$

And (if there are only two options in *M*)

$$P_{1/M} + P_{2/M} = 1 \tag{7.3}$$

Where λ_M is a dispersion parameter for the secondary split among the maritime modes. A similar λ_L dispersion parameter for the secondary split among the land-based modes is also defined. Equation 7.3 shows that if there are only two alternatives in the maritime nest, then all commodities selecting a maritime mode will be transported via maritime option 1 or 2. At this level, it is possible to have more options of a similar type (e.g. a third maritime option) which is assumed to follow an N-way structure (within the maritime nest) and thus share the same secondary dispersion parameter λ_M.

These secondary dispersion parameters can be calibrated as in the binary or N-way structure if the generalized cost of each option within the nest and its associated market share are known. Having estimated the secondary dispersion parameters, it is possible to estimate the so-called composite generalized cost GC_N. This composite cost GC_N for nest N is calculated through Eq. 7.4.

$$\mathrm{GC}_N = \frac{-1}{\lambda_N} \log\left(\sum_{j \in N} e^{-\lambda_N \cdot \mathrm{GC}_{j/N}}\right) \tag{7.4}$$

According to Eq. 7.4, if there is only one alternative j between a similar mode of type i, then the composite cost collapses into the generalized cost GC_j of that mode. In a similar manner, if there are only two modes of type i (e.g. L and M for land-based and maritime), and for each type, there is only one alternative, then the hierarchical model collapses into a binary logit model of only two options. Therefore, the described structure can be readily applied to all types of case studies affected by the SECA regulation. These can be:

- Routes that face no competition from land-based modes, but more than one shipping operators are serving
- Routes with a unique shipping operator and one land-based alternative
- A combination of the previous

It has to be noted that there could be a hierarchical structure with more than two nests (e.g. a maritime, a road and a rail nest), but this is not considered in this chapter.

2.2 Modeling Framework

A six-step modeling framework is created to identify the impacts of the regulation on SSS. The first step in this process is concerned with the data collection on the situation of the market before any changes (in this case the introduction of the new limit). After gathering the necessary data, it is possible to decide which for of logit model will be used for the examined shippers' options. The model is then calibrated in the second step, and the scale parameters are estimated. In the third step, the new generalized costs following the new limits are estimated for each travel option. For example, if the fuel price increases and no other change is introduced, it is possible to estimate the new freight rate for any maritime legs based on information on the bunker adjustment factor. Using the calibrated model, and having calculated the new generalized cost for each option, it is now possible to retrieve the new market share of each option. In the fourth step, the new environmental balance of the system (emissions from all modes in each option) and the new route profitability of each maritime link are estimated. If one of the two is not satisfactory, step 5 and step 6 are, respectively, examining whether an operators measure or a new policy can mitigate or reverse the unsatisfactory situation of step 4. The whole process is depicted in Fig. 7.5 below. The next sections of this chapter will present the effects of the operators and policy measures on a set of representative case studies.

2.3 Data Collection and Assumptions

For the model calibration of any modal split model, there are some necessary data that need to be collected. Most discrete choice models use the so-called revealed preference data to predict aggregate market behavior (Ben-Akiva et al. 1994) and require information on the key explanatory parameters (e.g. travel times, travel costs, number of interchanges, time of travel, weather condition, etc.). An alternative approach is the use of stated preference data, which revolve around observations on hypothetical choice behavior, typically collected through surveys, interviews or focus groups. Necessary data require the acquisition of information on the market share of each of the available options (e.g. how many users select each available option) to model the probability of making a selection. Subsequently, one has to decide which of all the explanatory variables should be used in the model during the calibration stage.

Most of the aforementioned discrete choice models used in transportation are focusing on the behavior of passengers or drivers and differentiate between the various transportation modes that are available to them. Ortúzar and Willumsen (2011) classify the factors influencing mode choice that include among others travel cost, travel time, number of transit changes, weather, comfort, vehicle availability, trip purpose, income and time of day. In contrast, for freight transport the shipper usually has to decide based on fewer influencing factors, mainly the total travel cost and overall time, taking into consideration the reliability of service. Nam (1997) considers as explanatory variables for freight transport the shipment weight, freight charge, commodity type, distance, travel time and frequency of service. These can all be transformed into cost and time for each option. In this work, the focus is on modeling the mode choice of shippers when one or more short sea shipping modes are available and compete with each other and with one or more land-based model. A key assumption here is that the shipper decides solely on information about total travel cost and time, as these are the explanatory variables that are heavily affected by changes in policies and the examined operators' measures.

The necessary steps for the data collection before calibrating the model are the following:

- Enumerate all competing modes (maritime and/or land-based).
- Select origin–destination (O-D) pairs for shipments that may use the Ro-Ro service.

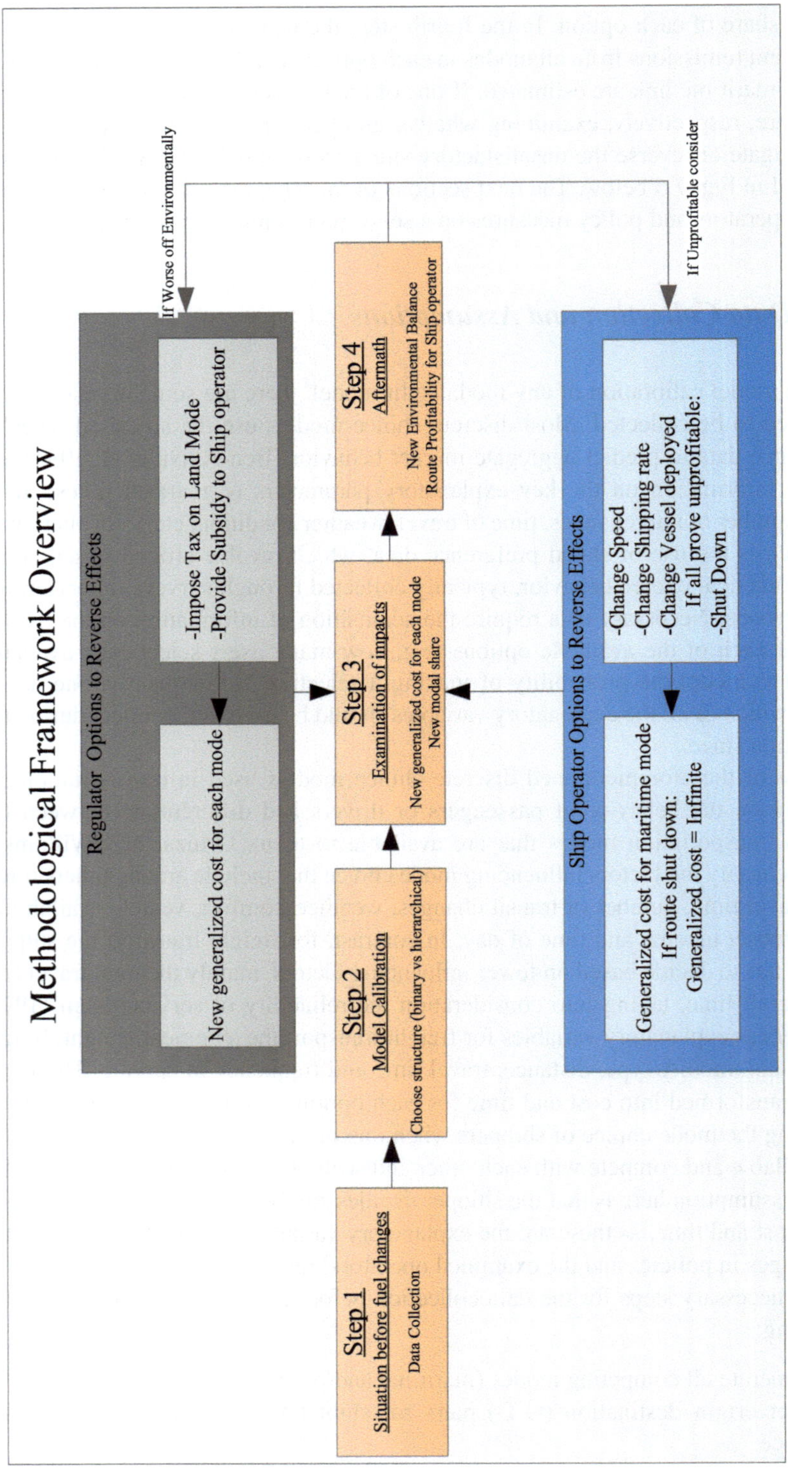

Fig. 7.5 The modeling framework to examine impacts of the new sulfur limits on SSS

- Estimate of the total travel time for each available option in the transportation system.
- Collect information on freight rates to estimate total transportation cost for each option.
- Gather information on transported volumes for each option as (%) market shares.

In an ideal scenario, all O-D pairs for all cargoes traveling through the available options will be collected. That info should contain the value and weight of cargo for each shipment, its depreciation rate which may vary significantly from product to product, the total travel time (including waiting times during mode changes), the travel cost as experienced by the shipper (which may vary if the shipper is a longtime customer or sends very big shipments), as well as any information on delayed or incomplete journeys. However, this sort of data are realistically impossible to obtain. Ro-Ro ship operators typically set their freight rates in € per lane meter (lm), which measures the volume of what the cargo takes on board the vessel. Information on the actual weight of the transported trailers are not controlled by the Ro-Ro operator, and only an estimation can be made based on the cargo type (if that is known). However for land-based options, the weight of the cargo is more important as it also has a more significant impact on the fuel consumption of the road vehicle. Typically freight rates are confidential and negotiated on an individual customer level (with provision of bulk discounts etc.). In this work, we use the average freight rate (€ per lane meter) that the shipping company was charging for each of the examined routes. Therefore, one weakness in our model is that during calibration, a drastic discount to a large supplier that may have prompted them to decide an all or nothing assignment between the available options is not considered. In the real world, apart from cost and travel time, a shipper may have a strong preference on a mode or service provider that cannot be explicitly modeled without direct information from the shipper. These shippers would therefore be inelastic and more reluctant to switch modes despite a small increase in the cost of one of the options. Due to the aforementioned limitations, this chapter presents a calibrated model that considered an aggregate case for the examined shipping network. However, the modeling framework could easily handle more disaggregate data should these become available in the future.

2.4 Selection Criteria of Routes and Model Calibration

This section presents the calibration results of seven routes in the North and Baltic Sea of a leading Ro-Ro operator. The selection of the seven routes was based on balancing the following criteria:

- Sailing distance
- Sailing frequency
- Deployed capacity
- Abatement technology (scrubbers vs low-sulfur fuel)

- Cargo types
- Vessel type (Ro-Ro, Ro-Pax, Pax)

The routes comprise of a mixture of cargo-only services (Ro-Ro), services that also carry passengers (Ro-Pax), and predominantly passenger services with limited cargo capacity (Pax). The Ro-Ro operator provided information on transported volumes (lanes per trip), freight rates (€/lm), passengers on-board (for Ro-Pax services), passenger fares (€/pax), on-board spending (€/lm), fleet deployment and fuel consumption for each vessel on each service for 2014 and 2015. The seven routes are a representative subset of the full Ro-Ro network that the operator is serving.

Based on these criteria, the examined services are the following:

- Gothenburg–Ghent (Ro-Ro – North Sea)
- Esbjerg–Immingham (Ro-Ro – North Sea)
- Rotterdam–Felixstowe (Ro-Ro – North Sea)
- Copenhagen–Oslo (Pax – North Sea)
- Klaipeda–Kiel (Ro-Pax – Baltic Sea)
- Klaipeda–Karlshamn (Ro-Pax – Baltic Sea)
- Dover–Calais (Ro-Pax – Cross Channel)

The modeled unit of transport in the problem is the lane meter of cargo for all competing modes in the analysis. It should be noted that information on the origin and final destination of the cargoes using the maritime links were not known to the collaborating ship operator for all cargoes. In addition, disaggregate level information on freight transport flows for the full European road network were also not possible to retrieve. To deal with these limitations, the calibration of the logit model was performed using a simulation approach where sensitivity analyses around central values for key characteristics (market shares, road distances, freight rates for land-based options) were conducted. The characteristics along with an aggregate data summary for 2014 and 2015 are shown in Table 7.2, showing only percentage changes between the 2 years (and not exact values) due to a confidentiality agreement with the shipping company that provided the datasets. These were the data for step 1 of Fig. 7.5.

Using these data, and the modeling framework presented in Sect. 2.2, the modal split model was calibrated for each route. For the information on the market shares of other options ranges around central values based on aggregate statistical information form Eurostat and the Shippax CFI journal were used. The market share information was also discussed with relevant experts in roundtable meetings in the duration of the project. The calibration results are shown in Table 7.3 considering only cargo flows for all routes, and the whole range of market shares during the simulation is shown for each route. The resulting scale parameters are the averages of the simulation process. For more information on the model calibration, the readers are referred to the paper of Zis and Psaraftis (2017) that explains this process in further detail.

Table 7.2 Overview of examined services and changes between 2014 and 2015

Route	Distance (NM)	Sailing speed (knots)	Frequency (sailings/week)	Fleet and abatement	Year	Trips total	ΔCargo volume (%)	ΔFreight rate change (%)	ΔRevenue (%)	ΔFuel cost (%)
Gothenburg–Ghent	577	18.1	6	3 scrubber	2014	553	6.06	−5.62	0.09	−52.89
					2015	569				
Esbjerg–Immingham	326	18.1	6	1 scrubber	2014	512	19.46	−0.5	18.85	−15.29
				1MGO	2015	580				
Rotterdam–Felixstowe	121	16.1	16	2 scrubber	2014	1514	15.13	00.5	15.71	−24.34
				1 MGO	2015	1637				
Copenhagen–Oslo	272	15.5	7	1 scrubber	2014	687	−5.82	1.58	4.28	−9.36
				1 MGO	2015	702				
Klaipeda–Kiel	397	18.4	6	2 scrubber	2014	611	−4.64	−7.71	−8.89	−30.05
					2015	615				
Klaipeda–Karlshamn	223	17.2	7	1 scrubber	2014	717	3.64	−2.32	3.73	−22.99
				1 MGO	2015	710				
Dover–Calais	26	15.3	99	2 MGO	2014	6210	−17.66	9.36	−18.04	−50.35
					2015	4994				

Source: Own compilation from data provided by operator. See also Zis and Psaraftis (2018a, b)

Table 7.3 Calibration results for the examined services

	Market share (%)			Scale parameter		
Route	Maritime	Maritime competitor	Land	λ	λ_I	λ_M
Gothenburg–Ghent	24–30	21–29	39–49	NA	0.027	0.025
Esbjerg–Immingham	60–70		30–40	0.08	NA	
Rotterdam–Felixstowe	30–40		60–70	0.14		
Copenhagen–Oslo	20–25	NA	75–80	0.108		
Klaipeda–Kiel	51–61	NA	39–49	0.019		
Klaipeda–Karlshamn	67–77	23–33	NA	0.08		
Dover–Calais	39–49	NA	51–61	0.015		

Source: Zis and Psaraftis (2017)

In the remainder of the chapter, with the examination of policy and operators measures, these average values of the scale parameters will be used for each case study. These values are in agreement with the dispersion parameters for freight transport of previous studies in the field (Lemper et al. 2009; Panagakos et al. 2014). Small values for λ indicate a low shift potential for the same change in the disutility function modeled. Therefore, if the generalized cost of all services increases by the same amount as a consequence of the mandatory use of low-sulfur fuel, then the route with the lower dispersion parameter will lose a smaller market share to competing modes and is more resilient to change. Therefore, routes with larger-scale parameters may be in need of more to be protected from potential service closures. The next section will analyze the effects of the proposed operators' measures.

3 Operators' Measures to Cope with Regulation

Following discussions with relevant stakeholders, a set of measures that ship operators could readily deploy in response to potential modal shifts if fuel prices increase again was developed. The measures were selected and tailored in a way to be transferable to other types of shipping, and not limited to only Ro-Ro operations. The set consists of the following measures which will be briefly presented here:

- New sailing speed
- Alterations in sailing frequency of a service
- Fleet reconfiguration with exchanges of ships between services
- Investments in abatement technologies (scrubbers)

In the next sections, each measure will be presented and its effects on the examined service and a summary of respective case studies will be shown. More detailed information can be found in Zis and Psaraftis (2018a, b). For some of the case studies, sensitivity analyses on the effects of fuel price are conducted. Three such case studies are considered:

- Case 1 considers the actual prices as experienced in 2015.
- Case 2 is a pessimistic scenario with high fuel prices as in early 2014.
- Case 3 is a hypothetical scenario during which HFO with 1% sulfur is still allowed as before the regulation, with the 2015 prices.

3.1 Effects of New Sailing Speeds in the Service

Even small reductions in sailing speed can bring significant fuel consumption economy in each journey. Therefore, for routes that are struggling with low traffic (reduced revenues), it may be an option to maintain a service financially viable. In contrast to other types of shipping, Ro-Ro services are faster and offer a high sailing frequency (multiple sailings per week, in certain short routes per day). This fact brings forth additional constraints on the viable ranges of sailing speeds. Due to the nature of the sector, most sailings are advertised (scheduled) to last an integer number of hours or integer multiples of 30-min periods. At the same time, typically departures and arrivals of most sailings are at sharp or half-past times. This facilitates the planning of cut-off times for the embarkation of goods and passengers. Finally, sailing speeds cannot be lowered at extreme levels in order to ensure a minimum turnaround time at each port for loading and unloading of vehicles.

For the examined services, we considered several different sailing speeds that would increase total sailing time by 30 min, 1 h, 2 h, up to 3 h for the longer routes. For one route (Klaipeda–Kiel), we also considered increasing the sailing speed as this is what actually was observed in 2016. The effects of speed changes in the fuel consumption of vessels deployed in the applicable services are shown in Table 7.4.

The average fuel consumption of the deployed vessels that were used in each route in the examined period is shown. The fuel consumption at the baseline sailing speeds was given from the ship operator, and estimations on the fuel consumption at other speeds were based on an in-house modeling tool that links speed with fuel consumption. Any changes in cargo volumes loaded are not considered in the fuel consumption. In reality, if due to the lower sailing speed the demand is reduced, this will result in a slightly lower fuel consumption due to the lower deadweight.

The next step of the analysis is to understand the effect of the lower sailing speed into modal choice, considering that no other change is introduced (e.g. the freight rates are remaining the same for all three fuel case scenarios as in the baseline). For speed reduction scenarios, a minor loss of cargo is observed, which is due to the very low effect that the little extra time has on the generalized cost of transport. However, it must be stressed that if a very high depreciation rate was used and/or cargoes of very high values, then the loss due to slow steaming would be more severe. An overall observation is that the revenue remains relatively unchanged, whereas the cost of fuel is changing dramatically for lower speeds for all fuel case scenarios. For cruise routes, a side benefit of a higher sailing times is an increase in revenue from on-board spending (passenger facilities such as casinos, restaurants, bars). In terms of capacity utilization, fuel price plays a crucial role in its value, whereas an

Table 7.4 The effects of a new sailing speed on fuel consumption

Route	Hours at berth	Hours sailing	Weekly fuel consumption (tonnes)	Change (%)
Gothenburg–Ghent	Baseline sailing speed 18.06 knots			
	38	130	286.9	NA
	Increase trip by 1 h, new sailing speed 17.3 knots			
	32	136	259.3	−10.7
	Increase trip by 2 h, new sailing speed 16.5 knots			
	26	142	235.5	−21.8
Esbjerg–Immingham	Baseline sailing speed 18.11 knots			
	60	108	292.4	NA
	Increase trip by 0.5 h, new sailing speed 17.6 knots			
	57	111	274.1	−6.3
	Increase trip by 1 h, new sailing speed 17.1 knots			
	54	114	257.4	−11.9
Copenhagen–Oslo	Baseline sailing speed 15.5 knots			
	45.5	122.5	271.3	NA
	Increase trip by 0.5 h, new sailing speed 15.1 knots			
	42	126	257.9	−4.9
	Increase trip by 1 h, new sailing speed 14.7 knots			
	38.5	129.5	245.9	−9.4
Klaipeda–Kiel	Baseline sailing speed 18.4 knots			
	26.5	129.5	321.1	NA
	Decrease trip by 1.5 h, new sailing speed 19.8 knots			
	35.5	120.5	373.8	+16.4
	Increase trip by 0.5 h, new sailing speed 18 knots			
	23.5	132.5	305.9	−4.7

increase in sailing time has a trivial effect. The effects on the capacity utilization are shown in Fig. 7.6.

In Fig. 7.6, the first bar for each route shows the baseline capacity utilization without a change in sailing speed. The next bars correspond to different tiers of speed changes as in Table 7.4. It can be observed that with lower sailing speeds, there is a reduction in the utilized capacity of the vessel (and thus revenues per trip); however, the total fuel savings as seen in Table 7.4 far outweigh these revenue losses.

3.2 *Altering Sailing Frequency*

For certain services where profitability may be hindered due to loss of cargo volumes, an option may be to reduce the number of weekly sailings. Instead of

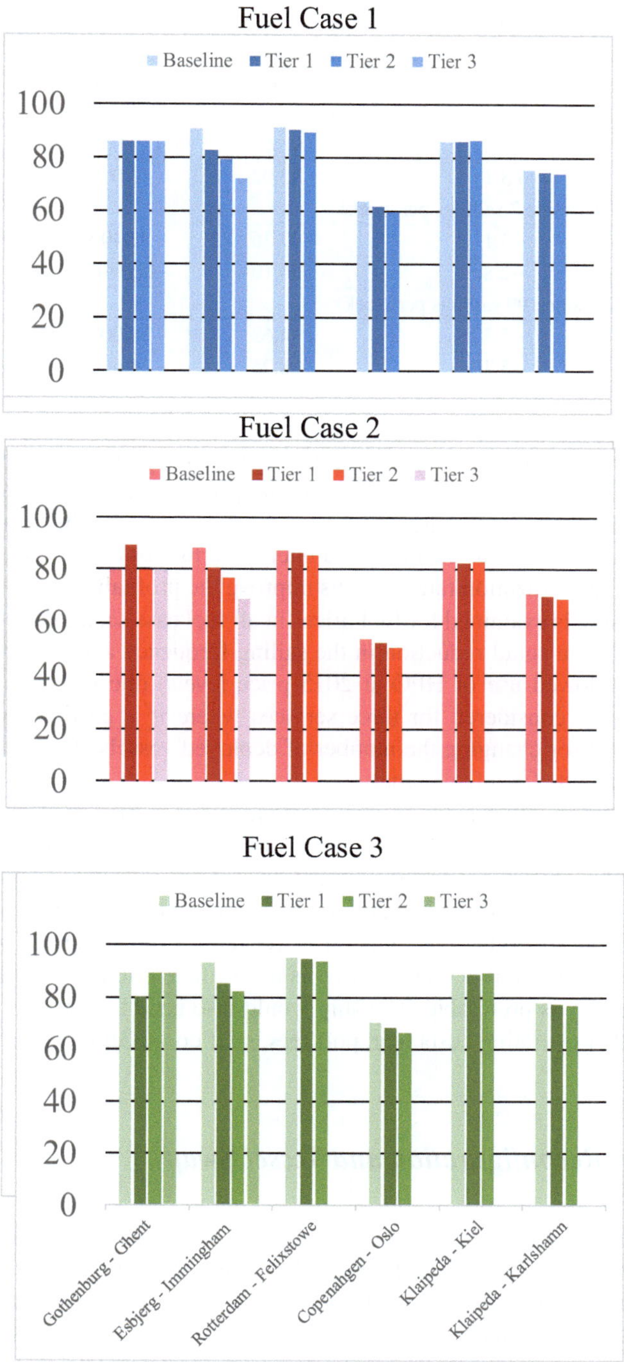

Fig. 7.6 Effects of speed changes on capacity utilization for different fuel price scenarios

Table 7.5 Effects of new sailing frequency

Fuel case	New frequency	Change in transported units (%)	Capacity utilization	ΔRevenue (€)	ΔFuel Cost (€)
Esbjerg–Immingham (baseline 6 sailings per week)					
2	5	−6	96.6	−112,000	−33,500
3	7	+3.1	82.02	40,000	16,600
Klaipeda–Kiel (baseline 7 sailings per week)					
1	6	−3.1	97.36	−32,400	−28,170
2	6	−2.8	96.19	−25,080	−57,090
Dover–Calais (baseline 99 sailings per week)					
1	75	−5.37	94.63	−56,000	−58,900
2	75	−5.39	88.25	−74,600	−119,300

shutting down a service completely, the sailing frequency may be adapted by either reducing the number of deployed vessels or simply reducing the number of weekly sailings. While the market share will drop in such an event (as this is increasing the average travel times due to the increased inter-departure times), it is expected that it will increase the utilization rate and thus improve the profitability of the route.

This measure is examined for fuel prices in all fuel case scenarios, where in Case 1 and 2, there is a small reduction in the sailing frequency and a higher frequency for Case 3 (allowed use of HFO at 2015 price levels) due to increased demand. This measure is considered for three services where it is easily applicable, with no requirement for changing the number of deployed vessels. Table 7.5 provides a summary of results for a few scenarios.

Table 7.5 shows that a lower sailing frequency for high fuel prices leads to significant increases in the utilization factor of the vessels, to a point where it would be undesirable. This is due to the assumption that the reduction would result in a small drop in transportation demand, as the service that would be canceled could be, for example, a service that runs on a weekend. Comparing the revenue from transport units and the fuel consumption is not sufficient to deduce whether a new sailing frequency is preferable. The ship would also need to carefully consider the change in other operating costs (e.g. port fees, staff costs, depreciation).

3.3 *Fleet Reconfiguration and Vessel Swaps*

This measure is essentially an adaptation of the sailing frequency option that the Ro-Ro operator has. Instead of changing the sailing frequency, the Ro-Ro operator can consider a different fleet assignment between routes served by swapping vessels according to their key technical characteristics in terms of capacity, speed and fuel consumption. There are certain constraints for the implementation of this measure. For example, vessels are assigned to existing services based on their type (pure cargo or cargo + passenger vessels), and thus vessels can be swapped only between

Table 7.6 Vessel swap between two North Sea services

Route	Gothenburg–Ghent		Esbjerg–Immingham	
Fuel case	Capacity utilization (%)	ΔFC (€)	Capacity utilization (%)	ΔFC (€)
1	92.08	−4660	94.32	−11,000
2	85.49	−9500	91.45	−22,400
3	95.36	−4500	96.59	−10,711

similar type services. Additional bureaucratic constraints may not allow the change of service for some of the vessels. For example, ships that were retrofitted with scrubber systems with subsidies from the European Commission were forced to run on the predetermined services.

The case study we examine in this section is the swapping of vessels between two Ro-Ro services in the North Sea (Gothenburg–Ghent and Esbjerg–Immingham). This swap can be considered if there is a drop in transport demand in one route and a smaller vessel is assigned to it. The results of the analysis are shown in Table 7.6.

For Gothenburg–Ghent, the benefit is small for the low fuel price scenarios. The swap is considered with a smaller vessel currently sailing in a different service of the operator (Gothenburg–Immingham). For Esbjerg–Immingham, the fuel savings can be important at high fuel prices as a more fuel-efficient vessel is moved from a less frequent service. The capacity utilization is not changing significantly due to the small difference in maximum capacity of the vessels. We assume that a vessel swap would not affect the transportation demand for the service (the shipper is unaware of which ship is moving their cargo), considering that the vessel would sail at the same sailing speed and frequency. For Fuel Case 3, the fuel cost benefit is marginal, and there is the added risk of having vessels that are loaded extremely close to the maximum capacity which could backfire due to potential losses of revenues for not picking up cargoes.

3.4 *Scrubbers vs Low-Sulfur Fuel*

The decision of using low-sulfur fuel has the advantage that it does not require a significant capital investment (for retrofits), at the expense of pricier fuel. Various cost-benefit analyses have been conducted in recent years, and the common denominator in these studies is that the fuel price differential is critical on whether the retrofit is a good option. More detailed information can be found in the literature (Jiang et al. 2014; Zis et al. 2016). The ship operator of the examined network already has a significant number of their vessels retrofitted with scrubber systems. In this case study, a hypothetical conversion of the most fuel consuming vessel (that currently uses MGO) is considered. We assume that the conversion would be performed on the first of January 2015. We use a typical retrofit cost of €250 per kW of installed main engine power, and thus the capital cost of investment lies in the region of approximately 4.8 M€. The total weekly fuel consumption for the vessel reaches 303 tons. Following an installation of scrubbers, the additional fuel consumption is assumed to be 3% to cover the scrubber's energy requirements (estimate on the high end of the range). The operating cost savings depends on the fuel price differential of HFO and MGO. At the highest fuel prices observed in the 2 years between 2014 and 2015, the investment in scrubber systems would seem as very promising with a payback period that was less than 2 years. In contrast, taking into account the lowest fuel prices observed in the end of 2015 (and very low fuel price differential), the payback period would increase to 4.3 years, e.g. 2020. By 2020, the global sulfur cap will be enforced, and potentially new technologies will be available that would constitute investing in scrubbers in 2016 less appealing. Considering these simplistic calculations, the age of the vessel should also be taken into account as if a vessel has less than 5 years of remaining service; investing in scrubbers may not make sense.

Other measures that the operator could use to cope with such regulations would be the potential use of Liquefied Natural Gas (LNG) as fuel (particularly for the post-2020 era of the global sulfur cap). The ship operator could also consider changing their Bunker Adjustment Factor (BAF) in order to better capture the impacts of the premium they have to pay on low-sulfur fuel, in such a way that the shippers are not tempted to change modes (or shipping company). Finally, in the section each of the measures in the case studies was presented on a one-by-one basis. However, a ship operator could do more than one at the same time. In the RoRoSECA project website,[2] a free software tool has been developed that allows the examination of more than one operational measures on the shippers choice, ship operators' profitability as well as the system's environmental performance (all available transport modes at each service scenario).

4 Policy Measures

The previous operators' measures may prove critical in the survival of certain services in the event of a re-emergence of high fuel prices in the near future. Even in the event that fuel prices remain at low levels, the operators may also need to fine-tune their services in order to maximize the financial performance of a route. However, during extreme fuel prices, the operators' measures may not be sufficient to revert the negative impacts of the SECA limits. This section of the book chapter presents potential policy measures that could be used to offset the potential modal backshifts that could result from an increase in fuel prices. More detailed information on the examined case studies is available in the RoRoSECA

[2] www.roroseca.transport.dtu.dk

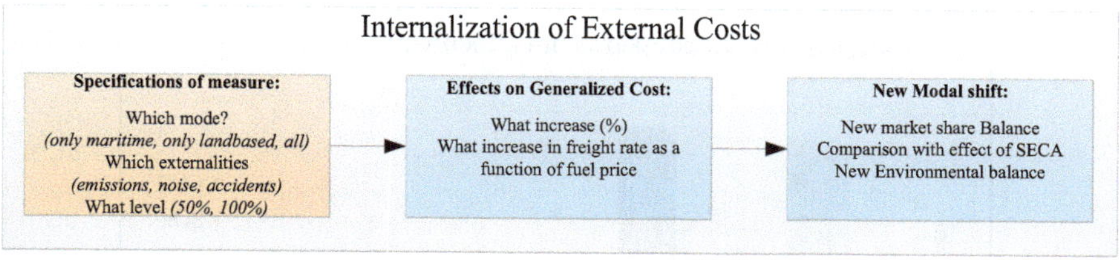

Fig. 7.7 The internalization of external costs as a potential policy measure

website in the deliverable on Task 3.2 entitled "Measures from policy makers" of the RoRoSECA project (Zis et al. 2017). This section draws heavily from that report.

4.1 *Internalization of External Costs*

This measure considers the full or partial internalization of external costs associated with the transportation of the examined routes and the various modes used. Transport generates negative externalities that involve a cost to society and economy. Internalizing these would encourage the use of safer, more silent and environmentally friendly vehicles.

In the examined case studies, the external costs are added to the transport cost element in the generalized cost formulation used in the modal shift models. Various different specifications can be considered for the internalization process. Figure 7.7 illustrates the process of internalization of external costs, as a measure to combat the negative effects of the low-sulfur fuel requirements.

A potential internalization of external costs may include all of these externalities, or merely a subset of them, and the effects of a partial (here we examine 50%) or full internalization will be considered. The second step considers the actual monetary effect of such a measure to the generalized cost of transportation for one unit of transport, in the different routes and for the different project specifications. The third and final step would be the use of the modal shift model (see Sect. 2 of this chapter) to estimate the new market shares of the competing transportation modes. The effects of this measure are then to be compared with the effects of the 0.10% sulfur limit on the short sea shipping sector and the impacts on emissions generation.

A critical decision in such measures is who would actually be responsible for the payment of the additional cost. In this section the assumption is that the shipper has to pay an additional contribution that is calculated as a function of the allocated emissions for the transportation of the cargo. It can be argued that in such a measure, the ship/freight operator would have to pay according to the emissions and transport work produced. The external costs used in this study are adapted from the DTU study on policy measures (Zis et al. 2017); more information on the derivation of these values is available in the deliverable report. Two main sources were used

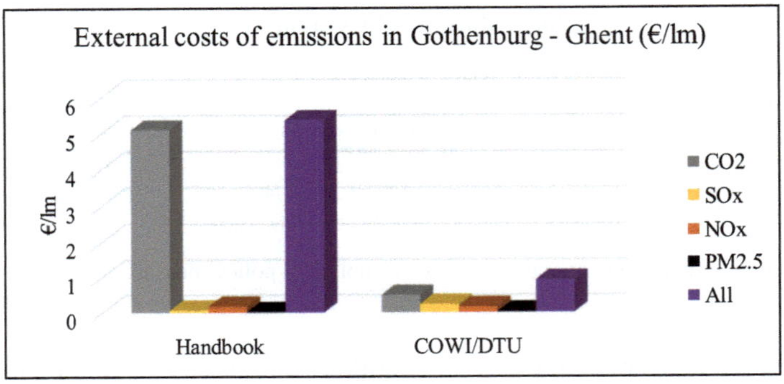

Fig. 7.8 Breakdown of external costs of transport for Gothenburg–Ghent. (Source: Zis et al. 2017)

for these values, the study of COWI/DTU (2016) and the 2014 Handbook for the European Commission (Ricardo-AEA 2014) on external costs.

These costs will be examined, but it can be anticipated that if the operator is paying for the externalities, these additional costs would in some manner be passed on to the shipper. There can also be an expected volatility in the values of these costs and a significant increase in the transportation costs for each mode. Consequently, it can be anticipated that such a measure could lead in an overall reduction in transport demand from an economics perspective.

For services that also carry passengers, the way the emissions would be allocated between cargo and passengers is not standard, and there can be great variability. For the sake of comparison across the routes, the next section assumes that all emissions are always attributed to the cargo even though that would not be realistic in an internalization scenario. For the maritime modes, the emissions at the port are considered for CO_2, SO_x, NO_x and $PM_{2.5}$ emissions, while at the sailing phases, only the CO_2 emissions are internalized. The values used in this chapter per kg of pollutant are the following (Zis et al. 2017):

- For CO_2 €0.11/kg (at sea and at port)
- For SO_x €12.96/kg (at port)
- For NO_x €13.47/kg (at port)
- For $PM_{2.5}$ €50.15/kg (at port)

It should be noted that this is only one set of possible values for the external cost per pollutant species, and there are other variations (low or higher estimates) per pollutant, and there are also variations depending on the region where the pollutants are emitted. More analytical information is shown in the project's report (Zis et al. 2017). For Gothenburg–Ghent, more analytical results are presented in Fig. 7.8 using different combinations of scenarios (full internalization of CO_2 only, or of all emissions, in combination with the different levels of the external cost values – low, medium, high).

Figure 7.8 shows that the main difference for both methodologies is due to the cost of carbon emissions. If all pollutants external costs are internalized, there is a

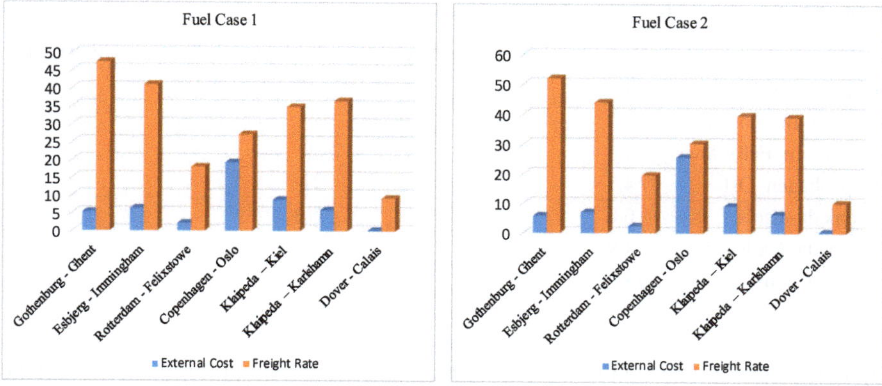

Fig. 7.9 Cost of internalization of emissions for each route. (Source: Zis et al. 2017)

significant increase in the total external cost per lane meter at this route. Figure 7.9 summarizes the cost of internalization per unit transport as compared to the freight rate of each route for the two fuel case scenarios, assuming full emissions allocation to the cargoes.

It is evident that for the cargo routes, the external costs are between 10% and 16% of the freight rate for both fuel case scenarios.

4.2 Easing the Port Dues of a Ship Operator

This measure is considering the option of subsidizing part of the port dues that the affected ship operators have to pay during their vessel calls. The rationale behind this measure is that since there are lower emissions as a consequence of the regulation (during the approach/departure phases), it may be reasonable to reduce the port fees by a certain extent. Retrieving information on the actual port fees for each vessel call is possible through the websites of the respective port authorities. For example, the port of Esbjerg charges 7.95 DKK (approximately €1.07) per GT per month for a visiting vessel that in the case of the two vessels accounts for approximately 0.85 million € a year. Esbjerg[3] is then providing a 90% refund of this cost in the next year if the vessel is a Ro-Ro ship. In contrast, the port of Immingham[4] has a tariff of £3.98 per NT just for the port and the examined vessels have a net tonnage of approximately 10,000. In addition, different ports have additional costs for mooring, waste charges, dock rents, fairway dues in certain countries, etc. While most of these costs are available online, the actual costs paid

[3]http://portesbjerg.dk

[4]http://www.humber.com

Table 7.7 Port fees vs fuel costs as % of operating costs

Route	Port cost as % of operating costs	Fuel costs as % of operating costs
Gothenburg–Ghent	4.6	30
Esbjerg–Immingham	4.2	39
Rotterdam–Felixstowe	4.5	30
Copenhagen–Oslo	4.7	21
Klaipeda–Kiel	6.8	NA
Klaipeda–Karlshamn	4.9	21
Dover–Calais	14.7	23

by frequent callers (as in the case of Ro-Ro services) are not public. For this reason, the actual port dues that the collaborating ship operator was paying was used in this analysis. These are shown in Table 7.7.

Table 7.7 shows that the port costs in the year are a very small component of the overall operating cost for each service, with the exception of Dover–Calais with a very high number of port calls a year. If a subsidy is provided to the ship operators toward their port dues, this could amount (for a full refund) to between 1.2 and 8 million € per year, which is similar for most routes to the cost of simply covering the BAF surcharges as seen in Sect. 4.4.

4.3 ECO-Bonus System

The first ECO-bonus system was authorized by the European Commission as a temporary state aid scheme in Italy for freight operators moving from road to sea. The main objective of ECO-bonus was to establish a mechanism to promote short sea shipping (Tsamboulas et al. 2015). The first implementation of the scheme considered the provisions of 20% toward the seaway tariffs of services (up to 30% for new services to be introduced after the system) while setting certain minimum limits (in terms of annual trips by the benefited operator). Due to limited resources and the ensuing recession, this scheme was operational only for a little bit over 2 years. However there are now new efforts attempting to replicate similar schemes.

In Norway, shipping lines that seek to establish new cargo services in the country can apply for grants from the Norwegian Coastal Administration. For this scheme, the aid goes directly to the ship operator and can be either the environmental benefit of the modal shift, the 30% of the operating costs of the service or a 10% of the transhipment equipment costs. In Italy, two additional state aid schemes were approved from the European Commission targeting modal shifts toward rail and sea. The maritime scheme is called Marebonus and will have a budget of €138 million. These subsidies will be used for the introduction of new services or upgrades to existing sea routes (Danesi and Longhi 2016). Finally, work on an ECO-bonus-like system is

Table 7.8 Cost and impacts on market share of a 20% subsidy to shippers (Source: Zis et al. 2017)

Route	Fuel case 1	Fuel case 2
Gothenburg–Ghent		
Modal shift	+5.93%	+6.18%
Total policy cost (€)	21 M	23 M
Esbjerg–Immingham		
Modal shift	+13.6%	+16%
Total policy cost (€)	14.6 M	14.9 M
Rotterdam–Felixstowe		
Modal shift	+10.92%	+10.95%
Total policy cost (€)	17.6 M	19.5 M
Copenhagen–Oslo		
Modal shift	+10.63	+9.97
Total policy cost (€)	4.5 M	5.4 M
Klaipeda–Kiel		
Modal shift	+3.2%	+3.67%
Total policy cost (€)	9.3 M	10.7 M
Klaipeda–Karlshamn		
Modal shift	+10.55	+12.4%
Total policy cost (€)	11.1 M	12.4 M
Dover–Calais		
Modal shift	+3.9%	+4.24%
Total policy cost (€)	14.2 M	15.7 M

currently conducted by the MED-Atlantic ecobonus project (co-funded by the EC) that seeks to increase the use of MoS in the Western Mediterranean and Atlantic markets.

In the context of this chapter, annual costs for such schemes are estimated for each of the seven routes. The arising modal shifts due to the induced lower generalized costs of the maritime transport options are estimated. We assume the provision of a 20% subsidy on the freight rate paid in each service for all customers (new and old) during 2015, and the impact this would have on the market shares of the examined service (additional %share captured) (Table 7.8).

The results show that such schemes could be very successful in attracting additional customers using the Ro-Ro links; however, the cost would be very high if applied to all users. In case a pilot implementation was considered, whereby the refund would be provided only to new users of the link, the cost would be proportionally lower. It is clear that such a policy would have an objective of increasing the users of maritime services, and not simply to reverse the negative effects of the low-sulfur regulation, as the monetary incentive exceeds the actual surcharge imposed on shippers because of low-sulfur fuel use. A subsidy of a different level could also be considered as a potential measure and could be the subject of interesting academic research.

Table 7.9 Impacts of subsidizing the BAF surcharges (Source: Zis et al. 2017)

Route	Gothenburg Ghent		Esbjerg Immingham		Klaipeda Kiel		Dover Calais	
Fuel case	1	2	1	2	1	2	1	2
BAF (€/lm)	1.37	5.13	1.19	4.3	1.76	6.34	0.33	1.2
Cost (M€)	2.5	10.1	1.96	7.82	2.27	8.48	2.35	9.0
Change in capacity utilization	+2.5%		+3.5%		+1.43%		+1.6%	

4.4 *Subsidies for the Bunker Adjustment Factor (BAF)*

The exact value of BAF depends on various service characteristics, including length, frequency, sailing speed and ship type. In this work, the BAF policy of the ship operator providing most of the data is used. The annual costs for the policy body are shown in Table 7.9.

If the fuel prices were as high as in early 2014, then the policy would cost approximately four times more for each route. The lower costs in Case 1 essentially represent the effects of the SECA limit on the shippers using this service. If the BAF was paid back to the shippers, then services would increase their market share and sail at increased utilization rates. However, a uniform policy to refund the shippers using the maritime mode shows that it will be very costly, considering that that the annual policy costs shown in Table 7.9 are for just one of the numerous affected services.

4.5 *Additional Road Tax to Reverse Modal Shifts*

This measure considers the identification of the necessary increase in the land-based freight rates that a shipper must pay, in order to negate the modal shift loss that is triggered by the low-sulfur fuel requirement. It is evident that this is a very case-specific measure, as the necessary increase per land-based transport work (in lm-NM units) will depend on the relative weight of the maritime costs in the generalized cost of the shipper. An explanatory analysis is conducted where the objective is to identify what percentage increase in the total monetary cost of land-based transport options will result in absorbing the modal backshift attributed to the low-sulfur fuel requirement. The necessary percentage increases are summarized in Table 7.10, for the two fuel price scenarios. For the Dover–Calais service, the percentage increase refers to the Eurotunnel cost. For Klaipeda–Karlshamn, this measure was not considered due to the lack of competition with land-based options.

Table 7.10 shows that the examined Ro-Ro services would be at considerable risk for high fuel prices. The necessary increase in the land-based option to offset the

Table 7.10 Necessary tax (% of freight rate) to reverse modal splits caused by 0.10% sulfur limit

Route	Fuel case 1	Fuel case 2
Gothenburg–Ghent	3.83	14.48
Esbjerg–Immingham	2.48	8.95
Rotterdam–Felixstowe	3.3	11.88
Copenhagen–Oslo	7.15	25.8
Klaipeda–Kiel	3.52	12.68
Klaipeda–Karlshamn	NA	
Dover–Calais	2.12	7.74

Source: Zis et al. (2017)

Table 7.11 Retrofit subsidy requirements

Route	Number of deployed vessels	Retrofit subsidy (M€)
Gothenburg–Ghent	3	6
Esbjerg–Immingham	2	3.9
Rotterdam–Felixstowe	3	6.6
Copenhagen–Oslo	2	4.7
Klaipeda–Kiel	2	4.8
Klaipeda–Karlshamn	2	4.3
Dover–Calais	2	4.4

Source: Zis et al. (2017)

effects of the higher BAF is increasing significantly in with higher fuel prices. The wide variance of the necessary land-based tax is evidence of the sensitivity of the total road lengths in the shippers' decision-making process. Therefore, suggesting a flat levy at 10% (e.g. in the form of an additional tax on petrol) would lead to net modal shifts toward maritime services for most routes.

4.6 Subsidies for Abatement Technologies

As presented in Sect. 3.4, one of the operators' measures that was examined was the investments in abatement technologies such as scrubbers or LNG engines. From the operator's perspective, the main question in such investments is the net present value and the length of the payback period. Companies that have already invested in such technologies can be considered as an early adapter to the technology in terms of size of investment (number of vessels). The European Commission, under the Motorways of the Sea (MoS) programme provided subsidies of 20% for the retrofitting of vessels, with indicative costs of 1.5 M€ per vessel. The assumption of this measure is that a policy body would cover 20% of the required investment costs for each retrofit. This analysis is only conducted to compare the total costs with the previously examined measures. The actual costs of a retrofit were taken based on published estimates as a function of total installed power (250€/kW). The costs are summarized in Table 7.11.

It can be seen that such a policy would require significant funds for the installation of scrubbers on all the available vessels. However, these costs are one-off (unlike other policies that could be annual) and in theory could be combined with a requirement that the benefitted ship operators would reduce the BAF surcharge since they could still use HFO.

5 Conclusions, Ongoing and Future Work

The previous sections presented the first implications of the low-sulfur regulation on ship operators, shippers and their effects on the environmental balance. The focus has been the lower limit within SECAs since 2015. Despite initial concerns that the limit would have devastating effects on the economy of the market, the unexpectedly low fuel prices actually boosted the sector. In terms of emissions reduction, the limit was a considerable success. This is evident in Fig. 7.10 where the relative emissions per transported work of cargo (assuming all emissions are allocated to cargo and the passengers are not included in the calculation) is shown in 2014 and 2015.

While absolute carbon emissions were increased (more trips performed in 2015, and in some routes, higher sailing speeds were used), the emissions intensity was improved. For sulfur emissions, the reduction had been proportional to the lowered limit. However, if fuel prices increase again, the emissions intensity may be increased. A thorough modeling framework has been constructed that allows the estimation of potential modal shifts of regulations such as the SECA limits but could easily also examine impacts of other regulations (e.g. if limits on other types of pollutants are placed, any bunker levies, speed limits or other MBM). Free software tools are available in the RoRoSECA project website that allow such analyses. A set of policy and operational measures were proposed that can be used to mitigate and reverse any negative effects of the regulation.

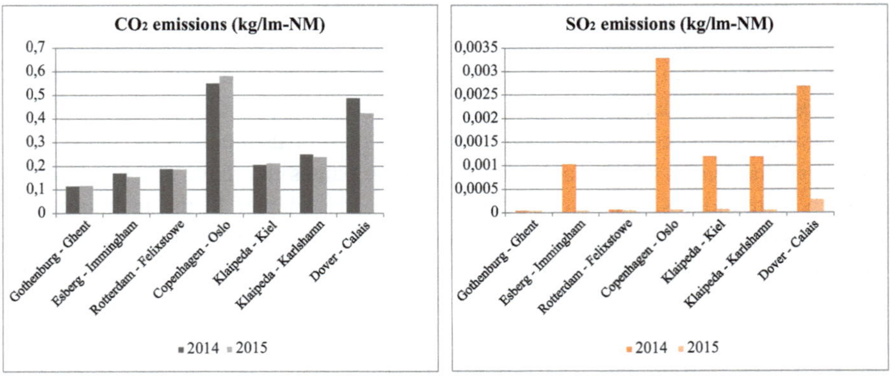

Fig. 7.10 Emissions intensity (kg/lm-NM) for CO_2 and SO_2 in the examined routes. (Source: Zis and Psaraftis 2017)

5.1 Early Adapters

Some ship operators started investing in scrubber systems before the new limit, at a time of high fuel prices and with concerns that low-sulfur fuel would be extremely expensive after 2015 due to the new demand. These early adapters took a major risk, as they committed significant capital to prepare for the regulation. The market for abatement technologies is increasing, and the technology is improving which will make it more affordable in the near future ahead of the global sulfur cap. In the European Union, early adapters were rewarded with subsidies toward the investment costs of the scrubber solution. However, certain constraints were put in place. For example, ship operators that benefited from such subsidies would need to sail on specified routes that limit their flexibility. Another risk is that new regulations may constitute the existing scrubber systems as non-compliant or simply require additional significant capital costs for the abatement of different pollutant species (e.g. black carbon). As a result, the option of not investing in "one-off" solutions, but instead relying on pricier low-sulfur fuel, may offer higher flexibility on ship operators. However technological progress can only occur if some operators start investing in such solutions, so that knowledge is generated and issues are resolved in the next versions of such solutions. It is therefore important to develop policies that help technological progress and also incentivize early adapters (ship operators).

5.2 The Global Sulfur Cap

It can be argued that the SSS operators caught a lucky break with the low fuel prices since 2014 until 2018, and the expected storm never hit them. Similar regulations that are targeting environmental impacts of the sector can be expected to affect key stakeholders in various ways. Indeed, the introduction of the Energy Efficiency Design Index (EEDI), the potential expansion of emission trading schemes (ETS) to cover emissions from the shipping sector, as well as the introduction of the global 0.5% sulfur cap, can all have the potential of being game changers. Particularly for scrubber systems, the coming of the global sulfur cap in 2020 is expected to raise the demand for such technologies, especially if the low-sulfur fuel availability is not enough to cover the necessary demand. However, a turn of the industry to scrubber technologies may in turn potentially increase the demand for HFO in contrast to what is anticipated and thus raise its price and reduce the fuel price differential with 0.5% fuel (and thus the benefit of investing in scrubber systems). Therefore, the impacts of environmental policies are much more complicated and difficult to predict, as it has been shown with the lowering of sulfur limits within SECAs. It is therefore vital to propose policies that mitigate the possible negative impacts of such regulation which will (post-2020) affect more severely the whole shipping sector and not just the relatively niche SSS. Therefore there are important research questions to be examined ahead of the global sulfur cap:

- How will the different levels of fuel availability affect the decision of ship operators to invest in scrubber systems?
- Will the global sulfur cap result in lower sailing speeds in the different shipping markets?
- How are the freight rates going to be affected, and will that lead into new modal shifts to other modes and new links like the One Belt One Road initiative?
- How can a level playing field be ensured?

5.3 The Challenge of Enforcing Sulfur Regulations

With regard to the last research question in Sect. 5.2, a new project at DTU is underway that seeks to answer this specific item. The new project is entitled SulphurGATE (Enforcement of Sulphur regulations; a Game Theoretic Approach). The purpose of the project is to develop a game theoretic modeling framework that improves the effectiveness of sulfur regulations enforcement. The existing legislative framework poses several challenges, stemming (mainly) from a highly non-homogeneous and spatially differentiated system, with cases where the penalty fines are as low as the benefit that the violator enjoyed from not complying. This project will examine the status quo of enforcement in different countries, where the regulation applies, and use a game theoretic approach for a uniform violation fine system. Such a system can help ship owners that currently have invested heavily in an abatement of options to comply with the sulfur regulations, by maintaining a level playing field among ship operators, while at the same time improve compliance rates and maximize societal environmental benefits. In anticipation of the global sulfur cap in 2020, it is expected that findings from the new project could form the basis for a new penalty system worldwide.

Acknowledgments Much of the work presented in this chapter is in the context of the project: "Mitigating and reversing the side-effects of environmental legislation on Ro-Ro shipping in Northern Europe" (also known as the RoRoSECA project) funded by the Danish Maritime Fund and the Orients Fund. We are grateful to Poul Woodall of DFDS whom we worked with in this project, for the provision of data and valuable input and discussions in producing this work.

References

Ben-akiva, M., Bradley, M., Morikawa, T., Benjamin, J., Novak, T., Oppewal, H., & Rao, V. (1994). Combining revealed and stated preferences data. *Marketing Letters, 5*(4), 335–349.

Buhaug, Ø., Corbett, J., Endresen, O., Eyring, V., Faber, J., Hanayama, S., Lee, D., Lindstad, H., Mjelde, A., Palsson, C., Wanquing, W., Winebrake, J., & Yoshida, K. (2009). *Second IMO greenhouse gas study*. London: International Maritime Organization.

COWI/DTUCOWI/DTU. (2016). Transport economic unit prices for use in socio-economic analyses – version 1.6 (in Danish). Accessed 14 Feb 2017.

Crowley, T. J. (2000). Causes of climate change over the past 1000 years. *Science, 289*(5477), 270–277.

Cullinane, K., & Bergqvist, R. (2014). Emission control areas and their impact on maritime transport. *Transportation Research Part D: Transport and Environment, 28*, 1–5.

Cullinane, K., & Cullinane, S. (2013). Atmospheric emissions from shipping: The need for regulation and approaches to compliance. *Transport Reviews, 33*(4), 377–401.

Danesi, A., & Longhi, M. (2016). Opening the southern door of the Italian inland waterway transport network. *Transport Logistics,* 16(41).

DMA. (2012). A feasibility study for an LNG filling station infrastructure and test of recommendations. North European LNG Infrastructure Project. Available at: https://www.dma.dk/Vaekst/EU/Documents/LNG_Full_report_20120524_optimerad.pdf. Accessed May 2017.

EMSA. (2010). European maritime safety. The 0.1% sulphur in fuel requirement as from 1 January 2015 in SECAs-An assessment of available impact studies and alternative means of compliance.

Environmental Protection Agency (EPA). (2013). Clean air markets: Allowance trading. Available at: http://www.epa.gov/airmarkets/trading/. Accessed Sep 2018.

European Commission. (2008). Directive 2008/50/EC of the European Parliament and of the Council of 21 May 2008 on ambient air quality and cleaner air for Europe. *Official Journal of the European Union.*

Eyring, V., Köhler, H. W., van Aardenne, J., & Lauer, A. (2005). Emissions from international shipping: 1. The last 50 years. *Journal of Geophysical Research: Atmospheres, 110*(D17), 1–12.

Henriksson, T. (2007). SO_x scrubbing of marine exhaust gases. *Wärtsilä Technical Journal, 2007*(2), 55–58.

Jiang, L., Kronbak, J., & Christensen, L. P. (2014). The costs and benefits of sulphur reduction measures: Sulphur scrubbers versus marine gas oil. *Transportation Research Part D: Transport and Environment, 28*, 19–27.

Lemper, B., Hader, A., Hübscher, A., Maatsch, S., & Tasto, M. (2009). *Reducing the sulphur conten of shipping fuels further to 0.1% in the North Sea and Baltic Sea in 2015: Consequences for shipping in this shipping area. Final report.* Bremen: Institut für Seeverkehrswirtschaft und Logistik.

MAN Diesel and Turbo. (2014). Guidelines for operation on fuels with less than 0.1% Sulphur. Copenhagen, Denmark. http://marine.man.eu/docs/librariesprovider6/test/guidelines-for-operationon-fuels-with-less-than-0-1-sulphur.pdf?sfvrsn=5. Accessed Dec 2017.

Nam, K. C. (1997). Study on the estimation and aggregation of disaggregate models of mode choice for freight transport. *Transportation Research Part E: Logistics and Transportation Review, 33*(3), 223–231.

Odgaard, T., Frank, C., Henriques, M., Bøge, M. (2013). *The impact on short sea shipping and the risk of modal shift from the establishment of a NOx emission control area in the North Sea*. Available at: http://eng.mst.dk/media/mst/9149808/theimpactonshortseashippingandtheriskofmodalshiftfromtheestablishmentofanecafina. Accessed Apr 2016.

OECD. (2017). Air and GHG emissions (indicator). https://doi.org/10.1787/93d10cf7-en

Ortúzar, J. d. D., & Willumsen, L. G. (2011). *Modelling transport* (4th ed.). Chichester: Wiley.

Panagakos, G. P., Stamatopoulou, E. V., & Psaraftis, H. N. (2014). The possible designation of the Mediterranean Sea as a SECA: A case study. *Transportation Research Part D: Transport and Environment, 28*, 74–90.

Pickering, K. T., & Owen, L. A. (1997). *An introduction to global environmental issues* (Vol. 2). London: Psychology Press.

Ricardo-AEA. (2014). *Update of the handbook on external costs of transport.* Report for the European Commission DG MOVE, London.

Tsamboulas, D., Chiappetta, A., Moraiti, P., Karousos, I. (2015). Could subsidies for maritime freight transportation achieve social and environmental benefits? The case of ecobonus. *Transportation Research Record: Journal of the Transportation Research Board,* (2479), 78–85.

Ward, P. L. (2009). Sulfur dioxide initiates global climate change in four ways. *Thin Solid Films, 517*(11), 3188–3203.

World Health Organization (WHO). (2005). Air quality guidelines for particulate matter, ozone, nitrogen dioxide and sulphur dioxide. Available at http://whqlibdoc.who.int/hq/2006/WHO_SDE_PHE_OEH_06.02_eng.pdf. Accessed Feb 2018.

Zis, T. (2015). The Implications and trade-offs of near-port ship emissions reduction policies.

Zis, T., & Psaraftis, H. N. (2017). The implications of the new sulphur limits on the European Ro-Ro sector. *Transportation Research Part D: Transport and Environment, 52*, 185–201.

Zis, T., & Psaraftis, H. N. (2018a). Operational measures to mitigate and reverse the potential modal shifts due to environmental legislation. *Maritime Policy & Management,* 1–16.

Zis, T., & Psaraftis, H. N. (2018b). Key performance indicators to assess and reverse the negative impacts of SECAs policies for Ro-Ro shipping. *FME Transactions, 46*(3), 347–354.

Zis, T., Angeloudis, P., Bell, M. G., Psaraftis, H. N. (2016). Payback period for emissions abatement alternatives: Role of regulation and fuel prices. *Transportation Research Record: Journal of the Transportation Research Board,* (2549), 37–44.

Zis, T., Panagakos, G., Kronbak, J., Psaraftis, H. N. (2017). Report on the outcome of Task 3.2 "Measures from policy makers". Available at: http://www.roroseca.transport.dtu.dk/dissemination

Chapter 8
Green Tramp Shipping Routing and Scheduling: Effects of Market-Based Measures on CO_2 Reduction

Xin Wang, Inge Norstad, Kjetil Fagerholt, and Marielle Christiansen

Abstract In this chapter we examine, from a tramp ship operator's point of view, how potential CO_2 emission reduction measures impact the operational decisions and their economic and environmental consequences. Two market-based measures (MBMs) are discussed, the bunker levy scheme and the emission trading scheme, and we show that both can be incorporated in a similar way into a typical tramp ship routing and scheduling model. We also demonstrate with a computational study the environmental benefits of these CO_2 reduction schemes.

List of Acronyms and Abbreviations

CO_2	Carbon dioxide
CO_2e	Carbon dioxide equivalent
COA	Contract of affreightment
dwt	Deadweight tonnes
ETS	Emission trading scheme
EU	European Union
EU ETS	European Union Emissions Trading Scheme
GHG	Greenhouse gas
IMO	International Maritime Organization
LNG	Liquefied natural gas
M	Nautical mile
MBM	Market-based measure
OR	Operations research
t	Tonnes

X. Wang · K. Fagerholt (✉) · M. Christiansen
Department of Industrial Economics and Technology Management, Norwegian University of Science and Technology, Trondheim, Norway
e-mail: xin.e.wang@tieto.com; kjetil.fagerholt@ntnu.no

I. Norstad
SINTEF Ocean, Trondheim, Norway

H. N. Psaraftis (ed.), *Sustainable Shipping*,
https://doi.org/10.1007/978-3-030-04330-8_8

1 Introduction

Traditionally for ship operators, the reduction of maritime greenhouse gas (GHG) emissions might just be a "happy side effect" of the increasing global competition in the shipping industry. While the thin profit margin generates the need to reduce bunker fuel consumption, through, e.g., better design of ship hulls, energy-saving engines, "slow steaming" (significantly reducing ship speed in response to depressed market conditions and/or high fuel prices (see Maersk 2011)), and more efficient deployment and operation of the fleet, it also contributes to less GHG produced, especially carbon dioxide (CO_2) emissions since they are directly proportional to fuel consumed.

However, as we marched into the second decade of the new millennium, there had been much discussion of stricter and more direct regulations on CO_2 emissions in the shipping sector (Buhaug et al. 2009; Shi 2016), due to the urgency of combating global warming and meeting the "two-degree goal" (Rajamani 2011). Among those proposed regulations are the so-called market-based measures (MBMs), including bunker levy, emission trading, and a variety of other schemes. We refer the readers to Chap. 11 for a discussion of MBMs in the shipping sector. A bunker levy scheme collects revenue from the sector in the form of a tax on fuel use, which may then be used to establish an international fund that invests in environmental causes. An emission trading scheme (ETS) sets a maximum quantity (cap) on emissions from the shipping sector and employs a trading mechanism to facilitate emission reductions. Although the effectiveness of any of these schemes has been controversial and the assessment and comparison of different MBMs are far from completion (Psaraftis 2012), it is useful to study how *ship operators* may react to different types of CO_2 emission reduction schemes.

In this chapter we examine, from a tramp ship operator's point of view, how potential CO_2 emission reduction schemes impact the operational decisions and their economic and environmental consequences. To the best of our knowledge, this is the first study in the literature that approaches this issue from an operations research perspective and in the tramp shipping context. We start by presenting the classic tramp routing and scheduling model in maritime transportation and extend the model to incorporate CO_2 reduction aspects under two scenarios: a bunker levy scenario and an ETS scenario. A computational study is then conducted on typical tramp shipping instances to show the effects of imposing these CO_2 reduction schemes.

The rest of this chapter is organized as follows. Section 2 introduces the tramp ship routing and scheduling problem and its mathematical models. Section 3 discusses the model extensions for incorporating two versions of CO_2 emission reduction scheme. Section 4 presents the computational study, and we conclude in Sect. 5.

2 Tramp Ship Routing and Scheduling

This section introduces the tramp ship routing and scheduling problem and its mathematical models. We start in Sect. 2.1 by discussing the operational characteristics of tramp shipping that distinguish itself as an important sector in maritime transportation. We then state the problem and present its mathematical models in Sect. 2.2.

2.1 *Operational Characteristics of Tramp Shipping*

In maritime transportation, ships are said to operate in the *tramp* mode if they do not have a fixed schedule or itinerary and do not expect repetition of voyages as a normal part of their operations. This is in contrast to the liner shipping business, characterized primarily by container shipping, which constitutes the provision of scheduled services with a fixed frequency over a predetermined route.

In tramp shipping, the sailings of a vessel follow the cargo commitments that vary with the vessel's employment (like taxicabs), usually catering to both mandatory contractual cargoes and optional spot ones. The mandatory cargoes are usually based on long-term agreements between the ship operator and cargo owners, or contracts of affreightment (COAs) in shipping parlance, where the ship operator is obliged to transport specified quantities of cargo between specified ports during a specified time period. Some contracts (in, e.g., oil trades) also demand repetitious voyages at a certain frequency, but unlike liner shipping, such voyages are usually not actively advertised, and the schedules are less strict. In addition to the mandatory contractual cargoes, a tramp operator often seeks optional cargoes from the spot market to better utilize their ship capacity and increase their revenue. Therefore, when planning for the routes and schedules of tramp ships in pursuit of maximized profits, the decisions regarding which optional cargoes to accept/reject are also nontrivial to the ship operator.

For the past two decades, there has been much work done in the operations research (OR) community toward the development of decision support tools in tramp shipping, where optimization theories and techniques are applied to achieve such better routes and schedules, optimized speeds, and improved composition of fleet. We refer the readers to Christiansen et al. (2004, 2013) for surveys in ship routing and scheduling problems and to Christiansen and Fagerholt (2014) for a review on tramp ship routing and scheduling in particular.

While it is common to distinguish between liner, industrial, and tramp (Lawrence 1972) when describing the mode of operation in maritime transportation, the line between the industrial and tramp shipping modes is narrow. A traditional industrial ship operator is considered to control its own "private" fleet that only provides transportation for its own cargoes. The recent trend, however, has been the shift from industrial to tramp shipping, as many companies previously involved in industrial shipping have outsourced their transportation, while others have become more engaged in the spot market during the process of industrial shipping operations being transformed from "cost centers" into "profit centers." From an OR perspective, the boundary between industrial and tramp shipping is even more obscure: essentially they are both defined around the principle of "following the available cargoes." While an industrial operator minimizes the cost of a somewhat closed system with a given number of ships and cargoes, the tramp ship routing and scheduling problem may be seen as a generalization of its industrial counterpart, where optional cargoes are also considered to generate additional revenue and the objective becomes profit maximization. Together, industrial and tramp shipping are responsible for the transportation of most of the bulk cargoes in global trades, including wet (oil and gas, chemicals, etc.) and dry bulk products (iron ore, coal, grain, etc.). In 2016, these products account for over 60% of the total weight transported at sea (UNCTAD 2017).

2.2 *The Tramp Ship Routing and Scheduling Problem*

In tramp shipping, as previously mentioned, the cargoes are the source of revenue, and the demands for transporting cargoes in a timely and efficient fashion are the main drivers for addressing tramp ship routing and scheduling problems. A cargo, mandatory or optional, represents the demand of a specified amount of product(s) to be loaded (picked up) at a specified origin port, transported, and unloaded (delivered) at a specified destination port. There usually is a time window associated with the pickup of each cargo during which the loading operations of the cargo must start. There sometimes are similar time windows for the delivery of the cargoes, but more often they are relatively wide (if any). Each optional cargo has a specified freight income rate that determines the revenue the ship operator will receive if the cargo is transported. The revenues from carrying the mandatory cargoes are also specified.

The ship operator controls a fleet of ships to service the cargoes. Such fleet is typically heterogeneous, in the sense that (a) the ships can be of different load capacity, speed range, fuel efficiency, and physical dimension (length, draft, etc.) and (b), more importantly, the initial locations of the ships are different, some ships may be at sea and others may be at dock in various sea areas and ports. A ship can sometimes carry multiple cargoes on board depending on the cargo sizes, although in several contexts, e.g., transporting major bulk commodities, a cargo is usually a full shipload. For various reasons there may also be compatibility constraints between ships and cargoes. For example, a small ship may not carry a cargo that is too heavy, and a large ship with deep draft may not carry a cargo because one of the associated ports of this cargo is too shallow.

In short, a typical tramp ship routing and scheduling problem is characterized by the simultaneous determination of acceptance/rejection of optional cargoes to service, assignment of cargoes to specific ships, the sequence and times of port calls for all ships, and, if variable speeds are applicable, the sailing speed during each voyage. The objective is to provide timely transportation services for all mandatory and accepted optional cargoes while maximizing profit which is computed as the revenue from all serviced cargoes subtracted by the variable transportation costs. These costs mainly consist of fuel costs, associated with sailing the ships; port and canal fees, dependent on the type and size of the ship when visiting a port and passing a canal; and sometimes also costs for spot charters (i.e., voyage/space charters from the spot market to service given cargoes).

The tramp ship routing and scheduling problem has many similarities with the so-called *multiple-vehicle one-to-one pickup and delivery problem with time windows* that arises in road-based transportation (Battarra et al. 2014). In the context of passenger transportation, it is often called the dial-a-ride problem (Doerner and Salazar-González 2014). In these land-based problems, each customer request also consists of transporting a load (goods or people) from one pickup vertex to one destination vertex. The differences are, however, equally significant. In tramp shipping the fleet is usually heterogeneous (even if the ships are of similar physical characteristics), the ships have different initial positions, and they generally do not have a common depot. In addition, since the transport distance is generally longer at sea than on land, the ships operate around the clock, and their voyages span days or weeks.

In the following we first give the mathematical formulation of a classic tramp ship routing and scheduling problem in Sect. 2.2.1, in which the sailing speed between a pair of ports for a given ship is fixed and the fuel consumption is not dependent on ship payload. The model takes the form of a mixed integer linear programming problem. We then show in Sect. 2.2.2 the nonlinear extension of the model that incorporates variable speeds and the dependency of fuel consumption on ship speed and payload. These models are based on Norstad et al. (2011) and Christiansen and Fagerholt (2014).

2.2.1 The Basic Linear Model

Let there be n cargoes that might be transported during the planning horizon. Let each of the n cargoes be represented by an index i. Associate to cargo i a loading port node i and an unloading port node $n + i$. Note that different nodes may correspond to the same physical port. Let $N^P = \{1, 2, \dots, n\}$ denote the set of pickup nodes and $N^D = \{n + 1, n + 2, \dots, 2n\}$ the set of delivery nodes. The set of pickup nodes is partitioned into two subsets, N^C and N^O, where N^C is the set of pickup nodes for the mandatory contracted cargoes and N^O is the set of pickup nodes for the optional cargoes.

Let V be the set of ships. A network (N_v, A_v) is associated with each ship v. Here, N_v is the set of nodes that can be visited by ship v, including the origin and an artificial destination for ship v, $o(v)$, and $d(v)$, respectively. Geographically, the origin can be either a port or a point at sea, while the artificial destination is the last planned unloading port for ship v. If the ship is not used, $d(v)$ will represent the same location as $o(v)$. From this, we can extract the sets $N_v^P = N^P \cap N_v$ and $N_v^D = N^D \cap N_v$ consisting of the pickup and delivery nodes that ship v may visit, respectively. The set A_v contains all feasible arcs for ship v, which is a subset of $N_v \times N_v$.

For each ship $v \in V$ and each arc $(i, j) \in A_v$, let T_{ijv}^S be the sailing time from node i to node j, while T_{iv}^P represents the service time in port at node i with ship v. The variable transportation costs C_{ijv} consist of the sum of the sailing costs from node i to node j and the port costs of node i for ship v. It is also assumed that a (contractual) cargo i can be serviced by a ship chartered from the spot market at a given cost, C_i^S. Further, let $[\underline{T}_{iv}, \overline{T}_{iv}]$ denote the time window for ship v associated with node i, where $\underline{T}_{iv}$ is the earliest time for start of service and $\overline{T}_{iv}$ is the latest. Each cargo i has a quantity Q_i and generates a revenue R_i per unit if it is transported. Let K_v be the capacity of ship v.

We also define the following decision variables. Let binary variable x_{ijv} be equal to 1 if ship v sails directly from node i to node j and 0 otherwise. Let t_{iv} represent the time for start of service for ship v at node i and l_{iv} the load (weight) on board ship v when leaving node i. To ease the reading of the model, we assume that each ship is empty when leaving the origin and when arriving at the artificial destination, i.e., $l_{o(v)v} = l_{d(v)v} = 0$. Let binary variable z_i be equal to 1 if cargo i is serviced by a ship from the spot market and 0 otherwise. Finally, let binary variable y_i be equal to 1 if optional cargo i is transported and 0 otherwise.

The basic tramp ship routing and scheduling problem can now be formulated as follows:

$$\max \sum_{i \in N^C} R_i Q_i + \sum_{i \in N^O} R_i Q_i y_i - \sum_{v \in V} \sum_{(i,j) \in A_v} C_{ijv} x_{ijv} - \sum_{i \in N^C} C_i^S z_i \tag{8.1}$$

subject to

$$\sum_{v \in V} \sum_{j \in N_v} x_{ijv} + z_i = 1, \qquad i \in N^C, \tag{8.2}$$

$$\sum_{v \in V} \sum_{j \in N_v} x_{ijv} - y_i = 0, \qquad i \in N^O, \tag{8.3}$$

$$\sum_{j \in N_v} x_{o(v)jv} = 1, \qquad v \in V, \tag{8.4}$$

$$\sum_{j \in N_v} x_{ijv} - \sum_{j \in N_v} x_{jiv} = 0, \qquad v \in V, i \in N_v \setminus \{o(v), d(v)\}, \tag{8.5}$$

$$\sum_{i \in N_v} x_{id(v)v} = 1, \qquad v \in V, \tag{8.6}$$

$$l_{iv} + Q_j - l_{jv} - K_v(1 - x_{ijv}) \leq 0, \qquad v \in V, (i, j) \in A_v | j \in N_v^P, \tag{8.7}$$

$$l_{iv} - Q_j - l_{n+j,v} - K_v(1 - x_{i,n+j,v}) \leq 0, \qquad v \in V, (i, n+j) \in A_v | j \in N_v^P, \tag{8.8}$$

$$\sum_{j \in N_v} Q_i x_{ijv} \leq l_{iv} \leq \sum_{j \in N_v} K_v x_{ijv}, \qquad v \in V, i \in N_v^P, \tag{8.9}$$

$$0 \leq l_{n+i,v} \leq \sum_{j \in N_v} (K_v - Q_i) x_{n+i,jv}, \qquad v \in V, i \in N_v^P, \tag{8.10}$$

$$t_{iv} + T_{iv}^P + T_{ijv}^S - t_{jv} - M_{ijv}(1 - x_{ijv}) \leq 0, \quad v \in V, (i, j) \in A_v, \tag{8.11}$$

$$\sum_{j \in N_v} x_{ijv} - \sum_{j \in N_v} x_{n+i,jv} = 0, \qquad v \in V, i \in N_v^P, \tag{8.12}$$

$$t_{iv} + T_{iv}^P + T_{i,n+i,v}^S - t_{n+i,v} \leq 0, \qquad v \in V, i \in N_v^P, \tag{8.13}$$

$$\underline{T}_{iv} \leq t_{iv} \leq \overline{T}_{iv}, \qquad v \in V, i \in N_v, \tag{8.14}$$

$$l_{iv} \geq 0, \qquad v \in V, i \in N_v, \tag{8.15}$$

$$x_{ijv} \in \{0, 1\}, \qquad v \in V, (i, j) \in A_v, \tag{8.16}$$

$$y_i \in \{0, 1\}, \qquad i \in N^O, \tag{8.17}$$

$$z_i \in \{0, 1\}, \qquad i \in N^C. \tag{8.18}$$

The objective function (8.1) maximizes the profit from operating the fleet. The four terms are the revenue gained by transporting the mandatory contracted cargoes, the revenue from transporting the optional cargoes, the variable transportation costs, and the cost of using spot charters. The fixed revenue for the contracted cargoes can be omitted, but is included here to obtain a more complete picture of the profit. Constraints (8.2) state that all mandatory contract cargoes are transported, either by a ship in the fleet or by a spot charter. The corresponding requirements for the optional cargoes are given by constraints (8.3). Constraints (8.4), (8.5), and (8.6) describe the flow along the sailing route used by ship v. Constraints (8.7) and (8.8) keep track of the load on board at the pickup and delivery nodes, respectively. Constraints (8.9) and (8.10) represent the ship capacity constraints at the loading and discharging nodes, respectively. Constraints (8.11) ensure that the time of starting service at node j must be greater than or equal to the departure time from the previous node i, plus the sailing time between the nodes. The big M coefficient in constraints (8.11) can be calculated as $M_{ijv} = \max(0, \overline{T}_{iv} + T_{iv}^P + T_{ijv}^S - \underline{T}_{jv})$. Constraints (8.12) ensure that the same ship v visits both loading node i and the corresponding discharging node $n + i$. Constraints (8.13) force node i to be visited before node $n + i$, while constraints (8.14) define the time window within which service must start. If ship v is not visiting node i, we will get an artificial starting time within the time windows for that (i, v) combination. The non-negativity requirements for the load on board the ship are given by constraints (8.15). Constraints (8.16), (8.17), and (8.18) impose the binary requirements on the flow, optional cargo, and spot charter variables, respectively.

Note that in the industrial shipping context (which may be seen as a special case of tramp shipping as discussed earlier), the objective will be to minimize the variable transportation costs, which correspond to the third and fourth terms in objective function (8.1), while constraints (8.3) and variable y_i are no longer required since in industrial shipping all cargoes are mandatory.

2.2.2 The Extended Nonlinear Model with Speed Optimization

Most of the earlier studies in the tramp ship routing and scheduling literature, e.g., Brown et al. (1987), Korsvik et al. (2010), Malliappi et al. (2011), and Lin and Liu (2011) among others, assume a fixed and known speed for every ship in the fleet, which usually is the service speed traditionally used when the shipping company makes its planning. In reality, the ship can of course sail at other speeds as well. Normally, a ship has a minimum and a maximum cruising speed which define the range of speeds at which it can actually travel. The option of speeding up affords the ship operator operational flexibility to absorb delays at ports and handle schedule disruptions. On the other hand, the shipping industry has seen significant economic savings by prevailing the practice of slow steaming in almost every commercial ship sector.

As shown by Ronen (1982), a cubic function provides a good estimation of the relationship between fuel consumption per time unit and speed for cargo ships. The impact of a change in ship speed on both fuel costs and emissions can therefore be quite dramatic. In fact, as a response to the growing awareness of the economic and environmental benefits brought by planning with variable speeds, in recent years many studies have been dedicated on speed optimization on given routes or have included speeds as decision variables in their routing and scheduling models (Psaraftis and Kontovas 2013, 2014). Some examples are Fagerholt et al. (2010), Gatica and Miranda (2011), Norstad et al. (2011), and Hvattum et al. (2013).

Another important but often overlooked consideration when determining the fuel costs along a ship route is that the payload of the ship varies, especially in pickup and delivery situations, and that the fuel consumption, other than being a nonlinear function of speed, is also a function of ship payload (Psaraftis 2017). According to Barrass (2004), a common approximation is that for a given speed, fuel consumption is proportional to $(l + L)^{2/3}$, where l is the payload and L is the lightship weight of the ship. Also as suggested in Psaraftis and Kontovas (2016), the difference between laden and ballast fuel consumption at the same speed for a specific ship type can be as high as 40%. It is therefore inspiring to see that, recently, some studies have taken payload dependency into account (only for laden and ballast conditions in some case) in the tramp routing and scheduling context, e.g., Wen et al. (2016) and

Vilhelmsen et al. (2017), and in other contexts too, e.g., Andersson et al. (2015) and Wen et al. (2017).

In the following we show that the basic model presented above in Sect. 2.2.1 can be modified to incorporate speed optimization, where the fuel consumption rate of each ship can be a specific function of its speed and payload. Let D_{ij} be the sailing distance from node i to node j. The variable s_{ijv} defines the speed of travel from node i to node j with ship v. Thus the time it takes to sail along arc (i, j) can be computed by D_{ij}/s_{ijv}. The nonlinear function $C_v(s, l)$, defined on the speed interval $[\underline{S}_v, \overline{S}_v]$, represents the variable transportation costs per unit of distance for ship v sailing at speed s with load l on board. The cost of sailing an arc (i, j) with ship v departing node i with load l_{iv} at speed s_{ijv} is then $D_{ij}C_v(s_{ijv}, l_{iv})$.

The model for the basic tramp ship routing and scheduling problem (8.1)–(8.18) can now be adjusted as follows:

$$\max \sum_{i\in N^C} R_i Q_i + \sum_{i\in N^O} R_i Q_i y_i - \sum_{v\in V} \sum_{(i,j)\in A_v} D_{ij}C_v(s_{ijv}, l_{iv})x_{ijv} - \sum_{i\in N^C} C_i^S z_i, \tag{8.19}$$

subject to (8.2)–(8.10), (8.12), (8.14)–(8.18) and

$$t_{iv} + T_{iv}^P + D_{ij}/s_{ijv} - t_{jv} - M_{ijv}(1 - x_{ijv}) \leq 0, \quad v \in V, (i, j) \in A_v, \tag{8.20}$$

$$t_{iv} + T_{iv}^P + D_{i,n+i}/s_{i,n+i,v} - t_{n+i,v} \leq 0, \quad v \in V, i \in N_v^P, \tag{8.21}$$

$$\underline{S}_v \leq s_{ijv} \leq \overline{S}_v, \quad v \in V, (i, j) \in A_v. \tag{8.22}$$

The objective function (8.19) has now become a nonlinear function because of the nonlinear relationships between fuel consumption and speed and payload. Constraints (8.20) and (8.21) correspond to constraints (8.11) and (8.13) in the original formulation. These constraints are also nonlinear because the sailing time depends on the speed variable. The new constraints (8.22) define the lower and upper bounds for the speed variables.

3 Modeling the Emission Reduction Schemes

In this section we present and discuss the model extensions for incorporating two versions of CO_2 emission mitigation strategy: a bunker levy scheme in Sect. 3.1 and an emission trading scheme (ETS) in Sect. 3.2. The bunker levy and ETS proposals are both market-based measures (MBMs) that can potentially help meet global climate goals through a more flexible approach than the traditional regulatory measures ("command-and-control," where public authorities mandate the performance to be achieved or the technologies to be used). On the one hand, the MBMs can be used to establish an international fund to invest into emission reduction projects outside the marine sector. They are also economic (or "price-based") instruments that, on the other hand, potentially provide the required incentives to ship owners for enhancing their energy efficiency and reducing "in-sector" emissions, through the adoption of long-term technological measures (e.g., more efficient engines or ships) and short-term logistical measures (e.g., slow steaming, optimal fleet management).

3.1 Model with Bunker Levy

Bunker levy, or "carbon tax," is a measure of collecting revenue from the shipping sector in the form of a tax on fuel use. The scheme may also be enforced as a percentage on fuel price. The bunker levy scheme has gained much favor with researchers compared with other emission mitigation solutions (European speed limit, ETS, etc.), mainly because it is easy to implement and provides price certainty in terms of increase in fuel costs to which shipping companies can respond proactively (Cariou and Cheaitou 2012; Psaraftis 2012; Kapetanis et al. 2014). There are also concerns on the resulting modal shifts and that the extra costs will only be passed along the supply chain (GSF 2012).

In a tramp ship routing and scheduling problem, the bunker levy scheme can be modeled as an extra charge on every tonne of fuel consumed. In terms of mathematical formulation, the model incorporating a bunker levy requires the following modifications. Similar to the nonlinear function $C_v(s, l)$ in Sect. 2.2.2., we let $F_v(s, l)$ denote the amount of fuel consumed (in tonnes) per unit of distance for ship v sailing at speed s with load l. The total fuel consumption, represented by FUEL, can then be written as

$$\text{FUEL} = \sum_{v \in V} \sum_{(i,j) \in A_v} D_{ij} F_v(s_{ijv}, l_{iv}) x_{ijv}. \tag{8.23}$$

Let LEVY be the tax imposed on every tonne of fuel consumed. The objective function (8.19) is then modified as follows to account for the extra fuel costs:

$$\max \sum_{i \in N^C} R_i Q_i + \sum_{i \in N^O} R_i Q_i y_i - \sum_{v \in V} \sum_{(i,j) \in A_v} D_{ij} C_v(s_{ijv}, l_{iv}) x_{ijv} - \sum_{i \in N^C} C_i^S z_i - \text{LEVY} \times \text{FUEL}. \tag{8.24}$$

3.2 The Emission Trading Scheme

To provide a basis for considering a potential ETS in our model on tramp shipping, we look at the European Union Emissions Trading System (EU ETS) for some details of the mechanism. The EU ETS has been in operation from 2005 and was the first large GHG emission trading scheme in the world. The scheme now covers more than 11,000 factories, power stations, and other installations in 31 countries – all 28 EU member states plus Iceland, Norway, and Liechtenstein. In 2012, the EU ETS was extended to the airline industry. In November 2017 the European Parliament and EU member states agreed on a revision of the EU ETS that excludes shipping for the time being, but "will include shipping in the trading system from 2023 if IMO progress in a CO_2 strategy is considered insufficient" (The Maritime Executive 2017).

The ETS functions under the "cap and trade" principle, where a maximum (cap) is set on the total amount of CO_2 that can be emitted by all participants in the system. "Allowances" for emissions are created equal to the size of the cap, which are measured in units where one unit corresponds to the right to emit one tonne of carbon dioxide equivalent (CO_2e). The allowances are allocated for free or auctioned off to the emitters and can subsequently be traded among them. If emission exceeds what is permitted by its allowances, an emitter must purchase allowances from others. Conversely, if an emitter has performed well at reducing its emissions, it can sell its leftover allowances. This potentially allows the participants of the system to find the most cost-effective ways of reducing emissions without significant government intervention.

To include an ETS mechanism in the model, we make the following modifications to the model presented in Sect. 2.2.2. Let H be the amount of CO_2 allowance (in tonnes) acquired by the shipping company from public authorities or auctions at price P^C per tonne (which may be zero or non-zero). Let P^S be the spot price of one tonne of CO_2 allowance trading in the secondary market, same for buying and selling. Note that H and P^C are input to our model, since the tramp ship routing and scheduling problem we address in this chapter typically focuses on decisions on the operational/tactical level. Also note that the assumption regarding CO_2 allowance trading price P^S in the spot market is based on the viewpoint of a single tramp shipping company; therefore, such spot price is assumed to be exogenous and constant during our planning horizon.

As in Sect. 3.1, we use $F_v(s, l)$ and FUEL to represent the amount of fuel consumed for every unit of distance sailed by ship v at speed s with load l and the total fuel consumption, respectively. There is a linear relationship between fuel burned and CO_2 produced, with the proportionality constant being known as the emission factor. The third IMO GHG study (Smith et al. 2015) indicates that such factor is between 3.11 and 3.21 (tonnes of CO_2 per tonne of fuel) independent of fuel type (for most common fuel types, emission factor for marine LNG is 2.75). Therefore, the total amount of CO_2 emitted can be expressed by $3.2 \times$ FUEL, using 3.2 as the emission factor.

The objective function (8.19) is then changed to

$$\max \sum_{i\in N^C} R_i Q_i + \sum_{i\in N^O} R_i Q_i y_i - \sum_{v\in V}\sum_{(i,j)\in A_v} D_{ij} C_v(s_{ijv}, l_{iv}) x_{ijv} - \sum_{i\in N^C} C_i^S z_i$$
$$- \left[\, P^C H + P^S(3.2 \times \text{FUEL} - H) \,\right] \tag{8.25}$$

where the expression in the brackets [·] represents the total costs for CO_2 emissions, including the costs of acquiring the initial CO_2 allowances H and the costs of buying additional allowances from the spot market (or the revenue of selling leftover allowances if the actual amount emitted is lower than H). The constraints remain unchanged compared with the model presented in Sect. 2.2.2.

Notice that by separating out terms that are constant values, we may rewrite the expression for total emission costs, i.e., the expression inside [·] in Eq. (8.25), into

$$P^S \times 3.2 \times \text{FUEL} + (P^C - P^S)H \tag{8.26}$$

in which the first term is the amount of total CO_2 emissions multiplied by the spot CO_2 allowance price P^S and the second term is a constant value independent of any decision variable in the model. Therefore, the maximization of objective function (8.25) is equivalent to solving

$$\max \sum_{i\in N^C} R_i Q_i + \sum_{i\in N^O} R_i Q_i y_i - \sum_{v\in V}\sum_{(i,j)\in A_v} D_{ij} C_v(s_{ijv}, l_{iv}) x_{ijv} - \sum_{i\in N^C} C_i^S z_i$$
$$- \; P^S \times 3.2 \times \text{FUEL}\,. \tag{8.27}$$

Therefore, compared with objective function (8.19) in the original model, incorporating an ETS implicates adding an *extra charge* $P^S \times 3.2$ on *every tonne of fuel consumed* independent of the amount and price of the CO_2 allowances initially received (provided that the amount and price of the initial allowance are both input), and such charge depends on the trading price of CO_2 allowance in a spot market. Also note that this objective function is analogous to objective function (8.24) in the bunker levy case.

It is important to emphasize again the caveats of this conclusion and that the underlying assumptions be comprehended. First, as mentioned earlier, the amount of allocated CO_2 allowances H and the average unit cost of acquiring these allowances P^C are input to our model due to the scope of a typical tramp shipping problem. These initial costs are therefore "sunk" and will not affect the ship routing, scheduling, or optional cargo selection decisions. In reality, H may also be a decision variable when the initial allowances held by the shipping company are acquired in part (or all) from auctions. In EU ETS, for example, over 50% of the total amount of allowances over the period 2013–2020 will be auctioned in the primary market (on average overall sectors covered by EU ETS; in the aviation industry, the proportion of auctioned allowances is 15%), while the remaining allowances are granted free and allocated to companies based on their historical emissions (European Commission 2015). Therefore, for those companies with expected allowance demand higher than their allocated amount, H is also a decision to be made as the company may buy allowances through auctioning (at prices usually comparable to the spot price at the time of auctioning) to avoid having to fulfill its obligations from a secondary market later where the spot price may fluctuate widely.

Second, the additional costs resulted from an emission reduction scheme may have large effects on many tactical/strategic decisions. For example, if the amount of free allowances received from public authorities is little, and the expected costs to fulfill its obligations through either allowance auctioning or the secondary market are significant, the shipping company might cut back on long-term contracts or reduce the size of its fleet which are all incentives for modal shifts that take cargoes off seaways (e.g., from short sea shipping to land-based transportation, which may also be a source of "carbon leakage" into sectors with less stringent climate policy). These are outside the scope of a typical tramp ship routing and scheduling problem but are significant issues that need further exploration.

Third, since the model is based on the viewpoint of one tramp shipping company, the trading price of CO_2 allowance in the spot market (P^S) is assumed to be exogenous and constant. In reality the spot market may exhibit an increasing marginal purchasing costs, i.e., the buying price of one unit of CO_2 allowance may increase when purchasing more, especially if the market is thin and if the shipping company is a major player in the business.

4 Computational Study

In this section we present a computational study to demonstrate the effects of implementing an emission reduction scheme in the form of a bunker levy. We only discuss the bunker levy scenario as in the previous section, we have shown that the imposition of an ETS also implicates an extra charge on fuel use (from the viewpoint of a typical tramp operator).

We use 16 test instances taken from the benchmark instances for industrial and tramp routing and scheduling problems (Hemmati et al. 2014). The tests are performed based on the model with variable speeds presented in Sect. 2.2.2, where we *increase the input fuel price* to imitate the implementation of a bunker levy. By doing so we can examine the impact of a bunker levy on the tramp operator's operational decisions and its total fuel consumption (and hence emissions).

The problems are solved on the commercial ship routing decision support system TurboRouter (Fagerholt 2004; Fagerholt and Lindstad 2007) from SINTEF Ocean, using the multi-start local-search heuristic method presented in Brønmo et al. (2007) and Norstad et al. (2011). Note that the particular algorithm used for solving the fixed-route speed optimization problems (which are subproblems in the multi-start heuristic) is based on the discretization arrival times at each route node (Fagerholt et al. 2010); the alternative "recursive smoothing algorithm," although shown to be more efficient in the case where fuel consumption depends only on speed, cannot be used here because of our inclusion of payload dependency (see discussions in Norstad et al. 2011).

4.1 Input Data and Test Instances

To represent realistic situations faced in tramp shipping, the test instances we use have characteristics combining two geographical settings, short-sea and deep-sea shipping, and two cargo settings, full-load and mixed-load cargoes. In deep-sea shipping, the cargoes are transported long distances and across at least one of the big oceans, for example, from Liverpool to Yokohama. In short-sea shipping, the cargo movements are within Europe, for example, from Gdansk to Dunkirk. For full-load instances, the cargo sizes are such that a cargo is a full shipload. And in mixed-load cases, some of the cargoes are of smaller size, and the ship capacity may accommodate several cargoes simultaneously.

We use 4 instances for each such combination of the above geographical and cargo settings, i.e., "Deep-Full," "Deep-Mix," "Short-Full," and "Short-Mix," which amount to 16 instances in total. Each instance is referred to in the format *setting*-Cx-Vy-z, where *setting* is the combination of geographical and cargo settings, x is the number of cargoes, y is the number of ships, and z is the z-th instance with the same setting and size. The complete list of instances is shown in Table 8.2, and these instances may be downloaded from http://home.himolde.no/~hvattum/benchmarks/.

Recall that in the model presented in Sect. 2.2.2, we use a nonlinear function $C_v(s, l)$, defined on the speed interval $[\underline{S}_v, \overline{S}_v]$, to represent the variable transportation costs per unit of distance for ship v sailing at speed s with load l on board. To describe such a function well, one must have a good approximation of the relationship between the ship's fuel consumption rate and its speed and payload, since the fuel costs make up most of the variable transportation costs.

For every specific ship, the fuel consumption (FC) rate is in tonnes (t) per traveled nautical mile (M) which is a function of speed s (knots) and payload ρ (% of total capacity). We use the following empirical function, $\text{FC} = (\text{A}s^2 + \text{B}s + \text{C}) \times (0.8 + 0.2\rho)$, where s is within the ship's feasible speed range and ρ takes its value between 0% (ballast) and 100% (fully loaded). The parameters A, B, and C are ship specific and obtained based on empirical fuel consumption values for each ship. The function also implies a linear relationship between payload and fuel consumption at a given speed, instead of a more sophisticated nonlinear version (see Sect. 2.2.2), but it gives an acceptable approximation of the fuel consumption between ballast and fully loaded (laden) states based on real data of these ships. Figure 8.1 depicts the fuel consumption characteristics for one of the ships used in the test instances, a Handymax bulk carrier of 56800 deadweight tonnes (dwt) capacity. The fuel consumption function in this case is $\text{FC} = (0.0019s^2 - 0.045s +$

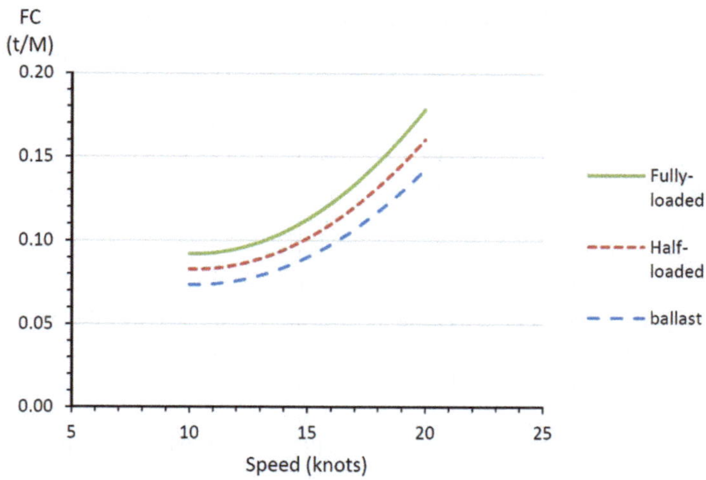

Fig. 8.1 Fuel consumption characteristics for a Handymax bulk carrier

0.3739) $\times$ $(0.8 + 0.2\rho)$, where the feasible speed range of s is between 10 and 20 knots. The fuel consumption curves for ballast, half-loaded, and fully loaded are shown in Fig. 8.1, which correspond to $\rho = 0\%$, 50%, and 100%, respectively.

4.2 Computational Results

We first use one instance, Deep-Full-C50-V20-1, as an illustrative example to demonstrate the impact of an increase in fuel price on the tramp operator's operational decisions and its fuel consumption and CO_2 emissions. This instance considers 50 cargoes and 20 ships and has the characteristics of deep-sea shipping and full-load cargoes.

Table 8.1 shows the results for instance Deep-Full-C50-V20-1 when altering the fuel price from \$200 to \$600 per tonne. It is clearly visible that as the fuel price goes up, the profit of the company decreases. The number of cargoes served also decreases from 46, when fuel price is \$400 and below, to 43 and 37, when fuel price is as high as \$500 and \$600, respectively. This is because the operator is rejecting more optional cargoes when fuel is expensive, so as to reduce its fuel consumption; see the "Fuel Consump.(t)" row.

One may notice that the number of cargoes served remains 46 for the first three columns, while fuel consumption in the low fuel price case is significantly higher. This is because when the fuel price is low, the operator sails the ships at faster speed to chase cargoes with higher income. Take \$200 fuel price, for example, compared to the \$300 case, only 1 cargo is different, while the other 45 cargoes are identical. When fuel is \$200 per tonne, the ship operator takes the cargo with higher income,

Table 8.1 Comparison under different fuel prices for instance Deep-Full-C50-V20

Fuel price/tonne	\$200	\$300	\$400	\$500	\$600
# Total cargoes	50	50	50	50	50
# Served cargoes	46	46	46	43	37
Income (mill \$)	34.51	34.21	34.39	32.37	28.71
Profit (mill \$)	24.57	20.54	17.05	13.58	10.66
Total days at sea	1594	1591	1611	1489	1285
Total mileage	425,947	423,724	425,950	381,135	314,651
Avg speed (knots)	11.14	11.09	11.02	10.67	10.20
Fuel consump.(t)	49,674	45,548	43,329	37,574	30,088
CO_2 emissions (t)	158,957	145,754	138,653	120,236	96,281
Tonne-miles (mill t-M)	13,073	13,073	13,068	12,129	10,370
CO_2/Tonne-mile (10^{-3} t/t-M)	0.0122	0.0111	0.0106	0.0099	0.0093

but also needs to operate a ship at a much higher speed to be able to serve this cargo within its stipulated time windows. The fuel consumption increases accordingly, but the ship operator can afford it because of low fuel price in this case. In fact, we see this trend across Table 8.1 when the fuel price increases from \$200 to \$600, namely, the ship operator gradually gives up those optional cargoes that are "harder" to service, so that ships can sail at lower speed and save more on fuel. This can be seen from the decreasing speed values from the "Avg speed (knots)" row, which shows the average sailing speed of all ships in the fleet.

We also show the amounts of CO_2 emissions in Table 8.1, calculated from multiplying the fuel consumption by the emission factor 3.2 (tonnes of CO_2 per tonne fuel). We then compute the total tonne-miles of all cargoes in each case. It can be seen that when the fuel price goes up, the amount of CO_2 emissions per tonne-mile of cargo transported decreases, meaning the ships are operated in a more "CO_2-efficient" way. However, this is achieved by giving up the "hard" optional cargoes, such as the ones with demanding time windows or at difficult locations that require long ballast sailings.

Table 8.2 shows the comparison under 2 fuel prices, \$300 and \$600 per tonne, for all 16 instances. Increasing the fuel price from \$300 to \$600 per tonne implies a bunker levy of 100%, which is not realistic in the near future. In addition, higher fuel prices due to taxation probably would also result in increased freight rates (and thus higher revenues from the same cargoes, since shipping companies cannot bear all the increased costs), while the rates in our study are assumed constant. Therefore, the results are intended, only for illustrative purposes, to show the effects of a bunker levy on a tramp operator's economic and environmental performances.

In Table 8.2, we summarize, for each instance and under two fuel prices, five important attributes, including the number of cargoes served, profit, average speed of the fleet, total amount of CO_2 emissions, and the amount of CO_2 emitted per

Table 8.2 Comparison of results for all instances under two fuel prices, \$300 and \$600 per tonne

Instance	# Cargoes served			Profits (mill \$)			Avg speed (knots)			CO_2 emissions (t)			CO_2/Tonne-mile (10-3 t/t-M)		
	\$300	\$600	%Δ	\$300	\$600	%Δ	\$300	\$600	%Δ	\$300	\$600	%Δ	\$300	\$600	%Δ
Deep–Full–C25–V7–1	23	14	−39.1	9.57	3.47	−63.7	12.4	11.6	−6.5	70,438	38,345	−45.6	0.0106	0.0094	−11.5
Deep–Full–C25–V7–2	25	20	−20.0	12.17	5.88	−51.7	12.0	11.0	−8.3	76,003	55,251	−27.3	0.0095	0.0084	−11.4
Deep–Full–C50–V20–1	46	37	−19.6	20.54	10.66	−48.1	11.1	10.2	−8.1	145,754	96,281	−33.9	0.0111	0.0093	−16.2
Deep–Full–C50–V20–2	49	39	−20.4	22.31	11.45	−48.7	10.9	10.4	−4.6	128,294	92,563	−27.9	0.0088	0.0078	−11.4
Deep–Mix–C30–V6–1	20	17	−15.0	10.26	5.92	−42.3	10.9	10.1	−7.3	45,667	36,541	−20.0	0.0138	0.0130	−5.8
Deep–Mix–C30–V6–2	21	20	−4.8	11.07	7.92	−28.5	12.0	11.1	−7.5	48,240	33,805	−29.9	0.0164	0.0127	−22.6
Deep–Mix–C35–V7–1	35	31	−11.4	41.50	22.03	−46.9	14.4	14.2	−1.5	227,949	174,845	−23.3	0.0059	0.0059	0.0
Deep–Mix–C35–V7–2	35	33	−5.7	50.85	34.78	−31.6	14.2	14.3	0.4	176,144	162,297	−7.9	0.0057	0.0055	−3.5
Short–Full–C25–V7–1	24	23	−4.2	2.07	1.21	−41.5	11.6	11.5	−0.9	9,277	8,630	−7.0	0.0206	0.0201	−2.4
Short–Full–C25–V7–2	25	24	−4.0	2.11	1.17	−44.5	11.1	11.6	4.5	10,218	9,325	−8.7	0.0205	0.0198	−3.4
Short–Full–C50–V20–1	50	50	0.0	5.30	3.58	−32.5	12.5	12.4	−0.8	19,123	18,726	−2.1	0.0146	0.0143	−2.1
Short–Full–C50–V20–2	50	48	−4.0	4.73	3.01	−36.4	12.5	12.4	−0.7	19,260	17,469	−9.3	0.0155	0.0149	−3.9
Short–Mix–C30–V6–1	30	28	−6.7	4.14	3.04	−26.6	12.8	12.4	−3.3	12,102	9,856	−18.6	0.0251	0.0248	−1.2
Short–Mix–C30–V6–2	30	29	−3.3	3.76	2.75	−26.9	11.9	11.7	−1.7	10,842	9,898	−8.7	0.0260	0.0263	1.2
Short–Mix–C35–V7–1	35	34	−2.9	4.32	3.28	−24.1	11.4	11.3	−1.3	12,732	11,245	−11.7	0.0235	0.0218	−7.2
Short–Mix–C35–V7–2	35	34	−2.9	5.03	3.91	−22.3	12.0	11.0	−8.0	11,136	10,800	−3.0	0.0242	0.0243	0.4
Average change			−10.2			−38.5			−3.5			−17.8			−6.3

tonne-mile. The "%Δ" columns indicate the relative changes when increasing the fuel price from \$300 to \$600. We observe that these changes are in general consistent with the trend found based on Table 8.1: when the fuel becomes expensive because of a levy, the tramp operator accepts fewer cargoes to transport, especially those that need ships sailing faster to meet their time windows. There are a few exceptions, e.g., Short-Full-C25-V7-2, the average fleet speed is increased by 4.5% in spite of expensive fuel. In this instance, the single cargo being dropped by the \$600 solution (compared to its \$300 counterpart) is relatively poorly paid and has a long transporting distance. In addition this cargo has an "easy" time window that allows the ship servicing it to sail a slow voyage which brings the average speed of the fleet down. When fuel becomes expensive, this slow and long voyage is dropped due to the low income of the corresponding cargo, leading to an overall increase in the fleet's average speed.

On average across all 16 instances, we observe in Table 8.2 that as a consequence of the high levy on fuel, the tramp operator accepts around 10% fewer cargoes and sails its fleet 3.5% slower. In addition, the ship operator's total profits are 38.5% lower, whereas its fuel consumption and hence CO_2 emissions are reduced by 17.8%. Moreover, the average "CO_2/tonne-mile" measure decreases by 6.3% when fuel is expensive, indicating that the ships are operated more efficiently in terms of CO_2 emitted for every tonne-mile of cargo transported. As was discussed earlier, such efficiency is achieved by dropping the "hard" optional cargoes, such as the ones with difficult time windows (e.g., loading needed rather soon or requiring fast transport) that demand high sailing speed or at difficult locations that require long ballast sailings. These potentially "inefficient" cargoes (from the single tramp operator's perspective) are accepted when fuel price is low, but when fuel becomes too expensive with added levy, the "hard" optional cargoes no longer make worthwhile contribution to the total profits. At a broader level, these cargoes may still find their way to their respective destinations in any case, perhaps by other shipping companies. However, the imposition of a levy may help the players in the market increase their CO_2 efficiency as a whole by redistributing the cargoes to their appropriate carriers.

5 Conclusion

This chapter has presented the typical tramp ship routing and scheduling model and discussed how market-based CO_2 reduction measures, including the bunker levy and ETS schemes, can be incorporated into the model. It has been shown that from the viewpoint of a tramp ship operator on the operational level, the implementation of an ETS implicates the addition of an extra charge on every tonne of fuel consumed, which is similar to a bunker levy. Such conclusion was obtained when assuming that the CO_2 allowances initially acquired are sunk costs. This assumption is consistent with the typical context of a tramp ship routing and scheduling problem, but the effects of an emission reduction scheme on a ship operator's tactical decisions need to be addressed and further studied. For example, the shipping company might cut back on long-term contracts or reduce the size of its fleet if the extra costs for CO_2 reduction are too expensive. These may lead to modal shifts from (short sea) shipping to land-based transportation modes and potential carbon leakage.

A computational study on 16 benchmark instances has been done to demonstrate the effects of implementing a bunker levy in the form of a tax based on fuel price. It has been shown that in response to a largely elevated fuel price, the ship operator will accept fewer optional cargoes, slow down the ships, and operate the fleet in a more "CO_2-efficient" way, i.e., emit less per tonne-mile of cargo transported.

Many perspectives remain open with respect to this study. First, we focused on decisions made by a tramp ship operator on an operational/tactical level. This scope can be expanded to include some important and directly relevant tactical/strategic decisions such as the composition of the fleet, i.e., to determine if the size and mix of the fleet need changing to adapt to new environmental regulations. Second, similar analysis from this work can also be done in other shipping sectors, such as container shipping.

References

Andersson, H., Fagerholt, K., & Hobbesland, K. (2015). Integrated maritime fleet deployment and speed optimization: Case study from RoRo shipping. *Computers & Operations Research, 55*, 233–240.

Barrass, B. (2004). *Ship design and performance for masters and mates*. Oxford: Butterworth-Heinemann.

Battarra, M., Cordeau, J.-F., & Iori, M. (2014). Pickup-and-delivery problems for goods transportation. In P. Toth & D. Vigo (Eds.), *Vehicle routing: Problems, methods, and applications* (MOS-SIAM series on optimization, 2nd ed., Chap. 6, pp. 161–191). Philadelphia: Society for Industrial and Applied Mathematics.

Brønmo, G., Christiansen, M., Fagerholt, K., & Nygreen, B. (2007). A multi-start local search heuristic for ship scheduling – A computational study. *Computers & Operations Research, 34*(3), 900–917.

Brown, G. G., Graves, G. W., & Ronen, D. (1987). Scheduling ocean transportation of crude oil. *Management Science, 33*(3), 335–346.

Buhaug, Ø., Corbett, J. J., Endresen, Ø., Eyring, V., Faber, J., Hanayama, S., Lee, D., Lee, D., Lindstad, H., Markowska, A., Mjelde, A., Nelissen, D., Nilsen, J., Pålsson, C., Winebrake, J., Wu, W., & Yoshida, K. (2009). *Second IMO GHG study 2009*. Technical report, International Maritime Organization (IMO), London.

Cariou, P., & Cheaitou, A. (2012). The effectiveness of a European speed limit versus an international bunker-levy to reduce CO_2 emissions from container shipping. *Transportation Research Part D: Transport and Environment, 17*(2), 116–123.

Christiansen, M., & Fagerholt, K. (2014). Ship routing and scheduling in industrial and tramp shipping. In P. Toth & D. Vigo (Eds.), *Vehicle routing: Problems, methods, and applications* (MOS-SIAM Series on Optimization, 2nd ed., chap. 13, pp. 381–408). Philadelphia: Philadelphia Society for Industrial and Applied Mathematics.

Christiansen, M., Fagerholt, K., Nygreen, B., & Ronen, D. (2013). Ship routing and scheduling in the new millennium. *European Journal of Operational Research, 228*(3), 467–483.

Christiansen, M., Fagerholt, K., & Ronen, D. (2004). Ship routing and scheduling: Status and perspectives. *Transportation Science, 38*(1), 1–18.

Doerner, K. F., & Salazar-González, J.-J. (2014). Pickup-and-delivery problems for people transportation. In P. Toth & D. Vigo (Eds.), *Vehicle routing: Problems, methods, and applications* (MOS-SIAM series on optimization, 2nd ed., chap. 7, pp. 193–212). Philadelphia: Society for Industrial and Applied Mathematics.

European Commission (2015). *EU ETS handbook*. Technical report, European Union.

Fagerholt, K. (2004). A computer-based decision support system for vessel fleet scheduling-experience and future research. *Decision Support Systems, 37*(1), 35–47.

Fagerholt, K., & Lindstad, H. (2007). TurboRouter: An interactive optimisation-based decision support system for ship routing and scheduling. *Maritime Economics & Logistics, 9*(3), 214–233.

Fagerholt, K., Laporte, G., & Norstad, I. (2010). Reducing fuel emissions by optimizing speed on shipping routes. *Journal of the Operational Research Society, 61*(3), 523–529.

Gatica, R. A., & Miranda, P. A. (2011). Special issue on Latin-American research: A time based discretization approach for ship routing and scheduling with variable speed. *Networks and Spatial Economics, 11*(3), 465–485.

GSF (2012). *Maritime emissions briefing note*. Technical report, Global Shippers' Forum.

Hemmati, A., Hvattum, L. M., Fagerholt, K., & Norstad, I. (2014). Benchmark suite for industrial and tramp ship routing and scheduling problems. *INFOR: Information Systems and Operational Research, 52*(1), 28–38.

Hvattum, L. M., Norstad, I., Fagerholt, K., & Laporte, G. (2013). Analysis of an exact algorithm for the vessel speed optimization problem. *Networks, 62*(2), 132–135.

Kapetanis, G. N., Gkonis, K. G., & Psaraftis, H. N. (2014). Estimating the operational effect of a Bunker Levy: The case of Handymax bulk carriers. In *Transport Research Arena 2014 Conference*, Paris.

Korsvik, J. E., Fagerholt, K., & Laporte, G. (2010). A tabu search heuristic for ship routing and scheduling. *Journal of the Operational Research Society, 61*(4), 594–603.

Lawrence, S. A. (1972). *International sea transport: The years ahead*. Lexington: Lexington Books.

Lin, D.-Y., & Liu, H.-Y. (2011). Combined ship allocation, routing and freight assignment in tramp shipping. *Transportation Research Part E: Logistics and Transportation Review, 47*(4), 414–431.

Maersk (2011). Slow steaming – The full story. https://www.yumpu.com/en/document/view/7803715/slow-steaming-the-full-story-maersk. Accessed Jan 1, 2018.

Malliappi, F., Bennell, J. A., & Potts, C. N. (2011). A variable neighborhood search heuristic for tramp ship scheduling. In *Computational Logistics. ICCL 2011* (Lecture notes in computer science, vol. 6971, pp. 273–285). Berlin/Heidelberg: Springer.

Norstad, I., Fagerholt, K., & Laporte, G. (2011). Tramp ship routing and scheduling with speed optimization. *Transportation Research Part C: Emerging Technologies, 19*(5), 853–865.

Psaraftis, H. N. (2012). Market-based measures for greenhouse gas emissions from ships: A review. *WMU Journal of Maritime Affairs, 11*(2), 211–232.

Psaraftis, H. N. (2017). Ship routing and scheduling: the cart before the horse conjecture. *Maritime Economics & Logistics*. https://doi.org/10.1057/s41278-017-0080-x.

Psaraftis, H. N., & Kontovas, C. A. (2013). Speed models for energy-efficient maritime transportation: A taxonomy and survey. *Transportation Research Part C: Emerging Technologies, 26*, 331–351.

Psaraftis, H. N. & Kontovas, C. A. (2014). Ship speed optimization: Concepts, models and combined speed-routing scenarios. *Transportation Research Part C: Emerging Technologies, 44*, 52–69.

Psaraftis, H. N., & Kontovas, C. A. (2016). Green maritime transportation: Speed and route optimization. In H. N. Psaraftis (Eds.), *Green transportation logistics: The quest for win-win solutions* (pp. 299–349). Cham: Springer.

Rajamani, L. (2011). The cancun climate agreements: Reading the text, subtext and tea leaves. *International and Comparative Law Quarterly, 60*(02), 499–519.

Ronen, D. (1982). The effect of oil price on the optimal speed of ships. *Journal of the Operational Research Society, 33*(11), 1035–1040.

Shi, Y. (2016). Reducing greenhouse gas emissions from international shipping: Is it time to consider market-based measures? *Marine Policy, 64*, 123–134.

Smith, T. W. P., Jalkanen, J. P., Anderson, B. A., Corbett, J. J., Faber, J., Hanayama, S., O'Keeffe, E., Parker, S., Johansson, L., Aldous, L., Raucci, C., Traut, M., Ettinger, S., Nelissen, D., Lee, D. S., Ng, S., Agrawal, A., Winebrake, J. J., Hoen, M., Chesworth, S., & Pandey, A. (2015). *Third IMO GHG study 2014*. Technical report, International Maritime Organization (IMO), London.

The Maritime Executive (2017). E.U. emissions trading system excludes shipping, for now. https://maritime-executive.com/article/eu-emissions-trading-system-excludes-shipping-for-now. Accessed Jan 1, 2018.

UNCTAD (2017). *Review of maritime transport 2017*. Technical report, United Nations Conference on Trade and Development, Geneva.

Vilhelmsen, C., Lusby, R. M., & Larsen, J. (2017). Tramp ship routing and scheduling with voyage separation requirements. *OR Spectrum, 39*(4), 913–943.

Wen, M., Ropke, S., Petersen, H., Larsen, R., & Madsen, O. (2016). Full-shipload tramp ship routing and scheduling with variable speeds. *Computers & Operations Research, 70*, 1–8.

Wen, M., Pacino, D., Kontovas, C., & Psaraftis, H. (2017). A multiple ship routing and speed optimization problem under time, cost and environmental objectives. *Transportation Research Part D: Transport and Environment, 52*, 303–321.

Chapter 9
Green Liner Shipping Network Design

Erik Hellsten, David Pisinger, David Sacramento, and Charlotte Vilhelmsen

Abstract *Green Liner Shipping Network Design* refers to the problems in *green logistics* related to the design of maritime services in liner shipping with a focus on reducing the environmental impact. This chapter discusses how to more efficiently plan the vessel services with the use of mathematical optimization models. A brief introduction to the main characteristics of Liner Shipping Network Design is given, as well as the different variants and assumptions that can be considered when defining this problem. The chapter also includes an overview of the algorithms and approaches that have been presented in the literature to design such networks.

Acronyms and Abbreviations

ECA	Emission Control Areas
IMO	International Maritime Organization
LNG	Liquid Natural Gas
LP	Linear Programming
LSNDP	Liner Shipping Network Design Problem
LSP	Liner Service Planning
MARPOL	International Convention for the Prevention of Pollution from Ships
MCFP	Multi-Commodity Flow Problem
MIP	Mixed Integer Programming
SSSCRP	Simultaneous Ship Scheduling and Cargo Routing Problem
TEU	Twenty-Foot Equivalent Units

E. Hellsten · D. Pisinger (✉) · D. Sacramento · C. Vilhelmsen
DTU Management Engineering, Technical University of Denmark, Kongens Lyngby, Denmark
e-mail: dapi@dtu.dk

H. N. Psaraftis (ed.), *Sustainable Shipping*,
https://doi.org/10.1007/978-3-030-04330-8_9

TSP	Traveling Salesman Problem
UNCTAD	United Nations Conference on Trade and Development
VNS	Variable Neighborhood Search
VRP	Vehicle Routing Problem

1 Introduction

The liner shipping industry is a vital part of the global economy, constituting one of the greenest modes of cargo transport. In full load, the new mega-vessels emit only 3 g of CO_2 for transporting 1 metric tonne of cargo 1 kilometer (Maersk 2017); in comparison, trains average on 18 g and flights on 560 g (see Fig. 9.1). Today, around 90% of the global trade, by volume, is carried out by seaborne transportation, a number which is expected to continue rising. During the last three decades, the volume of containerized cargo has grown by more than 8% per year, and more than 5.150 container vessels were in operation worldwide in 2017. The largest vessels carry more than 20.000 20-foot equivalent units (TEU), and during 2016, a container volume of around 140.000.000 TEU was estimated to pass through this vast network (Unctad 2017a,b). In this chapter we will show how optimization techniques can be used to design more efficient liner shipping networks in order to further decrease the environmental footprint of liner shipping.

The liner shipping industry is built up by so-called *services*. A service is a fixed cyclic itinerary, sailed by a number of similar vessels. Services usually have weekly or biweekly departures, to add consistency and regularity for the customers. The vessels are operated by shipping companies called carriers, where the largest carriers operate over 600 vessels. As larger vessels are more energy efficient (see Fig. 9.2), the trend is to build ever-larger vessels. To efficiently utilise those very large liner vessels, each region typically has a few larger ports, called *hubs*, where the liner ships pick up and deliver containers. From the hubs, the containers are then transported to other ports by smaller, more flexible vessels, called *feeder vessels*. *Transshipments* occur both between larger vessels and smaller vessels but also between larger vessels when no suitable service connects the origin and destination

Fig. 9.1 Estimated CO_2 emission for transporting 1 tonne of goods 1 kilometer for different transportation modes (Source: Maersk 2017)

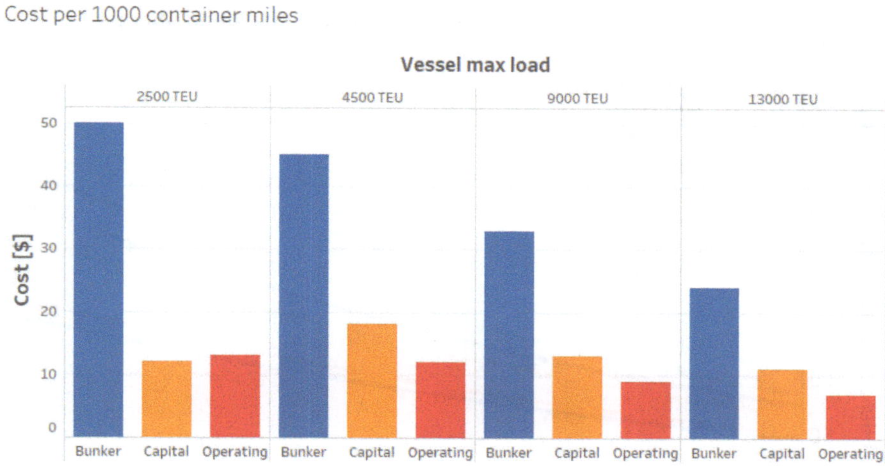

Fig. 9.2 Estimated cost per 1000 container miles for different vessel sizes. The vessels are assumed to sail at 19 knots and the bunker price is estimated as 750$/tonne. We see that bunker represents the largest cost and that transporting containers on larger vessels requires significantly less fuel (Source: Germanischer Lloyd)

hub. While transshipments add flexibility, they tend to be costly, as the cargo needs to be unloaded, stored until the arrival of the new vessel and then reloaded again.

Another major constraint in liner shipping is *cabotage rules*. To protect the national trade business, many countries forbid foreign carriers to ship cargo between two ports within the country. See Brouer et al. (2014a) for examples of cabotage rules.

The major costs for the carriers are vessel acquirement and bunker. But other costs, like canal fees, port costs and transshipment costs, are also highly significant. Most papers in the literature presumes that fuel consumption is frequently estimated as a cubic function of the speed (see Fig. 9.3). Psaraftis and Kontovas (2013) point out that the fuel consumption is given as a complex function depending on many ship parameters and that the cubic approximation on terms of the sailing speed is valid for tankers and bulk carriers, whereas higher exponents should be considered in liner shipping. As the speed has such an impact on the fuel consumption, *slow steaming* is often used to reduce the consumption, i.e., operating the container vessels at speeds, significantly lower than their maximum speed. Especially after the financial crisis in 2008, maritime shipping companies implemented slow steaming policies for cost-cutting purposes. The drawback with slow steaming is, however, that more vessels are required to keep the regularity with respect to weekly departures and also that transit times become longer, yielding a lower level of service for the customers. In general, services has two directions, *head-* and *backhaul*, where most of the cargo is transported in the head-haul direction. A good example of this is the trade between Europe and Asia, where most of the goods are delivered

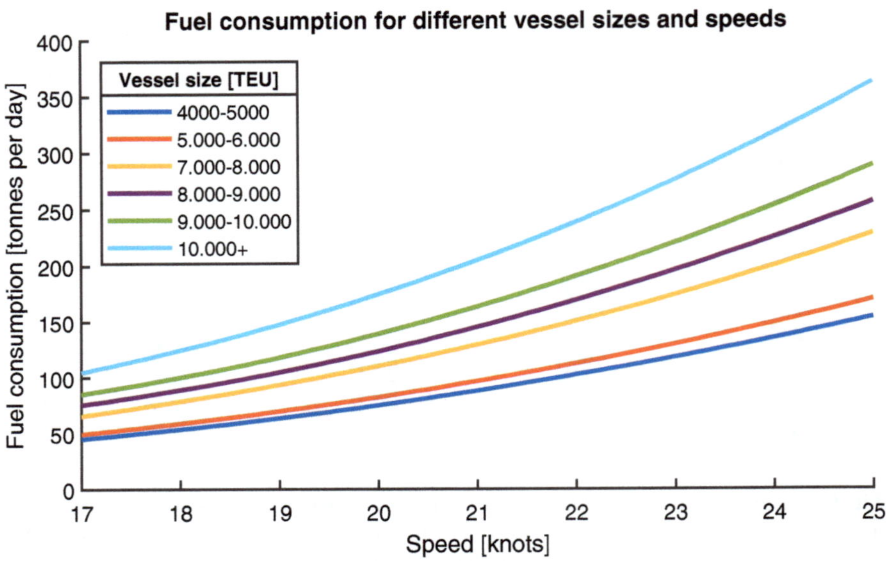

Fig. 9.3 Estimated fuel consumption as a function of steaming speed and vessel size (Source: Notteboom and Vernimmen 2009)

from Asia to Europe. In this case, vessels are slow steaming in the backhaul direction where less customers are affected by the increased transit time.

Due to the ability to transport large numbers of containers with each vessel, liner shipping is one of the most energy-efficient transportation forms. Nonetheless, due to the large volumes transported, the shipping industry contributes significantly to the global CO_2 emissions. According to the *IMO Green House Gas Study 2014*, in 2012 the international shipping industry was estimated to account for 2.2% of the global greenhouse gas emissions, of which approximately a quarter was caused by container vessels, which corresponds to around a billion tonnes of CO_2 annually. These emissions are further expected to increase between 50% and 250% in the next 30 years (IMO 2014).

Although liner shipping is the most efficient transportation mode in terms of CO_2, the vessels commonly operate using "dirty" fuel, emitting various pollutants which are harmful for the environment and the human health. In 2013, it was estimated that, in Europe, ships accounted for 18% of the nitrogen oxides (NO_x), 18% of the sulfur oxides (SO_x) and 11% of the particle matter ($PM_{2.5}$) of the total annual emissions, respectively (Wan et al. 2016). Measures to control SO_x and PM emissions are being applied through the International Convention for the Prevention of Pollution from Ships (also known as MARPOL) and Emission Control Areas (ECA). The emission percentages, of these gases, from seaborne trade are currently much higher than from other modes of transportation, such as rail or aviation. The maximum permitted level of sulfur content contained in marine fuels is currently 3.5%, but it will be reduced to 0.5% by 2020 (IMO 2014).

There are several measures which could be applied to counteract these polluting emissions in the maritime industry. Cleaner practices and maritime policies should be imposed, both by industry and by governments, to control the environmental impact. It is important, however, to emphasize that maritime companies follow long-term strategic plans, where the vessel fleet has a long life expectancy, around 25 to 30 years, since building new vessels is a huge investment. Therefore, it takes a long time before green innovations regarding engines or vessel design can be applied in practice.

Instead, one of the major roads toward a greener shipping industry must be through more efficient utilization of the current assets. If a more efficient service structure can be developed, the same vessels could transport the same amount of cargo while running at a lower speed. This far, the literature on pure green liner shipping network design is highly limited. However, as the bunker cost is one of the major costs for the carriers, reducing the cost is strongly correlated to reducing the fuel consumption. Hence, reducing the cost can, indirectly, be seen as contributing to the green objective. Further, increasing the level of service would likely result in that transportation changes to shipping from other modes. As the CO_2 emissions of liner shipping are lower, an increase in the level of service could also be expected to result in a more sustainable overall transportation system. All in all, to make a greener shipping industry, it would be of great value to develop models, solution algorithms and decision support tools for liner shipping network design.

1.1 Liner Shipping Network Design Problem

The Liner Shipping Network Design Problem (LSNDP) can informally be defined as follows: given a collection of ports, a fleet of container vessels and a group of origin-destination demands, construct a set of services for the container vessels such that the overall operational expenses are minimized while ensuring that all demands can be routed through the resulting network, respecting the capacity of the vessels.

In this section, we present some notation of the LSNDP. For a complete model, see Brouer et al. (2014a). The set of ports is denoted by N and represents the set of physical ports in the problem. The set of arcs A represents all possible sailings between two ports. To each port, there is a corresponding port call cost c_{P}^{i}, as well as a berthing time b_i, for the port call. The set of commodities to transport is denoted by K, and for each commodity $k \in K$, there is an origin port o_k, a destination port d_k and a quantity δ_k measured in TEUs. Finally, the set V denotes the set of *vessel classes* with the corresponding cargo capacities q_v, available quantity M_v, fuel consumption g_v per nautical mile and additional speed limitations. Furthermore, for convenience, the demand of the commodities in the ports are defined as:

$$\xi_i^k = \begin{cases} \delta_k & \text{if port } i \text{ is the origin port of commodity } k \\ -\delta_k & \text{if port } i \text{ is the destination port of commodity } k \\ 0 & \text{otherwise.} \end{cases} \tag{9.1}$$

There is a limited fleet of container vessels, but not all vessels need to be used. The deployment of a vessel has an associated *charter cost* c^v. Additionally, there exist other costs related to the resulting network, such as the *sailing cost* c_{ij}^v associated with each vessel and each arc, which is given as a combination of the port call cost c_{P}^j and the fuel consumption for the corresponding leg. Furthermore, handling costs of containers in the ports are considered as well, incurring a cost c_{L}^i (c_{U}^i) per unit of container (un)loaded in the port. Containers can also be transferred from one vessel to another in the ports, which incurs a unit *transshipment cost* $c_{\mathrm{T}}^{i,k}$.

One of the main traits of the liner shipping industry is the regular operation of services under a pre-established schedule. It is imposed that all services should have *weekly operations*, meaning that if a round trip takes 8 weeks to complete, then eight similar vessels need to be deployed to the service in order to ensure that each port is visited once a week. In addition, services must be cyclic, visiting a sequence of ports before returning to the original port. However, a service is allowed to be non-simple, meaning that a port can be visited several times, since this may improve transit times. Services where only one port is visited twice are called *butterfly services*, and the port which is visited twice is denoted the *butterfly port*.

The variants of the LSNDP, which have been studied in the literature, vary mainly in the following four aspects:

- *Transit time constraints* As described above, the transit time of each commodity has an associated time limit that must be respected. If the transit time is not respected, perishable goods may become unsalable. Many early models for LSNDP did not consider this constraint.
- *Transshipment costs* Several early models for LSNDP did not consider transshipment costs. However, the costs of transshipments are a significant part of the operational costs (Karsten 2015), so it is generally important to represent these costs properly in the model.
- *Rejected demands* Although the formulation of LSNDP states that all commodities must be flowed through the network, many models allow rejection of commodities and instead impose a penalty.
- *Speed optimization* There are three main categories of models regarding speed optimization: models which have constant speed for all services, models which choose a speed for each service and models which choose a speed on each individual leg in each service. As the fuel consumption depends non-linearly on the speed, it is common to choose between a number of discrete speed alternatives, each with a corresponding cost.

Most models for the LSNDP design a network without a specific schedule. Hence the route for each vessel is defined, but not the exact day of arrival/departure.

This is typically done in a later step, where port availabilities are negotiated and transshipment times at ports are adjusted.

For a detailed review of the research on liner shipping optimization problems, see the survey papers Ronen (1983, 1993), Christiansen et al. (2004, 2013), Meng et al. (2014), Brouer et al. (2016, 2017) and Lee and Song (2017).

1.2 Measuring and Calculating Transportation Emissions

The environmental effects associated with the maritime industry are becoming a major concern. The large amount of pollution produced by container vessels has not gone unnoticed, due to considerable emissions of various types of pollutants such as SO_x, NO_x, PM and CO_2. The International Maritime Organization (IMO) is investigating the possibility of reducing these emissions by establishing regulatory policies.

The maritime industry is an economy-dependent industry, and the minimization of the operational cost is paramount. As noted by Notteboom (2006), the price of fossil fuels is one of the largest in maritime transportation. Ronen (2011) estimates that the bunker cost makes up more than 75% of the total operating cost of a vessel. The fuel cost is strongly related to the operating speed of the vessels, where there exists an important trade-off. Based on this, the estimation of greenhouse gases such as CO_2 can be given by an energy approach, which can be obtained from the fuel consumption and an appropriate emission factor to convert carbon content of the fuel into CO_2 emissions. These conversion factors have been established by IMO according to the type of fuel used by the container vessel (IMO 2014). The default values are given on the basis of gram CO_2 per gram fuel, being 3.114 g CO_2/g for heavy fuel oil and 3.206 g CO_2/g for marine diesel and marine gas oils. An estimate E_{ijv} of the total CO_2 emissions for a vessel v in a leg trip between port i and j can be obtained as:

$$E_{ijv} = \sum_{z \in Z} \alpha_{v,z} \left[g_{\mathrm{S}}^{v,z} \left(\frac{s_{ij}^{v}}{s_{v}^{*}} \right)^{n} d_{ij}^{z} + g_{\mathrm{I}}^{v,z} b_{j}^{z} \right] \tag{9.2}$$

where Z is the set of bunker types, indexed by z; $\alpha_{v,z}$ is the corresponding conversion factor for vessel v according to the type of fuel z; $g_{\mathrm{S}}^{v,z}$ and $g_{\mathrm{I}}^{v,z}$ is the fuel consumption of vessel v when sailing and idle at the port with bunker type z, respectively; s_{ij}^{v} is the operational speed of the vessel between the ports; and s_{v}^{*} is the design speed of the vessel. The exponent n is usually approximated to be around 3, meaning that the fuel consumption varies cubically with the speed (Stopford 2009). Moreover, d_{ij}^{z} is the sailing distance between the ports in nautical miles, and b_{j}^{z} is the berthing time at the port j for the vessel with bunker type z. This estimate is a simple representation of how CO_2 emissions can be calculated for its incorporation into a mathematical model.

Although sustainable maritime transportation is gaining more importance in Operations Research, the literature is still very scarce. In the context of routing and scheduling, there are a few papers dealing with green maritime transportation. Kontovas (2014) presents different approaches that can be considered when incorporating environmental dimensions: through the minimization of total emissions, internalizing the external cost of emissions and adding constraints to limit the produced emissions. The author remarks that minimizing fuel consumption is not equivalent to minimizing the total emissions, since vessels are generally equipped with main and auxiliary engines, which usually use different types of fuel. Another way to reduce the greenhouse gas emissions is to introduce ECAs, which are predefined areas where vessels are not allowed to use fuels with high sulfur content. Fagerholt et al. (2015) and Dithmer et al. (2017) present mathematical formulations introducing these emission control regulations. In the latter case, in a similar way as described in Kontovas (2014), the authors also study the approach of internalizing the external costs of emissions, making it possible to analyze the routing and scheduling of the services if a tax system is implemented in the future.

1.3 The LINER-LIB Test Instances

In order to make it easier to compare algorithms for liner shipping network, (Brouer et al. 2014a) published the LINER-LIB benchmark suite. The test instances in LINER-LIB are based on real-life data from leading shipping companies along with several other industry and public stakeholders. The benchmark suite contains data on ports including port call cost; cargo handling cost and draft restrictions; distances between ports considering draft and canal traversal; vessel-related data for capacity, cost, speed interval and bunker consumption; and finally a commodity set with quantities, revenue and maximal transit time. The commodity data is intended to reflect the differentiated revenue associated with the current imbalance of world trade.

The LINER-LIB benchmark suite consists of seven instances described in Brouer et al. (2014a) and is available at http://www.linerlib.org. They range from smaller networks suitable for optimal methods to large-scale instances spanning the globe. Table 9.1 gives an overview of these instances.

Each of the instances can be used in a low, base and high capacity case depending on the fleet of the instance. For the low capacity case, the fleet quantity and the weekly vessel costs are adjusted to fewer vessels with a higher vessel cost. For the high capacity case, the adjustments are reversed.

Currently, most papers only report results for the six first instances, with (Krogsgaard et al. 2018) being the only to report results for the *WorldLarge* instance.

Table 9.1 The seven test instances included in LINER-LIB with indication of the number of ports ($|N|$), the number of origin-destination pairs ($|K|$), the number of vessel classes ($|V|$), the minimum (min v) and maximum number of vessels (max v)

Instance	Category	$\|N\|$	$\|K\|$	$\|V\|$	min v	max v
Baltic	Single-hub	12	22	2	5	7
West Africa (WAF)	Single-hub	19	38	2	33	51
Mediterranean	Multi-hub	39	369	3	15	25
Pacific	Trade-Lane	45	722	4	81	119
AsiaEurope	Trade-Lane	111	4000	6	140	212
WorldSmall	Multi-hub	47	1764	6	209	317
WorldLarge	Multi-hub	197	9630	6	401	601

1.4 Outline

This chapter is organized as follows. In Sect. 2 we discuss the challenges in designing an energy-efficient liner shipping network and show that algorithms roughly can be split into four different families, which are studied in Sects. 3, 4, 5, and 6. In Sect. 3 we give an overview of integrated MIP models, while Sect. 4 studies two-stage algorithms where the routes are constructed in a first step, and containers are flowed through the resulting network in the second step. Section 5 considers algorithms for selecting a subset of proposed candidate routes. In Sect. 6 we consider algorithms based on first flowing containers and then designing routes. Finally, Sect. 7 shows how speed optimization can be used to lower energy consumption in liner shipping. The chapter is concluded in Sect. 8 with a short discussion of future trends and challenges.

2 Overview of Algorithms

Designing a green liner shipping network is a difficult task, embracing several decisions: not only do we need to construct the individual routes, but we should also deploy vessels of the right size to each route and ensure that there is sufficient capacity in the network to transport all containers from their origin to their destination. Designing the individual routes is an NP-hard problem, as proved in Brouer et al. (2014a), but also routing the containers through a given network subject to time constraints for each container can be recognized as a time-constrained multi-commodity flow problem, which is NP-hard.

The problem is further complicated by the fact that ports often are visited several times in the same route. This is obviously the case for pendulum routes where a vessel is sailing back and forth along the same route, but multiple visits to a port (typically a hub) often take place to ensure faster transportation times. However, formulating the problem as MIP model becomes more difficult.

Finally, one should notice that transshipment costs represent the majority of the cost of routing the containers through the network according to Psaraftis and Kontovas (2015). It is therefore important to carefully model which containers are transshipped and at which costs. This adds further complexity to the problem and makes a graph formulation huge and difficult to solve.

Algorithms for liner shipping network design can roughly be divided into the following four groups:

- *MIP-based algorithms* These algorithms are based on a unified MIP model that designs routes and flows containers through the resulting network. In order to handle this task, two sets of variables are needed: variables to select edges in a route and variables to denote the flow on each edge. If multiple visits to a node are allowed (butterfly nodes), then an additional index is needed to indicate the visit number at each node. Several MIP-based models have been presented in the literature, including Álvarez (2009), Reinhardt and Pisinger (2012), Plum et al. (2014) and Wang and Meng (2014).
- *Two-stage algorithms* As the name suggests, these algorithms solve the problem in two steps: designing the routes and flowing containers through the resulting network. Frequently, these algorithms contain a feedback mechanism, where output from the second-stage flow model is used as input to improve the routes in the first stage. Successful applications of this approach include Agarwal and Ergun (2008), Álvarez (2009), Brouer et al. (2014a,b), Karsten et al. (2017b), Thun et al. (2017) and Neamatian Monemi and Gelareh (2017).
- *Subset of routes* Both Meng and Wang (2011b) and Balakrishnan and Karsten (2017) suggest a method for generating a network by having a list of candidate routes as input. The idea behind these algorithms is to use the experience from existing planners to design a large number of promising candidate routes. The algorithm then selects a subset of the candidate routes to form a network. Many shipping companies and customers do not want the network to be completely restructured, in which case proposing small variations to each route may be a useful method.
- *Backbone flow* The idea behind this approach is that it can be difficult to design the individual routes without knowing how the containers will flow through the network. Hence reverse the order of the subproblems in the two-stage algorithms, and start by finding an initial flow (a so-called backbone network) where cargo is flowed through a complete network with all connections between ports available. The connections are priced such that they are expensive at low loads and cheap at high loads, in order to make the cargo gather at fewer connections. The initial flow can be seen as an accomplishment of the *physical Internet* (Montreuil 2011) where point-to-point transport has been replaced by multisegment intermodal transport. A successful application of the backbone network idea was presented in Krogsgaard et al. (2018).

Many of the MIP-based algorithms can in principle solve the LSNDP to optimality. However, due to the intrinsic complexity, only smaller instances can be solved to proven optimality within a reasonable time frame; hence the algorithms will often return a suboptimal solution. The subset-of-routes-based algorithms also solve the problem to optimality given that only the proposed candidate routes are valid. In practice, however, there may be an exponential number of valid routes, and we cannot expect to get all routes as input. If only a subset of all valid routes is given as input, the found solution may be suboptimal. The two-stage algorithms and backbone-network algorithms are both heuristics, since they first solve one stage and then optimize the second stage with the first-stage decisions fixed.

3 Mixed Integer Programming Models

The design of a liner shipping network includes numerous decisions, such as the routing of containers, the fleet deployment and the service design. The design of shipping networks is beyond the limited capacity of human planners, and it requires the use of several complex decision support tools. Mixed Integer Programming and graph-based models will be used in the subsequent sections to define the network design problem mathematically. Several MIP formulations of the LSNDP have been proposed during the last decades. We will give an overview of some of the formulations and discuss their advantages and limitations.

3.1 Service Formulation for LSNDP

Liner Shipping is based on the operation of services, which are defined by a sequence of ports that are visited by the vessel under a previously established schedule. The main objective of LSNDP is to design the shipping network by selecting services for the vessels so that the demand can be flowed at minimum cost while the overall benefit is maximized. Considering this fact, the first mathematical formulation is introduced in this section, which models the problem based on a service formulation, i.e. where the set of all feasible services are predefined in the model.

Before introducing the mathematical models presented in the literature, we briefly introduce a simple mathematical model based on a service formulation for better understanding. We will consider the notation presented in the introduction in Sect. 1.1 but with a small extension. Let $G = (N, A)$ be a directed graph, where N is the set of ports and A is the set of arcs connecting the ports. We now define the set S as the set of all feasible services in the model. Notice that S may be exponentially large. Let c_s be the cost of operating service $s \in S$, c^k_{ij} the unit cost per commodity $k \in K$ for traversing arc $(i, j) \in A$, $v(s)$ the corresponding vessel class $v \in V$ for the service $s \in S$, $m_{v(s)}$ the required number of vessels of class $v(s)$ for operating service $s \in S$ and a^s_{ij} a binary parameter indicating if the arc $(i, j) \in A$ is traversed in service $s \in S$. Finally, let x^{ks}_{ij} be a continuous variable indicating the amount of commodity $k \in K$ transported in service $s \in S$ through the arc $(i, j) \in A$ and y_s a binary variable for the selection of service $s \in S$ in the network. Now, the service formulation of the LSNDP can be expressed as:

$$\min \sum_{s \in S} c_s y_s + \sum_{k \in K} \sum_{(i,j) \in A} c_{ij}^k \sum_{s \in S} x_{ij}^{ks} \tag{9.3a}$$

$$s.t. \sum_{s \in S} \sum_{j:(i,j) \in A} x_{ij}^{ks} - \sum_{s \in S} \sum_{j:(j,i) \in A} x_{ji}^{ks} = \xi_i^k \quad i \in N, k \in K \tag{9.3b}$$

$$\sum_{\substack{s \in S \\ v(s)=v}} m_{v(s)} y_s \leq M_v \quad v \in V \tag{9.3c}$$

$$\sum_{k \in K} x_{ij}^{ks} \leq q_{v(s)} a_{ij}^s y_s \quad (i, j) \in A, s \in S \tag{9.3d}$$

$$x_{ij}^{ks} \geq 0 \quad (i, j) \in A, k \in K, s \in S \tag{9.3e}$$

$$y_s \in \{0, 1\} \quad s \in S. \tag{9.3f}$$

The objective function (9.3a) minimizes the total operational cost of the network. The first term accounts for the total fixed cost of the selected services, whereas the second term accounts for the sailing cost of shipping the demand. Constraints (9.3b) are the flow conservation constraints, constraints (9.3c) ensure that the deployed vessels on the services do not exceed the available fleet, and the flow capacity of the selected services has to be respected, which is described by constraints (9.3d). Finally, the domain of the variables is defined by constraints (9.3e) and (9.3f).

A successful implementation, based on a service formulation, was presented by Álvarez (2009). Álvarez extends the previous formulation to define the Liner Shipping Network Design at the tactical level, where the formulation combines the routing and deployment of a fleet of container vessels. The formulation relies on the set of all feasible services, which are given as a combination of a vessel type, its corresponding speed and the route structure. Therefore, it is possible to accommodate services that are proposed externally by the planners as services generated internally by a solution algorithm, meaning that any type of non-simple services can be considered in the set S of services. However, as the size of the problem increases, the number of feasible services in the problem grows exponentially, making the model intractable to solve.

Moreover, for a better utilization of the capacity of the vessels, the model allows the rejection of cargo incurring a goodwill penalty, where continuous variables are defined to account for the demand that is delivered and rejected by the liner company. With reference to the above, Álvarez also defines continuous variables for the amount of cargo that is transported along an arc on a service as well as continuous variables for different operations of loading and unloading containers in ports on specific services. These variables can be used to identify the amount of containers that are transshipped between services. However the model is unable to accurately calculate the transshipment cost of non-simple services. Finally, the model considers the fleet deployment of the available fleet, using integer variables to control the amount of vessels deployed for a chosen service. The model includes many relevant parameters in the objective function to correctly represent the operational cost of the selected services over a tactical planning horizon, and it is one of the first formulations to consider transshipment when designing the shipping network.

3.2 *Arc Formulation for LSNDP*

The main problem with a service-based formulation is that generating all services S is non-trivial, due to the high number of combinatorial possibilities. This process can be inefficient and very time-consuming. Therefore, an alternative mathematical formulation is introduced in this section, which is based on an arc formulation. The set of services S is no longer considered in the problem, but the services are instead designed as part of the problem.

Next, we will present a simple mathematical model based on an arc formulation. For this we will again use the notation presented in the introduction, Sect. 1.1, with small extensions. Let $G = (N, A)$ be a directed graph, where N is the set of ports and A is the set of arcs connecting the ports. Moreover, let V be defined as the set of vessel classes and c^v the cost for deploying a vessel belonging to class $v \in V$. We will introduce the set S^v as the set of services for the vessel class $v \in V$. We also introduce t_{ij}^v and c_{ij}^v as the sailing time and cost by a vessel of type $v \in V$ traversing arc $(i, j) \in A$, respectively, and b_j the berthing time at port $j \in N$. Finally, let x_{ij}^{ks} be a continuous variable denoting the flow of a commodity $k \in K$ on an arc $(i, j) \in A$ in the service $s \in S^v$ belonging to vessel class $v \in V$, y_{ij}^{sv} a binary variable for the selection of an arc $(i, j) \in A$ in the service $s \in S^v$ operated by the vessel class $v \in V$, τ_i^s a continuous variable for the time in service $s \in S^v$ of a vessel class $v \in V$ arriving at port $i \in N$, and m_s^v an integer variable indicating the number of vessels from class $v \in V$ needed to be deployed to maintain the weekly frequency in the service $s \in S^v$. Now, the arc formulation of the LSNDP can be expressed as follows:

$$\min \sum_{v\in V}\sum_{s\in S^v} c^v m_s^v + \sum_{v\in V}\sum_{s\in S^v}\sum_{(i,j)\in A} c_{ij}^v y_{ij}^{sv} + \sum_{k\in K}\sum_{(i,j)\in A} c_{ij}^k \sum_{v\in V}\sum_{s\in S^v} x_{ij}^{ks} \quad (9.4a)$$

$$s.t. \sum_{v\in V}\sum_{s\in S^v}\sum_{j:(i,j)\in A} x_{ij}^{ks} - \sum_{v\in V}\sum_{s\in S^v}\sum_{j:(j,i)\in A} x_{ji}^{ks} = \xi_i^k \qquad i \in N, k \in K \quad (9.4b)$$

$$\sum_{j:(i,j)\in A} y_{ij}^{sv} - \sum_{j:(j,i)\in A} y_{ji}^{sv} = 0 \qquad i \in N, v \in V, s \in S^v \quad (9.4c)$$

$$\sum_{k \in K} x_{ij}^{ks} \leq q_v \cdot y_{ij}^{sv} \qquad (i,j) \in A, v \in V, s \in S^v \tag{9.4d}$$

$$\tau_j^s \geq (\tau_i^s + t_{ij}^v + b_j) y_{ij}^{sv} \qquad i,j \in N, v \in V, s \in S^v \tag{9.4e}$$

$$\sum_{(i,j) \in A} y_{ij}^{sv}(t_{ij}^v + b_j) \leq 24 \cdot 7 \cdot m_s \qquad v \in V, s \in S^v \tag{9.4f}$$

$$\sum_{s \in S^v} m_s^v \leq M_v \qquad v \in V \tag{9.4g}$$

$$x_{ij}^{ks} \geq 0 \qquad (i,j) \in A, k \in K, v \in V, s \in S^v \tag{9.4h}$$

$$y_{ij}^{sv} \in \{0,1\} \qquad (i,j) \in A, v \in V, s \in S^v \tag{9.4i}$$

$$m_s^v \in \mathbb{Z}^+ \qquad v \in V, s \in S^v \tag{9.4j}$$

$$\tau_i^s \geq 0 \qquad i \in N, v \in V, s \in S^v \tag{9.4k}$$

The objective function (9.4a) minimizes the cost of deploying the vessels and designing the services and the cost for transporting the commodities through the network. The flow conservation constraints for the cargo variables are given in constraints (9.4b), whereas the flow conservation constraints for the routing variables are given in constraints (9.4c). The flow of cargo on an edge (i, j) cannot exceed the capacity q_v of a vessel, as expressed in (9.4d). If the vessel is not used for the given edge, i.e., $y_{ij}^{sv} = 0$, then the capacity is zero. The subtour elimination constraints for the routing variables are given by the time variables in constraints (9.4e). Note that it is required to linearize these constraints, as they are non-linear. Moreover, the weekly frequency of the services and the deployment of the fleet is limited by constraints (9.4f). The availability of the fleet is limited by constraints (9.4g). Finally the domain of the variables is defined in constraints (9.4h), (9.4i), (9.4j) and (9.4k).

The model presented above is a simple representation of the arc formulation for the LSNDP. It is a fairly easy adaptation of a variant of the Vehicle Routing Problem (VRP) (Toth and Vigo 2015). However, this model can be extended to consider all the assumptions that can occur in Liner Shipping. Reinhardt and Pisinger (2012) proposed a MIP model based on an arc-flow formulation where the network design and the fleet assignment are combined; however, in this case, cargo rejection is not considered. As argued in Agarwal and Ergun (2008), transshipment is the core of liner shipping; hence, these operations should not be ignored when designing the network. Reinhardt and Pisinger (2012) include these operations into the formulation and accounts correctly for the transshipment cost in the intermediate ports. Moreover, one of the main considerations of the model is the inclusion of butterfly services, where it is allowed to visit a single port twice during the service. Due to the allowance of butterfly services, the model requires the definition of extra binary variables for the identification of the unique centre point, i.e. the hub port in the vessel route, and the finding of the first and last arc visiting the hub port, respectively. Similarly, as proposed by Miller et al. (1960), positive integer variables are defined for enumerating the arcs in the vessel route to avoid the formation of subtours in the services. The definition of these variables will be used to model the transshipment of cargo in hub ports. Furthermore, the model also considers the fleet deployment with a heterogeneous fleet. It is possible to define service-dependent capacities according to the time horizon and the frequency.

The high level of detail in the model allows a fairly realistic representation of the problem, making it possible to design efficient services reducing the overall operational costs and CO_2 emissions. Nonetheless, the model is NP-hard and also in practice very difficult to solve. The model includes several "big-M" constraints, resulting in weak bounds from LP relaxation. The proposed method to solve this problem is Branch-and-Cut, as it has presented good results to the VRP and other transportation network design problems. The idea is to solve the previous relaxed problem without the transshipment constraints and the connectivity constraints in butterfly nodes and, then, gradually add cuts to the formulation whenever they are violated. The implemented method is tested against the CPLEX MIP solver on a set of test instance with up to 15 ports. The results show that the developed branch-and-cut method clearly outperforms the solver, even though some test instances are not solved to optimality. This method is not suitable for solving real-life instances such as LINER-LIB; however it provides promising results for smaller feeder services in liner shipping network design problems.

3.3 *Port Call Formulation for LSNDP*

The majority of the models for LSNDP are defined using an arc formulation, but such formulations can be problematic when formulating non-simple services, as it requires the inclusion of many extra variables in the model, as seen in Reinhardt and Pisinger (2012). Alternatively, the problem can be defined with a service formulation, but the number of variables will increase exponentially with the size of the problem, as seen in Álvarez (2009). Plum et al. (2014) propose a new mathematical formulation based on a service formulation, where the set of all services S are defined beforehand. The set of services can handle several calls to the same port during the same route, i.e., it can include the non-simple routes, which better represents the services operated by liner shipping companies in the real world. In order to do that, the authors define the set of port calls B, and a service is defined to consist of a number of port calls. The model is defined with a series of continuous flow variables that represent the amount of demand that is transported on a certain port call leg of a service, among other flow variables which represent the flow of cargo from and to a specific port call in the different ports of the problem. This formulation allows the rejection of part of the demand, which is subtracted from the objective function incurring a penalty for not flowing the cargo. The model defines the decision variables in such a way that the flow of containers from one service to another can be considered correctly and transshipment can be modelled. Furthermore, the model imposes the services to have weekly frequency while limiting the fleet deployment according to the available fleet. Finally, the authors present an objective function where the operator's profit of the flowed cargo on the operating network is maximized, while the operational cost of the services and the cost for handling the cargo are minimized.

3.4 Outbound-inbound Principle with Transit Time Constraints

Wang and Meng (2014) incorporate the transit time constraints when designing the network in liner shipping problems. However, transshipment between services is excluded in this approach as they define the ship routes with the outbound-inbound principle. The problem is defined with a set R of geographical ship routes, which is an itinerary of port calls, using binary variables for the selection of these itineraries. The proposed model is a mixed-integer non-linear and non-convex programming model with an exponential number of decision variables, and it determines the network design and cargo routing of containers through the network. Binary variables are defined for assignment of arcs to the ship routes in order to construct feasible geographical ship routes. Furthermore, there exists a limited available fleet, and the fleet deployment is controlled by integer variables. Demand can be split among different vessels, and the model defines continuous variables for the amount of demand flowing through the arcs. These variables allow the model to define feasible patterns of the demand on the selected geographical ship routes. Additionally, binary variables are defined to ensure that the transit times of the cargoes are not violated. Finally, the model defines the port time as a function of the number of containers handled at the port. This is taken into account when the route length is enforced to have weekly frequencies. The problem is proved to be strongly NP-hard by reduction from the Bin Packing Problem, and Wang and Meng (2014) describe a column generation-based algorithmic scheme for its resolution. The approach efficiently finds high-quality solutions that can help planners to design better liner shipping networks.

4 Two-Stage Algorithms

The LSNDP consists of two tightly interrelated problems – the vessel service network design and the container flow problem. One of the most successful approaches so far, for finding good solutions to the LSNDP, has been to use heuristics exploiting this two-tier structure.

The idea, in general, is to first generate a set of services for the vessels and then to solve the container flow problem, given the set of services. It is then commonplace to use information from the container flow to update the services. This way a feedback loop is created, iteratively improving the services and solving the container flow. The different frameworks, in which this has been used, range from column generation and Benders' decomposition (Agarwal and Ergun 2008) to various matheuristics (Álvarez 2009; Brouer et al. 2014b). This section will discuss some of those methods. Various versions of the LSNDP will be featured, both with and without transshipment costs, transit time constraints and rejection of demand.

4.1 The Container Flow Problem

Before going into the full two-stage algorithms, let us briefly discuss the container flow problem, which is the lower-tier problem in the LSNDP two-tier structure. In general, for a given set of services, the container flow problem reduces to a *multi-commodity flow problem* (MCFP), with fractional flows allowed.

Let $G = (N, A)$ be a directed graph, where N represents the ports and A represents the arc set that connects the ports. Let K be the set of commodities with corresponding parameters as defined in Sect. 1.1. To each arc $(i, j) \in A$, further, define the corresponding cost, c_{ij}^k, of transporting one unit of commodity k through (i, j) and its corresponding flow capacity, u_{ij}. The arc set A and its corresponding costs c_{ij}^k and capacities u_{ij} are defined by the vessel services, designed in the upper-tier problem. Lastly, let x_{ij}^k be a continuous variable denoting the flow of commodity k through arc (i, j). The MCFP can then be expressed as:

$$\min \sum_{(i,j)\in A} \sum_{k\in K} c_{ij}^k x_{ij}^k \tag{9.5a}$$

$$s.t. \sum_{(i,j)\in A:i=p} x_{ij}^k - \sum_{(i,j)\in A:j=p} x_{ij}^k = \xi_p^k \qquad p \in N, k \in K \tag{9.5b}$$

$$\sum_{k\in K} x_{ij}^k \le u_{ij} \qquad (i, j) \in A \tag{9.5c}$$

$$x_{ij}^k \ge 0 \qquad (i, j) \in A, k \in K. \tag{9.5d}$$

Here, the objective, (9.5a), is to minimize the total cost. Constraints (9.5b) are the flow conservation constraints, constraints (9.5c) are the capacity constraints, and constraints (9.5d) define the domain of the variables x_{ij}^k.

When fractional flows are allowed, the MCFP is solvable in polynomial time, but for larger instances, it is still computationally demanding. As it generally has to be solved a multitude of times in the presented two-tier solutions to the LSNDP, efficient solution methods to the MCFP are essential.

One of the most common solution approaches is to exploit its block-angular constraint matrix and apply Dantzig-Wolfe decomposition (Ahuja et al. 1993; Karsten et al. 2015). First reformulate the problem to a path-flow formulation, where the goal is to allocate the commodities onto a number of flow paths from the commodity origins to their destinations while respecting the capacity constraints on the arcs. Let P^k be the set of all paths for commodity $k \in K$, from o_k to d_k, and let P_a^k be the set of paths for commodity k, which uses the arc a. For each path, p, define its cost $c_p = \sum_{a\in A:p\in P_a^k} c_a^k$, and a corresponding decision variable f_p, deciding the flow through path p. The path-flow formulation can then be expressed as:

$$\min \sum_{k\in K} \sum_{p\in P^k} c_p f_p \tag{9.6a}$$

$$s.t. \sum_{p\in P^k} f_p = \xi_p^k \qquad k \in K \tag{9.6b}$$

$$\sum_{p\in \bigcup_{k\in K} P_a^k} f_p \leq u_a \qquad a \in A \tag{9.6c}$$

$$f_p \geq 0 \qquad k \in K, p \in P^k. \tag{9.6d}$$

The objective function, (9.6a), is to minimize the cost. Constraints (9.6b) ensure that all commodities are delivered and constraints (9.6c) assert that the arc capacity is not exceeded. Lastly, constraints (9.6d) define the domain of the variables.

The path formulation has a very large number of variables, but generally, only a few of them are needed for the optimal solution. Using column generation, the problem can be restricted to only consider a limited amount of paths for each commodity, and new paths can then be generated dynamically. This way, the path formulation can generally be solved faster than the arc formulation, described above. The path formulation makes it relatively easy to implement transit time constraints as they can be handled in the pricing problem.

Another efficient method of solving the MCFP (without time constraints) is by using so-called interior point methods, as is done by Álvarez (2009). In contrast to the simplex method, which searches through the vertices of the solution space, interior point methods search through solutions in its interior.

4.2 Matheuristics Methods for the LSNDP

While the lower-tier container flow problem is solvable in polynomial time (when no transit time constraints are imposed), the upper-tier service selection problem is NP-hard, and just to calculate the objective value of a given solution, one has to solve the container flow problem. This makes the service selection problem difficult to solve optimally, and instead several matheuristics have been developed to find good solutions to larger instances. A matheuristic is a method that employs heuristics together with methods from linear and integer programming. In the case of the LSNDP, the most common procedure is to use linear programming tools to solve the MCFP and then various heuristics to update the vessel services.

The first two-stage algorithms for liner shipping network design were presented by Agarwal and Ergun (2008) that solved the *simultaneous ship scheduling and cargo routing problem* (SSSCRP) with a column generation and a Benders' decomposition heuristic. As the name implies, they also took the ship scheduling into account which has been more or less neglected since. They did not, however, account for transshipment costs. The column generation heuristic was designed such that the cargo routing was solved in the master problem, and the dual variables were then utilized to generate and choose new services for the vessels. As column generation solves only the LP relaxation of the problem, once no more services with negative reduced cost could be found, they used the generated columns to find an integer solution using branch-and-bound. In the Benders' decomposition heuristic, the container flow problem was solved in the subproblem to add optimality cuts for the service generation in the master problem. In both cases they found it most efficient to generate new services using a labeling algorithm. They reported good results for instances of up to 20 ports and 100 vessels.

Another prominent approach was presented by Álvarez (2009) that used a matheuristic which perturbed the services with a tabu-search scheme, solved the container flow problem using an interior point method and generated new services from the dual variables from the container flow solutions. Álvarez's model included the cost of transshipments and also allowed for butterfly routes. The moves considered in the tabu-search for the services were deletion, change in vessel speed and change in number of vessels assigned. To guide the search, from the solution of the commodity flow problem, information about which services were under-/overutilized was used to increase/decrease the number of vessels and the speed. The paper presents computational results for up to 100 available vessels and 120 ports.

Another tabu-search approach was presented by Brouer et al. (2015), which was later improved upon by Karsten et al. (2017b), by adding time constraints for the commodities. As it is computationally costly to solve the full cargo flow problem, both papers instead developed a method to estimate the impact of a change in the service structure. Their solution method is then based on an improvement heuristic, first presented by Archetti and Speranza (2014), in which in each iteration, an integer program is solved to update the current services.

Here follows a brief description of the algorithm from Brouer et al. (2015). Let $G = (N, A)$ be a complete directed graph, where N represents the ports and A represents the possible connections between ports. Let S denote the set of services, where each service, $s \in S$, visits a set of ports, $N_s \subseteq N$, and has a corresponding vessel class v_s, a number of assigned vessels m_s and a duration τ_s. The algorithm is initialized, using a greedy knapsack heuristic to generate an initial set of services. The change in revenue and time by including or excluding ports from the current services is estimated by solving a set of shortest path problems. Let us define r_{is}^{+} (r_{is}^{-}) to be the estimated revenue change and t_{is}^{+} (t_{is}^{-}) to be the estimated duration change from including (excluding) port $i \in N$ in (from) service $s \in S$. Denote the weekly cost of using a vessel of the vessel class v_s by c_s and the number of free vessels of the type v_s by M_s. Lastly, let us define the binary variables λ_{is}^+ and λ_{is}^-, which control the inclusion and removal, respectively, of port i from service s, and the integer variable ω_s, which denotes the number of vessels to add to/subtract from service s. We also define a maximum number of inclusions, γ_s^+, and removals γ_s^- and a number of locksets L_i. For each service $s \in S$, we can then define the following mixed-integer program:

$$\max \sum_{i \in N_s} r_{is}^+ \lambda_{is}^+ + \sum_{i \in N \setminus N_s} r_{is}^- \lambda_{is}^- - c_s \omega_s \tag{9.7a}$$

$$\text{s.t. } \tau_s + \sum_{i \in N_s} t_{is}^+ \lambda_{is}^+ + \sum_{i \in F_s} t_{is}^- \lambda_{is}^- \leq 24 \cdot 7 \cdot (m_s + \omega_s) \tag{9.7b}$$

$$\omega_s \leq M_s \tag{9.7c}$$

$$\sum_{i \in N_s} \lambda_{is}^+ \leq \gamma_s^+ \tag{9.7d}$$

$$\sum_{i \in F_s} \lambda_{is}^- \leq \gamma_s^- \tag{9.7e}$$

$$\sum_{j \in L_i} \lambda_{js}^- \leq |L_i|(1 - \lambda_{is}^+) \qquad i \in N_s \tag{9.7f}$$

$$\sum_{j \in L_i} \lambda_{js}^- \leq |L_i|(1 - \lambda_{is}^-) \qquad i \in N \setminus N_s \tag{9.7g}$$

$$\lambda_{is}^+ \in \{0, 1\}, \; i \in N_s \qquad \lambda_{is}^- \in \{0, 1\}, \; i \in N \setminus N_s \qquad \omega_s \in \mathbb{Z}, \tag{9.7h}$$

where the objective (9.7a) is to maximize the increase in revenue. Constraint (9.7b) ensures that there are enough vessels assigned to keep the weekly frequency, and constraint (9.7c) specifies that no more than the number of free vessels can be added to the service. Constraints (9.7d) and (9.7e) set a limit on the number of insertions and removals and (9.7f) and (9.7g) enforce the locksets L_i. Constraints (9.7d), (9.7e), (9.7f) and (9.7g) are defined to limit the amount of changes which can be applied, as the revenue and time change estimates are made for one or a few changes and deteriorate rapidly when multiple changes are applied. L_i are defined such that if a port i is to be inserted in between two ports, then neither of those are allowed to be removed and if inserting a new port means that a new commodity is transported, then the origin and destination nodes, of this commodity, are not allowed to be removed. Lastly, (9.7h) defines the domains of the variables.

The algorithm works such that each service, one by one, is updated according to the solution of the above-defined mixed-integer problem, and then the MCFP is solved to update the total revenue, and the effect of new changes is once again estimated with the shortest path procedure. To diversify the solutions, every tenth iteration the services with the lowest utilization are removed, and new services are created using the greedy construction heuristic.

Brouer et al. (2015) report satisfactory solutions for 6 out of 7 instances from the LINER-LIB benchmark set where the largest solved instance, the *world small*, contains 47 ports and 317 available vessels.

5 Subset of Routes

Balakrishnan and Karsten (2017) suggest a method for generating a network by selecting a subset of sailing services from an initial pool of candidate services given in advance by expert planners. The problem is therefore reduced from service design to service selection. Limits on the number of transshipments for each container are included in the model, and rejection of demand is allowed. This profit-maximizing problem is denoted the Liner Service Planning (LSP) problem.

The transportation network consists of a set of ports N indexed by i and j and a set of candidate services S where each service $s \in S$ has N_s port calls. Associated with each candidate service $s \in S$ is a set of sailing arcs $a \in A_s$ where each arc represents the part of a ship's itinerary between two successive ports on the service route. The fleet is composed of several vessel classes and V denotes the set of these classes. There are M_v available vessels of each class $v \in V$, and for each service $s \in S$ we let m_v^s denote the required number of vessels of class $v \in V$. Associated with each service $s \in S$ is also a cost c_s and for each arc $a \in A_s$ a capacity g_a.

K denotes the set of commodities where an origin port o_k, a destination port d_k and a demand δ_k are associated with each commodity $k \in K$. It is allowed to split the flow of each commodity, and a penalty cost c_k^r per container is used to penalize rejected demand of commodity k.

Given a commodity's route, a *sub-path* is defined as the part of the route in which the container travels on a single service. If this part is from port i to port j on service s, the sub-path is denoted $\langle i, j, s\rangle$. The set H_s denotes the full set of sub-paths for service s, i.e. the set contains one sub-path $\langle i, j, s\rangle$ for each combination of ports i and j included in service s. These sub-paths are used to introduce an augmented multi-commodity flow network in order to incorporate the limits on the number of transshipments and their associated costs. This modeling approach falls somewhere between the two more traditional modeling approaches of either using arc-flow, i.e. over sailing edges, or path-flows, i.e. origin-to-destination paths.

The augmented network contains one node for each port and one link for each sub-path of each service. The sub-path structure also extends to more complex routes, e.g. butterfly routes. A_{ij}^s denote the set of sailing arcs of service s included in sub-path $\langle i, j, s\rangle$. The cost of routing one container of commodity k on sub-path $\langle i, j, s\rangle$ is denoted c_{ijs}^k. Finally, h_k denote the maximum allowed number of sub-paths on which commodity k can travel. Note that h_k must be one larger than the maximum permitted number of transshipments to enforce this constraint.

Balakrishnan and Karsten (2017) present a multi-commodity model based on flows along sub-paths in the augmented network. The binary variable y_s is equal to 1 if service $s \in S$ is selected, and 0 otherwise. The flow of commodity k using sub-path $\langle i, j, s \rangle$ as the hth stage is described by the variable x_{ijs}^{hk} for $s \in S$, $\langle i, j, s \rangle \in A_s$ and $h = 1, 2, \ldots, h_k$. Finally, z_k is equal to the unmet demand (number of containers) for commodity $k \in K$.

The LSP problem can then be described by the following mixed-integer program:

$$\min \sum_{s \in S} c_s y_s + \sum_{k \in K} \sum_{s \in S} \sum_{h=1}^{h_k} c_{ijs}^k x_{ijs}^{hk} + \sum_{k \in K} c_k^r z_k \tag{9.8a}$$

$$s.t. \sum_{s \in S} \sum_{\langle o_k, j, s \rangle \in H_s} x_{o_k js}^{1k} + z_k = \delta_k \qquad \forall k \in K, \tag{9.8b}$$

$$\sum_{s \in S} \sum_{i: \langle i, j, s \rangle \in H_s} x_{ijs}^{hk} - \sum_{s \in S} \sum_{l: \langle j, l, s \rangle \in H_s} x_{jls}^{h+1,k} = 0 \qquad \forall k \in K, j \in N \setminus \{o_k, d_k\}, h = 1, \ldots, h_k - 1, \tag{9.8c}$$

$$\sum_{k \in K} \sum_{h=1}^{h_k} \sum_{\langle i, j, s \rangle \in H_s : a \in A_{ij}^s} x_{ijs}^{hk} \leq g_a y_s \qquad \forall s \in S, a \in A_s \tag{9.8d}$$

$$\sum_{s \in S} m_v^s y_s \leq M_v \qquad \forall v \in V, \tag{9.8e}$$

$$x_{ijs}^{hk} \geq 0 \qquad \forall k \in K, s \in S, \langle i, j, s \rangle \in H_s, h = 1, \ldots, h_k, \tag{9.8f}$$

$$z_k \geq 0 \qquad \forall k \in K, \tag{9.8g}$$

$$y_s \in \{0, 1\} \qquad \forall s \in S. \tag{9.8h}$$

The objective function (9.8a) minimizes total cost comprised of fixed costs for the selected services, the cost of transporting commodities along each sub-path and finally the penalties incurred for rejected demand. By including penalties, the problem is formulated as a cost minimization problem as opposed to a profit maximization problem where c_k^r would instead represent the revenue for transporting one unit of commodity k.

Constraints (9.8b) ensure that the flow of each commodity k is assigned to sub-paths incident to the corresponding origin port o_k. They also ensure that this flow out of the origin port in combination with the unmet demand for commodity k adds up to the total demand for commodity k. Constraints (9.8c) are flow-balancing constraints for intermediate ports. Together with constraints (9.8b), these constraints ensure that for each commodity k the demand subtracted any unmet demand will arrive at the destination port using at most h_k sub-paths, i.e. fulfilling the constraint on a maximum number of transshipments.

Constraints (9.8d) impose capacity constraints on the sailing arcs and ensure that only sub-paths from the selected services can be used. Constraints (9.8e) ensure that no more than the available vessels are used. Finally, constraints (9.8f), (9.8g) and (9.8h) impose non-negativity and binary restrictions on the respective decision variables.

The LSP model formulation is flexible enough to allow incorporation of several practical container routing issues such as cabotage rules, regional policies and embargoes. The incorporation of many of these constraints can be handled during preprocessing simply by removing sub-paths that are no longer permitted.

Balakrishnan and Karsten (2017) show that the LSP problem is NP-hard. A problem reduction procedure to eliminate or combine variables is outlined, and valid inequalities for increasing the lower bounds of its linear programming (LP) relaxation are described.

5.1 Optimization-Based Heuristic Procedure

Balakrishnan and Karsten (2017) propose an optimization-based heuristic algorithm to generate good initial solutions. The heuristic iteratively solves the LP relaxation of the problem and fixes service selection variables, y_s, that are integer in the corresponding solution, and rounds service selection variables, y_s, that are fractional. The highest or lowest fractional variable is selected in each iteration and rounded up or down correspondingly. The heuristic procedure first rounds down low y-values before rounding up high y-values. Thereby, unattractive services are eliminated early in the process. If rounding a variable up causes a violation of the fleet availability constraint, the variable is instead set to zero. The LP relaxation is then re-solved. When all y_s variables assume binary values, the procedure stops.

Balakrishnan and Karsten (2017) test their solution method on four data sets from the LINER-LIB benchmark suite with at most two transshipments per container. The initial pool of candidate services was generated using the matheuristic from Brouer et al. (2014b). The LP-based heuristic yields solutions that are close to optimality in relatively short time. This method can therefore be used as a stand-alone tool or to warm-start an exact solution procedure.

6 Backbone Flow

The main idea in a backbone flow algorithm, as presented by Krogsgaard et al. (2018), is to reverse the order of two-phase algorithms by first flowing the containers and then constructing services that cover the flow.

In order to find the backbone flow, an artificial network $G = (N, A)$ is used where N is the set of ports and A is a complete, directed graph. There are no capacities associated with the edges, but the cost of using an edge (i, j) depends on how many containers in total are flowing on the edge. This can be expressed as a concave function $f(x)$ of the flow x reflecting the economy of scale for flowing more containers: there is a large cost associated with opening an arc (i.e. deploying a vessel), while the cost per container decreases as the flow (and hence vessel size) is increased. See Fig. 9.2 for an illustration of the costs. The cost function implicitly aims at aggregating the flow on fewer arcs. Sun and Zheng (2016) also use a concave function to optimize the container flow.

Let the set of commodities K and demands ξ_i^k be defined as in (9.1), and let x_{ij}^k denote the flow of commodity k on edge (i, j). Then the backbone flow problem becomes a non-linear MCFP as given by

$$\min \sum_{(i,j)\in A} f(\sum_{k\in K} x_{ij}^k) \tag{9.9a}$$

$$s.t. \sum_{(i,j)\in A} x_{ij}^k - \sum_{(j,i)\in A} x_{ji}^k = \xi_i^k \qquad i \in N, k \in K \tag{9.9b}$$

$$x_{ij}^k \geq 0 \qquad (i, j) \in A, k \in K. \tag{9.9c}$$

As before, the objective, (9.9a), is to minimize the total cost, and constraints (9.9b) are the flow conservation constraints. Constraints (9.9c) define the domain of the variables.

Since the model is non-linear, (Krogsgaard et al. 2018) solve the problem heuristically through a randomized greedy algorithm. As the arc costs depend on previously flowed containers, the result of the flow will be very dependent on the order in which containers are flown. Generally, the first containers are more decisive for the arcs used heavily in the final solution than the last containers flown. It is thus necessary to run several iterations of the problem, with a random order of the containers, to achieve a reasonable *average* picture of the backbone flow. Running ten iterations for the demand matrix of the *WorldSmall* instance gives the average arc loads shown in Fig. 9.4. The figure clearly shows that only a fraction of the possible arcs is used in the solution.

6.1 Greedy Heuristic for Generating Services

Having found a backbone flow, Krogsgaard et al. (2018) present a greedy heuristic for generating services. The idea is to add one arc at a time to a service until all services have reached their maximum duration.

To generate a service, the unserved arc with the largest flow is selected as the first arc in the service, and a return arc is added to close the service. While the service is at or below the desired duration, a new arc is added to the service to expand it, and this arc replaces the return arc. The new arc is the unserved arc with the largest demand that either starts at the same port as the return arc, which is to be replaced, or ends at the same port as the return arc. A new return arc is added to close the service. The selection process continues until it is not possible to add a new arc

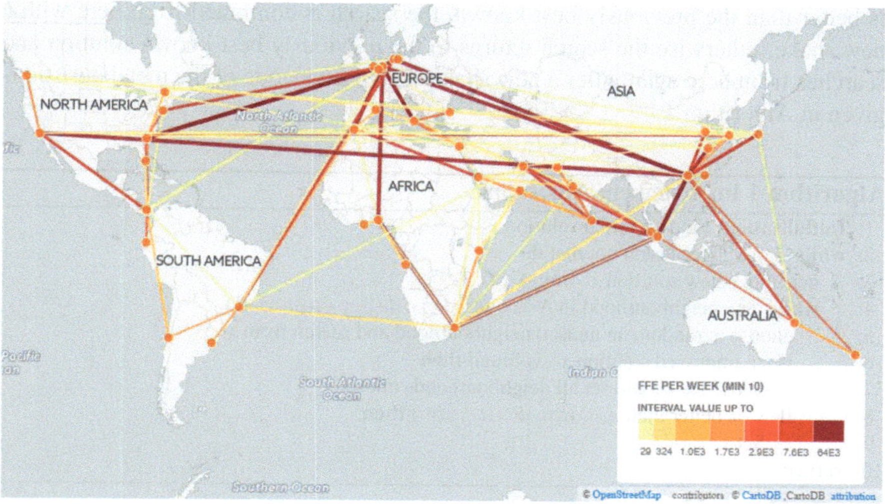

Fig. 9.4 Typical backbone flow for the *WorldSmall* instance (Source: Krogsgaard et al. 2018)

without exceeding the maximum duration of the service. After this, the creation of the next service starts.

To obtain a number of different start solutions to select from, the algorithm is repeated a number of times with random settings on the maximum service length for every service. The length is selected in a predefined interval depending on the size of the vessel, such that larger vessels, typically traveling between continents, get longer services than smaller vessels doing feeder service. For every service generated, a duration is selected in the interval at random, and the service is constructed. This is repeated until all available resources have been exhausted.

In the computational study by Krogsgaard et al. (2018), it is shown that usable solutions can be found in reasonable time. Using the *WorldSmall* instance, the authors generate 20 different sets of services by running the above algorithm where the containers are flown in random order. This can be done in about 80 s and results in profitable solution, although the resulting network is far from optimal.

6.2 Network Optimization

In order to improve the initial services found by the greedy heuristic, Krogsgaard et al. (2018) use a Variable Neighborhood Search (VNS) algorithm to reach a high-quality network. The general idea in VNS, as presented by Hansen and Mladenovic (2014), is to apply different neighborhood structures throughout the search to exploit the benefits from neighborhood changes. When a local optimum is encountered, it is escaped by doing a random move, a `shake`, from the best known solution and do hill climbing from here until a new local optimum is reached. If this solution is better than the previously best known, the search is continued from here with a new shake; otherwise the search returns to the previously best known solution and searches from here again after a new shake. The pseudocode of the metaheuristic is given in Algorithm 1.

Algorithm 1 Improvement Algorithm

```
Initialisation: Find an initial solution x
while stopping criterion not met do
    generate a new solution x' from x (shake)
    while any neighbourhood in N is unused (local search) do
        choose at random an unused neighbourhood and search from x'
        if an improved solution x'' is found then
            Set x' := x'' and set all neighbourhoods unused
        if x' is better than x (test solution) then
            Set x := x'
return x
```

As can be seen from the pseudocode, the `local search` procedure terminates after all neighborhoods have been tested without yielding an improving solution, as a local optimum with respect to all neighborhoods must then have been encountered. The `shake` procedure is applied less frequently than in a standard VNS framework. A lower degree of randomness is preferable here, because the evaluations are relatively expensive. It is thus desirable to search directly for a local optimum with respect to all neighborhoods before randomly altering the solution.

Although the local search only accepts moves that have an expected improvement, some moves may turn out to be degrading when calculating the real objective function. These moves are nevertheless kept on to progress the search. This can, however, lead to cycling in the local search, as it might both be expected to be an improvement to first insert a port and to remove it afterward. To break such cycles, only 20 loops are allowed in the `local search` part, after which the algorithm must continue to `test solution`.

If a cycle is encountered or a local optimum has been reached, the `shake` procedure is applied to progress the search from another point in the solution space. The procedure must change the solution sufficiently to escape the local optimum, but should, on the other hand, not destroy good characteristics of the solution. Preliminary studies show that there is a high risk of changing the solution too much to be able to return to a good solution, and a relatively modest shake procedure is thus implemented. This procedure modifies a number of services by either inserting or removing a port randomly, without considering the effect on objective value. To avoid inserting an obviously irrelevant port, a distance requirement is enforced such that only ports relatively close to the service can be inserted. The number of modified services is 10% of the total number of services and a least one.

In each iteration of `local search`, a neighborhood is randomly selected, and one or more services are altered through that neighborhood. Six different neighborhoods are applied: *Insert port*, *Service omission*, *Service unserved port*, *Remove port*, *Simple remove port* and *Create feeder services*.

In order to select the best move, delta evaluation is used to avoid time-consuming evaluations of the multi-commodity flow for the entire graph. Instead, a small graph (*rotation graph*) is constructed, covering only the rotation currently being altered. As this graph is much faster to evaluate, more moves can be tested before one is selected for implementation. See Krogsgaard et al. (2018) for a detailed description of the neighborhoods.

Promising computational results using the LINER-LIB instances are reported in Krogsgaard et al. (2018). The *WorldLarge* instance can be solved within 1 hour, while the smaller instances have much tighter CPU time limits. The authors report that they can improve the solutions of Brouer et al. (2014a,b), for instances *WestAfrica*, *WorldSmall* and *WorldLarge*. Perhaps the most important result is that the number of transshipments in general is very low, being below 1.14 per commodity. For the smaller instances, the number of transshipments is below 0.5 per commodity. Fewer transshipments mean shorter port stays, and hence vessels are not as likely to be restricted by the maximum transit times.

7 Speed Optimization

As described in Brouer et al. (2017), a key tool in achieving lower fuel consumption in liner shipping is to reduce the sailing speed between the serviced ports. However, a lower sailing speed will increase the transit times for containers, and more vessels are needed to transport the same amount of cargo as it takes longer time to complete a rotation.

Bunker consumption for a vessel profile is often modelled as a cubic function of speed, but in practice it depends not only on the speed of operation but also on wind and currents, the vessel type, the draft of the vessel, the time since the hull was cleaned and the number of reefer containers powered by the vessel's engine. During a round trip, the vessel may sail at different speeds between ports. The vessel may *slow steam* to save bunker fuel or increase speed to meet a crucial transit time. Hence speed optimization is a complex trade-off between these two criteria. A good strategy is to speed up when the vessel is fully loaded (and hence many containers need to meet their transit time), while slow steaming can be used when the load is low.

Several recent papers study speed optimization with increasing complexity and integration with routing decisions. Ronen (2011) presents a simple model where the speed and number of allocated vessels are optimized to minimize cost on a single predefined service and a single speed for the full service is assumed. Meng and Wang (2011a) also work with a single service but use a more detailed model, taking, for example, transit times into account. Further, the speed is optimized for each individual leg. Wang and Meng (2012) consider a liner shipping network with multiple predefined services and present a non-linear MIP model to optimize the ship deployment and speed of those services and the container routing through this network. The sailing speeds are optimized on each leg individually. In this model, however, no transit time constraints are considered. Kim (2013) presents a model to determine the speed and bunkering ports for a single vessel on a predetermined path. There are no transit time constraints, but a cost is imposed for each day a container is on the ship.

Another appearance of speed optimization is in the paper by Álvarez (2009), in which it is integrated with liner shipping network design. However, the model assumes a single speed for the full service, and the model has no transit time constraints.

Reinhardt et al. (2016) present a model for speed optimization of an existing liner shipping network which adjusts berthing times to minimize the overall bunker consumption. It is assumed that all services, as well as the number of deployed vessels, are fixed and that all containers are flowed along the same route as before speed optimization. Moreover, transit time constraints are taken into account. When rescheduling the berthing times, the overall transit time of a demand may change. Hence a constraint is imposed for each commodity ensuring that the transit time is within an acceptable range. A penalty is paid for each change in port calls to keep the schedule similar to the original one. Reinhardt et al. (2016) report that the model is able to save around 2% of bunker consumption while keeping all transit times unchanged. If transit times can be extended by up to 48 h, a saving of around 6–7% can be achieved.

Karsten et al. (2018) present a more advanced speed optimization model, where the services are fixed, but the speed on each leg is allowed to vary, and hence commodities may take a different route if speed changes allow for a cheaper or faster route than currently available. The problem is solved using Benders' decomposition, and results indicate that the flow changes significantly when the speed on the individual legs is changed.

Finally, Karsten et al. (2017a) consider a complete network design problem with speed optimization on individual legs, by extending the matheuristic from Karsten et al. (2017b). The leg speeds are iteratively calculated for each single service based on the current flow of containers. The method adjusts speed to the required transit times of the current container routings throughout the round trip. The individual leg speeds are calculated by solving a MIP model with the objective of minimizing the bunker consumption. A piece-wise linear function is used to approximate the cubic bunker consumption function.

8 Conclusion

Liner shipping is the backbone of international trade; hence it is important to develop decision support tools that can help designing more energy-efficient routes and balance several objectives. This includes finding the right trade-off between speed, transportation times, number of transshipments and operational costs.

Slow steaming together with larger vessels has proven to be an efficient tool for reducing energy consumption. However, slow steaming decreases the capacity of vessels, since they cannot transport as much cargo per time unit as before. Hence, more vessels are needed in order to maintain the same capacity, straining the environment. Bigger vessels tend to be more energy efficient per container, but the increased capacity results in longer port stays, making it necessary to speed up between the port stays. It is therefore necessary to design routes such that fewer transshipments are needed while still ensuring a good utilization of the mega-vessels.

Although liner shipping generally is one of the most energy-efficient modes of transportation per kilometer, the shipping industry emits large quantities of SO_x and NO_x.

In the future we will see container vessels operating with new, greener, propulsion types. Electric vessels may operate shorter routes, while liquid natural gas (LNG) may be used for operating longer routes. The new propulsion types will make it necessary to completely rethink route net design, since refueling/recharging will be more complicated, and vessels will have a more limited range of operation.

Nearly every vessel will be delayed in one or more ports during a round trip. Instead of just speeding up (and hence using more energy), advanced disruption management tools need to be developed that can ensure timely arrival to the end customer with the lowest possible energy consumption. Some studies along this path include Brouer et al. (2014a) and Li et al. (2015), but more work needs to be done in this area.

Vessel sharing agreements are an important tool for making it possible to operate larger and more energy-efficient vessels. In a vessel sharing agreement, two or more companies share the capacity of a vessel throughout the full rotation or on certain legs. Vessel sharing agreements, however, substantially increase the complexity of designing a network, since some legs and capacities are locked according to the agreement.

References

Agarwal, R., & Ergun, Ö. (2008). Ship scheduling and network design for cargo routing in liner shipping. *Transportation Science, 42*(2), 175–196.

Ahuja, R. K., Magnanti, T. L., & Orlin, J. B. (1993). *Network flows: Theory, algorithms, and applications*. Englewood Cliffs: Prentice Hall.

Álvarez, J. F. (2009). Joint routing and deployment of a fleet of container vessels. *Maritime Economics & Logistics, 11*(2), 186–208.

Archetti, C., & Speranza, M. G. (2014). A survey on matheuristics for routing problems. *EURO Journal on Computational Optimization, 2*(4), 223–246.

Balakrishnan, A., & Karsten, C. V. (2017). Container shipping service selection and cargo routing with transshipment limits. *European Journal of Operational Research, 263*(2), 652–663.

Brouer, B. D., Álvarez, J. F., Plum, C. E. M., Pisinger, D., & Sigurd, M. M. (2014a). A base integer programming model and benchmark suite for liner-shipping network design. *Transportation Science, 48*(2), 281–312.

Brouer, B. D., Desaulniers, G., & Pisinger, D. (2014b). A matheuristic for the liner shipping network design problem. *Transportation Research Part E: Logistics and Transportation Review, 72*, 42–59.

Brouer, B. D., Desaulniers, G., Karsten, C. V., & Pisinger, D. (2015). A matheuristic for the liner shipping network design problem with transit time restrictions. In F. Corman, S. Voß, & R. R. Negenborn (Eds.), *Computational logistics* (pp. 195–208). Cham: Springer.

Brouer, B. D., Karsten, C. V., & Pisinger, D. (2016). Big data optimization in maritime logistics. In A. Emrouznejad (Ed.), *Big data optimization: Recent developments and challenges* (Studies in big data, Vol. 18, pp. 319–344). Cham: Springer.

Brouer, B. D., Karsten, C. V., & Pisinger, D. (2017). Optimization in liner shipping. *4OR, 15*(1), 1–35.

Christiansen, M., Fagerholt, K., & Ronen, D. (2004). Ship routing and scheduling: Status and perspectives. *Transportation Science, 38*(1), 1–18.

Christiansen, M., Fagerholt, K., Nygreen, B., & Ronen, D. (2013). Ship routing and scheduling in the new millennium. *European Journal of Operational Research, 228*(3), 467–483.

Dithmer, P., Reinhardt, L. B., & Kontovas, C. A. (2017). The liner shipping routing and scheduling problem under environmental considerations: The case of emissions control areas. In *International Conference on Computational Logistics* (pp. 336–350). Springer.

Fagerholt, K., Gausel, N. T., Rakke, J. G., & Psaraftis, H. N. (2015). Maritime routing and speed optimization with emission control areas. *Transportation Research Part C: Emerging Technologies, 52*, 57–73.

Germanischer Lloyd. (2013). http://www.balkanlloyd.com/news/96-germanischer-lloyd-has-conducted-research-showing-that-new-and-efficient-4-500-teu-panamax. [Online; Accessed 20 Feb 2017].

Hansen, P., & Mladenovic, N. (2014). Variable neighborhood search. In E. K. Burke & G. Kendall (Eds.), *Search methodologies – Introductory tutorials in optimization and decision support techniques* (pp. 313–337). New York: Springer.

IMO. (2014). Third imo greenhouse gas study 2014. Technical report, International Maritime Organization.

Karsten, C. V. (2015). *Competitive liner shipping network design*.Ph. D. Thesis, DTU Management Engineering.

Karsten, C. V., Pisinger, D., Røpke, S., & Brouer, B. D. (2015). The time constrained multi-commodity network flow problem and its application to liner shipping network design. *Transportation Research Part E: Logistics and Transportation Review, 76*, 122–138.

Karsten, C. V., Brouer, B. D., & Pisinger, D. (2017a). Competitive liner shipping network design. *Computers & Operations Research, 87*, 125–136.

Karsten, C. V., Brouer, B. D., Desaulniers, G., & Pisinger, D. (2017b). Time constrained liner shipping network design. *Transportation Research. Part E: Logistics and Transportation Review, 105*, 152–162.

Karsten, C. V., Røpke, S., & Pisinger, D. (2018). Simultaneous optimization of container ship sailing speed and container routing with transit time restrictions. *Transportation Science, 52*(4), 739–1034.

Kim, H. J. (2013). A lagrangian heuristic for determining the speed and bunkering port of a ship. *Journal of the Operational Research Society, 65*(5), 747–754.

Kontovas, C. A. (2014). The green ship routing and scheduling problem (GSRSP): A conceptual approach. *Transportation Research Part D: Transport and Environment, 31*, 61–69.

Krogsgaard, A., Pisinger, D., & Thorsen, J. (2018). A flow-first route-next heuristic for liner shipping network design. *Transportation Science, 72*(3), 358–381. Special Issue on Emerging Challenges in Transportation Planning.

Lee, C.-Y., & Song, D.-P. (2017). Ocean container transport in global supply chains: Overview and research opportunities. *Transportation Research Part B: Methodological, 95*, 442–474.

Li, C., Qi, X., & Lee, C.-Y. (2015). Disruption recovery for a vessel in liner shipping. *Transportation Science, 49*(4), 900–921.

Maersk. (2017). The world's largest ship. https://www.maersk.com/en/explore/fleet/triple-e/environment

Meng, Q., & Wang, S. (2011a). Optimal operating strategy for a long-haul liner service route. *European Journal of Operational Research, 215*(1), 105–114.

Meng, Q., & Wang, S. (2011b). Liner shipping service network design with empty container repositioning. *Transportation Research Part E: Logistics and Transportation Review, 47*(5), 695–708.

Meng, Q., Wang, S., Andersson, H., & Thun, K. (2014). Containership routing and scheduling in liner shipping: Overview and future research directions. *Transportation Science, 48*, 265–280.

Miller, C. E., Tucker, A. W., & Zemlin, R. A. (1960). Integer programming formulation of traveling salesman problems. *Journal of the ACM (JACM), 7*(4), 326–329.

Montreuil, B. (2011). Towards a physical internet: Meeting the global logisitcs sustainability grand challenge. *Logistics Research, 3*, 71–87.

Neamatian Monemi, R., & Gelareh, S. (2017). Network design, fleet deployment and empty repositioning in liner shipping. *Transportation Research Part E, 108*, 60–79.

Notteboom, T. E. (2006). The time factor in liner shipping services. *Maritime Economics & Logistics, 8*(1), 19–39.

Notteboom, T. E., & Vernimmen, B. (2009). The effect of high fuel costs on liner service configuration in container shipping. *Journal of Transport Geography, 17*,(5), 325–337.

Plum, C. E. M., Pisinger, D., & Sigurd, M. M. (2014). A service flow model for the liner shipping network design problem. *European Journal of Operational Research, 235*(2), 378–386.

Psaraftis, H. N., & Kontovas, C. A. (2013). Speed models for energy-efficient maritime transportation: A taxonomy and survey. *Transportation Research Part C: Emerging Technologies, 26*, 331–351.

Psaraftis, H. N., & Kontovas, C. A. (2015). Slow steaming in maritime transportation: Fundamentals, trade-offs, and decision models. In C.-Y. Lee & Q. Meng (Eds.), *Handbook of ocean container transport logistics: Making global supply chains effective* (pp. 315–358). Heidelberg: Springer.

Reinhardt, L. B., & Pisinger, D. (2012). A branch and cut algorithm for the container shipping network design problem. *Flexible Services and Manufacturing Journal, 24*(3), 349–374.

Reinhardt, L. B., Plum, C. E. M., Pisinger, D., Sigurd, M. M., & Vial, G. T. P. (2016). The liner shipping berth scheduling problem with transit times. *Transportation Research Part E: Logistics and Transportation Review, 86*, 116–128.

Ronen, D. (1983). Cargo ships routing and scheduling: Survey of models and problems. *European Journal of Operational Research, 12*(2), 119–126.

Ronen, D. (1993). Ship scheduling: The last decade. *European Journal of Operational Research, 71*(3), 325–333.

Ronen, D. (2011). The effect of oil price on containership speed and fleet size. *Journal of the Operational Research Society, 62*(1), 211–216.

Stopford, M. (2009). *Maritime economics (3rd ed.)*. London: Taylor & Francis Ltd. ISBN:9780415275583.

Sun, Z., & Zheng, J. (2016). Finding potential hub locations for liner shipping. *Transportation Research Part B: Methodological, 93*, 750–761.

Thun, K., Andersson, H., & Christiansen, M. (2017). Analyzing complex service structures in liner shipping network design. *Flexible Services and Manufacturing Journal, 29*(3–4), 535–552.

Toth, P., & Vigo, D. (2015). *Vehicle routing: Problems, methods and applications* (Vol. 18). Philadelphia: SIAM.

Unctad. (2017a). Review of maritime transport. Technical report, United Nations Conference on Trade and Development.

Unctad. (2017b). UnctadSTAT. http://unctadstat.unctad.org/wds/TableViewer/tableView.aspx. [Online; Accessed 9 Feb 2017].

Wan, Z., Zhu, M., Chen, S., & Sperling, D. (2016). Pollution: Three steps to a green shipping industry. *Nature, 530*, 275.

Wang, S., & Meng, Q. (2012). Sailing speed optimization for container ships in a liner shipping network. *Transportation Research Part E: Logistics and Transportation Review, 48*(3), 701–714.

Wang, S., & Meng, Q. (2014). Liner shipping network design with deadlines. *Computers & Operations Research, 41*, 140–149.

Chapter 10
Speed Optimization for Sustainable Shipping

Harilaos N. Psaraftis

Abstract Among the spectrum of logistics – based measures for sustainable shipping – this chapter focuses on speed optimization. This involves the selection of an appropriate speed by the vessel, so as to optimize a certain objective. As ship speed is not fixed, depressed shipping markets and/or high fuel prices induce slow steaming which is being practiced in many sectors of the shipping industry. In recent years the environmental dimension of slow steaming has also become important, as ship emissions are directly proportional to fuel burned. Win-win solutions are sought, but they will not necessarily be possible. The chapter presents some basics, discusses the main trade-offs and also examines combined speed and route optimization problems. Some examples are presented so as to highlight the main issues that are at play, and the regulatory dimension of speed reduction via speed limits is also discussed.

Abbreviations

AIS	Automatic identification system
BRI	Belt and Road Initiative
CBO	(US) Congressional Budget Office
CIF	Cost insurance freight
CO_2	Carbon dioxide
CSC	Clean Shipping Coalition
DWT	Deadweight ton
EEDI	Energy Efficiency Design Index
ECA	Emissions Control Area
FMC	(US) Federal Maritime Commission
GHG	Greenhouse gas

H. N. Psaraftis (✉)
DTU Management Engineering, Technical University of Denmark, Kongens Lyngby, Denmark
e-mail: hnpsar@dtu.dk

H. N. Psaraftis (ed.), *Sustainable Shipping*,
https://doi.org/10.1007/978-3-030-04330-8_10

HFO	Heavy fuel oil
IMO	International Maritime Organization
MBM	Market-based measure
MEPC	Marine Environment Protection Committee
MSC	Mediterranean Shipping Company
NGO	Nongovernmental organization
Ro/Ro	Roll on/Roll off
Ro/Pax	Ro/Ro passenger
SECA	Sulfur emissions control area
SO_x	Sulfur oxides
TEU	Twenty-foot equivalent unit
USD	United States dollar
VLCC	Very large crude carrier
VSRP	Vessel speed reduction programme
WS	World scale (index)

1 Introduction

The spectrum of *logistics-based* problems in maritime transportation is broad. Already Chaps. 8 and 9 of this book gave a flavor of such problems, by exploring green solutions for the tramp and liner shipping markets, respectively. Chapter 7 of the book can also be considered to fall in the same category, by investigating possible modal shifts due to sulfur regulations. Logistics-based problems in shipping can be broken down in the categories broadly shown in Table 10.1 below. Some related references are also shown in the table (neither list is encyclopedic).

With the exception of network design and fleet size and mix problems, which are typically defined at the *strategic* level (planning horizon of several years), and of weather routing problems, which are typically defined at the *operational* level (planning horizon of a few hours to a few days), most of logistics-based problems in maritime transportation are defined at the *tactical* planning level (planning horizon of a few days to a few months).

This chapter will examine ship speed optimization from various angles, including combined speed and routing scenarios. In that context, some basics will be outlined, the main trade-offs will be analyzed, and some decision models will be presented. The examples presented will highlight the main issues that are at play. Material of this chapter is mainly drawn from various papers and other work by the authors and their colleagues. These include, among others, Gkonis and Psaraftis (2012), Kapetanis et al. (2014), Psaraftis and Kontovas (2013, 2014), Wen et al. (2017) and more recently Giovannini and Psaraftis (2018).

It is important to note that, in much of the maritime logistics literature, environmental criteria such as emissions reduction are scarce, traditional economic criteria such as cost reduction being the norm. Sometimes such economic criteria map directly into environmental criteria: if, for instance, *fuel cost* is the criterion, as

Table 10.1 Selected logistics-based problems and related references in maritime transportation

Problem category (selected)	Related references (selected)
Ship routing and scheduling (general)	Ronen (1982)
	Christiansen et al. (2013)
Ship routing and scheduling (tramp)	Andersson et al. (2011)
	Fagerholt et al. (2010)
	Lin and Liu (2011)
	Chapter 8 of the book
Ship routing and scheduling (offshore supply)	Halvorsen-Weare and Fagerholt (2011)
	Norlund and Gribkovskaia (2013)
Fleet deployment (liner)	Powell and Perakis (1997)
	Meng and Wang (2011)
	Andersson et al. (2014)
Fleet size and mix (liner)	Alvarez et al. (2011)
	Zeng and Yang (2007)
Modal split (Ro/Ro)	Panagakos et al. (2014)
	Zis and Psaraftis (2017, 2018)
	Chapter 7 of the book
Network design (liner)	Agarwal and Ergun (2008)
	Reinhardt and Pisinger (2014)
	Brouer et al. (2013)
	Chapter 9 of the book
Weather routing (general)	Perakis and Papadakis (1989)
	Lo and McCord (1998)
Transshipment (liner)	Hsu and Hsieh (2005)
	Wang and Meng (2012)
Terminal management	Moccia et al. (2006)
	Goodchild and Daganzo (2007)
	Stahlblock and Voss (2008)
Speed optimization	Norstad et al. (2011)
	Hvattum et al. (2013)
	Fagerholt and Ronen (2013)
	Zis et al. (2015)
	Chapters 7, 8 and 9 of the book
	This chapter

it is directly proportional to *emissions*, if fuel cost is to be minimized as an objective, so will emissions, and the solution is win-win.

However, for other objectives, this direct relationship may cease to exist, and one would need to look at environmental criteria in their own right. Even though such criteria were not very common in the past, the body of knowledge that includes such criteria is growing in recent years. Among the set of maritime logistics problems which are important as regards both economic and environmental criteria, perhaps *speed optimization* is the most important.

In conceptual terms, if x is a vector of the decision variables of the problem at hand, $f(x)$ is the fuel cost associated with x, $c(x)$ is the cost other than fuel and $m(x)$ are the associated maritime emissions (CO_2, SO_x or others), then a generic optimization problem is the following:

$$\text{Minimize } \alpha(f(x)+c(x))+\beta m(x) \ s.t. x \in X$$

where α and β are user-defined weights (both ≥ 0) representing the relative importance the decision-maker assigns to cost versus emissions and X represents the feasible solution space, usually defined by a set of constraints.

One can safely say and without loss of generality that if $d(x)$ is the amount of fuel consumed, p is the fuel price and e is the emission coefficient (kg of emissions per kg of fuel), then $f(x) = pd(x)$ and $m(x) = ed(x)$. Therefore $f(x) = km(x)$ with $k = p/e$, as both $f(x)$ and $m(x)$ are proportional to the amount of fuel consumed $d(x)$. The cases that different fuels are used onboard the ship, for instance, in the main engine and the auxiliary engines, or if fuel is switched from high to low sulfur along the ship's trip, represent straightforward generalizations of the above formulation.

Then the above problem can also be written as

$$\text{Minimize } \alpha c(x)+(\alpha k+\beta)m(x) \ s.t. x \in X$$

The following special cases of the above problem are important:

(i) The case $\alpha = 0$, $\beta > 0$, in which the problem is to minimize emissions
(ii) The case $\alpha > 0$, $\beta = 0$, in which the problem is to minimize total cost
(iii) The case $c(x) = 0$, in which fuel cost is the only component of the cost

A solution x^* is called *win-win* if both case i and case ii have x^* as an optimal solution. It is important to realize that such a solution may not necessarily exist.

It is also straightforward to see that in case iii, cost and emissions are minimized at the same time and we have a win-win solution. It is clear that $c(x)=0$ is a *sufficient condition* for a win-win solution. But this is *not* a *necessary* condition, as it is conceivable to have the same solution being the optimal solution under two different objective functions. An interesting question is to what extent policy-makers can introduce either (a) a market-based measure (MBM) and/or (b) a set of constraints that would make win-win solutions possible. MBMs are examined in Chap. 11 of this book.

As alluded to earlier, a significant part of the recent literature on green maritime logistics deals with *speed optimization*. This is because an appropriate selection of ship speed is an important measure to achieving both fuel cost reduction and emissions reduction, therefore it is potentially a win-win proposition. Already Chaps. 7, 8 and 9 of this book look at variable ship speeds in their formulations from various angles.

Even though ships travel slower than the other transportation modes, a basic premise has always been that there is value in ship speed. As long-distance trips may typically last 1–2 months, the benefits of a higher speed may be significant: they mainly entail the economic added value of faster delivery of goods, lower inventory costs and increased trade throughput per unit time.

The need for higher speeds in shipping was mainly spurred by strong growth in world trade and development and in turn was made possible by significant technological advances in maritime transportation in a broad spectrum of areas, including hull design, hydrodynamic performance of vessels, engine and propulsion efficiency, to name just a few. By extension, developments in cargo handling systems and supply chain management and operation have also contributed significantly to fast door-to-door transportation. However, the above basic premise is being challenged whenever shipping markets are depressed and whenever fuel prices are on the increase. In such situations, ships tend to slow down.

Perhaps the most significant factor that is making a difference in recent years is fact that a ship has to be environmentally friendly as regards air emissions. Because of the nonlinear relationship between speed and fuel consumption, it is obvious that a ship that goes slower will emit much less than the same ship going faster.

The importance of ship speed on ship emissions can be seen in Fig. 10.1, which breaks down CO_2 emissions from the world commercial fleet by ship type-size combination (Psaraftis and Kontovas, 2009a). The data of Fig. 10.1 is from the IHS Fairplay database, and the base year is 2007 (45,620 commercial ships accounted for).

According to this analysis, containerships are the top CO_2 emitters in the world fleet. This is perhaps something to be expected, given the relatively high design speeds of these vessels (20–26 knots) as opposed to those carrying bulk

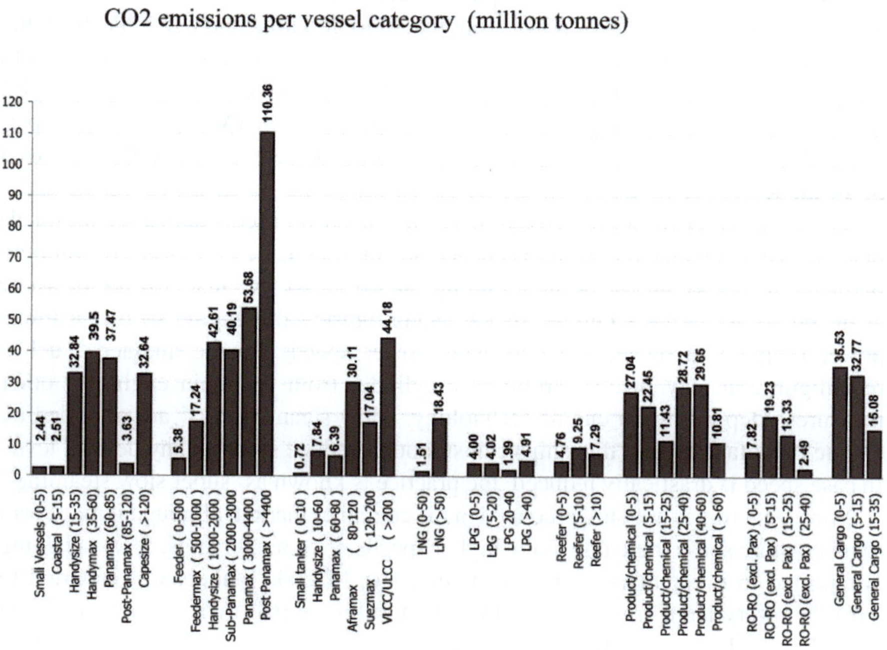

Fig. 10.1 CO_2 emissions, world fleet, 2007. (Source: Psaraftis and Kontovas (2009a))

cargoes (13–15 knots) and given the nonlinear relationship between speed and fuel consumption and hence emissions.

What is perhaps not so obvious to expect and can be seen in Fig. 10.1 is that just the top-tier category of container vessels (712 vessels of 4400 TEU and above) is seen to produce 110.36 million tonnes of CO_2 emissions, which is higher than the 106 million tonnes produced by the entire crude oil tanker fleet (2028 vessels). This means that if ship speed were to be reduced, perhaps uniformly across the board, or even selectively for some categories of vessels, emissions would be reduced too, perhaps drastically. Reducing speed could also have important side benefits: cost reduction is one, and helping a depressed market in which shipping overcapacity is the norm these days is another. In that sense, reducing ship speed may conceivably be a 'win-win' proposition.

We note here that since the above analysis refers to the 2007 fleet, today's picture may be different. However, the 2007 picture is also in line with later results. In the third greenhouse gas (GHG) study of the International Maritime Organization (IMO 2014), which refers to 2012 fleet data, the containership class of vessels was identified as the top CO_2 emitter of the world fleet.

If one starts with the simple way to reduce fuel costs (and by extension emissions) by reducing speed, this can be done at two levels. One level is the technological one, that is, build future ships with reduced installed horsepower so that they cannot sail faster than a prescribed speed. The first cellular containerships of the late 1960s and early 1970s that went up to 33 knots in the late 1960s when fuel was cheap are gone forever. Maersk's flagship 'Triple-E' fleet of 18,000 TEU containerships has a design speed of 17.8 knots, down from the 20 to 26 knots range that has been the industry's norm, and will emit 20% less CO_2 per container moved as compared to the *Emma Maersk*, the Triple-E predecessor as flagship in the Maersk fleet, and 50% less than the industry average on the Asia-Europe trade lane (Maersk 2013). Triple-E stands for economy of scale, energy efficiency and environmentally improved performance.

The other level of speed reduction is the logistics-based (tactical/operational) one. At that level, an existing ship can sail slower than its design speed. In shipping parlance, this is known as 'slow steaming' and may involve just slowing down or even 'derating' a ship's engine, that is, reconfiguring the engine so that a lower power output is achieved, so that even slower speeds can be attained. Such a reconfiguration may involve dropping a cylinder from the main engine or other measures. Depending on engine technology, 'slow steaming kits' are provided by engine manufacturers so that ships can smoothly reduce speed at any desired level. In case speed is drastically reduced, the practice is known as 'super slow steaming'.

Slow steaming is not only practiced in the container market, although it may seem to make more sense there due to the higher speeds of containerships. Slow steaming is reported in every market. In December 2010, Maersk Tankers was reported to have their very large crude carriers (VLCCs) sailing at half their speed. The speed of 16 knots (design speed) was reduced to less than 10 knots on almost one third of its ballast legs and between 11 and 13 knots on over one third of its operating days. For example, a typical voyage from the Persian Gulf to Asia normally takes

42 days (at 15 knots laden and 16 knots in ballast). Maersk Tankers decreased speed to 8.5 knots on the ballast leg, thus increasing round trip time to 55 days and saving nearly USD 400,000 off the voyage's bunker bill (TradeWinds 2010).

The fact that slow steaming is being practiced in periods of depressed market conditions can be confirmed by the fact that whatever fleet overcapacity existed has been virtually absorbed. Since early 2009, the total containership capacity absorbed due to the longer duration of total round trip time for long-haul services has reached 1.27 million TEU in October 2013 (taking early 2009 as a starting point), based on Alphaliner's estimates (Alphaliner 2013). More recently, UNCTAD (2016) documented a continuing *sluggish demand* challenged by an accelerated massive global *expansion* in container supply capacity, estimated at 8% in 2015 – its highest level since 2010. A similar situation pertains in the tanker and dry-bulk markets (Devanney 2011). Even more recently, the two largest container carriers, Maersk and MSC, have agreed to further slow steam to cut costs, with some speeds as low as 13 knots (Lloyds List 2018). Moreover, and according to the third GHG study of the IMO, the reduction of global maritime CO_2 emissions from 885 million tonnes in 2007 to 796 million tonnes in 2012 is mainly attributed to slow steaming due to the serious slump in the shipping markets after 2008 (IMO 2014).

At the same time, and even though win-win solutions may look as natural consequences of speed reduction, the practice may have other ramifications which may not be beneficial. For instance, in the long run, more ships will be needed to produce the same transport throughput, and this will entail some costs, some of them financial and some environmental, such as lifecycle emissions due to shipbuilding and recycling (see Chatzinikolaou and Ventikos (2016) for a discussion on the lifecycle approach and Chap. 6 of this book on the issue of ship recycling).

Also, in-transit inventory costs will generally increase, due to the increased transit time of the cargo. These inventory costs are proportional to the value of the cargo, so if a ship hauls high-value goods, sailing at a lower speed may entail significant costs to the shipper (we shall come back to this point later in the chapter).

Yet another side effect of speed reduction is that in the short run, freight rates will go up once the overall transport supply shrinks because of slower speeds. Reducing speed may help a depressed market, but it is the shippers who will suffer and in fact they will do so in two ways: they will pay more and receive their cargo later. For a discussion how tanker spot rates may be impacted as a result of slow steaming see Devanney (2007).

Psaraftis and Kontovas (2009b) investigated, among other things, the option to slow down in Sulfur Emissions Control Areas (SECAs) to reduce the quantity of SO_X produced. Realizing that a reduced speed cannot alter the percentage of SOx emissions in a ship's exhaust, it was shown that if the ship speeds up outside the SECA to make up for lost time within the SECA, more emissions will be produced overall, including SO_X. Fagerholt et al. (2015) and Fagerholt and Psaraftis (2015) examined route-speed alternatives for ships operating in and out of emissions control areas (ECAs), and Magirou et al. (2015) developed stochastic optimal control schemes for speed optimization in a dynamic setting.

Another possible side effect concerns effects that speed reduction may have on other modes of transportation, to the extent these are alternatives to sea transportation. This is the situation mostly as regards short sea trades, in Europe but also in North America. If ships are made to go slower, shippers may be induced to prefer land-based transportation alternatives, mostly road, and that may increase overall GHG emissions. Even in long-haul scenarios such as the Far East to Europe trade, some cargoes may be tempted to use the rail alternative (via the Trans-Siberian railway) if the speed of vessels is slow enough (see Psaraftis and Kontovas (2010) for a discussion). Such considerations may also be relevant as regards the recent Belt and Road Initiative (BRI), which aims to promote Chinese trade to Europe via a combination of land-based and maritime corridors.

In short sea shipping, possible modal shifts due to speed reduction and other measures were investigated in Zis and Psaraftis (2017, 2018) in the context of European SECAs and in the Ro/Ro sector. The impact of adjusting ship speed as a mitigation measure was also examined (more on this in Chap. 7 of this book).

Psaraftis and Kontovas (2015), among other things, provided a discussion on the possible impact of slow steaming on *port operations*. If a port is congested, it would clearly make no sense to sail there at full speed, wasting money on fuel and producing emissions that can be avoided if ship speed were slower. A recent initiative is the so-called virtual arrival, which has been used in order to manage the vessels' arrival time based on the experience of congestion at some discharging ports. This initiative recognizes known inefficiencies in the supply chain, such as waiting to discharge because of port delays, and reduces fuel consumption and, consequently, emissions by implementing a mutually agreed reduction in a vessel's speed in order to achieve an agreed arrival time at a port. After the agreement of both parties, the ship slows to the economic speed based on the revised arrival time. Once the voyage is completed, demurrage is calculated based on the original plans, and bunker savings are split between the parties. At the same time, Californian ports have been offering monetary incentives for ocean-going vessels that reduce speed down to 12 knots in the proximity of the port as an emissions reduction measure (vessel speed reduction programme – VSRP) which has seen great participation rates (Zis et al. 2014). In separate but related initiatives, Golias et al. (2010) and Du et al. (2011) developed models that combined optimizing berth allocation with reducing associated vessel emissions.

The rest of this chapter is organized as follows. Section 2 presents some basics of speed optimization. Section 3 discusses factors that may impact fuel consumption and the possible impact of inventory costs. Section 4 deals with combined speed and routing decisions. Section 5 investigates speed optimization in a liner shipping context, and Sect. 6 (together with Appendix A) discusses the speed limit issue. Last but not least, Sect. 7 presents this chapter's conclusions.

2 Ship Speed Optimization Basics

Before we see how ship speed can be optimized, we present some basics. We do this so as to clear possible misconceptions and highlight some issues which we find important.

The first basic is that *ships do not trade at fixed or predetermined speeds.*

In the charter (tramp) market, those who pay for the fuel, that is, the ship owner whose ship trades on the spot market, or the charterer if the ship is on time or bareboat charter, will typically choose ship speed as a function of two main input parameters: (a) the fuel price and (b) the market freight rate. In periods of depressed market conditions, as is the typical situation in recent years, ships tend to slow steam. The same is the case if bunker prices are high. Conversely, in boom periods or in case fuel prices are low, ships tend to sail faster.

An exception to the above is in case the ship is *on spot charter* (rental of the ship for a single voyage) and its speed is prescribed in the charter party contract, either explicitly (speed is, say, 15 knots) or implicitly (cargo pickup and delivery dates are prescribed). In spot charters the fuel is paid for by the ship owner. Agreeing on a prescribed speed in the charter party involves in most cases only the laden part of the trip, with the owner free to choose his speed on the ballast return leg. The speed that is agreed upon for the laden leg may or may not be the speed that the ship owner would have freely chosen if no explicit agreement were in place. If it is higher, the ship owner may ask for a higher rate than the prevailing market spot rate, understanding of course that in this case he may lose the customer to a competitor ship, with whom the charterer can obtain more favorable terms. For a discussion of possible distortions and additional emissions that can be caused by charter party speed agreements, see Devanney (2011a).

A similar situation plays out in the liner market. Container and Ro/Ro operators typically operate a mixed fleet of vessels, some of which are owned vessels and some are chartered from independent owners who are not engaged in liner logistics. In either case, fuel is paid for by the liner operator. The operator receives income from the multitude of shippers whose cargoes are carried on the ship and the rates charged to these shippers can be high or low depending on the state of the market. As in the charter market, high fuel prices and/or depressed market conditions imply lower speeds for the fleet. More on how this can be manifested in the liner sector can be found in Sect. 5 of this chapter.

In spite of the above, many of the models found in the maritime logistics literature assume fixed and known ship speeds. See, for instance, Rana and Vickson (1991), Agarwal and Ergun (2008), Hwang et al. (2008), Grønhaug et al. (2010) and Song and Xu (2012), among others. In these models, ship speed is typically considered not as a decision variable but as a *fixed input* to the problem. Most of the time, this input is *implicit*, in the sense that it is used to compute various other *explicit inputs* that depend on speed, such as sailing times, due dates for cargo pickup and delivery and ship operating costs, of which fuel cost is an important component.

Assuming fixed ship speeds is typically also the case for models that compute shipping emissions worldwide, even though these do not belong to the maritime logistics literature. See, for instance, the second IMO GHG study (IMO 2009) and Psaraftis and Kontovas (2009a), among others. In their calculations, these models typically take as input *design speeds* extracted from commercially available ship databases, such as those maintained by IHS Fairplay, among others. Such information may be inaccurate and does not necessarily represent actual ship speeds. The third IMO GHG study (IMO 2014) is more advanced in that it uses actual ship speeds in its calculations. Actual ship speeds were taken from the ship's automatic identification system (AIS) data.

Coming back to maritime logistics models, it is clear that *not* considering speed as a decision variable may render solutions suboptimal. This is because doing so ignores the economic trade-off between (a) the lower voyage and cargo inventory costs associated with a higher speed and (b) the higher fuel costs associated with such higher speed. Assuming a fixed speed precludes the balancing of such trade-offs.

A speed that is assumed fixed may also in some cases remove flexibility in the overall decision-making process. For problems that include port capacity constraints, berth occupancy constraints, time window constraints or other constraints that preclude the simultaneous service of more than a given number of vessels (see, e.g. Cordeau et al. (2005) and Halvorsen-Weare and Fagerholt (2013), among others), satisfying such constraints would conceivably be easier to meet were it not for the assumed constancy in ship speed. The same is the case for problems that analyze disruptions of service due to weather or other unpredictable events. It is clear that removing the flexibility to adjust ship speed in such scenarios would render any response to the disruption suboptimal.

Still, dealing with speed is not new in the maritime transportation literature, and this body of knowledge is rapidly growing. In Psaraftis and Kontovas (2013), some 42 relevant papers were reviewed, and a taxonomy of these papers according to various criteria was developed. An amended taxonomy, consisting of 51 papers, was presented in Psaraftis and Kontovas (2016); however many additional papers dealing with ship speed appeared after the 2013 paper was published. The 2013 paper's Google Scholar citations as of October 2018 stood at 207, and even included papers in seemingly unrelated journals such as *Meat Science* (Mills et al. 2014). The growing number of references indicates a strong interest of researchers in this topic.

Another basic property of optimal speeds is not immediately obvious. It applies mainly to the charter (tramp) market and compares, for a specific ship and a specific route, the speed optimization problem of its ship owner and that of a time charterer who may charter the same ship. The ship owner wants to maximize average profit per day, and the charterer wants to minimize average cost per day. Even though these two optimization problems appear at first glance different, the optimal ship speed for both problems turns out to be the same. For a proof in a rudimentary tramp scenario, see Devanney (2010).

Figure 10.2 is adapted from Gkonis and Psaraftis (2012) and captures the impact of both freight rate and bunker price on optimal speed for a specific very large

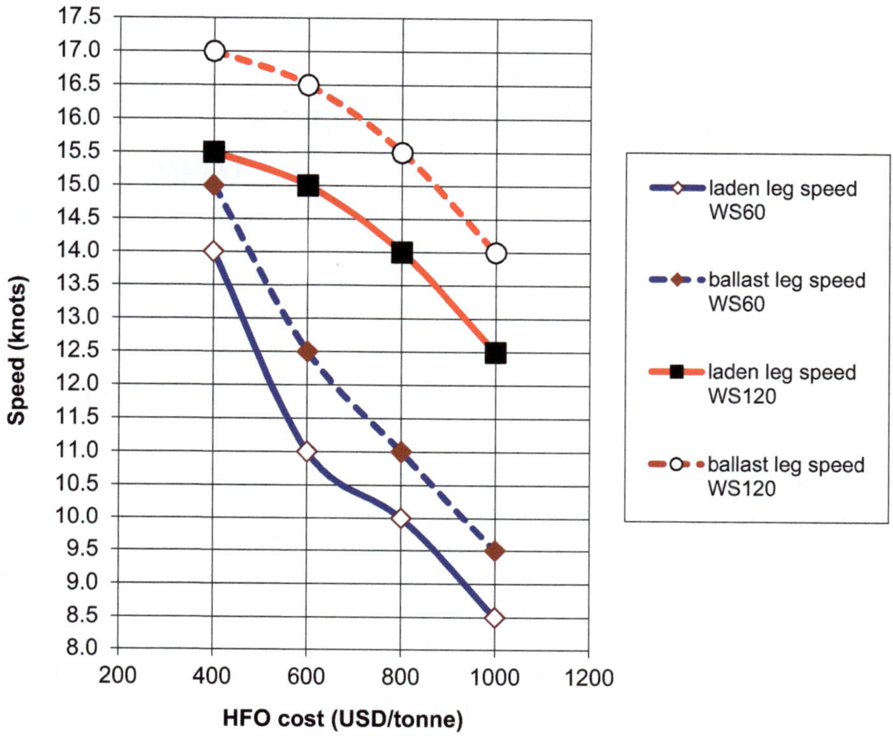

Fig. 10.2 Optimal VLCC speed as a function of spot rate and bunker price. WS is the Worldscale index. (Adapted from Gkonis and Psaraftis (2012))

crude carrier (VLCC) trading from the Persian Gulf to Japan. Optimal here means maximize average per day profit for the ship owner, and speeds are optimized in both laden and ballast conditions. Two market conditions are shown for the spot rate, one at Worldscale (WS) 60 and one at WS120.[1] Bunker prices (HFO, heavy fuel oil) range from USD 400 to USD 1000 per tonne. It can be observed that the impact of both freight rate and bunker price on optimal speed can be quite dramatic and that the range of optimal speeds can be very broad, depending on the combination of values of these two input parameters. It can be also observed that ballast speeds are typically higher than laden speeds by 1.0 knot in the lower rate scenario and by 1.5 knots in the higher rate scenario.

Figure 10.3 shows annual CO_2 emissions for the same VLCC as a function of bunker price and spot rate. It can be seen that CO_2 emissions can be reduced significantly if fuel price goes up. This points out to the possible importance of a

[1]WS is a nondimensional index measuring the spot rate and is exclusively used in the tanker market. For a specific route, WS is proportional to the spot rate on that route (in USD/tonne) and is normalized by the 'base rate' on that route. See Stopford (2007) for a detailed definition.

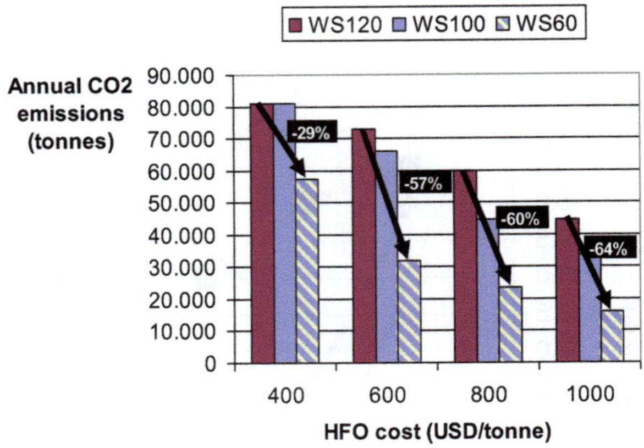

Fig. 10.3 Annual CO_2 emissions (single VLCC tanker) as a function of fuel price and spot rate. WS is the Worldscale index. (Source: Gkonis and Psaraftis (2012))

bunker levy as a tool to reduce CO_2 emissions. The figure also shows that emissions will be reduced (sometimes significantly) whenever fuel prices are up and/or the state of the market is down. Such a reduction is attributed to slow steaming.

A similar situation plays out in the liner market, in spite of obvious differences in the logistical scenario. Liner ships tend to slow down in periods of high fuel prices and/or depressed freight rates and speed up if the opposite is the case. Yet, and even though the impact of a high or a low fuel price is captured by many models in the liner shipping literature, the possible impact of *the state of the market* on liner ship speed is typically considered as outside the scope of much of that literature. Section 5 of this chapter addresses this issue by presenting a recently developed model that includes the impact of fuel price, state of the market and cargo inventory costs on liner ship speed.

3 Factors That Affect Fuel Consumption and Impact of In-Transit Cargo Inventory Costs

3.1 Fuel Consumption Function

It is known from basic naval architecture that fuel consumption depends nonlinearly on both ship sailing speed and ship payload. Many papers assume that fuel consumption per day is a *cubic function* of ship speed. The cubic approximation is reasonable for some ship types, such as tankers, bulk carriers or ships of small size, but may not be realistic at slow or near-zero speeds and for some other ship types such as high-speed large container vessels. Even at zero speed the ship consumes some fuel, as its auxiliary engines are typically on to produce electricity. An exception is if electricity is provided to the ship by shore-side supply (also known as 'cold ironing'), but this is currently an exception rather than the rule.

In-port fuel consumption is proportional to overall total port residence time. See Chap. 12 of this book for more on the subject of green ports.

A more serious assumption in many related models is that no dependency between fuel consumption and ship payload is considered. This assumption is reasonable in case ship payload is constant or does not change much. Cruise vessels, passenger vessels, and sometimes Ro/Ro carriers and Ro/Pax vessels belong to this category. However, if this assumption is not valid, it can cause serious under- or overestimation of fuel costs. Ship resistance and hence fuel consumption at a given speed can be drastically different if the ship is full, empty or at an intermediate loading condition.

In tankers and bulk carriers, we have a 'binary' situation, as the ship is typically either full or empty, and the difference in fuel consumption between these two extreme conditions can be quite substantial. In container vessels the ship is typically intermediately laden most of the time, but ships in some trunk routes (e.g. Far East to Europe) are mostly full in one direction and mostly empty in the opposite. This can come close to a binary situation, and one would expect non-trivial differences in fuel consumption as a result.

In general, if a ship's loading condition varies along the legs of a ship's route (which is typical in pickup and delivery scenarios in which the ship is not fully laden all of the time), it is important that the dependency between ship load and fuel consumption along that route be realistically modeled. In an optimization setting, it would not make sense to claim solutions within, say, 1, 2 or 5% from the optimal solution, or even solutions at the exact optimum, if the fuel consumption function and hence fuel costs are misrepresented by 10, 20 or 30%.

In order to capture this dependency, it is useful to extend the previous formulation of the daily 'at-sea' fuel consumption of the ship and assume that it is a known function $f(v, w)$ of both v, the ship's speed, and w, the ship's payload, which may actually vary along the ship's route. Function $f(v, w)$ depends on the ship and essentially on the hull geometry-engine-propeller configuration. It can even be defined for $v = 0$ (ship in port) and $w = 0$ (ship going on ballast), and it need not be assumed in closed form but could be given as a point/wise function, as a table or even as the output of a relevant subroutine. Strictly speaking, f must also take into account the reduction of the ship's total displacement due to fuel being consumed along the ship's route. However, since displacement would not change much as a result of that consumption, one can practically assume f independent of en-route fuel consumption.

A realistic closed-form approximation of f that takes both v and w into account is $f(v, w) = k(p+v^q)\ (w+A)^{2/3}$ with k, p and q constants such as $k > 0$, $p \geq 0$ and $q \geq 3$. A is the 'lightship weight', that is, the weight of the ship if empty including fuel and other consumables (modified admiralty formula). The rationale for such a formulation is that fuel consumption is proportional to the wetted surface of the ship, which is crudely proportional to the displacement of the ship $\Delta = w + A$, raised to the power of 2/3; see also Barrass (2005).

As said earlier, most papers in the literature assume a cubic function, that is, $p = 0$ and $q = 3$ and no dependency on payload.

The fuel consumption function also depends on the prevailing weather conditions along the route, which may actually vary in time and space. The way weather conditions are treated in the literature ranges from nontreatment (implying that the *average* weather conditions the ship expects along its route are implicitly factored into the function f, perhaps by a *sea margin* coefficient) to more sophisticated approaches in which f depends on the specific weather conditions along the ship's route, including wave height, wave direction, wind speed, wind direction, sea currents and possibly others. These factors, most of which may be stochastic, can significantly influence both wave and wind resistance and hence fuel consumption and cost. Weather routing models typically take the more sophisticated approach, whereas all other models including ship routing and scheduling, fleet deployment, and other models typically follow the simpler one.

Last but not least, hull condition can also be an important factor that influences the frictional resistance of a ship, and, as a result, its fuel consumption. A foul hull from seaweed and other sea organisms would entail a higher resistance (and hence fuel consumption) than a clean hull, and efforts are being made (via antifouling paints and hull cleaning at regular intervals) to maintain a clean hull. To our knowledge, no maritime logistics model takes into account such factor, all assuming an average hull condition. An interesting problem would be to determine the optimal timing of hull cleaning, assuming an average rate of hull decay and a corresponding increase of fuel consumption through time.

3.2 In-Transit Cargo Inventory Costs

Many of the models that include speed do not include *in-transit cargo inventory costs* as part of the cost function. These are inventory costs that accrue while the ship is in transit, and they can be a non-trivial component of the cost that the owner of the cargo bears if the ship will sail at a reduced speed. They can be important if timely delivery of the cargo is significant. They can also be important if the voyage time and/or the quantities to be transported are non-trivial. This can be the case in long-haul problems.

It is clear that in-transit inventory costs are important for the charterer, assuming that he is the owner of the cargo. These costs are also important for the ship owner, as a charterer will prefer a ship that delivers his cargo earlier than another ship that sails slower. Thus, if the owner of the slower ship would like to attract that cargo, he may have to rebate to the charterer the loss due to delayed delivery of cargo. In that sense, the in-transit inventory cost is very much relevant in the ship owner's profit equation, as much as it is relevant in the charterer's cost equation.

The same is the case if the charterer does not move his own cargo but uses the ship to move somebody else's cargo. This is a typical situation in liner trades, where a significant part of a liner company's fleet consists of chartered ships, owned by independent ship owners but operated by the liner company. As the cargo owner will prefer a ship that moves his cargo faster, his in-transit inventory costs are again very much part of the chartered ship's cost equation.

If we call γ the per day and per tonne in-transit inventory cost of the cargo, it is straightforward to see that $\gamma = PR/365$, where P is the CIF value of the cargo (value of cargo at destination) and R the cargo owner's cost of capital. This represents the revenue that is lost due to a delayed delivery of 1 tonne of the cargo by 1 day. This means (as expected) that expensive cargoes are more costly than cheaper cargoes in terms of inventory cost. This also explains why expensive cargoes tend to get hauled by liner ships that go faster, whereas cheaper (bulk) cargoes go by tramp ships that go slower. Conversely, it also means that in periods of low interest rates this cost component is less important.

Cargo inventory costs can be important in the liner business which involves trades of higher valued goods than those in bulk trades. The unit value of the top 20 containerized imports at the Los Angeles and Long Beach ports in 2004 varied from about USD 14,000/tonne for furniture and bedding to USD 95,000/tonne for optic, photographic and medical instruments (CBO 2006). Delaying 1 tonne of the latter category of cargo by 1 week because of reduced speed would cost some USD91 if the cost of capital is 5%. For a 80,000 tonne payload, this would amount to some USD 7.25 million. This may or may not be greater than the economic benefit of a reduced speed.

We shall come back to the impact of in-transit cargo inventory costs as regards liner shipping in Sect. 5.

4 Combining Speed and Routing Decisions

Speed optimization can be extended into combined ship routing and speed scenarios. We divide the discussion into tactical level problems and operational level problems. As will be seen, these two levels are very different, even though many of the examined issues are similar.

4.1 Tactical Level Problems

At the tactical planning level, where the problem at hand is basically a distribution problem, a number of papers in the literature have looked at combined ship routing and speed scenarios. See, for instance, Hvattum et al. (2013) and Fagerholt and Ronen (2013), among others.

Psaraftis and Kontovas (2014) examined combined single-ship scenarios in which the fuel consumption function depended on both ship speed and payload and in which fuel price, charter rate and inventory costs were also taken onboard. By increasing order of complexity, these scenarios included:

- Fixed-route scenarios: A ship going from port A to port B, or even on a multiple leg route in which the sequence of port visits is already determined at a higher level, but ship payload varies along the route.
- Feeder scenarios: A feeder ship collecting cargoes from several ports and bringing them into a hub port or vice versa.
- Combined pickup and delivery scenarios: A ship picking up cargoes from distinct origins and delivering them to distinct destinations. The route and sequence of pickups and deliveries has to be determined, along with the ship speed at each leg of the route.

The latter scenario was actually a generalized version of the feeder scenario and included several sub-scenarios itself, depending on whether each port has one or multiple pickup cargoes, to be delivered to one or several delivery ports.

Several alternative objective functions were examined, including a minimum time objective, a minimum emissions objective and a minimum total cost objective. It was seen that fuel costs, freight rates and cargo inventory costs generally have an impact not only on the speed decision, but also on the choice of the route.

Wen et al. (2017) extended the above work to a multiple ship setting. A branch and price algorithm and a constraint programming model were developed that considered (a) fuel consumption as a function of payload, (b) fuel price as an explicit input, (c) freight rate as an input and (d) in-transit cargo inventory costs. The alternative objective functions that were examined were minimum total trip duration, minimum total cost and minimum emissions.

A result of this work, which is to be expected, is that more expensive cargoes induce higher ship speeds and therefore more CO2. This is due to the impact of in-transit inventory costs. Table 10.2 shows a five-leg (fixed) ship route sailed by a 16,000 DWT ship in which the payload of the ship varies along the route and the value of the cargo varies from 0 to 25,000 USD/tonne.

As much as problem inputs generally influence both speed and route selection, another result of these combined speed/routing scenarios is perhaps counter-intuitive. It was found that sailing the minimum distance route at minimum speed does not necessarily minimize emissions. This may be so whenever the minimum distance route involves a heavier load profile for the ship. A heavier load profile would result in a higher fuel consumption (and emissions) overall, even though the route may be shorter. So in this case what would intuitively seem like an optimal policy is actually suboptimal. For some examples confirming the above, see Psaraftis and Kontovas (2014) and Wen et al. (2017).

4.2 Operational Level Problems

At the operational planning level, a separate but very important class of the combined speed/route class of problems concerns *weather routing* scenarios. The important difference vis-à-vis the types of problems described in Sect. 4.1 is

Table 10.2 Variation of optimal speed with value of cargo

Value of cargo (USD/tonne)			0	5000	10,000	15,000	20,000	25,000
		Payload (000 tonnes)	Speed (knots)					
Leg	0–1	0	13.54	13.54	13.54	13.54	13.54	13.54
	1–2	5	11.61	12.12	12.58	13.02	13.43	13.81
	2–3	6	11.36	11.96	12.49	12.99	13.45	13.88
	3–4	8	10.95	11.70	12.36	12.96	13.51	14.00
	4–5	11	10.46	11.42	12.24	12.96	13.61	14.00
Fuel cost (USD)			39,751	44,433	48,808	52,945	56,890	59,854
Charter cost (USD)			79,502	75,324	72,136	69,580	67,461	65,996
Inventory cost(USD)			0	13,542	25,480	36,310	46,318	56,189
Total cost (USD)			119,253	133,299	146,424	158,835	170,669	182,039
CO_2 emitted (tonnes)			206.04	230.31	252.99	274.43	294.88	310.24
Trip time (days)			5,30	5,02	4,81	4,64	4,50	4,40

Adapted from Psaraftis and Kontovas (2014)

that weather routing problems are typically *path problems* defined as trying to optimize a ship's track from a specified origin to a specified destination, under a prescribed objective and under time varying and maybe also stochastic weather conditions. Decision variables include the selection of the ship's path and the speeds along the path, and typical objectives include minimum transit time and minimum fuel consumption. Several constraints such as time windows or constraints to accommodate a feasible envelope on ship motions, vertical and transverse accelerations and ship loads such as shear forces, bending moments and torsional moments can be introduced. The influence of currents, tides, winds and waves, which may be varying in both time and space, should be taken into account.

A variety of methods have been developed to solve the weather routing problem. Below is a non-exhaustive exposition (see Psaraftis et al. 2017 for more details).

The so-called isochrone method was originally proposed by James (1957) and was modified by Hagiwara (1989). Calculus of variations was originally proposed by Haltiner et al. (1962). Papadakis and Perakis (1990) developed the method further and could find the routes and the vessel's power setting. Perakis and Papadakis (1989) also extended the method to be valid in a time-dependent environment.

Bekker and Schmid (2006) investigated the use of Dijkstra's algorithm and developed a genetic algorithm to achieve practical strategies and a method in which the two optimization techniques interact to provide a safe route considering the risk of both the sea mine and the environment and making it applicable to sea mine avoidance. Padhy et al. (2008) also developed an application of Dijkstra's algorithm.

Azaron and Kianfar (2003) accommodated environmental variable at each node and arc logistics independent variables to find the shortest path from upstream to downstream by applying stochastic dynamic programming. Bauk and Kovac (2004) proposed a neural networks approach for determining the optimal route.

Elbeltagi et al. (2005) introduced evolutionary algorithms based on the natural biological evolution and the social behavior of species so as to arrive at near optimum solutions in large-scale optimization problems for which traditional techniques may fail; they also compared five recent evolutionary algorithms: the genetic algorithm, the memetic algorithm, particle swarm, ant colony system and frog leaping.

Kumar and Kumar (2010) implemented a genetic algorithm to find the set of optimal routes to send traffic from source to destination; he also discussed about static and dynamic routing.

Tsou and Hsueh (2010) achieved the objectives such as warning and pre-collision preparation by using the concept of e-navigation and path planning for positioning collision avoidance and applied an ant colony algorithm in the field of artificial intelligence for constructing a collision avoidance model that imitates optimization behavior in real-life application.

Ko (2009) focused the international multimodal transport to connect one or more adjoining countries for delivering cargo with assurance of delivery reliability while minimizing transit time and costs and also considered the economic standard that varies from country to country, which can affect the routes' performance, due to lack of interconnections, interchangeability and legal framework while delivering cargo.

Last but not the least, in Psaraftis et al. (2017), a proof-of-concept analysis was performed for a ship weather routing system using satellite altimetry to provide data on ocean currents, including an assessment of the potential benefits of such a system under several alternative scenarios. This was in the context of the BlueSIROS project funded by the European Space Agency (ESA).[2]

5 Impact of Freight Rates, Bunker Prices and Inventory Costs on Liner Ship Speeds and Frequencies

In Giovannini and Psaraftis (2018), a simple model was developed for a fixed route liner shipping scenario which, among other things, incorporates the influence of freight rates, along with that of fuel prices and cargo inventory costs into the overall speed optimization process. The objective to be maximized is the line's average daily profit. Departing from convention, the model was also able to consider *flexible service frequencies*, to be selected among a broader set than the standard assumption of one call per week. It was shown that this may lead to better solutions and that the cost of forcing a fixed frequency can be significant. Such cost is attributed either to additional fuel cost if the fleet is forced to sail faster to accommodate a frequency that is higher than the optimal one, or to lost income if the opposite is the case. The impact of the line's decisions on CO_2 emissions was also examined.

[2] https://business.esa.int/projects/blue-siros

The model assumes without loss of generality a fleet of N identical containerships deployed on a given fixed route. Inputs to the problem include:

- The route geometry, represented by a set of ports J and a set of legs I representing the route
- The length L_i of each leg i of the route
- The freight rate F_{zx} of transporting a TEU from a port z on the route to another port x on the route, for all relevant port pairs (in general $F_{xz} \neq F_{zx}$)
- The demand in TEUs c_{zx} from a port z on the route to another port x on the route, for all relevant port pairs (in general $c_{xz} \neq c_{zx}$)
- The bunker price P
- The daily operating costs E of each vessel, other than fuel
- The daily at-sea fuel consumption function $f(v)$ as a function of ship speed v
- The daily auxiliary engine at port fuel consumption A
- The average monetary value W_i of ship cargo on each leg i of the route
- The operator's annual cost of capital, R
- The time G_j spent at each port j
- The cargo handling cost H per TEU handled
- The total cargo C_i on the ship along leg i of the route
- The minimum and maximum allowable ship speeds, $v_{\min}$ and $v_{\max}$, respectively

The problem's main decision variables are:

- The ship speeds v_i along each leg i of the route, allowed to be different in each leg
- The service period t_0, which is the inverse of the service frequency and which is also allowed to vary
- The number of ships N deployed on the route

If we define $\alpha_i = \frac{R\ W_i}{365}$ (daily unit cargo inventory costs on leg i of the route), the problem formulation is as follows:

$$\dot{\pi}=\text{Max}_{v_i,t_0,N}\left\{\frac{1}{t_0}\left(\sum_x\sum_z F_{zx}\, c_{zx} - P\sum_i f\,(v_i)\,\frac{L_i}{24\, v_i}-\text{PA}\sum_j G_j-\sum_i \alpha_i\, C_i\,\frac{L_i}{24\, v_i}\right.\right.$$
$$\left.\left.-H\sum_j D_j\right) - N\,E\right\} \tag{10.1}$$

subject to the following constraints:

$$v_{\min} \leq v_i \leq v_{\max} \qquad i \in I \tag{10.2}$$

$$Nt_0 = \sum_i \frac{L_i}{24\, v_i} + \sum_i G_j \tag{10.3}$$

and

$$N \in \mathbb{N}^+ \tag{10.4}$$

Constraints (10.2) are the upper and lower bounds on ship speed for each leg of the route, constraint (10.3) links the three decision variables of the problem (number of ships, service period and ship speeds) together, and constraint (10.4) is the integrality constraint. Non-negativity constraints could also be added for t_0, but they are redundant because of (10.3).

In (10.1), a difference versus other formulations in which the objective function is defined on a *per route* basis is that the objective function is defined on a *per unit time* basis. In that sense, the objective function in (10.1) – to maximize operator's average daily profit – is the *maximization of a ratio*, that of total route profit divided by the total duration of the route. Both numerator and denominator of the ratio are *nonlinear* functions of speed, and of course so is the ratio itself. Constraint (10.3) is also nonlinear. Last but not the least, another difference from other models is that the service period t_0 is not fixed but flexible.

A *constrained* version of the above problem (and in fact this is a common assumption in most liner shipping formulations) is if one of the three decision variables, the service period t_0, is fixed, that is, it is considered an exogenous input and cannot vary freely. In fact for liner services, it is typically expected that the service period t_0 can take on only prespecified values, the most common of which is 7 days for a weekly service. In theory, other values of t_0 can also be considered (e.g. $t_0 = 14$ corresponds to a biweekly service, and $t_0 = 3.5$ is a service twice a week), but this is not very common. Almost unheard of is the case that t_0 may take on other values, such as 6, 8 and 9, or even fractional values. However, as liner services schedules are published in each carrier's web site and other media well in advance, there is really nothing fundamental that prevents a carrier from setting up a service with t_0 equal to any prescribed value, if these 'unconventional' service frequencies happen to achieve better results for the carrier. Whatever it is, it is obvious that the constrained version of the problem (t_0 fixed and equal to 7) will not achieve better results vis-à-vis the case in which t_0 is allowed to vary freely, or is restricted to a wider range or set of values. In that sense, a fixed t_0 will generally come at a price.

The nonlinear optimization problem as formulated above was solved by linearizing the objective function, coding the model in MATLAB and using an Excel spreadsheet solver.

The following three actual liner routes were examined (see also Table 10.3):

- *AE2* – North Europe and Asia: such service links Asia to North Europe and is provided by Maersk. The same service is also provided by MSC under the name SWAN. Indeed, both Maersk's ships and MSC's ships are deployed along this route.
- *TP1* – North America (West Coast) and Asia: the route connects Asia to the West Coast of North America. Maersk offers this service; however the same service is

Table 10.3 Ports in the routes under study

Ports					
AE2		TP1		NEUATL1	
Felixstowe	1	Vancouver	1	Antwerp	1
Antwerp	2	Seattle	2	Rotterdam	2
Wilhelmshaven	3	Yokohama	3	Bremerhaven	3
Bremerhaven	4	Busan	4	Norfolk	4
Rotterdam	5	Kaohsiung	5	Charleston	5
Colombo	6	Yantian	6	Miami	6
Singapore	7	Xiamen	7	Houston	7
Hong Kong	8	Shanghai	8	Norfolk	8
Yantian	9	Busan	9		
Xingang	10				
Qingdao	11				
Busan	12				
Shanghai	13				
Ningbo	14				
Yantian	15				
Tanjung Pelepas	16				
Algeciras	17				

Source: Giovannini and Psaraftis (2018)

also provided by MSC, and it is called EAGLE. As for the AE2 service, along the TP1 route are deployed Maersk's vessels as well as MSC's vessels.

- *NEUATL1* – North Europe and North America (East Coast): the NEUATL1 lane links North Europe to the US East Coast. The service is furnished by MSC or similarly by Maersk under the name TA1.

For these routes, three different cases were analyzed:

First case: the service frequency is constant, and the number of ships is variable. Therefore the main decision variables in such case are two, the speeds and the number of deployed vessels.

Second case: the number of ships is constant, and the frequency is variable, the service period t_0 being allowed to take on the following values: 3.5, 4, 5, 6, 7, 8, 9, 10, and 14 days. Hence the main decision variables are again two, the speeds and the service frequency.

Third case: both the frequency and the number of ships are variable, in which case the main decision variables are three. However, in this case the number of ships is bounded from above. This bound is imposed because otherwise the optimal number of ships may reach unrealistic values.

Hereby we show a sample of the results of the above cases. The number of ships concerning the 'base scenarios' is the actual number of ships employed on the route involved in the examined routes; these are ten ships for the AE2 route and five ships for each of the TP1 and NEUATL1 routes.

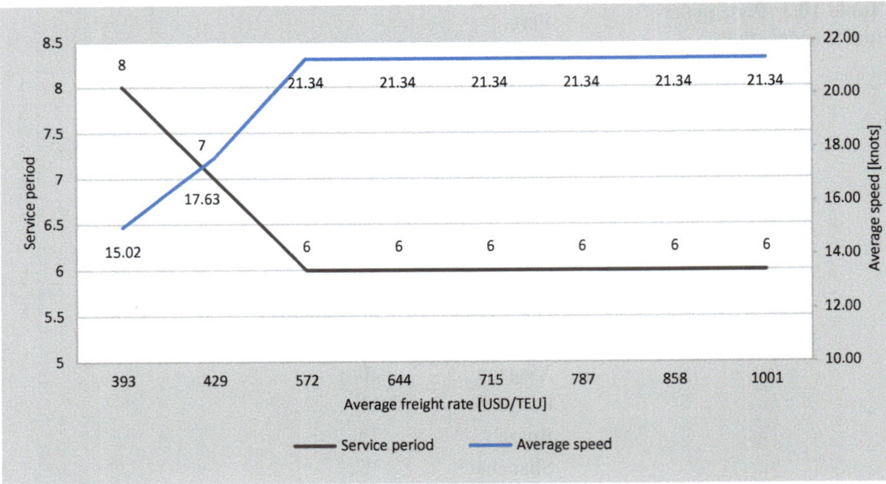

Fig. 10.4 Fixed number of ships scenario, optimal service period and optimal average speed at different average freight rates (route TP1). (Source: Giovannini and Psaraftis 2018)

Table 10.4 Cost of forcing $t_0 = 7$ days in a fixed number of ships scenario

Instance	Average freight rate (USD/TEU)	Optimal t_0 (days)	Δ (USD/day)
1	393	8	4132
2	429	7	0
3	572	6	15,717
4	644	6	35,029
5	715	6	54,341
6	787	6	73,653
7	858	6	92,965
8	1001	6	131,590

Source: Giovannini and Psaraftis (2018)
Speed if $t_0 = 7$ days is 17.63 knots

Figure 10.4 depicts a fixed number of ships' scenarios and shows service frequency's trend and the average speed's trend at eight different freight rate values for the route TP1. It can be seen that if the freight rate is low enough, a service period of 8 days is better, whereas for higher rates a service period of 6 days is better.

The above means that if we force $t_0 = 7$, the solution will be suboptimal in seven out of the eight instances. We can compute the fleet-level difference Δ in the objective function between the optimal solution and the solution in which t_0 is forced to be equal to 7. This is shown in Table 10.4 above.

At the low end of the freight rate spectrum (instance 1), the model chooses an 8-day service period as optimal and a low corresponding average speed, 15.02 knots. If one forces a higher frequency (and specifically a call every 7 days) and the number of ships is constant, this would only be achievable if the average speed increases to 17.63 knots. The higher frequency would increase the amount of cargo transported and associated revenue, but as the freight rate is low the additional revenue cannot match the increased cost due to the higher speed, hence daily profit for the fleet is lower by 4132 USD/day.

The situation at the high end of the freight rate spectrum (instance 8) is the opposite but its effect the same. At the last instance, the high average rate of 1001 USD/TEU suggests a 6-day service period as optimal and a high corresponding average speed, 21.32 knots. If one forces a lower frequency (a call every 7 days) and the number of ships is constant, this would only be achievable by a lower ship speed, again 17.63 knots. The lower frequency would decrease the cargo transported, but given the freight rate is high, the associated loss of revenue would be greater than the savings in fuel cost due to the lower speed and hence again a lower daily profit (in this instance lower by 131,590 USD/day for the entire fleet). The situation in instances three to seven is similar.

Figure 10.5 is an example of the bounded above number of ships' scenario and shows the bunker price effect on the average speed and the service period for route AE2. Bunker prices range from 146 to 583 USD/tonne.

This example confirms that, under certain circumstances, service periods different from 7 days may achieve better results for the operator and that speed generally is a nondecreasing function of the freight rate. Also, a higher bunker price makes the high service frequency disadvantageous since this would entail deploying more ships and increasing the average speed, hence a higher fuel expenditure (and more emissions).

As one can see in the objective function, expression (1), given a specific service period and a specific number of ships, the optimal sailing speeds along the legs v_i depend essentially on two factors, the bunker price and the cargo inventory costs.

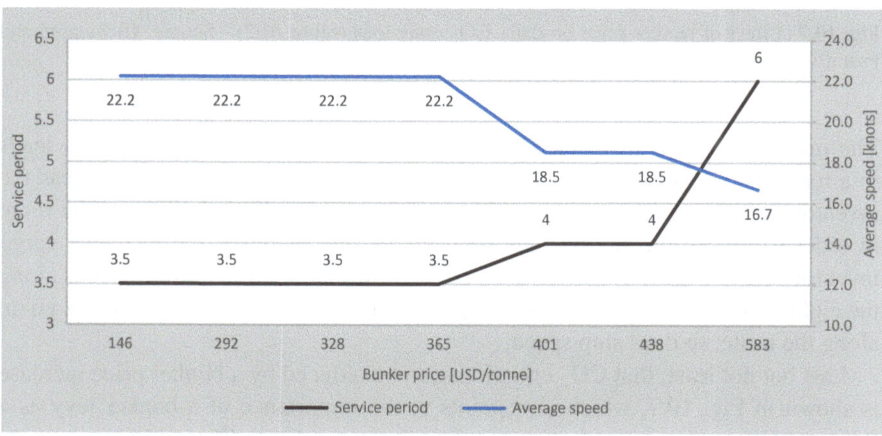

Fig. 10.5 Number of ships bounded above scenario, optimal service period and optimal average speed at different bunker prices (route AE2). (Source: Giovannini and Psaraftis 2018)

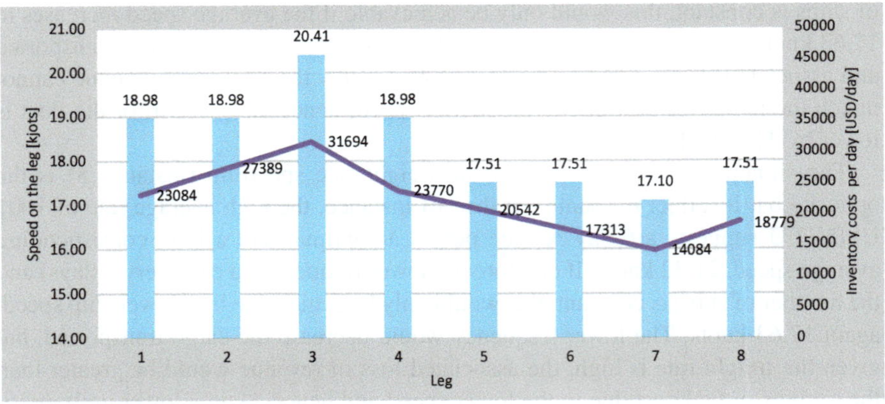

Fig. 10.6 Effect of inventory costs on the speeds along the legs (route NEUATL1). The figure refers to a base scenario in which $N = 5$ and $t_0 = 6$. (Source: Giovannini and Psaraftis 2018)

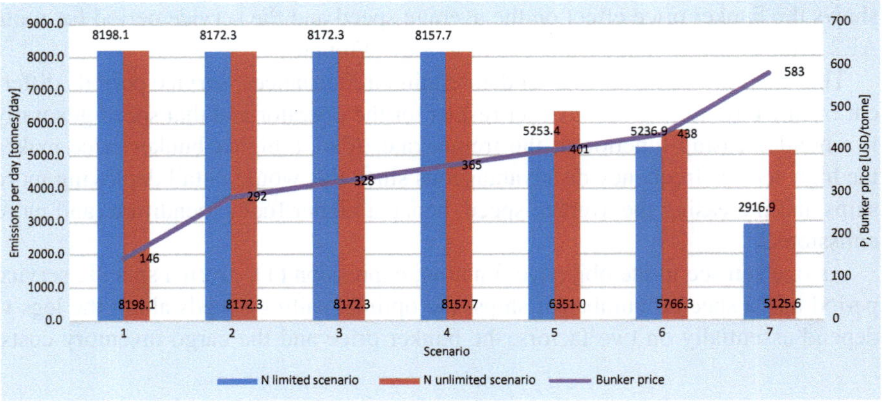

Fig. 10.7 Effect of bunker price on daily CO_2 emissions (route AE2). (Source: Giovannini and Psaraftis 2018)

The influence of these two factors is opposite: the fuel consumption factor leads to a reduction of the speeds v_i, so as to respect the service frequency, whereas the inventory cost factor leads to an increase of the speeds along the legs of the route in order to reduce the sailing time on each leg and therefore the in-transit cargo inventory costs. Figure 10.6 shows the effect of inventory costs on ship speeds along the eight legs of the NEUATL1 route. It is seen that whenever inventory costs go up along the route, so does ship speed.

Last but not least, that CO_2 emissions can be reduced by a bunker price increase is shown in Fig. 10.7, which also points to the importance of a bunker levy as a potential CO_2 emissions reduction measure (more on this in Chap. 11).

More scenarios and results of this analysis can be found in Giovannini and Psaraftis (2018).

6 The Speed Limit Debate

At the latest IMO/MEPC 72 landmark decision to aim for at least 50% GHG reductions by 2050 (IMO 2018), some countries in South America (and most notably Chile and Peru) objected to the use of the term 'speed reduction' as a possible emissions reduction measure, on the ground that this may constitute a barrier to their exports to Asia (and particularly to those that involve perishable products such as agricultural products and others). They suggested the use of 'speed optimization' instead. In a compromise solution, both wordings were included in the IMO decision text. However, what is meant by 'speed optimization' in that text is far from clear and hence is subject to different interpretations.

Irrespective of such legitimate wishes of South American and possibly other countries that have similar concerns, speed directional imbalances have been manifested in several trades worldwide and have been reported in several publications (see, e.g. Cariou (2011), Cheaitou and Cariou (2012) and FMC (2012), among others). Ongoing research by this author and his colleagues in the context of the ShipCLEAN project[3] has actually confirmed that liner cargo from South America to Asia moves at a much slower average speed than cargo in the opposite direction and that at current market conditions (spring and summer 2018), slow steaming was being practiced (see also Vilas (2018)). This difference in average speeds is also manifested in the trades between Asia and Europe, with cargoes from Asia to Europe moving faster than cargoes going in the opposite direction. This imbalance is surely due to commercial considerations that take into account, among other things, the difference in the values of the cargo between the two trade directions and the implied difference in in-transit cargo inventory costs. The optimization model presented in Sect. 5 of this chapter is seen to be able to capture such differences, among other things, and reflect them in the speed profile along the route's legs.

As ship speeds are the decisions of the carriers and not of the shippers, it is not immediately clear what can be done by South American countries who want their cherries and other agricultural or perishable products to China shipped faster. Our understanding is that these countries are mostly concerned by the conceivable imposition of *speed limits*, which is one of the most controversial (in our opinion) measures that are currently on the table as potential measures to reduce GHG emissions.

In fact, a recurrent measure that has been and is being promoted by various nongovernmental organizations (NGOs) is mandating *direct speed limits*. Since GHG emissions can be reduced by reducing speed, can someone achieve the same desirable outcome by imposing speed limits? This is an argument that is being heard frequently over the last several years. Among various lobbying groups, the Clean Shipping Coalition (CSC), an NGO, advocated at IMO/MEPC 61 that "speed reduction should be pursued as a regulatory option in its own right and not only

[3]https://www.chalmers.se/en/projects/Pages/ShipCLEAN%2D%2D-Energy-efficient-marine-transport-through_1.aspx

as possible consequences of market-based instruments or the EEDI." However, that proposal was rejected by the IMO at the time. In spite of this decision, lobbying for speed limits has continued by CSC and other groups, and speed limits have been discussed at IMO/MEPC 72 and have succeeded in being included in the roster of potential *short term* measures toward the 50% GHG emissions reduction target (IMO 2018).

Noting that the speed limit debate is still ongoing, our own position on this issue is that such speed limits are not a good idea. In addition to difficulties in enforcing such a rule, or in deciding what should be the speed limit as a function of ship type and size, it is clear that slow steaming and speed limits are two different things: the first is *a voluntary response* to market conditions which can dynamically change, and the second is *a mandated measure*. If the speed limit is above the optimal speed that is voluntarily chosen, then it is superfluous. This may conceivably alleviate the concerns of Chile and Peru, as no reasonable speed limit is likely to be below actual ship speeds to Asia under the current market circumstances. The question is what happens if or when the speed limit is below the optimal ship speed, as is likely to happen in a boom period. If this happens, a speed limit may cause distortions in the market and costs that may exceed the benefits of speed reduction. A likely short-term effect would be an increase in freight rates due to the contraction of the fleet's annual tonne-km supply curve. This may conceivably render the measure agreeable to some ship owners; however shippers would be hit twice: they would pay more for their cargo and also suffer increased transit times and increased in-transit inventory costs.

We have seen no comprehensive analysis of the possible market distortions of a speed limit. A discussion of some of the issues is in Devanney (2011b). Also we note that Cariou and Cheaitou (2012) investigated policy options contemplated by the European Commission and compared speed limits versus a bunker levy as two measures to abate GHGs, with a scenario from the container trades. They concluded that the former measure is counterproductive because it may ultimately generate more emissions and incur a cost per tonne of CO_2 which is more than society is willing to pay and because it is suboptimal as compared to results obtained if an international bunker levy were to be implemented.

A bunker levy belongs to the class of market-based measures (MBMs) for reducing GHG emissions which are examined in Chap. 11 of this book. It is interesting to note that for any given bunker levy that achieves a specific GHG emissions reduction in a given ship route scenario, *an equivalent speed limit* can be calculated that achieves exactly the same GHG emissions reduction. However, other attributes of the solution are different. A rudimentary scenario that is described in Appendix A of this chapter compares the two options and provides an indication that a bunker levy is preferable to a speed limit.

As this chapter was being finalized, the speed limit option was among the set of short-term options being considered by the IMO/MEPC in the quest to reduce maritime GHG emissions, and the fate of this option remained by and large unknown. It was also being considered by the European Commission, among other possible measures to reduce maritime GHG emissions, and again the fate of this option remained unknown.

7 Conclusions

This chapter has examined speed optimization in maritime transportation from various angles, including its interaction with route optimization. It has confirmed, among other things, that solutions for optimal environmental performance are not necessarily the same as those for optimal economic performance and that a win-win scenario is not necessarily obvious. Also policies that may seem at first glance optimal from an environmental viewpoint may actually be suboptimal. As a private operator would most certainly choose optimal economic performance as a criterion, if policy-makers want to influence the operator in his decision so as to achieve results that are good from a societal point of view, they could either play with parameters that would internalize the external costs of emissions produced and move the solution closer to what is deemed more appropriate for the environment and for the benefit of society or introduce constraints that would aim to achieve a similar outcome.

Acknowledgments Work reported in this chapter was funded in part by various sources. Early work was supported in part by the Lloyd's Register Foundation (LRF) in the context of the Centre of Excellence in Ship Total Energy-Emissions-Economy at the National Technical University of Athens (NTUA), the author's former affiliation. Later sources include an internal grant by the President of the Technical University of Denmark (DTU) and an internal grant at the DTU Department of Management Engineering, Management Science Division; the BlueSIROS project at DTU, funded by the European Space Agency (DTU Space leader); and the ShipCLEAN project at DTU, funded by the Swedish Energy Agency (Chalmers University project leader). Three recent DTU MSc theses, by Juan Morales, Massimo Giovannini and Fabio Vilas, have also contributed to the chapter (in Sects. 4.2, 5, and 6, respectively).

Appendix A

Comparison Between a Bunker Levy and Speed Limits

With all the discussion on speed limits at the IMO, the purpose of this appendix is to investigate the issue of a bunker levy vs speed limits. Both measures would cause speed reduction and hence a reduction in CO_2 and other emissions (GHG and non-GHG). A bunker levy would induce speed reduction, and a speed limit would mandate it. Below we attempt to compare the two measures, in terms of emissions reduction and other attributes.

To investigate the issue, we use a rudimentary scenario in the container sector. A generalization to more realistic scenarios or other shipping markets is straightforward. The example is taken from the 'cart before the horse' paper (Psaraftis 2017).

Table 10.5 Assumed inputs

Input	Value
Q	10,000 TEU
L	20,000 nm
R (base case)	1500 USD/TEU
u	0.6
p	500 USD/tonne
v_{min}	16 knots
v_{max}	26 knots
X	15,000 USD/day

A containership of capacity Q (TEU) shuttles between port A and port B, whose interport distance is L (nautical miles, nm). The ship's speed is v (nm/day)[4] which is within the bracket v_{min} and v_{max}.

Assume that the ship is semi-full in both directions and that the freight rate received by the ship owner is R (USD/TEU), assumed the same in both directions. The assumed load factor of the ship is u ($0 \leq u \leq 1$), again assumed the same in both directions. R is assumed to be on a *per loaded TEU* basis, meaning that if the ship is 75% full ($u = 0.75$), its per (one way) trip income will be $0.75RQ$. Assume that the fuel price is p (USD/tonne) and that the fuel consumption function is $FC = kv^3$ (tonnes/day) with k being a constant. Assume finally that miscellaneous/other operating expenses are X (USD/day) and that port turnaround times are ignored. Q, L, R, u, p, k, v_{min}, v_{max} and X are assumed known inputs (see Table 10.5 above for an example), and the sole decision variable is the ship's speed v.

We also note that this analysis assumes that R is an exogenous variable outside the line's control, and we do not attempt to estimate R as a function of container capacity supply and demand. In that sense, it is expected that slow steaming or speed reduction, if applied for all ships sailing the given route, will generally increase R; however this is not captured in our model.

Finally k is such that $FC = 144$ tonnes/day when $v = 22$ knots. The value of k for which this is the case is $9.7827*10^{-7}$ (again, v in the formulas is in nm/day).

In this scenario, we can compute various attributes of the round trip, such as:

Round trip time $T = 2\,L/v$ (days)
Round trip TEU throughput $H = 2uQ$ (TEU)
Round trip cost $C = T(pkv^3+X) = 2(pkLv^2+LX/v)$ (USD)
Round trip income $I = 2uRQ$ (USD)
Round trip profit $P = I - C = 2(uRQ - pkLv^2 - LX/v)$ (USD)
Average per day profit $P' = P/T = uRQv/L - pkv^3 - X$ (USD/day)
Average per day TEU throughput $H' = 2uQ/T = uQv/L$ (TEU/day)

[4]This is 24 times the speed in knots. The reason we use nm/day instead of knots *in the formulas* is to avoid having the number 24 in the equations. However, *in the tables and results*, knots will be used.

Table 10.6 Optimal speed as a function of freight rate R, individual ship

R (USD/TEU)	500	1000	1500 (base case)	1800	2000
v_{opt} (knots)	16.00	18.84	23.07	25.28	26.00
T (days)	104.17	88.47	73.23	65.94	64.10
H (TEU)	12,000	12,000	12,000	12,000	12,000
C (USD)	4,447,589	5,326,987	7,083,480	8,189,078	8,579,871
I (USD)	6,000,000	12,000,000	18,000,000	21,600,000	24,000,000
P (USD)	1,552,451	6,673,013	10,916,520	13,410,922	15,420,129
P' (USD/day)	14,904	75,430	151,131	203,385	240,554
H' (TEU/day)	115.20	135.65	166.13	181.99	187.20
CO_2 (tonnes/day)	172.27	281.24	516.67	679.18	739.22

If the objective of the line is to maximize *average per day profit*, that is, P', the optimal speed can be shown to be as follows.

$$v_{opt} = v_{min} \text{ if } v_{min} > v_0$$

$$v_{opt} = v_0 \text{ if } v_{min} \le v_0 \le v_{max}$$

$$v_{opt} = v_{max} \text{ if } v_{max} < v_0$$

with $v_0 = (uRQ/3pkL)^{1/2}$

Then CO_2 emissions per unit time (tonnes/day) for this ship are equal to

$$CO_2 = fkv_{opt}{}^3$$

with f being the carbon coefficient (assumed here equal to 3.11).

For an individual ship, Table 10.6 above shows the optimal speed and other solution attributes for the above inputs and for selected values of the freight rate R ranging between 500 USD/TEU to 2000 USD/TEU, with a base case value of 1500 USD/tonne.

One can see in general that a higher state of the market (higher R) induces a higher speed and hence higher CO_2 emissions for the ship, and vice versa. It should also be noted that in this particular example and for the two extreme cases $R = 500$ and 2000 USD/TEU, the optimal speed hits the speed's lower and upper bounds, respectively.

To lower CO_2 emissions, one contemplates either a levy q on fuel or a speed limit equal to V, with q and V being user inputs. Either of those would generally result in a lower speed. The question is: Which of these alternatives achieves lower CO_2 emissions? The answer of course depends on the values of q and V. Depending on

Table 10.7 Comparison between the speed limit and levy cases, individual ship

	Speed limit case	Levy case
v_0	$(uRQ/3pkL)^{1/2}$	$(uRQ/3(p+q)kL)^{1/2}$
v_{opt}	$v_{opt} = v_{min}$ if $v_{min} > v_0$	$v_{opt} = v_{min}$ if $v_{min} > v_0$
	$v_{opt} = v_0$ if $v_{min} \leq v_0 \leq V$	$v_{opt} = v_0$ if $v_{min} \leq v_0 \leq v_{max}$
	$v_{opt} = V$ if $V < v_0$	$v_{opt} = v_{max}$ if $v_{max} < v_0$
P'	$uRQv_{opt}/L - pkv_{opt}{}^3 - X$	$uRQv_{opt}/L - (p+q)kv_{opt}{}^3 - X$
CO_2	$fkv_{opt}{}^3$	$fkv_{opt}{}^3$

Table 10.8 Reductions of CO_2 and other attributes as a function of the speed limit V, constant throughput

V (knots)	18.00	20.00	22.00
v_{opt} (knots)	18.00	20.00	22.00
T (days)	92.59	83.33	75.76
C (USD)	5,040,279	5,757,889	6,590,909
I (USD)	18,000,000	18,000,000	18,000,000
P (USD)	12,959,721	12,242,111	11,409,091
r	1.28	1.15	1.05
P' (USD/day)	179,418	169,483	157,951
CO_2 (tonnes/day)	314.43	388.18	469.70
ΔCO_2 (tonnes/day)	202.24	128.49	46.97

these values, a levy can achieve lower, the same or higher CO_2 emissions reductions vis-à-vis those achieved by a speed limit.

Note also that for this comparison to make sense, constant average per day TEU throughput should be maintained, even though speed is reduced. This would necessitate deploying additional ships.

If the initial speed before the levy or the speed limit is v_1 and the final speed after the levy or the speed limit is v_2 ($<v_1$), we define as the 'throughput factor' the ratio $r = v_1/v_2$ (>1). A ratio $r = 1.20$ means that r-1 (in this case 20%) more ships should be deployed on the route so as to maintain the same average per day TEU throughput. These additional ships would generate additional profit and additional CO_2, both of which should be taken into account. To do so, the average per day profit and the average per day CO_2 emissions should be multiplied by r, vis-à -vis those for an individual ship.

To further investigate the issue, we assume that $v_{min} \leq V \leq v_{max}$ because if V is outside that range, then either the speed limit is superfluous ($V > v_{max}$) or the problem is infeasible ($V < v_{min}$).

The two cases are compared in Table 10.7 above as follows.

The superfluous speed limit case occurs if $V \geq (uRQ/3pkL)^{1/2}$, which for our case and for the base case for R means $V \geq 23.07$ knots. If this is the case, v_{opt} is also 23.07 knots.

The non-superfluous (binding) speed limit case occurs if $V < (uRQ/3pkL)^{1/2} =$ 23.07 knots.

Table 10.8 shows the results for the base case R and for selected values of the speed limit V ranging from 18 to 22 knots.

Table 10.9 Reductions of CO_2 and other attributes as a function of the levy q, constant throughput

q (USD/tonne)	100	300	500
v_{opt} (knots)	21.06	18.24	16.32
T (days)	79.13	91.37	102.15
C (USD/rtrip)	7,186,893	7,370,506	7,532,272
I (USD/rtrip)	18,000,000	18,000,000	18,000,000
P (USD/rtrip)	10,813,107	10,629,494	10,467,728
r	1.10	1.27	1.41
P' (USD/day)	149,723	147,169	144,880
CO_2 (tonnes/day)	430.62	322.94	258.27
ΔCO_2 (tonnes/day)	86.05	193.73	258.40

The last row in the table shows the reductions of CO_2 (in tonnes/day) that can be achieved as a function of the speed limit, vis-à-vis the 'no speed limit' case (516.67 tonnes/day). Note that the figures for P', CO_2 and ΔCO_2 have factored in the effect of the throughput factor r.

In turn, we can investigate what happens if we impose a levy q on bunker fuel. Table 10.9 shows these results (again base case for R) for selected values of q ranging from 100 to 500 USD/tonne.

Again, the last row in the table shows the reductions of CO_2 (tonnes/day) that can be achieved as a function of the levy, vis-à-vis the 'no levy' case (516.67 tonnes/day). As before, the figures for P', CO_2 and ΔCO_2 have factored in the effect of the throughput factor r.

Tables 10.8 and 10.9 are not directly comparable, in the sense that from these tables no direct conclusions can be drawn as to what is preferrable, a speed limit or a levy. To draw such conclusions, we ask the following question: For a given levy q, what is the value of the speed limit V so that the results are *the same* in terms of CO_2? And once this happens, what are the other differences between the two cases?

It turns out that the speed limit V for which the optimal speed is the same as that with a levy q is as follows.

$$V = (uRQ/3\,(p+q)\,kL)^{1/2}$$

Then the optimal speed is equal to V in both cases.

In this case, and for an individual ship, daily CO_2 is also the same and equal to $fkV^3 = 2fk(uRQ/3(p+q)kL)^{3/2}$

However, daily profit P' is different. With a levy q, it is $P^{'} = uRQV/L - (p+q)kV^3 - X$.

With an equivalent speed limit V, and no levy, it is $P'' = uRQV/L - pkV^3 - X\ (>P^{'})$.

The difference in daily profit is $\Delta P^{'} = qkV^3$.

The above are for an individual ship. To maintain the same TEU throughput, the effect of the throughput factor r has also to be taken into account.

This means that for the ship owner, and if the same speed (and hence the same CO_2 emissions) reduction are to be achieved, a speed limit is more profitable than a

Table 10.10 Equivalent speed limit V, CO_2 and $\Delta P'$ as functions of levy q, constant throughput

q (USD/tonne)	100	300	500
V (knots)	21.06	18.24	16.32
r	1.10	1.27	1.41
CO_2 (tonnes/day)	430.62	322.94	258.27
$\Delta P'$ (USD/day)	13,902	31,275	41,409

bunker levy. The ship owner will sail the ship at the same speed as that with a levy, but without paying the levy.

Table 10.10 shows the values of V, CO_2 and ΔP' for values of q between 100 and 500 USD/tonne.

This cuts both ways. The difference in daily profit $\Delta P'$, which is positive for the ship owner and which possibly reflects an external cost of CO_2 pollution that is not internalized, is a net cost to society. It is money not collected which could be used to achieve out-of-sector emissions reductions[5] or for other noble causes (e.g. financial aid to developing countries, research and development, etc.). In that sense, and from a societal point of view, a levy is better than an equivalent speed limit.

An equally serious problem with a speed limit is that for ships of different size, a common and uniform levy q will result in different optimal speeds. A larger ship would in general imply a higher optimal speed, everything else being equal. Therefore, achieving equivalence such as the above by a common and uniform speed limit V will be impossible. To do so, one would have to set *size-specific* (or maybe even *ship type-specific* or *route-specific*) speed limits, which will make the whole exercise an administrative nightmare.

Conversely, if a common and uniform speed limit V is imposed, the limit may be superfluous for some ship sizes and binding for some others, depending on the state of the market, the price of fuel and a host of other parameters. Having the same speed limit in boom market periods and in depressed market periods could create all sorts of distortions. In depressed market periods, the limit may be superfluous, and in boom market periods, the limit would force some ships (likely at the high end of the scale) to slow down, whereas others do not. A speed limit may also be superfluous in one route direction (e.g. from Europe to the Far East, where ships go slower anyway) and binding in the other direction (ships go faster from the Far East to Europe).

Last but not least, a speed limit would be difficult or impossible to enforce, even if it is the same for all ship sizes or types, and it would hardly serve as an incentive to economize and improve the energy efficiency of ships.

[5]Out-of-sector emissions reductions (or offsets) are emissions reductions that can be realized by investing the monies that are collected by a bunker levy into emissions reduction projects outside the maritime sector, for instance, by developing a wind farm in New Zealand or a solar farm in Indonesia. See Chap. 11 for more details.

For at least the above reasons, a conjecture that we can safely make is that a bunker levy is a preferable instrument (as compared to a speed limit) if one wants to reduce maritime emissions.

References

Agarwal, R., & Ergun, Ö. (2008). Ship scheduling and network design for cargo routing in liner shipping. *Transportation Science, 42*(2), 175–196.

Alphaliner. (2013). Extra and super slow steaming help absorb 7.4% of fleet. *Alphaliner Weekly Newsletter, 2013*(44), October 2013.

Alvarez, J. F., Tsilingiris, P., Engebrethsen, E. S., & Kakalis, N. M. (2011). Robust fleet sizing and deployment for industrial and independent bulk ocean shipping companies. *INFOR: Information Systems and Operational Research, 49*(2), 93–107.

Andersson, H., Duesund, J. M., & Fagerholt, K. (2011). Ship routing and scheduling with cargo coupling and synchronization constraints. *Computers & Industrial Engineering, 61*(4), 1107–1116.

Andersson, H., Fagerholt, K., & Hobbesland, K. (2014). Integrated maritime fleet deployment and speed optimization: Case study from RoRo shipping. *Computers & Operations Research* 2015, *55*, 233–240.

Azaron, A., & Kianfar, F. (2003). Dynamic shortest path in stochastic dynamic networks: Ship routing problem[J]. *European Journal of Operational Research, 144*(1), 138–156.

Barrass, C. B. (2005). *Ship design and performance for masters and mates*. UK: Butterworth-Heinemann.

Bauk, S., & Kovac, N. (2004). Modeling ship's route by the adaptation of Hopfield-Tank TSP neural algorithm[J]. *Journal of Maritime Research, 1*(3), 45–64.

Bekker, J. F., & Schmid, J. P. (2006). Planning the safe transit of a ship through a mapped minefield[J]. *ORiON, 22*(1), 1–18.

Brouer, B. D., Alvarez, J. F., Plum, C. E., Pisinger, D., & Sigurd, M. M. (2013). A base integer programming model and benchmark suite for liner-shipping network design. *Transportation Science, 48*(2), 281–312.

Cariou, P. (2011). Is slow steaming a sustainable means of reducing CO_2 emissions from container shipping? *Transportation Research Part D, 16*(3), 260–264.

Cariou, P., & Cheaitou, A. (2012). The effectiveness of a European speed limit versus an international bunker-levy to reduce CO_2 emissions from container shipping. *Transportation Research Part D, 17*, 116–123.

CBO. (2006). *The economic costs of disruptions in container shipments*. Washington, DC: U.S. Congress, Congressional Budget Office.

Chatzinikolaou, S.D., Ventikos, N. P., (2016), Critical analysis of air emissions from ships: Lifecycle thinking and results, in Psaraftis, H,N. (ed.), *Green Transportation Logistics: in Search for Win-Win Solutions*, Springer.

Cheaitou, A., & Cariou, P. (2012). Liner shipping service optimisation with reefer containers capacity: An application to northern Europe–South America trade. *Maritime Policy & Management, 39*(6), 589–602.

Christiansen, M., Fagerholt, K., Nygreen, B., & Ronen, D. (2013). Ship routing and scheduling in the new millennium. *European Journal of Operational Research, 228*(2013), 467–483.

Cordeau, J.-F., Laporte, G., Legato, P., & Moccia, L. (2005). Models and Tabu search heuristics for the berth-allocation problem. *Transportation Science, 39*(4), 526–538.

Devanney, J. W. (2007). Solving elastic transportation networks. Center for tankship excellence [online]. Available at: www.c4tx.org.

Devanney, J. W. (2010). The impact of bunker price on VLCC spot rates. Proceedings of the 3rd International Symposium on Ship Operations, Management and Economics. SNAME Greek Section, Athens, October.

Devanney, J. W. (2011a). The impact of charter party speeds on CO_2 emissions. Center for tankship excellence [online]. Available at: www.c4tx.org.

Devanney, J. W. (2011b). Speed limits versus slow steaming, center for tankship excellence [online]. Available at: www.c4tx.org.

Du, Y., Chen, Q., Quan, X., Long, L., & Fung, R. Y. K. (2011). Berth allocation considering fuel consumption and vessel emissions. *Transportation Research Part E, 47*, 1021–1037.

Elbeltagi, E., Hegazy, T., & Grierson, D. (2005). Comparison among five evolutionary-based optimization algorithms[J]. *Advanced Engineering Informatics, 19*(1), 43–53.

Fagerholt, K., & Psaraftis, H. N. (2015). On two speed optimization problems for ships that sail in and out of emission control areas. *Transportation Research Part D, 39*, 56–64 2015.

Fagerholt, K., & Ronen, D. (2013). Bulk ship routing and scheduling: Solving practical problems may provide better results. *Maritime Policy and Management, 40*(1), 48–64.

Fagerholt, K., Laporte, G., & Norstad, I. (2010). Reducing fuel emissions by optimizing speed on shipping routes. *Journal of the Operational Research Society, 61*, 523–529.

Fagerholt, K., Gausel, N., Rakke, J., & Psaraftis, H. (2015). Maritime routing and speed optimization with emission control areas. *Transportation Research Part C, 52*, 57–63.

FMC. (2012). *Study of the 2008 repeal of the liner conference exemption from European Union Competition Law, Bureau of trade analysis*. Washington, DC: Federal Maritime Commission.

Giovannini, M., & Psaraftis, H. N. (2018). The profit maximizing liner shipping problem with flexible frequencies: Logistical and environmental considerations. *Flexible Services and Manufacturing Journal*. https://doi.org/10.1007/s10696-018-9308-z.

Gkonis, K. G., & Psaraftis, H. N. (2012). "Modelling tankers' optimal speed and emissions," Archival Paper, 2012 SNAME Transactions, Vol. 120, 90–115, (Annual Meeting of the Society of Naval Architects and Marine Engineers, Providence, RI, USA, Oct 2012).

Golias, M., Boile, M., Theofanis, S., & Efstathiou, C. (2010). The berth-scheduling problem: Maximizing berth productivity and minimizing fuel consumption and emissions production. *Transportation Research Record: Journal of the Transportation Research Board, 2166*, 20–27.

Goodchild, A. V., & Daganzo, C. F. (2007). Crane double cycling in container ports: planning methods and evaluation. *Transportation Research Part B: Methodological, 41*(8), 875–891.

Grønhaug, R., Christiansen, M., Desaulniers, G., & Desrosiers, J. (2010). A branch-and-price method for a liquefied natural gas inventory routing problem. *Transportation Science, 44*(3), 400–415.

Hagiwara, H. (1989). Weather routing of(sail-assisted) motor vessels[D]. Technische Universiteit Delft.

Haltiner G J, Hamilton H D, 'Arnason G. Minimal-time ship routing[J]. Journal of Applied Meteorology, 1962, 1(1): 1–7.

Halvorsen-Weare, E. E., & Fagerholt, K. (2011). Robust supply vessel planning. In *Network optimization* (pp. 559–573). Berlin/Heidelberg: Springer.

Halvorsen-Weare, E. E., & Fagerholt, K. (2013). Routing and scheduling in a liquefied natural gas shipping problem with inventory and berth constraints. *Annals of Operations Research, 203*(1), 167–186.

Hsu, C. I., & Hsieh, Y. P. (2005). Direct versus terminal routing on a maritime hub-and-spoke container network. *Journal of Marine Science and Technology, 13*(3), 209–217.

Hvattum, L. M., Norstad, I., Fagerholt, K., & Laporte, G. (2013). Analysis of an exact algorithm for the vessel speed optimization problem. *Networks, 62*(2), 132–135.

Hwang, H.-S., Visoldilokpun, S., & Rosenberger, J. M. (2008). A branch-and-price-and-cut method for ship scheduling with limited risk. *Transportation Science, 42*(3), 336–351.

IMO. (2009). Second IMO GHG study. Co authored by Buhaug, Ø., Corbett, J. J., Endresen, Ø., Eyring, V., Faber, J., Hanayama, S., et al. IMO document MEPC59/INF. 10.

IMO. (2014). Third IMO GHG study 2014, Co-authored by Smith, T. W. P., Jalkanen, J. P., Anderson, B. A., Corbett, J. J., Faber, J., Hanayama, S., O'Keeffe, E., Parker, S., Johansson, L.,

Aldous, L., Raucci, C., Traut, M., Ettinger, S., Nelissen, D., Lee, D. S., Ng, S., Agrawal, A., Winebrake, J., Hoen, M., Chesworth, S., Pandey, A., International Maritime Organization (IMO) London, UK, June.

IMO. (2018). Resolution MEPC.304(72) (adopted on 13 April 2018), Initial IMO strategy on reduction of GHG emissions from ships, IMO doc MEPC 72/17/Add.1, Annex 11.

James, R. W. (1957). *Application of wave forecasts to marine navigation. U.S.* Washington, DC: Naval Oceanographic Office.

Kapetanis, G. N., Gkonis, K., & Psaraftis, H. N. (2014). Estimating the operational effects of a bunker levy: The case of handymax bulk carriers," TRA 2014 conference, Paris, France, April 2014.

Ko, H. J. (2009). A DSS approach with Fuzzy AHP to facilitate international multimodal transportation network. *KMI International Journal of Maritime Affairs and Fisheries, 1*(1), 51–70.

Kumar, R., & Kumar, M. (2010). Exploring genetic algorithm for shortest path optimization in data networks. *Global Journal of Computer Science and Technology, 10*(11), 8.

Lin, D.-Y., & Liu, H.-Y. (2011). Combined ship allocation, routing and freight assignment in tramp shipping. *Transportation Research Part E: Logistics and Transportation Review, 47*(4), 414–431.

Lloyds List. (2018). MSC boxships slow down as chief executive Diego Aponte overhauls network, 13 June.

Lo, H. K., & McCord, M. R. (1998). Adaptive ship routing through stochastic ocean currents: General formulations and empirical results. *Transportation Research Part A: Policy and Practice, 32*(7), 547–561.

Maersk. (2013). Building the world's Biggest Ship. Available online at: http://www.maersk.com/innovation/leadingthroughinnovation/pages/buildingtheworldsbiggestship.aspx.

Magirou, E. F., Psaraftis, H. N., & Bouritas, T. (2015). The economic speed of an oceangoing vessel in a dynamic setting. *Transportation Research Part B, 76*, 48–67.

Meng, Q., & Wang, S. (2011). Optimal operating strategy for a long-haul liner service route. *European Journal of Operational Research, 215*, 105–114.

Mills, J., Donnison, A., & Brightwell, G. (2014). Factors affecting microbial spoilage and shelf-life of chilled vacuum-packed lamb transported to distant markets: A review. *Meat Science, 98*(1), 71–80.

Moccia, L., Cordeau, J. F., Gaudioso, M., & Laporte, G. (2006). A branch-and-cut algorithm for the quay crane scheduling problem in a container terminal. *Naval Research Logistics (NRL), 53*(1), 45–59.

Norstad, I., Fagerholt, K., & Laporte, G. (2011). Tramp ship routing and scheduling with speed optimization. *Transportation Research Part C, 19*, 853–865.

Norlund, E. K., & Gribkovskaia, I. (2013). Reducing emissions through speed optimization in supply vessel operations. *Transportation Research Part D, 23*, 105–113.

Padhy, C. P., Sen, D., & Bhaskaran, P. K. (2008). Application of wave model for weather routing of ships in the North Indian Ocean[J]. *Natural Hazards, 44*(3), 373–385.

Panagakos, G., Stamatopoulou, I. V., & Psaraftis, H. P. (2014). The possible designation of the Mediterranean as a SECA: A case study. *Transportation Research Part D, 28*, 74–90.

Papadakis, N. A., & Perakis, A. N. (1990). Deterministic minimal time vessel routing[J]. *Operations Research, 38*(3), 426–438.

Psaraftis, H. N., & Kontovas, C. A. (2016). Green maritime transportation: Speed and route optimization. In H. N. Psaraftis (Ed.), *Green transportation logistics: In search for win-win solutions*. Cham: Springer.

Perakis, A. N., & Papadakis, N. A. (1989). Minimal time vessel routing in a time-dependent environment. *Transportation Science, 23*(4), 266–276.

Powell, B. J., & Perakis, A. N. (1997). Fleet deployment optimization for liner shipping: An integer programming model. *Maritime Policy and Management, 24*(2), 183–192.

Psaraftis, H. N. (2017). Ship routing and scheduling: the cart before the horse conjecture. *Maritime Economics and Logistics, 17*(2), 1–14.

Psaraftis, H. N., & Kontovas, C. A. (2009a). CO_2 emissions statistics for the world commercial fleet. *WMU Journal of Maritime Affairs, 8*(1), 1–25.
Psaraftis, H. N., & Kontovas, C. A. (2009b). Ship emissions: Logistics and other tradeoffs. Proceedings of10th International Marine Design Conference. Trondheim, Norway, 26–29 May.
Psaraftis, H. N., & Kontovas, C. A. (2010). Balancing the economic and environmental performance of maritime transportation. *Transportation Research Part D, 15*(8), 458–462.
Psaraftis, H. N., & Kontovas, C. A. (2013). Speed models for energy-efficient maritime transportation: A taxonomy and survey. *Transportation Research Part C: Emerging Technologies, 26*, 331–351.
Psaraftis, H. N., & Kontovas, C. A. (2014). Ship speed optimization: Concepts, models and combined speed-routing scenarios. *Transportation Research Part C: Emerging Technologies, 44*, 52–69.
Psaraftis, H. N., & Kontovas, C. A. (2015). Slow steaming in maritime transportation: Fundamentals, trade-offs, and decision models. In C.-Y. Lee & Q. Meng (Eds.), *Handbook of ocean container transportation logistics: Making global supply chains effective*. Cham: Springer.
Psaraftis, H. N., Morales Llamas, J., Ding, L., Nehammer, J. (2017). BlueSIROS project WP3, proof of concept. BlueSIROS project technical report, Technical University of Denmark.
Rana, K., & Vickson, R. G. (1991). Routing container ships using lagrangean relaxation and decomposition. *Transportation Science, 25*(3), 201–214.
Reinhardt, L. B., & Pisinger, D. (2014). A branch and cut algorithm for the container shipping network design problem. *Flexible Services and Manufacturing Journal, 24*(3), 349–374.
Ronen, D. (1982). The effect of oil price on the optimal speed of ships. *Journal of the Operational Research Society, 33*, 1035–1040.
Song, D. P., & Xu, J. J. (2012). CO2 emission comparison between direct and feeder liner services: A case study of Asia-Europe services interfacing with the UK. *International Journal of Sustainable Transportation, 6*(4), 214–237.
Stahlbock, R., & Voss, S. (2008). Operations research at container terminals: A literature update. *OR Spectrum, 30*(1), 1–52.
TradeWinds. (2010). Slow spur for Maersk VLCCs. TradeWinds magazine, 13 December.
Tsou, M. C., & Hsueh, C. K. (2010). The study of ship collision avoidance route planning by ant colony algorithm[J]. *Journal of Marine Science and Technology, 18*(5), 746–756.
UNCTAD. (2016). Review of maritime transport 2016, United Nations Conference on Trade and Development, *United Nations*, New York.
Vilas, R. F (2018). Container shipping performance: A case study on a transpacific service, M.Sc. thesis, Technical University of Denmark, July 2018.
Wang, S., & Meng, Q. (2012). Liner ship route schedule design with sea contingency time and port time uncertainty. *Transportation Research Part B, 46*(5), 615–633.
Wen, M., Pacino, D., Kontovas, C., & Psaraftis, H. N. (2017). A multiple ship routing and speed optimization problem under time, cost and environmental objectives. *Transportation Research Part-D, 52*, 303–321.
Zeng, Q., & Yang, Z. (2007). Model integrating fleet design and ship routing problems for coal shipping. In *Computational science–ICCS 2007* (pp. 1000–1003). Berlin/Heidelberg: Springer.
Zis, T., & Psaraftis, H. N. (2017). The implications of the new sulphur limits on the European Ro-Ro sector. *Transportation Research Part-D, 52*, 185–201.
Zis, T., & Psaraftis, H. N. (2018). Operational measures to mitigate and reverse the potential modal shifts due to environmental legislation. *Maritime Policy & Management*. https://doi.org/10.1080/03088839.2018.1468938.
Zis, T., North, R. J., Angeloudis, P., Ochieng, W. Y., & Bell, M. G. H. (2014). Evaluation of cold ironing and speed reduction policies to reduce ship emissions near and at ports. *Maritime Economics & Logistics, 16*(4), 371–398.
Zis, T., North, R. J., Angeloudis, P., Ochieng, W. Y., & Bell, M. G. (2015). Environmental balance of shipping emissions reduction strategies. *Transportation Research Record: Journal of the Transportation Research Board, 2479*, 25–33.

Chapter 11
Reducing GHGs: The MBM and MRV Agendas

Harilaos N. Psaraftis and Poul Woodall

Abstract The purpose of this chapter is to introduce the concept of market-based measures (MBMs) to reduce greenhouse gas (GHG) emissions from ships and review several distinct MBM proposals that were under consideration by the International Maritime Organization (IMO). The chapter then moves on to discuss the concept of monitoring, reporting and verification (MRV) of CO_2 emissions and the distinct mechanisms set up by the European Union (EU) and the IMO for MRV. The reason the MBM and MRV subjects are treated in the same chapter is twofold: (a) the MRV discussion essentially started when the MBM discussion was suspended in 2013, and (b) MRV is a critical step for any eventual MBM implementation in the future.

Abbreviations

BAU	Business As Usual
BDN	Bunker Delivery Note
CBDR-RC	Common But Differentiated Responsibilities and Respective Capabilities
CH_4	Methane
CO_2	Carbon dioxide
DCS	Data collection system
DNV	Det Norske Veritas
DoC	Document of compliance
DWT	Deadweight
EEDI	Energy Efficiency Design Index

H. N. Psaraftis (✉)
DTU Management Engineering, Technical University of Denmark, Kongens Lyngby, Denmark
e-mail: hnpsar@dtu.dk

P. Woodall
DFDS Seaways, Copenhagen, Denmark
e-mail: pouwo@dfds.com

H. N. Psaraftis (ed.), *Sustainable Shipping*,
https://doi.org/10.1007/978-3-030-04330-8_11

EIS	Efficiency Incentive Scheme
EIV	Estimated Index Value
EMSA	European Maritime Safety Agency
ER	Emission Report
ESSF	European Sustainable Shipping Forum
ETS	Emissions Trading System
EU	European Union
GHG	Greenhouse gas
GRT	Gross registered tons
HFO	Heavy Fuel Oil
IMAREST	Institute of marine engineering, science and technology
IMO	International Maritime Organization
IOPCF	International Oil Pollution Compensation Fund
IPTA	International Parcel Tankers Association
ISM	International Ship Management
ISO	International Standards Organization
IUCN	International Union for the Conservation of Nature
KPI	Key Performance Indicator
LDC	Least developed country
LIS	Leveraged Incentive Scheme
LNG	Liquefied natural gas
MAC	Marginal abatement cost
MBM	Market-based measure
MDO	Marine Diesel Oil
MEPC	Marine Environment Protection Committee
MP	Monitoring plan
MRV	Monitoring, Reporting and Verification
NGO	Nongovernmental organization
NO_x	Nitrogen oxides
PWC	PricewaterhouseCoopers
RoPax	Roll-on/roll-off passenger vessel
SECA	Sulfur Emission Control Area
SECT	Ship Efficiency and Credit Trading
SEEMP	Ship Energy Efficiency Monitoring Plan
SIDS	Small Island Developing State
UK	United Kingdom
UNFCCC	United Nations Framework Convention on Climate Change
USA	United States
USD	United States Dollar
VES	Vessel Efficiency System
VLCC	Very Large Crude Carrier
WSC	World Shipping Council
WTO	World Trade Organization

1 Introduction

It has been customary to break down the spectrum of measures to reduce maritime emissions into basically three major classes.

First, *technological* measures include more efficient (energy-saving) engines, more efficient ship hulls and designs, more efficient propellers, cleaner fuels (low carbon content, liquefied natural gas – LNG), alternative fuels (fuel cells, biofuels, etc.), devices to trap exhaust emissions (scrubbers, etc.), energy recuperation devices (exhaust heat recovery systems, etc.), 'cold ironing' in ports, various kites and others. Chapter 2 of the book gives a flavor of such technologies. Compliance with the Energy Efficiency Design Index (EEDI, Chap. 3), which is a *design index*, will mainly induce technological measures.

Second, we have *logistics-based* (tactical and operational) measures, which include speed optimization, optimized weather routing, optimal fleet management and deployment, efficient supply chain management and others that impact the logistical operation. Chapters 8, 9, and 10 (and to a lesser extent Chap. 7) of the book provide more insights on such measures.

Third, we have what we call *market-based measures* or MBMs. These include Emissions Trading Systems (ETS), an international fund based on a contribution imposed on fuel, and a variety of others, as will be explained later.

We note that the partition into the above three categories is, in many respects, artificial. This is so because an MBM may induce the ship owner to adopt (a) logistics-based measures in the short run and (b) technological measures in the long run. Both sets of measures would result in emission reductions.

This chapter focuses on the third category of measures, the MBMs, and specifically for greenhouse gases (GHGs), even though it will also touch upon the other two categories whenever warranted. As an extension of the MBM debate, this chapter also reports on the issue of monitoring, reporting and verification (MRV) of CO_2 emissions, which is a related but critical element in the quest of reducing such emissions. The reason the MBM and MRV subjects are treated in the same chapter is twofold: (a) the MRV discussion essentially started when the MBM discussion was suspended in 2013, and (b) MRV is a critical step for any eventual MBM implementation in the future.

To obtain some insights into the possible role of MBMs, consider the practice of slow steaming, widely applied in recent times mainly to reduce fuel costs and help sustain a fragile market by absorbing excess shipping capacity. From basic naval architecture, the dependency of fuel consumption on ship speed is at least cubic. GHG emissions being directly proportional to fuel consumed, a simple way to reduce these emissions, perhaps drastically, is for a ship to slow down.

By making a ship owner pay for his ship's CO_2 emissions, an MBM is an instrument that implements the 'polluter pays' principle. In that sense, it helps internalize the external costs of these emissions. In addition, monies raised by an MBM can be used to reduce CO_2 emissions *outside* the marine sector, for instance, by purchasing what are known as 'offsets'. Such offsets could be used to invest in projects such as, for instance, a wind farm in New Zealand, a solar cell farm in Indonesia or others and so contribute to GHG reduction outside the marine sector. These are known as 'out of sector' reductions.

It should be noted that earlier discussions on MBMs have been presented in Psaraftis (2012, 2016), and Sects. 2, 3, and 4 of this chapter draw from these references. The reader may also look at Gkonis and Psaraftis (2012) and Kapetanis et al. (2014) for analyses of the impact of a bunker levy on tankers and Handymax bulk carriers (respectively) and Kosmas and Acciaro (2017) for a discussion of bunker levy schemes in international shipping. Last but not least, Chap. 8 of this book examines the possible impact of MBMs to reduce CO_2 in a tramp shipping routing and scheduling context.

The rest of this chapter is organized as follows. Section 2 discusses some basic concepts including that of the marginal abatement cost (MAC) curve. Section 3 outlines the MBM proposals that were submitted to the International Maritime Organization (IMO). Section 4 comments on the modeling effort to evaluate the MBM proposals and outlines how the MBM discussion was suspended in 2013. Sections 5, 6, 7, and 8 introduce the concept of MRV of CO_2 emissions and discuss the two distinct regimes that are at play, the EU MRV and the IMO DCS schemes. Finally Sect. 9 presents the conclusions of the chapter.

2 Basic Concepts

Before we proceed, some basic concepts are in order.

2.1 Carbon Coefficients

There is a linear relationship between fuel burned and CO_2 produced, with the proportionality constant being known as the 'carbon coefficient'. The first IMO GHG study of 2000 used a coefficient of 3.17 (tonnes of CO_2 per tonne of fuel) independent of fuel type, but its 2009 update (Buhaug et al. 2009) used slightly lower coefficients, which ranged from 3.021 for Heavy Fuel Oil (HFO) to 3.082 for Marine Diesel Oil (MDO). The factor that IMO uses for EEDI reference line calculations is 3.11, for both main engine and auxiliaries. For alternative fuels such as LNG, the carbon coefficient can range from 2.6 to 2.8. This feature makes LNG more attractive than fossil fuels for propulsion, among other advantages, such as lack of sulfur and other substances and producing more energy per unit weight than fossil fuels. However, a disadvantage of LNG is the so-called 'methane slip', as some methane (CH_4) is released by LNG use. CH_4 is a GHG that is many times more potent than CO_2 (see Chap. 13 for details).

2.2 CO_2 Produced by International Shipping

The third IMO GHG study (Smith et al. 2014) provided updated estimates of CO_2 emissions from international shipping from 2007 to 2012. The 2012 figure, estimated by a 'bottom-up' method, was 796 million tonnes, down from 885 million (updated figure) in 2007, or 2.2% of global CO_2 emissions. It should be noted that the equivalent percentage in the second IMO GHG study (Buhaug et al. (2009) was 2.7% for 2007 fleet data. CO_2 from all shipping was estimated at 940 million tonnes in 2012, down from 1100 tonnes in 2007. The reduction from 2007 to 2012 was mainly attributed to slow steaming due to depressed market conditions after 2008.

2.3 Marginal Abatement Costs

The concept of the marginal abatement cost (MAC) may be critical in the context of MBMs. Let us call *A* a well-defined available *technological* measure to avert CO_2. For instance, measure *A* can be a more efficient hull form, a more fuel-efficient engine, a more efficient propeller, a device such as a kite that can save energy or others (see Chap. 2 for a discussion of such technologies). Suppose that we implement measure *A* on a given ship and we compare the ship without measure *A* to the ship with measure *A*, ceteris paribus.

Define also:

- $\Delta GCOST(A)$: the total *gross* cost in implementing measure *A*, defined as the gross difference in annualized costs of the ship with measure *A*, minus those costs without measure *A*, gross meaning excluding fuel costs
- $\Delta FUEL(A)$: the total annual fuel consumption averted by implementing measure *A*, for the same ship
- $\Delta NCOST(A)$: the total *net* cost in implementing measure *A*, defined as the net difference in annualized costs of the ship attributed to measure *A*, net meaning including fuel costs
- $\Delta CO_2\,(A)$: the total tonnes of CO_2 averted by measure *A*
- *PFUEL:* the average price of fuel over a year
- *F*: the carbon coefficient (between 3.02 and 3.11)

Then the marginal abatement cost (*MAC*) of measure *A* is defined as follows:

$$MAC(A) = \Delta NCOST(A)/\Delta CO_2\,(A) \tag{11.1}$$

Given that

$$\Delta NCOST(A) = \Delta GCOST(A) - \Delta FUEL(A) * PFUEL \text{ and that}$$

$$\Delta CO_2\,(A) = \Delta FUEL(A)^* F$$

it follows that

$$MAC(A) = \Delta GCOST(A)/\Delta CO_2(A) - PFUEL/F \tag{11.2}$$

The negative term in the right-hand side of (11.2) reflects the savings in costs (per tonne of CO_2 averted) *due to fuel consumption reduction*. This also means that for any measure *A*, *MAC*(*A*) can be negative if the price of fuel is high enough. Measures for which the ratio $\Delta GCOST(A)/\Delta CO_2(A)$ is low are also more likely to have a negative *MAC(A)* than other measures for which these costs are high.

A negative MAC means that the ship owner would have an economic incentive to implement the respective measure. Doing so would increase his profits, and, as an important side-effect, would also reduce CO_2. It would be a win-win proposition and would not need a regulation mandating the measure.[1] Conversely, if the MAC of a measure is positive, then the ship owner would have no incentive to adopt it. The measure would have to be mandated in order to be implemented.

2.4 MAC Curves

If one examines a set of feasible measures to reduce CO_2 and compute the MACs for such measures applied to the world fleet, one comes up with what is known as the *MAC curves*. Several attempts to construct MAC curves are known; see, for instance, DNV (2009), Eide et al. (2010) and IMAREST (2011). MAC curves are supposed to be constructed for the entire set of possible measures to reduce CO_2. The horizontal axis of a MAC curve measures the total amount of CO_2 averted and the vertical axis measures the corresponding MAC.

In economic terms, the MAC curve is really a *supply curve*, in the sense that measures are rank-ordered by non-decreasing order of MAC. This assumes that before a certain measure is implemented, all other measures with a MAC lower than the MAC of this measure have been implemented. In practice this may not necessarily be the case, as some measures may be mandated and therefore given priority over others, and in these situations a MAC curve may not be monotonically increasing. Also the MAC curve assumes no interdependencies among measures which also may not be the case.

Figure 11.1 shows a typical MAC curve, taken from the Expert Group report on MBMs (IMO 2010) and carried out by (former) Norwegian classification society DNV, which was commissioned by the IMO for the task. The data used for this analysis was not made publicly available. One can see here that some of the MAC curves are not monotonic, meaning that some measures may take precedence over other measures even though their MAC is higher.

---

[1] This would also assume the absence of barriers that would make the adoption of the measure difficult or impossible.

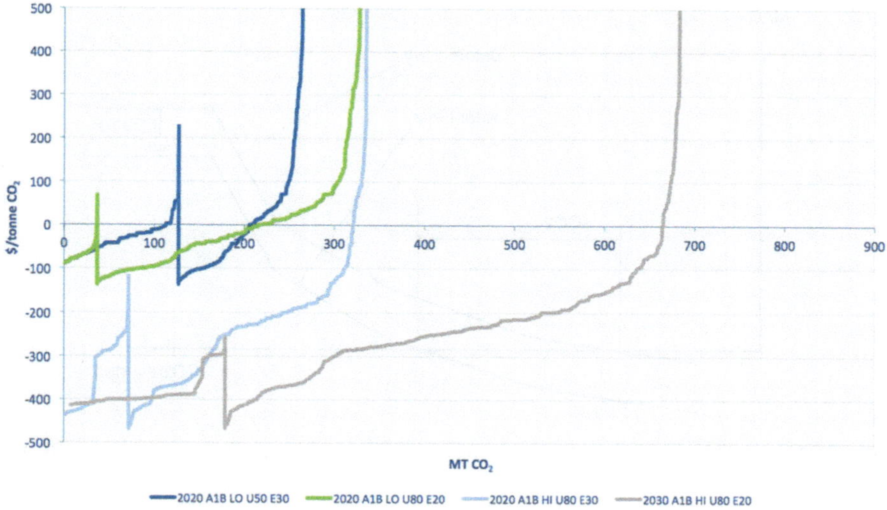

Fig. 11.1 Sample MAC curves by DNV. (Source: IMO 2010)

It is also important to realize that the MAC curves directly depend on the projected price of fuel, as Eq. (11.2) above stipulates. The MAC curve will shift up and down depending on what *PFUEL* will be. It will shift down by about USD 100/tonne for each increase of USD 300/tonne in the price of fuel. In DNV (2009), DNV assumed a fuel price of USD 350/tonne for a standard bunker oil and USD 500/tonne for a high-quality bunker oil for 2030. If these prices change, the MAC curves will change. In IMO (2010), DNV and the MBM Expert Group examined a variety of scenarios on projected future parameters including fuel prices. More on this in Sect. 4.

2.5 *Effect of a Bunker Levy on MAC Curves*

The MAC curve can be useful if one wants to evaluate the effect of a bunker levy (or tax) on the amount of CO_2 emissions. Figure 11.2 below shows how. The figure shows two MAC curves. The one on top is before a levy is applied, and the one below is after the levy. Applying a levy equal to *LEVY* means that the price of fuel will increase from PFUEL to *PFUEL+LEVY.*

Assuming that *LEVY* > *0* and that the gross costs of each the various measures to reduce emissions (*ΔGCOST*) do not change as a result of the levy, the MAC curve will uniformly go down by an amount equal to *LEVY/F*. Note that this is a first-order approximation, as the new fuel price may result in speed reduction and a corresponding reduction of fuel consumption and hence emissions. So a second-order effect may move the MAC curve also to the left and not only down.

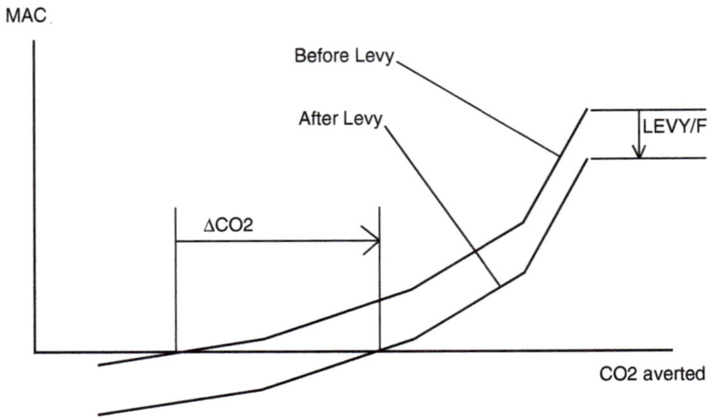

Fig. 11.2 Using MAC curves to determine the effect of a bunker levy. (Source: Psaraftis 2012)

Ignoring for the moment this second-order effect, a first-order approximation of the CO_2 reduction ΔCO_2 can be estimated if we assume that for every fuel price/levy, the equilibrium CO_2 emissions will be at the point where $MAC = 0$.

We can even make a crude estimate of ΔCO_2 as equal to $LEVY/(F*s)$, where s is the slope of the MAC curve at $MAC = 0$. However one has to be careful as from Fig. 11.1 one can see that this slope can vary widely.

3 The MBM Proposals Before the IMO

Following the update of the IMO GHG study in 2009 (Buhaug et al. 2009), IMO activity on GHGs was largely on two 'parallel' tracks. The first track mainly concerned EEDI (see also Chap. 3). The second track concerned MBMs. It is interesting that discussion on these two tracks was conducted with no apparent connection between the two, even though both tracks concerned the same objective (reduce GHG emissions from ships). In reality these tracks are not disjoint, as some of the proposed MBMs embedded EEDI in their formulation.

For MBMs, an Expert Group was appointed by the IMO's Secretary General after solicitation of member states and was tasked to evaluate as many as ten (10) separate MBM proposals, submitted by various member states and other organizations. All submitted MBM proposals described programmes and procedures that would target GHG reductions through either 'in-sector' emission reductions from shipping or 'out-of-sector' reductions via the collection of funds to be used for mitigation activities in other sectors that would contribute toward global reduction of GHG emissions.

The IMO formulated the following nine (9) criteria for evaluation of GHG reduction measures, including MBMs:

1. Environmental effectiveness
2. Cost-effectiveness and potential impact on trade and sustainable development
3. The potential to provide incentives to technological change and innovation
4. Practical feasibility of implementing MBM
5. The need for technology transfer to and capacity building within developing countries, in particular the least developed countries (LDCs) and the small island development states (SIDS)
6. The relation with other relevant conventions (United Nations Framework Convention on Climate Change, UNFCCC; Kyoto Protocol; and World Trade Organization, WTO) and the compatibility with customary international law
7. The potential additional administrative burden and the legal aspects for National Administrations to implement and enforce MBM
8. The potential additional workload, economic burden and operational impact for individual ships, the shipping industry and the maritime sector as a whole, of implementing MBM
9. The compatibility with the existing enforcement and control provisions under the IMO legal framework

Brief descriptions of each of the ten original IMO MBM proposals are as follows (see IMO (2010) for more details):

1. The International Fund for Greenhouse Gas emissions from ships (GHG Fund) originally proposed by Cyprus, Denmark, the Marshall Islands, Nigeria and the International Parcel Tankers Association – IPTA (Denmark 2010)

Liberia and the Republic of Korea were later added as cosponsors of this MBM. This Fund would establish a global reduction target for international shipping, set by either the UNFCCC or the IMO. Emissions above the target line would be offset largely by purchasing approved emission reduction credits. The offsetting activities would be financed by what the proposers called a 'contribution' paid by ships on every tonne of bunker fuel purchased. It was envisaged that contributions would be collected through bunker fuel suppliers (Option 1) or via direct payment from ship owners (Option 2). The contribution rate would be adjusted at regular intervals to ensure that sufficient funds are available to purchase project credits to achieve the agreed target line.

It should be noted that this MBM was essentially a levy on bunker fuel, even though its proposers carefully avoided the use of that word (or of the word 'tax').

2. The Leveraged Incentive Scheme (LIS) to improve the energy efficiency of ships based on the International GHG Fund proposed by Japan (2010)

This resembled the GHG Fund scheme with an important difference: The concept of the Leveraged Incentive Scheme is that a part of the GHG Fund contributions, which are collected on marine bunker, is refunded to ships meeting or exceeding agreed efficiency benchmarks and labeled as 'good performance ships'. To that effect, the EEDI index is the main instrument in determining which ships are efficient and should receive the refunds. In that sense, this MBM was a hybrid one, as it included EEDI as part of its formulation.

3. Achieving reduction in greenhouse gas emissions from ships through Port State arrangements utilizing the ship traffic, energy and environment model, STEEM (PSL) proposal by Jamaica (2010)

Under this MBM, member states participate in levying a uniform emissions charge on all vessels calling at their respective ports based on the amount of fuel consumed by the respective vessel on its voyage to that port (not bunker suppliers). The proposal is directly aimed at reducing maritime emissions of CO_2 without regard to design, operations or energy source. The Port State Levy would be structured to achieve the global reduction targets for GHG and could be leveraged in a manner as proposed by Japan to reward vessels exceeding efficiency targets.

4. The US proposal to reduce greenhouse gas emissions from international shipping, the Ship Efficiency and Credit Trading (SECT) (USA 2010)

This MBM is designed to focus emission reduction activities just in the shipping sector. Under SECT, all ships, including those in the existing fleet, would be subject to mandatory energy efficiency standards, rather than a cap on emissions or a surcharge on fuel. As one means of complying with the standard, SECT would establish an efficiency-credit trading programme. Similar to the EEDI, these efficiency standards would be based on a reduction from an established baseline and would establish efficiency standards for both new and existing ships. As the LIS MBM by Japan, the SECT MBM by the USA was a hybrid MBM, as it embedded EEDI within its formulation. However, the mechanism is different. Under SECT, ships would trade on EEDI. A 'good EEDI' ship would sell credits to a 'bad EEDI' ship.

5. Vessel Efficiency System (VES) proposal by World Shipping Council (2010)

VES would establish mandatory efficiency standards for both new and existing ships. Each vessel would be judged against a requirement to improve its efficiency by X% below the average efficiency (the baseline) for the specific vessel class and size. Standards would be tiered over time with increasing stringency. Both new build and existing ships would be covered. New builds must meet the specified standards or they may not operate. Existing ships may comply by improving their efficiency scores through technical modifications that have been inspected and certified by the administration or recognized organizations. Existing ships failing to meet the required standard through technical modifications would be subject to a fee applied to each tonne of fuel consumed. The total fee applied (non-compliant ships only) would vary depending upon how far the vessel's efficiency (as measured by the EEDI) falls short of the applicable standard. This was another example of a hybrid MBM, as it embedded EEDI within its formulation.

6. The Global Emissions Trading System (ETS) for international shipping proposal by Norway (2010)

This MBM would set a sector-wide cap on net emissions from international shipping and establish a trading mechanism to facilitate the necessary emission reductions, be they in-sector or out-of-sector. The use of out-of-sector credits allows for further growth of the shipping sector beyond the cap. In addition the auction revenue would be used to provide for adaptation and mitigation (additional emission reductions) through UNFCCC processes and R&D of clean technologies within the maritime sector. A number of allowances (Ship Emission Units) corresponding to the cap would be released into the market each year. It was proposed that the units would be released via a global auctioning process. Ships would be required to surrender one Ship Emission Unit, or one recognized out-of-sector allowance or one recognized out-of-sector project credit, for each tonne of CO_2 they emit. The Norwegian ETS would apply to all CO_2 emissions from the use of fossil fuels by ships engaged in international trade above a certain size threshold. The proposal also indicated that limited exemptions could be provided for specific voyages to Small Island Developing States.

7. Global Emissions Trading System (ETS) for international shipping proposal by the United Kingdom (2010)

This was very similar in most respects to the global ETS proposal by Norway. Two aspects of the UK proposal that differ from the Norwegian ETS proposal were the method of allocating emissions allowances and the approach for setting the emissions cap.

8. Further elements for the development of an Emissions Trading System (ETS) for International Shipping proposal by France (2010)

This MBM set out additional detail on auction design under a shipping ETS. In all other aspect, the proposal was similar to the Norwegian proposal for an international ETS.

9. Market-based instruments: a penalty on trade and development proposal by the Bahamas (2010)

This MBM did not set explicit standards or reductions to be achieved in the shipping sector or out-of-sector for GHG reductions. In that sense, it was a 'do nothing' MBM. The proposal clearly set forth that the imposition of any costs should be proportionate to the contribution by international shipping to global CO_2 emissions. Bahamas indicated that it was assuming that mandatory technical and operational measures would be implemented such as the EEDI.

10. A Rebate Mechanism (RM) for a market-based instrument for international shipping proposal by the International Union for the Conservation of Nature (IUCN) (2010)

This MBM focused on a Rebate Mechanism to compensate developing countries for the financial impact of a MBM. This Rebate Mechanism would 'piggy back' any of the other MBMs. A developing country's rebate would be calculated on the basis of their share of global costs of the MBM, using readily available data on a developing country's share of global imports by value as a proxy for that share (or another metric such as value-distance if data becomes available).

4 Continuation and Suspension of the MBM Discussion

The following developments took place after the above ten MBMs were submitted:

(a) A German ETS proposal (Germany 2010) that was not included in the original MBM list for administrative reasons was reinstated as part of the MBM roster.
(b) The LIS and VES proposals were combined into what was relabeled the Efficiency Incentive Scheme (EIS) (Japan and WSC 2011).
(c) The Bahamas submitted what they called an evolution of their original proposal (Bahamas 2011), which they subsequently withdrew altogether.
(d) The US MBM proposal was radically restructured.

After considerable discussion, a 300+ page report (IMO 2010) evaluating the MBM proposals was prepared by the MBM Expert Group and was presented and discussed at MEPC 61 (September 2010). The report went at length in assessing each MBM according to the evaluation criteria, in modeling future scenarios and in assessing the impact of MBMs on trade and developing countries. However, the report contained no horizontal comparison of MBMs and no recommendation as to which MBMs should be further pursued.

The Expert Group's modeling effort, which also involved the work of external consultants, was to develop and apply a model to make quantitative estimates of emission reductions, revenues generated, costs and other attributes of each MBM proposal. Modeling scenarios included:

- Two growth rates (1.65% and 2.8%)
- Three targets lines/caps for GHG Fund and ETS (0%, 10% and 20% below 2007 level)
- Twenty-eight percent revenue used for mitigation for Rebate Mechanism and 25%, 50% and 75% revenue refunded for LIS
- Low, medium and high stringency standards for VES and SECT
- Two carbon price scenarios (medium and high) and two fuel price scenarios (reference and high)

Projections of emissions and remaining proceeds were also made.

Even though this effort was certainly worthy of note, reservations are expressed herewith for some of the modeling assumptions. As an example (which is one of several), a key assumption was made that an increase in fuel prices of 100% over the long-term will result in a 4% reduction in emissions below the so-called 'Business As Usual' (BAU) scenario. However, this percentage (4% or others) critically depends on the slope of the marginal abatement cost (MAC) curve at the point it crosses the x-axis. As illustrated by the DNV MAC curves for the 72 scenarios examined (see Fig. 11.1), that slope can vary widely from very low to very high, projected future fuel price being the main determinant. Therefore the fixed 4% assumption is not necessarily correct. In that sense, strong reservations are expressed on all the numerical results of this model, which are sometimes difficult to follow and, at a minimum, should be interpreted with caution.

The same applies to the numerical results that pertain to a variety of estimates for each MBM proposal, such as in-sector and out-of-sector emission reductions, revenues generated, costs and a variety of others. Even estimates of CO_2 reductions with or without EEDI enacted were produced. All of these results critically depend on a variety of assumptions, which cannot be fully substantiated.

It should also be pointed out that the data and models on the MAC curves supplied to the IMO MBM Expert Group were not fully available to scrutiny, and this significantly limits their usefulness. Both the data and the models are subject to confidentiality clauses. The problem here is if the models are not made available for scrutiny by the experts or anybody else (remaining virtually a black box), then obviously the correctness of their results cannot be confirmed.

In Table 11.1 below we present a comparison of the GHG Fund and the ETS proposal in terms of the evaluation criteria. All ETS proposals have been combined in the table. Comments on criteria 6 (compatibility to UNFCCC and other international laws) and 9 (compatibility with existing IMO framework) have been omitted as they are covered by the Expert Group report (IMO 2010). We also include some additional criteria.

Providing more detail, Psaraftis (2012) included all ten MBMs in a horizontal assessment according to the nine evaluation criteria. This has been, to our knowledge, the only comparison of these MBM proposals to date. The Expert Group report (IMO 2010) contained no recommendation on which MBM should be chosen, and discussion on MBMs at the IMO level after 2010 was pretty non-productive. In March 2011, an Intersessional Meeting of the Working Group on GHG emissions from ships took place, with a view to making progress toward the ultimate selection of an MBM, but did not arrive at a similar recommendation either. The same was the case at MEPC 62 (July 2011), as discussion there was entirely devoted to EEDI. The period immediately after the adoption of EEDI was focused on practical matters involving its implementation, and there was little discussion on MBMs. A proposal by Greece in 2012 (who had submitted no MBM proposal of its own) for the IMO to decide on a short list of MBMs (Levy and ETS) was rejected, apparently on the ground of not wanting to displease the MBM proposers. The same happened to a proposal by the Chairman of the MEPC in 2012 to conduct an impact assessment study, as political considerations and lack of agreement between developed and developing countries prevented a decision on the matter.

In fact, reception of the proposed MBMs at the IMO has been mixed at best. In addition to the lack of consensus among MBM proposers, the group of developing countries, such as China, India, Brazil and others, were as much against any MBM as they were against EEDI. This was mainly on the ground that MBMs are not compatible with the principle of *Common But Differentiated Responsibilities and Respective Capabilities (CBDR-RC).*

CBDR-RC (or simply CBDR, as it was known earlier) has been the main political argument of a group of developing countries (see above) to resist GHG emission reduction, not just for shipping but across the board, on the ground that this would

Table 11.1 Comparison of GHG Fund and ETS proposals

Main criterion	GHG fund (Denmark et al.)	ETS (Norway, the UK, France, Germany)
1. Environmental effectiveness (how certain is MBM to achieve a specific reduction target)	There is less certainty of CO_2 reductions than ETS, but MAC curves of DNV can give an estimate. If price is same, CO_2 reductions are same with ETS*. Offsets can contribute to meeting a cap. See also criterion 2 below	There is higher certainty of CO_2 reduction, but reduction target is arbitrary (or very difficult to determine). Plus, enforcing the cap can be difficult and carbon price may skyrocket if we are close to the cap
	**Assuming equal cost-effectiveness which not the case*	Significant carbon leakage risks exist (for instance, if not all ships are covered, some countries like LDCs excluded, etc.)
2. Cost-effectiveness	High. Costs are known as price is known. Simplest scheme (except Bahamas). According to several studies, Levy is most efficient way to reduce emissions	Low. High administrative costs, very unpredictable carbon prices
3. Incentives to technological change	High. Investors will respond to known price	Low. Investors will not know what future prices they will encounter and will pay high administrative costs
4. Practical feasibility	Reasonable. Can be modeled from the International Oil Pollution Compensation Fund (IOPCF)	Questionable. All GHG Fund (Option 2) processes, plus auction permits, monitor allowance market, enforce compliance, identify fraud, etc.
5. Impact on LDCs and SIDS	Neutral. From a revenue perspective, if prices are same, revenue is same as ETS	Distortions likely, as traffic to LDCs-SIDS countries is exempted, which may lead to traffic being diverted through these countries
6. Compatibility to UNFCCC and other international laws	See IMO (2010)	See IMO (2010)
7.National administrative burden	Reasonable. Tracking bunkers is not trivial but burden is lower than all other schemes (except Bahamas)	Significant. High administrative costs to track, monitor, enforce, avoid evasion and fraud, etc. if all ships in the scheme, impossible to implement
8. Administrative burden on industry	Same as above	Same as above
9. Compatibility to IMO framework	See IMO (2010)	See IMO (2010)
Other criteria		
Impact in slow steaming	Taken care of automatically	Difficult to impossible to implement or assess
Impact on safety	Neutral	Neutral
Risk of fraud	Average. Low at refinery level	High-documented cases in EU ETS and elsewhere
Money collected	Limited to in-sector contributions. Depends on level of Levy	If GHG Fund Levy and ETS carbon price are same, amount of money collected for ETS is same as GHG Fund minus difference in administrative costs

Source: Psaraftis (2016)

impede their economic development. In that sense, the stance of these countries was that their obligation to reduce GHGs should be less stringent than that of developed countries. It is however clear that this would be incompatible to the principle that any measure for GHG reduction should be non-discriminatory, so as to maintain a level playing field. At least in shipping, a sector which is based on the notion of free and fair competition, this principle is of paramount importance.

Another issue of disagreement has been how monies collected by the MBM would be used for the benefit of developing countries (capacity building, technology transfer, etc.). Among industrial stakeholders, the International Chamber of Shipping, BIMCO and several ship owners associations have come out against an ETS, on the ground that it would be unworkable for the shipping industry. Interestingly enough, these include the German and Norwegian ship owners associations, even though their national maritime administrations are for ETS.

Then in May of 2013, the MEPC decided to suspend discussion on MBMs altogether, at least for the time being. This reflected a channeling of the discussion toward the subject of monitoring, reporting and verification (MRV) of CO2 emissions. Sections 5, 6, 7, and 8 of this chapter report on the various aspects and issues of MRV.

A late twist in the MBM saga came with the February 2017 vote of the European Parliament (EP) to include shipping into the EU ETS as of 2023, in case no global agreement is reached by 2021. This followed a recommendation of the EP's Environment (ENVI) Committee to that effect in December 2016. As mentioned earlier, this caused serious concern among industry stakeholders that such a *regional* MBM would create serious distortions, not to mention that it might not necessarily reduce maritime CO_2 emissions. As an example, a ship calling at Kaliningrad, Russia, might be able to avoid the EU ETS. If so, one might see that Baltic port establishing itself as a regional hub, creating distortions in intermodal flows and ultimately more CO_2 in the supply chain. The same may be true for African or other non-EU ports in the Mediterranean or elsewhere. We know of no analysis of such possible distortions or other side-effects.

In November 2017, and after some negotiations between the EP and the EU Council of Ministers, it was agreed to align the EU with the IMO process and essentially refrain from taking action on ETS before seeing what the IMO intends to do on GHGs. Industry circles, concerned with the effects of an early EU ETS, welcomed this development. However, the European Commission will closely monitor the IMO process, starting from what is agreed on the initial strategy in 2018 and all the way to 2023. Whether or not this latest agreement at the EU level might put some pressure on the IMO to resume the suspended discussion on MBMs and adopt a global MBM before the EU moves on ETS is unclear at this time. And even though the ETS looks like the default scenario for the EU if progress at the IMO is not deemed satisfactory, precisely what action the EU will take and when that action will be taken is equally unclear.

In April 2018 the IMO/MEPC reached the landmark decision to adopt an initial strategy for GHG reductions that sets (among other things) a target of GHG reductions of at least 50% by 2050, vis-à-vis 2008 levels (IMO 2018). Chapter 13 comments on that decision. A list of potential measures has also been proposed; however at this point in time, no prioritization among measures exists. MBMs are included in the set of *medium term* measures but only obliquely. To quote from the decision text, 'new, innovative emission reduction mechanism(s), possibly including Market Based Measures (MBMs), to incentivize GHG reduction'. No measures have been specified to be adopted until 2023 at the earliest, year when the strategy is supposed to be finalized. Other than the above wording, and as this chapter was being finalized, there is nothing in the IMO process that would reopen the MBM discussion anytime soon. Some industry associations seem to favor a Levy; however this is off the record, and they would not (at least as things stand) go as far as officially propose such a measure to the IMO.

5 The 2011 EU White Paper: The Origin of the EU MRV

As early as 2011, the EU had adopted the White Paper on Transport (EU, 2011) that was to lay the foundation for the community's future transport policy.

Among other issues, the White Paper addressed the global warming predicament of aiming to limit global temperature to below 2 centigrade increase from pre-industrial levels while ensuring growth and prosperity to the community at large. The White Paper set out the roadmap for achieving a 60% reduction in CO_2 emissions from community transport; this included shipping. It also made it clear that curbing mobility or trade was not an option in achieving this goal.

Specifically on international and long-haul maritime transport, the White Paper established the following ambition:

> *In maritime, the need for a global level-playing field is equally pronounced8. The EU should strive – in cooperation with IMO and other international organisations – for the universal application and enforcement of high standards of safety, security, environmental protection and working conditions, and for eliminating piracy. The environmental record of shipping can and must be improved by both technology and better fuels and operations: overall, the EU CO_2 emissions from maritime transport should be cut by 40% (if feasible 50%) by 2050 compared to 2005 levels.*

The White Paper summarized the policies into ten specific goals, the tenth and final one establishing the principle of full application of the 'polluter pays' principle. We will later see how this point is reflected in subsequent documents and legislation. But we note that this is a point that to a large extent has gone unnoticed by the industry in general.

A general problem with international maritime GHG emissions is that they do not belong to any country. They are therefore outside the national statistics, and as a result of that, they are quite uncontrollable. Should one have elected to allocate vessels GHG emissions to the flag states emissions, we would see some very small countries with very large emissions. For instance for Denmark, the figure for CO_2 equivalent emissions would increase by 71% (2016) if Danish operated ships were to be added to the domestic figure.

The next milestone in the process came on June 28, 2013, when the European Commission released a proposal for a Regulation on MRV (EU 2013a). This proposal was a direct consequence of IMO not progressing the maritime CO_2 emissions agenda sufficiently fast to satisfy the EU Commission ambitions (see suspension of MBM discussion in May 2013, as per Sect. 4 of this chapter). The proposal contained a number of quite noticeable statements.

First was an acceptance that the GHG emissions from maritime transport was largely an unknown factor.

Second was a statement that a robust MRV system would save up to 2% of emissions or €1.2 billion in cost by 2030. Obviously, how this calculation can be made if one does not know the emissions in the first place is left somewhat to the readers to find out.

Whereas the main document of the proposal included the claim that the CO_2 emissions are largely unknown, the supplementing document of the impact assessment stated that the intended target was '210 tonnes of CO_2 in 2020 (+8% compared to 2005)' (EU 2013b).

The third key point was that an EU MRV system could be the model for an IMO global system. That was a clear hint for the IMO to get on with the work.

The scope of the proposed regulation was vessels of 5000 GRT and above, so as to limit the reporting to the larger tonnage, while at the same time ensuring it covers the majority of the total maritime emissions. The scope was CO_2 only and not any other GHGs such as CH_4, and finally the proposal established that the annual cost of this coming legislation would be € 26 million.

Whereas there appears to be no documentary evidence to this effect, among industry it was clear that PricewaterhouseCoopers (PWC) had assisted the Commission in drafting the MRV Regulation documentation. In doing so, PWC had a very good eye to the corresponding system covering the European Airline system. That the EU MRV thus was a virtual copy/paste of an airline system should prove to be one of the main errors when constructing this set of rules.

From an industry perspective, the ambition was to ensure the fine print of the Regulation would ensure an easy administration of the future reporting and that rules would be applied equally to all so not to disturb the competitive environment. There was not really any belief that this process in itself would contribute to any viable fuel reduction, in and of itself.

6 The EU MRV Regulation: EU 2015/757

6.1 Scope

On April 29, 2015, the EU Parliament approved Regulation EU 2015/757 also known as the MRV Regulation. The Regulation entry into force date was July 1, 2015. On October 28, 2016, the scope was extended to include EEA countries. The timeline for the process is shown in Table 11.2.

Section one of the Regulation deals with scope and definitions. Here the scope is interesting, as it includes the nine European outermost regions. So, not only are places like Canary Islands and Madeira included but also exotic locations such as Martinique, Mayotte, Guadeloupe, French Guiana and Réunion. Given that only one port of a voyage has to be located within the scope, this should eventually make an interesting appendix to the reporting and the analysis of the results.

Section two deals with which principles are allowed and how they should be documented. The key element here is the Monitoring Plan, which is a document that details how data is captured, stored and reported, who has the responsibility for which elements in the chain and also on procedures for filling any data gaps that may occur in the process.

Section three covers the verification process, that is, who is allowed to verify the data and how the process for monitoring plan assessment and the Emission Report is verified.

In the following, we will look into some of the key elements of the Regulation text.

Article 1 attempts to define the voyages that are within scope, '. . . from ships arriving at, within or departing from ports under the jurisdiction of a Member State,..'. Together with the wording in Article 2: *voyages from their last port of call to a port of call under the jurisdiction of a Member State and from a port of call under the jurisdiction of a Member State to their next port of call, as well as within ports of call under the jurisdiction of a Member State.* The definition of a port of call is in Article 3, where it is clarified that stops to bunker or crew change are not considered port calls.

Table 11.2 The EU MRV timeline

Date	Milestone
29/04/2015	MRV Regulation agreed
01/07/2015	Entry info force
31/08/2017	MMonitoring reports finalized
31/12/2017	Monitoring reports assessed
01/01/2018	Start 1st reporting period
31/12/2018	End 1st reporting period
30/04/2019	1st reporting period ready for assessment
30/06/2019	Assessment completed/DoC issued

Let us look at an example. A vessel departs Singapore having loaded cargo there, stops in Suez to take bunkers, then in Algeciras to for crew change and arrives at Rotterdam for cargo discharge. The entire voyage from Singapore to Rotterdam is 'in scope' as the two stops for bunkering, respectively, crew change are not considered port calls. So one needs to report the distance sailed, time at sea and the fuel consumed for the voyage from Singapore to Rotterdam. In order to do this correctly, one needs to record the port time and consumption during the two port stays, as these must be subtracted from the total in order to report the correct voyage details. So even with the ports calls being out of scope, the same data collection is required as if they had been within scope.

6.2 Data Collection

Moving on to Article 4, this deals with how CO_2 emissions need to be measured and establishes that there need be separate measurements/reporting for when a vessel is at sea and when in port. Article 4.4 is quite specific on what is required:

> *Companies shall obtain, record, compile, analyse and document monitoring data, including assumptions, references, emission factors and activity data, in a transparent manner that enables the reproduction of the determination of CO2 emissions by the verifier.*

It is interesting to note the obligation to also analyze the data. What this analysis should result in remains unclear.

Article 5 is a key element as it describes (through a reference to Annex 1) the permissible ways of obtaining CO_2 data. The first paragraph in Annex 1 establishes that CO_2 equals fuel consumption multiplied with the emission factor for the relevant fuel. This is straightforward. The emission factors for most standard fuels have been established by IMO, and these values are applicable for MRV reporting. In hindsight, one may have wished for a more simplified approach to this. CO_2 conversion factors for the most common fuel types – diesel, gasoil, light fuel oil and heavy fuel oils – as defined by ISO 8217 range from 3.1144 to 3.206. Given the uncertainties later agreed that would be allowed with MRV reporting, having one value of 3.15 for all these fuel types, could have made matters a lot simpler.

For many operators, 2020 will bring new concerns. With the global sulfur cap being implemented, a number of 0.50% fuels will enter the market that initially will not have an ISO 8217 spec. Then these emission factors will no longer be available in a standard table.

The second paragraph in Article 5 establishes four possible ways this data can be obtained:

(a) *Bunker Delivery Note (BDN) and periodic stocktakes of fuel tanks*
(b) *Bunker fuel tank monitoring on board*
(c) *Flowmeters for applicable combustion processes*
(d) *Direct CO_2 emissions measurements*

Let us look at these four methods one by one. Reporting of fuel consumption and CO_2 under MRV has to be in mass, typically metric tons.

In order to determine fuel consumption during a voyage, one needs to take one measurement at departure and one measure at arrival. Assuming there is no bunkering and/or debunkering, subtracting one figure from the other will give the consumption for the voyage. Any consumption during 'out of scope' port calls, such as a call for bunkering, will need to be subtracted. This procedure needs to be applied for each type of fuel consumed. A similar process needs to be done to establish fuel consumption during a port call.

(a) *Bunker Delivery Note (BDN) and periodic stocktakes of fuel tanks*

Before going deeper in to this, one needs to acknowledge that stocktake for MRV reporting is different from stocktake to verify fuel bunkered/(debunkered). The former is to ascertain what enters the combustion units and the latter to verify what you are paying for. When stocktaking for consumption purposes, one would need to sound or measure from the day tank and perhaps from any tanks from which one is replenishing the day tank. As part of this process, the quantity mentioned on the BDN really is not part of the equation. Where the BDN adds value is because it mentions the density and as most tank soundings are volumetric one needs the BDN to compute the correct mass.

(b) *Bunker fuel tank monitoring on board*

This is really no different from option (a). The omission of the reference to BDN makes sense if one either uses the default density values allowed, established density from other sources or has a tank measuring system that records mass.

In reality for most modern vessels, taking tank soundings is performed by reading the computer in the engine control room. Often these computers have a built-in conversion factor to convert volume to mass. To be aligned with the Regulation, one needs to ensure these conversion factors are correctly updated.

(c) *Flowmeters for applicable combustion processes*

Flowmeters come in two versions: volume flowmeters and mass flowmeters. The latter is the easiest but also the most expensive type. To follow the procedure correctly, one needs a flowmeter on each feedline to the engines and another one on the return line. The difference between the two represents fuel combustion consumption. In case of the use of volume flowmeters, one needs a temperature recording combined with the density figure (from the BDN) to calculate mass. Additional complications occur if one has mixed two or more fuel deliveries with different densities into the day tank.

As with the tank sounding options, one needs to take these measurements at the time of berthing and unberthing.

(d) *Direct CO_2 emissions measurements*

The final option is quite different, at it assumes measurement of the CO_2 mass in the exhaust. To do this one needs to measure the concentration of CO_2 in the exhaust and the total mass of the exhaust. How to measure the latter is still a topic for discussion. Add to this that a vessel may likely have more than one exhaust pipe, so multiple systems will have to be installed. A further complication arises if the exhaust in a pipe is a combination of exhaust from two or more engines running different types of fuel. The MRV regulation requires one to report the consumption of the various types of fuel, which is impossible to do by a backward calculation from the CO_2 emissions.

Even though the MRV Regulation is for CO_2, it is interesting to note that the sulfur content of the fuel is also recorded. According to Annex I of the Regulation, when performing fuel measurements according to either method (a) or (b) (stocktaking), 'the fuel type and the sulphur content need to be *specified*'. However if method (c) is used, 'the fuel type and the sulphur content need to be *monitored*'. Such information may conceivably prove useful in the quest of enforcing the global sulfur cap from 2020 on, even though there are currently no plans to use it to that effect. It is also noted that in either method, there is no place in the mandatory reporting template to insert this data.

Articles 6 and 7 deal with the Monitoring Plan (MP) and Articles 8–12 deal with the actual reporting requirements.

6.3 *Verification*

Chapter III, Articles 13–16 deal with the verification process.

Verifiers must be accredited by an EU National accreditation agency (Article 16) and must be independent 'from the company or from the operator of a ship' (Article 14).

The verifiers have two distinct tasks. The first is to *assess* the Monitoring Plan (MP). In other words, ensure that it complies with the requirements laid down in the Regulation, especially these mentioned in Articles 6 and 7. The second task is to *verify* the Emission Report (ER). This entails to ensure the data is collected and stored according to what is described in the MP and then check for the correctness of the submitted data.

Now that the initial deadline for submission and assessment of the monitoring reports has passed (August 31, 2017 and December 31, 2017, respectively), the MP assessment task will be an ongoing one. This will include reassessment of existing MPs where changes are made. Such changes could be either due to new technology being installed or to new data sources or to a change of responsibilities as laid down in the MP. Assessment of new MPs will be required for vessels entering EU for the first time either because of trading pattern now includes a port in scope or in the case of newbuilding's being delivered.

The ER verification has a fixed window every year from January 1 to April 30 when all needs to be submitted to the Commission – in reality to the 'Thetis-MRV' system (see Sect. 6.7) and to the respective flag states. One must anticipate that these 4 months will be quite hectic. The yearly report must include all voyages that have commenced in the year in question. A tanker sailing in ballast from North Europe to the Arabian Gulf departing late December and going south of Africa will take about 45 days to complete the voyage. Then it may anchor up outside of the port area waiting for cargo for a period before it berths. This voyage needs to be included in the annual report and needs to be verified and submitted by April 30. It will be a close call to achieve this.

Assuming there are 12,000 vessels under obligation to report to the EU MRV and that shipping companies take the first 2 months of the following year to conclude ongoing voyages and compile the ERs, that leaves the verifier community 60 days to audit them all, or some 200 audits per day. Time will show whether this is feasible.

6.4 Control and Communication

A distinct wish from the outset was to ensure transparency of the data collected under the MRV scheme. To ensure this the Regulation introduces a Document of Compliance (DoC) that must be on board every ship as from June 30, 2019. This DoC is issued by the verifier and confirms that the vessel has submitted verified MRV data for the preceding year. The DoC is valid for 18 months to ensure an overlap from period to period. Without this DoC on board a vessel is technically in non-compliance and subject to penalties should a port-state inspection observe this. Article 19 instructs member states to ensure that vessels calling in their territory are checked and that a proper penalty system is in place. It is beyond the power of the EU to set levels of penalties, this being a national prerogative, but member states are encouraged to ensure these are 'effective, proportionate and dissuasive' (Article 20). Article 20 gives some specific guidance with respect to repeated non-compliance. Here nations are guided to issue expulsion order preventing the offending vessel from entering EU waters.

Article 21 provides more clarity of what this will entail. The summarized annual data per vessel will be available as of June 30 of the following year. The data and level of details is quite specific:

(a) *The identity of the ship (name, IMO identification number and port of registry or home port)*
(b) *The technical efficiency of the ship (EEDI or Estimated Index Value- EIV, where applicable)*
(c) *The annual CO2 emissions*
(d) *The annual total fuel consumption for voyages*
(e) *The annual average fuel consumption and CO2 emissions per distance travelled of voyages*
(f) *The annual average fuel consumption and CO2 emissions per distance travelled and cargo carried on voyages*
(g) *The annual total time spent at sea in voyages*
(h) *The method applied for monitoring*

(i) *The date of issue and the expiry date of the document of compliance*
(j) *The identity of the verifier that assessed the emissions report*
(k) *Any other information monitored and reported on a voluntary basis in accordance with Article 10*

Exactly how the Commission intends to publicize this data for all ships (and on the same date as the submission deadline) remains to be seen.

Finally the Regulation has a section, putting the obligation on the European Commission to review an international agreement that has a similar objective and amend the MRV Regulation accordingly. This section is clearly aimed at the IMO. The IMO committee that deals with these matters – the Maritime Environment Protection Committee, MEPC – is aiming at launching its own data collection system (DCS) starting January 01, 2019. The Commission therefore has an obligation to review this, a process that started in the fall of 2017. It is important to note the obligation is to 'review' only, nothing more. The key differences between these two data collection systems are being dealt with in Sect. 7 of this chapter.

6.5 The End Result

Annex II of Commission implementing Regulation (EU) 2016/1927 contains the template for the annual vessel-specific emission report.

First of all one must provide from various identification data such as vessel, company and verifier details. Most of this makes good sense. There is however also a mandatory requirement to mention the Energy Efficiency Design Index (EEDI) or the Estimated Index Value (EIV) as appropriate. This provides a bit of a problem. Whereas the EEDI is a mandatory figure for all vessels delivered after July 1, 2015, providing the EIV is a bit more tricky as it does not apply to all ships and cannot be calculated. EIV was established according to Resolution MEPC.215(63) in 2012 and covers a certain number of ships built between January 1, 1999, and January 1, 2009. EIV is calculated using the data extracted from the IHS Fairplay ship database at *that time*. So EIV as per the official definition is a static figure for certain ships that cannot be reproduced.

Annex II provides good guidance on the data to be reported. Initially it is about the fuel and CO_2.

(a) *Amount and emission factor for each type of fuel consumed in total*
(b) *Total aggregated CO_2 emitted within the scope of this Regulation, expressed in tonnes CO_2*

This is where the 'direct emission monitoring' method becomes problematic. Even if one could calculate the CO_2 directly from the exhaust by measuring the CO_2 concentration in combination with the total mass, to be able to split that CO_2 into fuel types if more than one fuel type is used and mixed in the exhaust would be a mathematically impossible task.

The aggregated CO_2 in (b) above then needs to be specified in to four elements:

1. Internal EU voyages
2. Voyages leaving the EU
3. Voyages entering the EU
4. Port calls

There are further some special conditions for RoPax vessels and some voluntary data for different kinds of trading and/or vessels.

The following elements stipulate the need to mention total distance and time at sea.

Then we get in to the section for 'transport work'. This is quite vessel specific depending on the vessel type but generally cargo tons loaded multiplied with the distance. It is this condition that cannot be met for vessels applying the exception for bulk reporting (over 300 voyages p.a. as per Regulation article 9.2). If all voyages are not of identical length, one cannot calculate this figure correctly unless one calculates each individual voyage.

Also one must provide are all the calculated performance measurements.

- Fuel consumption per mile sailed
- Fuel consumption per transport work unit

Should one so desires, there are further voluntary figures one may provide. For all the voluntary data, it is important to remember that although data is voluntary, if they are submitted, they are subject to official verification.

6.6 MRV and the ESSF

The European Sustainable Shipping Forum (ESSF) was established leading up to the implementation of the Sulfur Emission Control Areas (SECAs) in the Baltic, the North Sea and the English Channel with effect from January 1, 2015. It is a forum where stakeholders – industry, nongovernmental organizations (NGOs) and other interested parties – can offer technical advice to the EU Commission and the national authorities. The structure is with a plenum that meets twice a year and then a number of topic-specific subcommittees. Whereas the ESSF was originally set up to deal with issues related to the Sulfur Directive, it has subsequently been expanded to a wider scope including different maritime environmental issues related to the maritime industry.

In connection with the decision to implement the MRV regulation, two subgroups under ESSF were established.

- The MRV monitoring subgroup
- The MRV verification and accreditation subgroup

The task of these subgroups has been to assist the Commission with drafting delegation acts and guidance papers.

6.7 *Thetis-MRV*

Throughout the MRV process, the European Maritime Safety Agency (EMSA) has assisted the Commission with practicalities around the entire process, not least carrying out a lot of work assisting the ESSF workgroups. EMSA was also tasked with setting up the database, and accompanying systems enable participating entities to manage the reporting and associated tasks. EMSA already operated a data system called Thetis, and for managing the CO_2 reporting process, a subsystem was developed and named Thetis-MRV.

The Thetis-MRV system basically has four types of interested parties:

1. Companies
2. Verifiers
3. Flag states
4. EU Commission

The systems key feature is to control certain sets of mandatory data. A company needs to record for which ships it is responsible to submit MRV data. The companies are also required to submit the annual ER.

The verifiers need to register in the system and must use the system to communicate on any Document of Compliance they have issued. The process also necessitates that once the companies have entered an emission report for a ship, the verifier needs to verify these data entries in the system.

The Thetis-MRV system is not fully developed at time of writing, and therefore it has not been possible to test all its functionality yet.

6.8 *Practical Implementation*

This section addresses a limited sample of practical issues related to implementation. As it is mainly based on the second author's involvement in the Ro/ro sector, it may not necessarily be applicable universally.

The first question one asks when starting on the MRV voyage is 'Which vessels am I responsible for?' A company may operate own and chartered tonnage, and some of the owned vessels are chartered out to other entities. Like most shipping entities, the legal owner of a ship is not necessarily the operator, and the technical and commercial operator may well be different entities. So the question is very relevant. Unfortunately, as touched upon, earlier the MRV definition of 'Company' does not provide full clarity on this. So the first important decision was to establish one's own policy. A policy might be 'If we are the International Ship Management (ISM) responsible, we are the MRV responsible'. This is what most of the industry is going with. As one has to register the vessels under a company's responsibility in the Thetis-MRV, there is no risk that the same vessel will be subject to two companies being responsible for its MRV. The opposite is however not so certain that a vessel will end up with no company assuming responsibility for the MRV process.

Once one has established his fleet, he needs to decide whether all ships are in scope. Vessels with no planned voyage to the EU can be eliminated immediately, but what about vessels where ISM responsibility has expired before year end 2017? Where do they belong to? What about new buildings being delivered during the 2nd half of 2017; what is the deadline for completing the MP for such vessels? This is when one comes to appreciate the total lack of official control built in to the MRV process. One can trade a vessel in Europe right up until June 30, 2019, without having a monitoring plan and without doing the CO_2 monitoring, and no one will ever notice. Only if one does not have the DoC on board to show during a port inspection post June 30, 2019, will the system recognize any non-compliance.

The next step in the process is to match what systems and data recording one has today and how that fits in to the MRV requirements for the future. Having the ability to reuse systems and data is naturally much sought after. Shipping companies generally have a system for documenting fuel consumption, so on the surface of it here is ground for reuse of data. Most performance systems are in place for financial controls. For instance, has one received the bunkers they have paid for and how much fuel cost should be allocated to each voyage? That is not necessary the same as how much fuel was fed into the engines.

When reporting MRV for a RoPax vessel, one needs to report one set of figures for the passenger segment and one for the freight segment, so the initial problem is to split a vessel's total CO_2 emissions into these two subsegments. Annex II, A, 1, (e) of the Regulation provides the first bit of assistance. One may use the principles in ISO 16258, Annex B. Note that one is free to choose another option for this. Should one elect the ISO 16258 as the way forward, Annex B provides two options, the mass method or the area method. The 'area method' is open for interpretation on what to include or not.

And all this is just for the initial splitting of a vessel's CO2 emissions into 'passenger emissions' and 'freight emissions'. Does this make much practical impact on the emissions? To illustrate this we have taken a vessel and applied the two ISO 16258 principles below to the following sample voyage:

Passengers on board: 460
Freight on board: 1750 tons
Voyage CO_2: 90 tons
Area split (area method): passengers 34%, freight 66%

Then, according to the method used, the emissions allocation is shown in Table 11.3 below.

Table 11.3 CO_2 emissions allocation to passengers and freight (tons)

	Mass method	Area method
Passengers CO_2	2.3	20.6
Freight CO_2	77.7	59.4

The fact that one can, within the parameters of the Regulation, get such a spread of permissible data to submit should warn anyone of using this type of information for assessing of any type of tax or levy system.

7 The IMO Data Collection System (DCS)

With Resolution MEPC.278(70), the IMO in October 2016 introduced their version of a CO_2 reporting system. This is the data collection system for fuel oil consumption of ships, or in short DCS.

Generally mapped on the EU MRV Regulation, there are some significant differences and one that is strange. Let us take the latter first. The IMO system applies for vessels of 5000 GT and above. So any vessel of exactly 5000 GT will only have to comply with one system.

The DCS system is implemented as an amendment to MARPOL Annex VI and by expanding the Ship Energy Efficiency Monitoring Plan (SEEMP) with a second chapter covering this procedure. By doing it in this way, the responsibility for compliance ends up with the vessel's flag state, who then also must ensure verification of the submitted data.

Unlike the EU system, the IMO system covers *all* voyages, and the rather complicated computation of transport work in the EU MRV is replaced by simply using the vessel's deadweight (DWT) instead. This will make the IMO system much simpler to administer.

The annual data must be submitted to the flag state who in turn has an obligation to report this to the IMO. Under the IMO system, individual ships will not be identifiable as vessel data will be anonymized before publication.

In the autumn of 2017, the EU commenced the process of finding out whether the IMO DCS could replace the EU MRV. This review process was a requirement laid down in the formal text of Regulation 2015/757. The initial step was to do a round of consultation with the various stakeholders.

As the time of writing, this process is not finalized but it is doubtful that the EU will find the IMO DCS sufficiently detailed and transparent to satisfy the Commission's desires. The EU has a seat at MEPC where the DCS was and is discussed, and throughout the drafting process, the EU delegates have argued for a closer alignment of the IMO system to that of the EU. But such an alignment eventually could not gather sufficient support.

The areas where the EU stance are quite firm are firstly the need for the transport work inclusion. Using a DWT figure as per the DCS says nothing about how efficient a vessel is in going about its business. The IMO system will collect data; the EU wants this data qualified in relation to productive work done.

In the area of publication or transparency, there is also a wide gap between what EU and the IMO wants. EU wants the public to be able to look at a specific ship's performance, something the majority of MEPC voters are against.

8 What Is Next?

All this data gathering is of little value unless it is put to something useful. Realistically one would need minimum 2 but preferably 3 years of reporting before the cumulative data in the MRV database can be analyzed and conclusions reached. It will be therefore 2021 before anyone can start looking and analyzing the data. From the outset the MRV was modeled to pave the way for a duty system on CO_2 emissions. It is very doubtful whether the first 3 years of data will be of such a quality and consistency that a levy/tax system based on a Key Performance Indicator (KPI) of 'CO_2 per transport work–mile' will have any meaning whatsoever. The total emitted CO_2 figure should however be fairly robust. If this is eventually the conclusion, then we are not far from what the IMO will collect according to their DCS, and a compromise may be achievable that could entail one global system.

Another issue that has not yet been discussed would be to use the MRV system to collect other relevant emission data. With the MRV system in place and routines established, it will not take much to expand it to cover other harmful emissions. The first such target could very well be nitrogen oxides (NOx).

9 Conclusions

This chapter has presented an overview of the main issues associated with MBMs and MRV. As mentioned earlier, the two issues are connected since the MRV discussion essentially started when the MBM discussion was suspended in 2013 and since MRV is a critical step for any eventual MBM implementation in the future.

It is clear that MRV by itself cannot directly lower CO_2 emissions, even though increased awareness of a ship's fuel consumption may induce the ship owner to adopt measures to reduce it. More importantly, MRV can be the first necessary step for subsequent measures to effectively reduce emissions. In that sense, the suspended discussion on possible MBMs can only resume whenever an efficient and effective global MRV system is established. The same is the case for any other emission reduction measures that may be implemented at the operational level. This means that any MRV system will have to be designed with a longer term view on what will be the next step, after the MRV is established. In our opinion, it is clear that the next step will be an MBM, whose nature would actually depend very much on the nature of the MRV system that will be adopted.

How can MBMs help with the recently adopted ambitious IMO GHG reduction targets? Irrespective of the fact that MBMs seem not to be up for discussion in the foreseeable future, one idea that might be worth considering would be to impose a *significant* bunker levy on a global level. By significant we mean not 10 or 20 USD per tonne of oil, as is being occasionally contemplated by industry, but *at least one order of magnitude higher*. This would induce both technological changes in the long run and logistical measures in the short run. In the long run, it would lead to changes in the global fleet toward vessels and technologies that are more energy efficient, more economically viable and less dependent on fossil fuels than those today. In MAC terms, it would make negative the MAC of many technologies that currently have a positive MAC, thus inducing ship owners to adopt them. In the short run, a bunker levy would lead to slow steaming, which would reduce fuel costs and emissions at the same time.

To understand the link between fuel price and technology used, a parallel to the automotive industry can be made: it is clear that the significant fuel price difference among the USA on the one hand and Europe and Japan on the other hand (ratio of approximately 1–2) is reflected in a similar major difference in these countries' automobile fleet profiles, as well as GHG emissions performance, which for the USA is way behind what it is in Europe and Japan (An and Sauer 2004). There is no serious incentive to build or use fuel-efficient cars if fuel prices are low, and hybrid and electric cars would have no such market penetration today were it not for the considerable state subsidies granted to them. Such subsidies are in fact MBMs, and without them we would not see either the development or the use of such technologies in the automotive sector. That this story has not yet found a parallel in the maritime sector is intriguing.

A maritime bunker levy could also collect monies that could be used to achieve *out-of-sector* GHG emission reductions. However, it would seem self-evident that out-of-sector GHG emission reductions (or *offsets*) should only be seen as ancillary reductions, in the sense that the shipping industry would eventually have little or no control over them. As far as what the industry can influence is concerned, *in sector* reductions seem far more relevant.

How much CO_2 can be reduced by a substantial global bunker levy? Devanney (2010) estimated that with a base HFO price of USD 465/tonne, a USD 50/tonne bunker levy would achieve a 6% reduction in total Very Large Crude Carrier (VLCC) emissions over their life cycle and that for a USD 150/tonne levy the reduction would be 11.5%. Some estimates of CO_2 reductions for tankers and Handymax bulk carriers, and for several bunker levy scenarios, were made in Gkonis and Psaraftis (2012) and in Kapetanis et al. (2014), respectively. These estimates showed CO_2 reductions of more than 50% for a single VLCC if fuel price rises from 400 to 1000 USD/tonne. However, the long-term fleet-level impacts of substantial levies are by and large unknown.

It should be obviously realized that any move in the above direction, even at the study level, would generate strong protests from many stakeholders. For instance, and at today's fuel prices, who would possibly entertain a global bunker levy so that total fuel cost becomes 800 or 1000 USD/tonne? Would the US administration support it, for instance? Could an appropriate legal regime be instituted on a global level? We consider the political prospects of such a measure extremely unlikely. The scheme may also have side effects in specific segments of the market, for instance, in short sea shipping higher fuel prices at sea may potentially shift cargo to land-based modes, ultimately increasing GHG emissions overall. It may also adversely impact the trade of LDCs and SIDS. Such potential side-effects ought to be examined carefully.

References

An, F., & Sauer, A. (2004). Comparison of passenger vehicle fuel economy and greenhouse gas emissions standards around the world. Report for the Pew Center on Global Climate Change, USA.

Bahamas. (2010). Market-Based Instruments: a penalty on trade and development proposal by the Bahamas, IMO doc. MEPC 60/4/10.

Bahamas. (2011). How technical and operational measures are the only direct and effective means to deliver cuts in CO2 emissions, IMO doc. GHG WG 3/2.

Buhaug, Ø., Corbett, J. J., Endresen, Ø., Eyring, V., Faber, J., Hanayama, S., Lee, D. S., Lee, D., Lindstad, H., Markowska, A. Z., Mjelde, A., Nelissen, D., Nilsen, J., Pålsson, C., Winebrake, J. J., Wu, W. Q., Yoshida, K. (2009). Second IMO GHG study 2009; IMO document MEPC59/INF.10.

Denmark. (2010). The International Fund for Greenhouse Gas emissions from ships (GHG Fund) proposed by Cyprus, Denmark, the Marshall Islands, Nigeria and IPTA, IMO doc. MEPC 60/4/8.

Devanney, J. W. (2010), The impact of EEDI on VLCC design and CO2 emissions, Center for Tankship Excellence, USA. www.c4tx.org.

DNV. (2009). Pathways to low carbon shipping. DNV report, 2009.

Eide, M. S., Endresen, Ø., Skjong, R., Longva, T., & Alvik, S. (2010). Cost-effectiveness assessment of CO2 reducing measures in shipping. *Maritime Policy & Management, 36*(4), 367–384.

EU (2013a). Proposal for a regulation of the European parliament and of the council on the monitoring, reporting and verification of carbon dioxide emissions from maritime transport and amending regulation (EU) No 525/2013, COM(2013) 480 final, Brussels, 28.6.2013.

EU. (2013b). Executive summary of the impact assessment, Brussels, 28.6.2013 SWD(2013) 236 final.

France. (2010). Further elements for the development of an Emissions Trading System (ETS) for International Shipping proposal by France, IMO doc. MEPC 60/4/41.

Germany. (2010). Impact assessment of an emissions trading scheme with a particular view on developing countries, IMO doc. MEPC 60/4/54.

Gkonis, K. G., & Psaraftis, H. N. (2012). Modelling tankers' optimal speed and emissions, Archival Paper, 2012 SNAME Transactions, Vol. 120, 90–115.

IMAREST. (2011). Marginal abatement costs and cost effectiveness of energy-efficiency measures, IMO doc. MEPC 62/INF.7.

IMO. (2010). Full report of the work undertaken by the expert group on feasibility study and impact assessment of possible market-based Measures, IMO doc. MEPC 61/INF.2.

IMO. (2018). Resolution MEPC.304(72) (Adopted on 13 April 2018) Initial IMO Strategy on Reduction of GHG Emissions from Ships, IMO doc. MEPC 72/17/Add.1, Annex 11.

IUCN. (2010), A Rebate Mechanism (RM) for a market-based instrument for international shipping proposal by IUCN, IMO doc. MEPC 60/4/55.

Jamaica. (2010). Achieving reduction in greenhouse gas emissions from ships through Port State arrangements utilizing the ship traffic, energy and environment model, STEEM (PSL) proposal by Jamaica, IMO doc. MEPC 60/4/40.

Japan. (2010). The Leveraged Incentive Scheme (LIS) to improve the energy efficiency of ships based on the International GHG Fund proposed by Japan, IMO doc. MEPC 60/4/37.

Japan and WSC. (2011). Consolidated proposal of "Efficiency incentive scheme" based on the Leveraged Incentive Scheme and the Vessel Efficiency System," IMO doc. GHG WG 3/3/2.

Kapetanis, G. N., Gkonis, K. G, Psaraftis, H. N. (2014). Estimating the operational effects of a bunker Levy: The case of Handymax Bulk Carriers, TRA 2014 conference, Paris, April 2014.

Kosmas, V., & Acciaro, M. (2017). Bunker levy schemes for greenhouse gas (GHG) emission reduction in international shipping. *Transportation Research Part D, 57*, 195–206.

Norway. (2010). The Global Emission Trading System (ETS) for international shipping proposal by Norway, IMO doc. MEPC 61/4/22.
Psaraftis, H. N. (2012). Market based measures for green house gas emissions from ships: A review. *WMU Journal of Maritime Affairs, 11*, 211–232 2012.
Psaraftis, H. N. (2016). Green maritime transportation: Market based measures, chapter. In H. N. Psaraftis (Ed.), *Green transportation logistics: In search for win-win solutions*. Cham: Springer.
Smith, T. W. P., Jalkanen, J. P., Anderson, B. A., Corbett, J. J., Faber, J., Hanayama, S., O'Keeffe, E., Parker, S., Johansson, L., Aldous, L., Raucci, C., Traut, M., Ettinger, S., Nelissen, D., Lee, D. S., Ng, S., Agrawal, A., Winebrake, J. J., Hoen, M., Chesworth, S., & Pandey, A. (2014). *Third IMO GHG Study 2014*. London: International Maritime Organization (IMO) June.
UK. (2010). Global Emissions Trading System (ETS) for international shipping proposal by the United Kingdom, IMO doc. MEPC 60/4/26.
USA. (2010). The United States proposal to reduce greenhouse gas emissions from international shipping, the Ship Efficiency and Credit Trading (SECT), IMO doc. MEPC 60/4/12.
WSC. (2010). Vessel Efficiency System (VES) proposal by World Shipping Council, IMO doc. MEPC 60/4/39.

Chapter 12
Green Ports

Thalis P. V. Zis

Abstract Sustainable shipping involves not only ships but ports as their extension. This chapter examines the issues associated with a green port operation. These include technologies such as cold ironing; market-based practices such as differentiated fairway dues, speed reduction, and noise and dust abatement; and others. The legislative framework in various countries is explained, and various environmental scorecards are discussed. This chapter starts with a brief review on recent academic research in the field of environmental management of ports and presents the status quo in leading ports around the world. The chapter emphasizes on the implementation of speed reduction programmes near the port, the use of cold ironing at berth, and the effects of fuel quality regulation, considering the perspectives of the port authority and the ship operator. The emerging environmental and economic trade-offs are discussed. The aim of this chapter is to be a starting point for researchers seeking to work on green ports. Insights of this chapter may also be useful for stakeholders seeking to select the best emissions reduction option depending on their unique characteristics.

Abbreviations

AGV	Automated guided vehicles
AMP	Alternative marine power
BPA	British Ports Association
CO	Carbon monoxide
CO_2	Carbon dioxide
ESPO	European Seaport Organization
ETS	Emissions trading system
EU	European Union

T. P. V. Zis (✉)
DTU Management Engineering, Technical University of Denmark, Kongens Lyngby, Denmark
e-mail: tzis@dtu.dk

H. N. Psaraftis (ed.), *Sustainable Shipping*,
https://doi.org/10.1007/978-3-030-04330-8_12

IAPH	International Association of Ports and Harbors
IMO	International Maritime Organization
ITS	Intelligent transport systems
LNG	Liquefied natural gas
NO_x	Nitrogen oxides
OPS	Onshore power supply
POLA	Port of Los Angeles
POLB	Port of Long Beach
RMG	Rail mounted gantry
RTG	Rubber-tired gantry
SO_x	Sulfur oxides
UNCTAD	United Nations Conference on Trade and Development
VSRP	Vessel Speed Reduction Programme

1 Introduction

Shipping is considered the most efficient mode of transport in economic and environmental terms. Due to economies of scale, it can offer the lowest cost per ton-km transported. The sector's contribution to global CO_2 emissions accounted for 2.2% in 2012 (Smith et al. 2014) down from 2.7% in 2007 (Buhaug et al. 2009) for international shipping. In absolute terms, the CO_2 emissions were reduced from 885 million tonnes in 2007–796 million tonnes in 2012 (Smith et al. 2014). At the same time, maritime transport moves approximately 90% of the world's trade, with increasing trends for transported cargo volumes.

However, its impacts on climate change through greenhouse gas emissions and on human health from air pollutants released near residential centers cannot be ignored. Over the last decades, regulatory bodies have been developing policies that seek to further improve the sector's environmental performance, and at the same time, new technologies improve the efficiency of vessels. Operational practices of ship operators and port authority initiatives are also relieving the sector's impacts.

While there has been significant research on the environmental impacts of maritime transport, there has been relatively little work focusing on the effects of maritime activity in the proximity and at ports. The majority of academic research in the environmental impacts of maritime transport has focused on its overall contribution. However, effects near ports have not been extensively researched, with the majority of relevant studies being technical reports of port authorities focusing on a very broad level of environmental concerns.

1.1 Background

Ports are areas on a shore or coast that contain one or more harbors where ships can call and transfer cargo or people to and from the land. Ports serve as intermodal nodes connecting water and various land modes while also providing other useful services such as shipbuilding, maintenance, and bunkering facilities to the maritime industry. Each port has unique characteristics in terms of operation, layout, volumes handled, geography, and organizational structure. Classification can vary according to the aforementioned characteristics. Alderton (2013) classifies ports into major groups by function (cargo interface, ship/shore interface) or geography (coastal, tidal, artificial, inland, and river). Classification can also be based on size and capability to handle large ships. This chapter will focus on container terminals and their implications to the environment.

1.2 Main Terminal Types and Overall Growth of the Sector

Terminals are essentially facilities within the port that provide several berths to handle vessels and the exchange of cargo goods and/or passengers. A port may have many terminals of different types (and sizes), and each terminal has a primary operator that is in charge of the various operations and is under the control of the port authority. Terminals comprise of the wet and dry infrastructure, superstructure, cargo handling equipment, and human resources for its operations. The wet infrastructure is defined as the harbor basin where one or more berths are in place to receive vessels. The storage area pavement, the roads inside the terminal, and the foundations for the crane tracks and drainage systems are part of the dry infrastructure. The superstructure is referring to the buildings, sheds, and all other covered storage spaces within the terminal. Cargo handling equipment and human resources vary depending on the terminal type and size. The main terminal types can therefore be distinguished into the following:

- Ro-Ro terminals
- Liquid bulk terminals (LNG, crude oil, chemical products)
- Dry bulk terminals (grain, coal, ore)
- Ferry terminals
- Multipurpose terminals
- Container terminals

For all terminal types, there are some services that are common; these include the loading/unloading of vessels, the temporary storage of cargo in the terminal, the processing of cargo (certain types), and the loading/unloading of cargo to the next transportation stage (e.g., before moving to the hinterland). From an environmental perspective, the emission intensity of each activity varies. However, in line with the continuous growth of seaborne trade (as seen in Fig. 12.1), ports are also increasing in size in order to handle the additional throughput and cater for larger vessels calling. Figure 12.1 presents the growth of international seaborne trade during the last decade as reported from data of UNCTAD.

This continuous growth has resulted in larger vessels being constructed and the requirement for ports to handle additional throughput each year. In 2006 the Emma Maersk was introduced as the largest containership ever built with a maximum capacity of 14770 TEUs. Eleven years later the largest containership was the OOCL

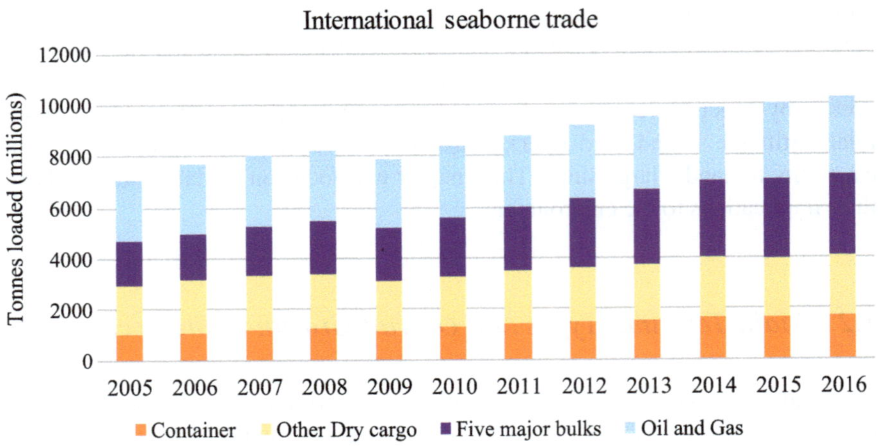

Fig. 12.1 Growth of international trade in recent years. (Data source: UNCTAD 2017)

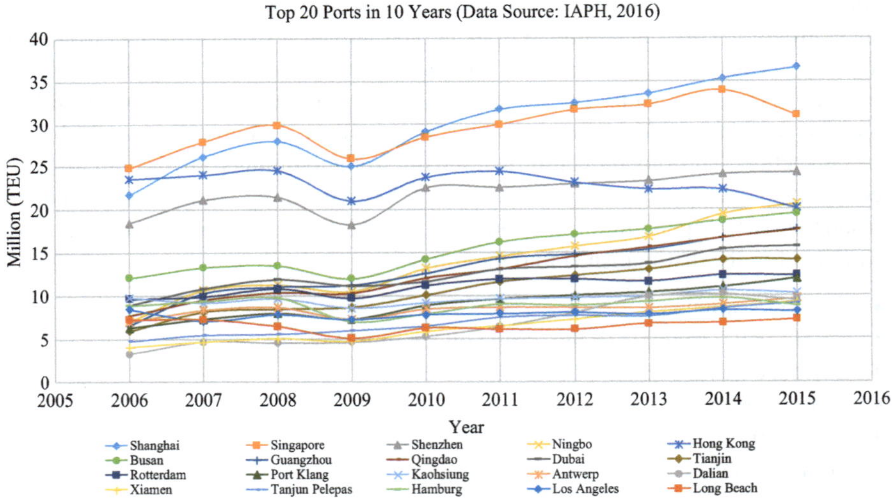

Fig. 12.2 Annual throughput in the top 20 container ports in the world (Data source: IAPH 2016)

Hong Kong with a capacity of 21413 TEUs. Such vessels cannot call at all ports, and as a result port authorities need to invest in additional dredging operations and install ship to shore cranes capable of handling such vessels. Along with the growth of the sector, and the introduction of bigger vessels, container terminals are every year required to handle larger volumes of containers. This is depicted in Fig. 12.2, where the growth in port throughput for the 20 biggest ports in the world is shown.

Figure 12.2 shows that for the majority of the biggest container ports in the world, the handled throughput was increasing between 2006 and 2015, with a notable exception of Hong Kong that has lost volumes, and the 2009 year which showed a drop for all ports, in line with the reduction of the maritime sector due to the financial crisis of 2008. The additional volumes handled will result in a higher environmental burden in the local environment, and it is important that port authorities create green agendas to reduce the negative environmental impacts of their growth.

1.3 What Is a Green Port

Passet (1979) proposed a three-pillar framework of societal, economical, and environmental development to describe sustainable development. The World Commission on Environment and Development (1987) defined sustainable development as meeting the needs of the present, without compromising the ability of future generations to meet their own needs. In transportation, the term sustainable or green transport is also based on a similar framework whereby the right balance of environmental, societal, and economical performance is sought after. Greene and Wegener (1997) note the importance of emissions, fatalities due to accidents, as well as the importance of satisfying the transportation demands of modern economies. Davarzani et al. (2016) conduct a bibliometric analysis on research related to green ports and maritime logistics but do not provide a definition for a green port. Nikitakos (2012) proposes the zero-emission port where any energy consumption within the port's operations is to be covered by in-port renewable energy sources (RES) generation, for example wind turbines or a small photovoltaic park. Arguably, a definition of a green port as being a zero-emission port is very exclusive as the energy demands of ports are quickly increasing with port throughput. In the context of this chapter, a "green port" is a port that has either developed a strategy to reduce emissions, energy consumption in their operations, and water pollution or has invested in new technology with improved environmental performance and in short is trying to become a "greener" port.

1.4 Structure of the Rest of This Chapter

The next section of this chapter presents recent research in academic literature in the field of green ports, ways of measuring the environmental performance, and the basic port operations that have an impact. In the third section, relevant legislation that may affect port performance from the IMO, the EU, or other regulatory bodies is discussed. The fourth section presents a summary of the different options port authorities may select to improve their environmental performance. The chapter concludes with the need for additional academic research in order to optimize the performance of said options.

2 The Environmental Angle of the Port System

The impacts of port operations on the surrounding area can be attributed to three main categories: maritime operations, in-port operations, and generated traffic outside the port's gates. The mechanisms through which each of these contributes to the environmental footprint of a specific port differ in each case, as do the potential mitigation measures. Due to the size of marine engines operated on board for propulsion and electricity requirements of each vessel, the fuel consumption of a large ship can result in massive emissions in each phase of the journey. Of particular concern are the emissions near the shoreline, as the generated emissions of marine engines contain pollutants with severe health effects. Certain port authorities monitor the emissions from each type of operations. Perhaps the most noteworthy example is the Port of Los Angeles (POLA) and its annual emissions inventory. Figure 12.3 presents the breakdown in the busiest container port of the USA.

It is evident that for pollutants with a local environmental impact, ocean going vessels (OGV) are by far the highest contributors with the notable exception of carbon monoxide (CO). It is interesting to notice that despite the very low sulfur limit allowed within the port due to environmental regulations from CARB, and the designation of the US ECA zone, in terms of SO_x emissions, the OGV are still the highest contributors at 93.5% of the total. Cargo handling equipment is the highest contributor in CO terms, which can be attributed to extended times of idling at a port that results in incomplete combustion in the diesel engines powering this equipment. Finally, heavy duty vehicles are the most important contributor in CO_2 terms and an important part of NO_x emissions. Not surprisingly, Californian ports have placed a lot of attention in reducing emissions from trucks, as California has had several problems with very high NOx emissions.

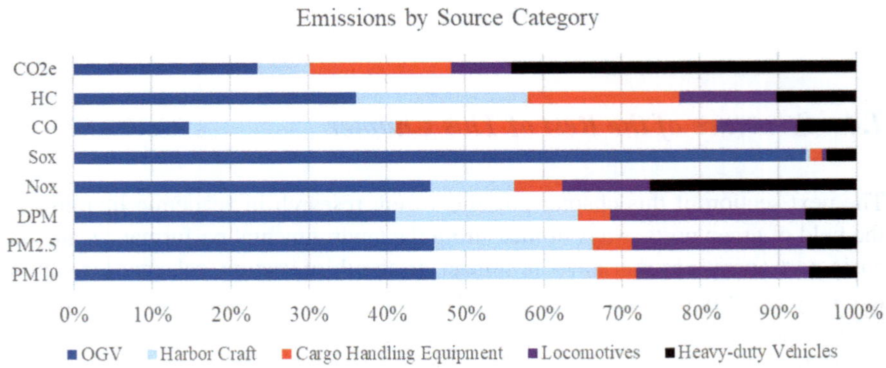

Fig. 12.3 Emissions breakdown by source in the Port of Los Angeles in 2016. (Data source: POLA 2017)

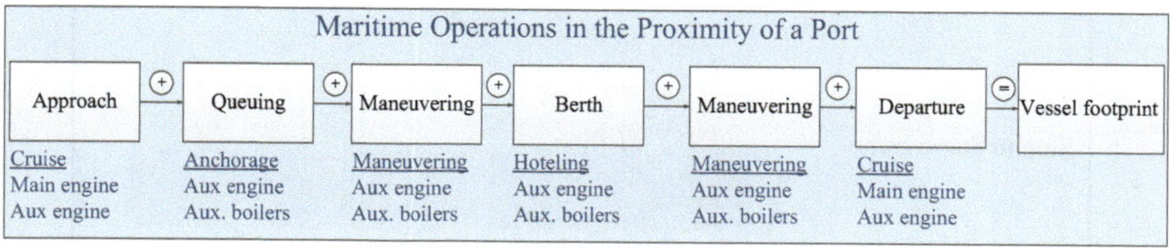

Fig. 12.4 The maritime operations of a vessel calling at a port and the machinery operating during each activity phase

2.1 *Maritime Operations*

For the construction of emissions inventories, the vessel activities near and at the port that are of interest include the approach, maneuvering, hoteling, and departure. These activities are shown in Fig. 12.4 along with the type of machinery that is operating during each phase.

Summing over all activity phases and their respective fuel consumption for each vessel calling at a port will result in an estimation of the environmental footprint of the maritime operations of the port. Such data can either be provided by a port authority or AIS data services that collect the position and speed of vessels. The main activity can be described by the pattern of vessel arrivals at the port and the duration of berth at each call. Ports tend to publish reports on their short-term expected traffic, patterns which combined with a comprehensive dataset of visiting vessels and berth durations could be used to obtain a thorough analysis of emissions within a port. In the event that these data are not retrievable, ship arrivals are usually modeled through Poisson processes which provide a good fit (El-Naggar 2010).

2.2 *Yard Operations*

Once a vessel is at berth, a number of operations take place at the port for the loading and unloading of cargoes and the embarkation/disembarkation of passengers depending on terminal and vessel types. When it comes to Ro-Ro terminals, vehicles need to quickly move from the ship to the yard and vice versa, while trailers need to be moved via either specialized yard equipment or via a truck-trailer combination. The yard operations are more complex in the case of container terminals due to the requirement for much more yard equipment. A typical layout of a container terminal and the three main areas of containers exchange is shown in Fig. 12.5.

Stopford (2009) defines the quayside as being comprised of several berths each serviced by one or more ship-to-shore (quay) cranes able of lifting containers weighing up to 40 tons. These cranes are generally rail mounted to move along the quay for positioning at the required place with respect to the berthed ship. They

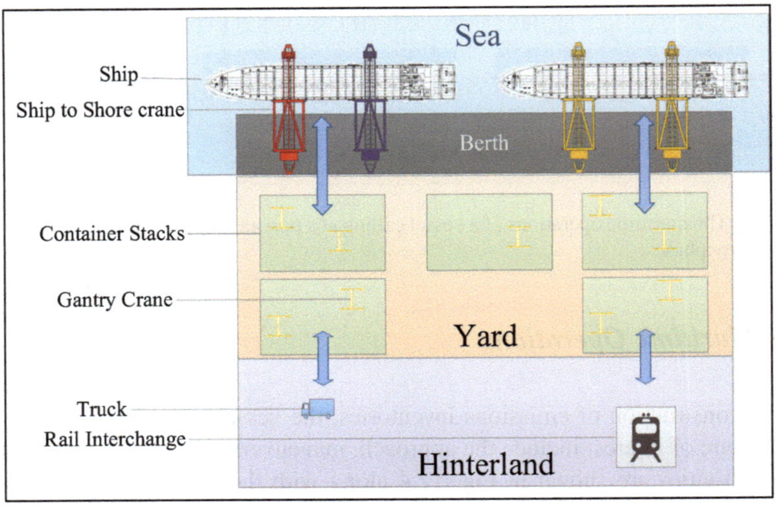

Fig. 12.5 Container terminal layout. (Adapted from Zis 2015)

are classified by lifting capacity and the maximum size of a container ship they may handle. The main categories are Panamax cranes which can handle a ship of 12–13 containers wide, the post-Panamax (18 containers), and the super-post-Panamax which have a reach that reaches 25 containers to handle the largest containerships. A super-post-Panamax crane may weigh up to 2000 tonnes and cost up to 14 million USD (Port Everglades bought three such cranes for a total of 41.4 million in June 2017). Due to the vast weights of the cranes, the quay needs to be strengthened to tolerate loads. Quay cranes can be the cause of a major bottleneck in the terminal's operation slowing down fast ship handling operations and increasing turnaround time (Imai et al. 2008). As technology improves, the quay cranes become faster able of lifting two containers at the same time and increase the number of maximum moves per hour. Energy efficiency also increases by taking advantage of hybrid technologies and energy regeneration when the cargo is lowered.

Container terminals require large storage spaces for the containers which may stay at the port several days. The stacks where the containers are placed ideally should be near the berth for fast unloading of the vessel. This area is typically called the yard where containers are stored in multi-tiered stacks which for ports with very limited area resources can reach up to 12 container tiers (e.g., Hong Kong). The transportation between the quay and the yard differs from port to port depending on size, throughput handled, and resources available. The most common machinery used are forklift trucks, reach stackers, chassis-trailers, straddle carriers, and automated guided vehicles (AGV). These vehicles pick up the container once the quay crane has unloaded it from the vessel (in the chassis and AGV case, the container is placed on top of their platform) and move it close to the storage stacks and vice versa for outgoing containers. Cargo handling equipment is required for the horizontal and vertical movements of containers at the stack (reshuffling of containers). The typical machinery involves:

- Rubber-tired gantries (RTG) which are flexible but cause high loads on the pavement.
- Rail mounted gantries (RMG) that are more appropriate for larger stacks but are more expensive.
- Automated stacking cranes, which are expensive to acquire and maintain, reduce labor costs.

Containers that are destined for the hinterland will have to be moved from the stacks in the storage yard to the stacks in the hinterland side before being boarded on the locomotives or heavy goods vehicles. These movements inside the yard are usually performed by stacking cranes. Energy losses are often observed due to relocation of containers or inefficient transportation due to congestion problems in the yard (Steenken et al. 2004). There has been significant research that aims to minimize the number of shuffle movements of containers at the yard, as well as on inland intermodal terminals facing the same problem (Colombaroni et al. 2017).

2.3 Hinterland Side

The final (or first process) occurs at the gate where export containers leave the port for their inland destination while import containers arrive, respectively. The busiest terminals use advanced information technology to reduce congestion at the gate and waiting times for trucks. As shown earlier in Fig. 12.3, the operations at the gate are a very significant contributor in most pollutant species generation.

In all of the aforementioned processes where containers are moved, significant energy is required. The source of this energy varies depending on the equipment used and whether this consumes fossil fuel (e.g., diesel engines), relies on electricity provided by the grid, or is a hybrid system. An estimation of the energy needs can be performed through analytical calculations based on the horizontal and vertical movements of containers from one place to another inside using basic energy models (equipment specifications, mass of container, speed of movement, and height differences are necessary inputs) or using simulation tools. The next section summarizes the main environmental challenges that ports are facing nowadays.

2.4 Environmental Challenges in Ports

The negative environmental effects of port operations are increasing with the growth of handled throughput. Port operations have both direct and indirect environmental impacts that regulators, shippers, and port authorities have been trying to address in recent years. The major environmental impacts are air and water pollution, depletion of fossil fuel due to the energy requirements of port operations, noise, and optical intrusion (Talley 2009).

Discharge of ballast water, dredging operations at the port, waste disposal, and oil spillage may all contribute in water pollution near the port. Large vessels carry massive amounts of water in their ballast tanks that is used to stabilize the ship. When cargo is removed, the ship pumps in water to compensate for the change in cargo weight distribution. When the cargo is loaded, the ballast water is discharged. The environmental concerns with ballast water treatment occur when it is discharged in different areas (pumped in in one port, released in a different port); it can lead to the unintentional invasion of nonindigenous species. These microorganisms can damage aquatic ecosystems and create health issues (Mooney 2005). A similar problem may occur with the transportation of nonindigenous through hull fouling of a vessel (Drake and Lodge 2007). The aquatic environment can also be negatively affected when dredging operations to increase the port's depth are taking place. Finally waste generated onboard a vessel has to be disposed in non-harmful ways, and ports are expected to be able to provide waste disposal solutions. Oil spillages can occur anywhere along the journey of a vessel including near the port with severe environmental consequences.

The visual intrusion or aesthetic pollution is the result of the vessels, cargo handling equipment, and port superstructure altering the appearance of the environment around the port. Together with the noise generated during port operations, and the lighting pollution during night-time operations, these have a severely negative effect on nearby residents particularly in terms of sleep deprivation and increase of stress. Noise is a serious concern these days for transportation, with a particular focus on noise from airplanes. Various strategies have risen to address the issue of noise from airport operations. For example, alterations in the approach of aircrafts to the airport, steeper descents to minimize exposure to residents, and adaptation of new technologies on the aircraft engines have been utilized. Parallels to the maritime sector exist; however, for ports the main source of noise pollution is the yard and hinterland operations and not so much the vessels themselves. A very different environmental concern for port operations is the effects of noise to marine mammals from maritime transport.

Air pollution in ports is the result of vehicle and cargo movements (ships, cargo handling equipment) and has both local and global consequences. Various different pollutant types are emitted, some of which affect the local air quality, while others are climate change forcing agents. Currently dealing with air pollutants is the most pressing issue port authorities, shippers, and regulators are trying to address with the majority of existing policies and port initiatives. The next section conducts a literature review on academic studies in the field of environmental impact of ports, focusing mainly on emissions.

2.5 *Emissions in Ports in Academic Literature*

A limited number of studies have specifically focused on emissions in ports and their surrounding regions. The review of Davarzani et al. (2016) suggests that the topic of green ports is at a very early stage, but it will continue to grow as practitioners and governments continue to face challenges that research can solve. There are various academic studies that construct emissions inventories in specific ports. Saxe and Larsen (2004) modeled NO_x and PM emissions in three Danish ports. They also model the dispersion of the pollutants and the maximum concentrations in nearby areas using meteorological air quality models. Marr et al. (2007) used a network of emissions monitoring stations in the harbor of Aberdeen to identify the most important pollutants and create an emissions inventory in the area for all transport activity. For ship emissions (mainly ferries), they sampled emissions from the ship funnels to model emissions.

De Meyer et al. (2008) use a bottom-up activity-based model to estimate emissions from international shipping in the Belgian part of the North Sea and four major Belgian ports. They compare their results to the national inventories of CO_2, SO_2, and NO_x emissions in Belgium and find that for the latter two, the contribution is very high (30 and 22%, respectively). Liao et al. (2009) compare the CO_2 emissions generation from trucking transportation with intermodal coastal shipping that incorporates in-port emissions in Taiwan. They show that a shift toward maritime modes will lead to carbon emissions reduction. However, they do not consider other pollutant species in which the maritime sector is less eco-friendly.

Tzannatos (2010) constructs an emissions inventory for NO_x and SO_2 in the port of Piraeus using a bottom-up approach with average load factors for each ship activity mode. He concludes that the port of Piraeus is responsible for 1.2 and 2.5%, respectively, of the total national transportation contribution in Greece. He also calculates the external costs of emissions and concludes that the majority of these are stemming from coastal passenger shipping due to high speed ferries. Berechman and Tseng (2012) construct comprehensive emissions inventories for all ship types and trucks in the port and conclude that tankers, container ships, and bulk carriers are the most polluting ship types. Ng et al. (2013) created an emissions inventory for the port of Hong Kong, based on AIS data for ship movements. Song (2014) did similar work for the port of Yangshan in Shanghai and additionally evaluated the social costs of ship emissions near the port. More recently, Dragovic et al. (2018) focused on near-port emissions from cruise vessels and arising externalities in the cruise ports of Dubrovnik and Kotor. Cullinane et al. (2016) used a bottom-up methodology to estimate emissions at berth from containerships in Taiwan and suggested emissions reduction actions to quantify their potential.

Regarding yard operations, the majority of academic literature is focusing on simulation of said operations or in optimization problems. For more information on the current status of in-yard operations on container terminals, the seminal literature paper of Steenken et al. (2004) provides a good overview of OR problems. Carlo et al. (2014) conduct a more recent literature review focusing on storage yard operations and suggest new research topics. There are certain research papers that focus on the interchange between quay and yard, focusing mainly on berth scheduling and quay-crane allocation. Zhou and Kang (2008) minimize the waiting time of vessels at berth in a stochastic environment. Zeng and Yang (2009) utilize a hybrid simulation/optimization approach for the container scheduling problem in yard operations considering quay and yard cranes. The seminal work of Golias et al. (2010) maximizes berth productivity and considers fuel consumption and arising emissions from vessels at berth. On purely yard operations, Angeloudis and Bell (2010) present a dispatch algorithm that minimizes delays of AGVs and increases port productivity, which will have indirect environmental benefits due to the minimization of energy consumption at the yard.

On the hinterland side, most papers focus on ways of minimizing queues at the gates or contemplate reward systems for booking slots. Aregall et al. (2018) recently conducted a literature review that focuses on the landside of port operations. Their paper is among the first to present the current status of green agendas in ports around the world with a focus on hinterland operations and categorize common measures as technological, infrastructure, or monitoring of activities. Chen et al. (2013) reduce truck emissions at the terminal by optimizing the arrival patterns. Of course, an improvement in one type of operations may result in benefits in other areas as well. Zhao and Goodchild (2010) show that by relieving bottlenecks caused at the port's gate through improved planning, the turnaround time of vessels can also be reduced.

Cao and Golias (2013) evaluated the effects of gate strategies on emissions reductions at marine container terminals. They developed a traffic simulation model capable of measuring the impact of various gate strategies on congestion at terminal gates. The proposed model was used to quantify both travel time and delay, and emission levels at terminal gates before and after gate strategies have been implemented. Each terminal was modeled as a series of tolls that were part of the network. This approach allowed a more accurate estimation of entrance and exit gate delays, equipment inspection delays, and wait time before the gates open and lane restrictions.

This section presented the main environmental problems that port operations are causing and showed the main research areas seeking to address these problems. The next section of the chapter will present the relevant regulation that affects port operations.

3 Relevant Regulation

In response to the growing concerns on the environmental impacts of transportation, a number of regulations and policies have been developed. This section presents the most important legislation affecting port operations. The most important associations of port authorities and their efforts are also discussed to set the scene for necessary research in the coming years in the field of green ports.

3.1 *The International Maritime Organization*

The primary regulator of maritime transport is the International Maritime Organization (IMO). In 1973 the IMO formed the Marine Environmental Protection Committee (MEPC) to address matters concerned with marine pollution. In the same year, MEPC adopted the International Convention for the Prevention of Pollution from Ships, known as the MARPOL Convention. Its aim is to prevent air pollution and address sewage, waste, garbage, and oil spillage and is applied to 99% of the world's merchant tonnage. The MARPOL Convention has been amended by two Protocols in 1978 and 1997. In this book, oil pollution is discussed in Chap. 5.

Emission Control Areas (ECAs) and in particular Sulfur ECAs (SECAs) are discuss in Chap. 7 of this book. From a port authority's perspective, the designation of SECAs is something that will result in less emissions during the approach and departure of the vessel. During the hoteling activities of the vessel, typically MGO or MDO is used from the vessels that are low on sulfur. However, in theory the SECAs could result in a loss of throughput for the ports, as some ship operators may opt to call at ports that are not within a SECA.

3.2 *European Union*

The European Union has long considered ports as vital for economic growth. In 2012 74% of the European trade was shipborne (ESPO 2012), while Eurostat estimates that 37% of the total intra-EU exchange of goods passes through some of the EU ports. Despite the importance of the port sector, the EU faces significant challenges including bottlenecks due to hinterland congestion and investment requirements to accommodate future growth. The environmental implications of sulfur in fuel in Europe were first considered through the Directive 93/12/EEC of March 1993 which regulated the sulfur content of certain liquid fuels. The Directive prohibited marketing of fuel up to 0.2% and 0.05% sulfur content (by weight) for fuel in all transport modes by October 1994 and October 1996, respectively. Vessels sailing between a member state and a third country were excluded from this regulation.

In 1999 this directive was amended through the Council Directive 1999/33/EC which essentially changed the limit of sulfur to 0.1% by the year 2008. The amended Directive required for the first time that from January 2003 heavy fuel oil with sulfur content exceeding 1% would be banned from use within the territory of a member state. The Directive would provide a period of no more than 6 months with a higher limit of sulfur for certain member states. These are the ones that could not apply the limits due to complications in the supply chain of crude oil and petroleum products.

The first effort of the EU to specifically address sulfur emissions from shipping came through Directive 2005/33/EC. It acknowledged the importance of the SO_x ECAs designated by the IMO and placed a limit of a maximum of 0.1% sulfur by weight fuel used by inland waterway vessels and ships at berth in community ports. Furthermore, it banned the use of heavy fuel oils exceeding 3% sulfur content in the territorial seas of each member state. Territorial waters are defined internationally as 12 nautical miles from the baseline of a coastal state under the 1982 United Nations Convention on the Law of the Sea – UNCLOS.

Placing sulfur limits within inland waterways and on vessel activity at berth signifies how important the EU considers the SO_x emissions to be near residential areas. In order to ensure proper use of fuel, the Directive requires all fuel switching operations to be recorded in ships' logbooks. In addition, the Directive allows the use of either shoreside electricity while at berth or alternative emission reduction technologies that would result in at least equivalent reductions to those achieved with the use of low-sulfur fuel. While there is currently no cold ironing targeting regulation, the 2005/33/EC as well as the will of the EU to promote the use of renewable energy sources should facilitate the use of AMP in European ports. An additional step was Directive 2014/94/EU that stipulates that from the 31st of December 2015, all EU ports will be required to have some capability of providing shore power.

3.3 California Air Resources Board (CARB)

The California Air Resources Board (CARB) is a part of the California Environmental Protection Agency and was setup in 1967 to attain and maintain healthy air quality. California had arguably the worst air quality and the highest levels of air pollution due to the largest number of cars in the USA. The two largest container terminals by volume in the USA are the ports of Los Angeles (POLA) and Long Beach (POLB). CARB has developed regulations targeting specifically transport activity in these ports. While the coasts of California are in the North American ECA, there were already stricter limits in place for maximum sulfur fuel content allowed for OGVs. There were two phases in the fuel requirements for OGVs in California. Phase 1 had an upper limit of 1.5% for MGO and 0.5% for MDO effective from July 2009 which would then be changed to 1% and 0.5%, respectively, from August 2012 (CARB 2012). The regulation in lieu (Phase 2) became effective in January 2014 and restricted the use of fuel with sulfur content by weight exceeding 0.1% by any machinery onboard a vessel within 24NM of the Californian coast.

The fact that both POLA and POLB are under the same regulation facilitates the operation of the ports. CARB has also promoted the use of alternative technology and in particular the provision of shore power. In 2007 the "Airborne toxic control measure for auxiliary diesel engines operated on Ocean-Going Vessels At-Berth in a California Port" Regulation (widely known as At-Berth Regulation) was approved. The regulation targets passenger, container, and refrigerated cargo ships berthing in any of the Californian ports. It currently dictates that ships must reduce by 70% the at-berth emissions from auxiliary engines for at least 70% of their calls in

Californian ports. This can be achieved either by turning engines off and connecting to other source of power or by using alternative control techniques that achieve similar reductions for PM and NOx emissions. It applies to POLA, POLB, and the ports of Oakland, San Diego, San Francisco, and Hueneme, for fleets with more than 25 annual visits (more than five for passenger vessels). This percentage will increase to 80% by 2020.

An important challenge with regard to AMP is the difficulty of accessing the AMP-ready berth which can already be in use by another vessel. Fleets are not in control of their allocated berths, while there are still compatibility issues faced between the dock facility and the ship. For some ship owners, the use of alternatives to shore power may be preferable economically as an emissions reduction method considering retrofitting costs for the vessel.

3.4 Port Associations

There are various port associations globally with the task of representing port authority members, the most famous of which is the International Association of Ports and Harbors (IAPH). Others include the American Association of Port Authorities (AAPA), the European Sea Port Organization (ESPO), and the British Ports Association (BPA). Such associations share the common objectives of representing their members, providing guidance toward more efficient operations and promoting the exchange of experience on successful green strategies developed by port authorities around the world. Unfortunately, the majority of objectives are monitored in a very qualitative manner usually revolving around the suggestion of good practice guides and are not backed by quantitative procedures to verify the potential in environmental improvement.

The IAPH has launched the World Ports Climate Initiative (WPCI) targeting GHG emission reductions for its members. The WPCI supports ports to monitor and reduce their CO_2 footprint through working groups that provide practical information on emissions reduction methods online. The IAPH has additionally designed a tool box that showcases successful implementation of port initiatives and clean air programmes for all operations taking place in a terminal. Finally, in March 2018 the IAPH launched the World Ports Sustainability Program to guide port members on how to achieve progress on the Sustainable Development Goals (SDG) of the United Nations (UN). The American Association of Port Authorities promotes the reduction of GHG from port-related activities and urges the need for IMO to set global standards for GHG emissions targets from vessels. The AAPA is also a strategic partner of the World Ports Sustainability Program.

The European equivalent of IAPH is ESPO that is also a strategic partner of the World Ports Sustainability Program. ESPO has developed the Self-Diagnosis Method (SDM) framework for port authorities within the EcoPorts network. A port in Europe or Norway may join this network by attaining the EcoPorts status as soon as its authority completes the SDM checklist. This is meant to provide insight on problematic areas within the port that should be prioritized for environmental improvement. ESPO has also published a green guide for the systematic port environmental management and designed the Port Environmental Review System (PERS). PERS complements the SDM and assists port authorities to introduce environmental management systems (ESPO 2012). EcoPorts members are expected to review their progress through the SDM checklist annually. The British Port Association (BPA) has adapted the ESPO environmental review and code of practice and holds annual meetings for the environmental managers of member ports. While tools such as PERS and the SDM are useful to provide a qualitative indication of improvement over the years, they are not sufficient. A quantitative estimate of actual reductions in energy use, emissions generated, or other environmental issues is necessary to ensure that each port is able of tracking its progress. This lack of quantitative evidence in the agendas of port associations around the world raises the issue of efficiently estimating, monitoring, and mitigating emissions near and at ports.

The main policies and regulations affecting maritime transport and its environmental impact were presented. However, there are also decisions that stakeholders may make which can also affect their emissions and environmental repercussions, even if the initial motivation is to minimize operating costs. The options span from rewarding clean practices of visiting fleet (either vessels or trucks in the hinterland) to major investments in equipment renewal with a focus on container terminals. These will be presented in the next section.

4 Toward a Green Port

This section will present the main options that ship operators, port authorities, and truck operators currently have at their disposal to improve their environmental performance.

4.1 Operational Practices

The first operational practice that has an impact on port operations was the constant increase in vessel sizes. Vessel sizes are increasing due to the arising economies of scale offering improved cost-efficiency per ton-NM (Cullinane and Khanna 2000). From a terminal operator's perspective, this means handling larger vessels but more containers per call. The energy intensity at the yard per call will increase (more cranes assigned to the larger vessels and more moves per call), while the vessel emits more in comparison to a smaller vessel. What is of interest is whether the ship emissions per TEU handled is lower. A simplistic calculation follows.

The fuel consumption during sailing in the proximity of the port (only main engines and auxiliary engines are active) is estimated by Eq. 12.1, while the fuel consumption FC_{berth} of a vessel at berth using auxiliary engines and boilers for its hoteling demands can be estimated using Eq. 12.2:

$$FC_{nearport}\,(ton) = 10^{-6}\,(SFOC_{main} \cdot EL_{main} \cdot EP_{main} + SFOC_{boiler} \cdot BP_{boiler}) \cdot \frac{D}{V_s} \tag{12.1}$$

$$FC_{berth}\,(ton) = 10^{-6}\,(SFOC_{aux} \cdot EL_{aux} \cdot EP_{aux} + SFOC_{boiler} \cdot BP_{boiler}) \cdot t_{berth} \tag{12.2}$$

Where

SFOC (g/kWh) is the specific fuel oil consumption of the machinery
EL(%) is the engine load at which the machinery operates
EP (kW) is the nominal installed power of the machinery
BP_{boiler} (kW) is the power demand of the boilers
D is the sailing distance from the port that we model
Vs is the approach/departure speed of the vessel
t_{berth} is the total time the vessel spends at berth.

Assume the vessels with the following technical specifications in Table 12.1 and that the maneuvering takes place in the first/last 1 NM from the port lasting 1 h in total. The distance of interest *D* is within 20 NM of the port, and for each vessel, it is assumed that 60% of its capacity is loaded and unloaded at the port (to estimate the time at berth). Under these assumptions, and using Eqs. 12.1 and 12.2, it is possible to estimate the total CO_2 emissions per call for each vessel. These are depicted in Fig. 12.6 broken down per activity phase (tonnes per call) and also per TEU handled (kg/TEU) at the port.

Figure 12.6 shows that as an individual call the ULCV emits more than the Panamax (particularly at berth where it spends more time), but broken down per TEU handled, the larger vessel is more efficient. From a terminal operator's perspective, this will mainly depend on the assigned number of cranes for each boat. In the example in this chapter, twice the cranes were assigned for the larger vessel, which had a more than three times TEU handling demand than the smaller vessel. If the number of cranes assigned was proportional, the ULCV would offer a further improved efficiency. Of course, this could lead in an increase of the total throughput handled at the port (more vessel calls per period) and thus increase its environmental impact in absolute terms.

The next operational practice of ship operators is the gradual fleet renewal where new builds are more fuel efficient. Their engines have a lower *SFOC* which reduces the total fuel consumption at each activity phase. The improved vessel designs will also result in reduced hydrodynamic resistances and thus lower *EL* or necessary *EP*. In recent years, the practice of slow steaming resurfaced due to the depressed market conditions and the relatively high fuel prices (until 2013). This practice has been proved to reduce CO_2 emissions despite the potential deployment of additional

Table 12.1 Technical specifications of containerships and port call/berth information

Vessel class	Technical specifications						
	EP_{main}(kW)	EP_{aux}(kW)	$SFOC_{main}$(g/kWh) at Vs	$SFOC_{aux}$(g/kWh)	EL_{main}(%) at Vs	EL_{aux}(%) at Vs	BP_{boiler} (kW)
Panamax	35,600	7800	178	220	85	30	525
ULCV	81,000	22,000	172	220	85	30	630
Vessel class	Port call berth information						
	V_s	Capacity (TEU)	Import/export traffic	Assigned cranes	Crane productivity (lifts per hour)	TEU factor	$FC_{maneuvering}$ (tonnes per call)
Panamax	17	5000	60%	5	25	1.7	0.92
ULCV	21	16,000		10			1.32

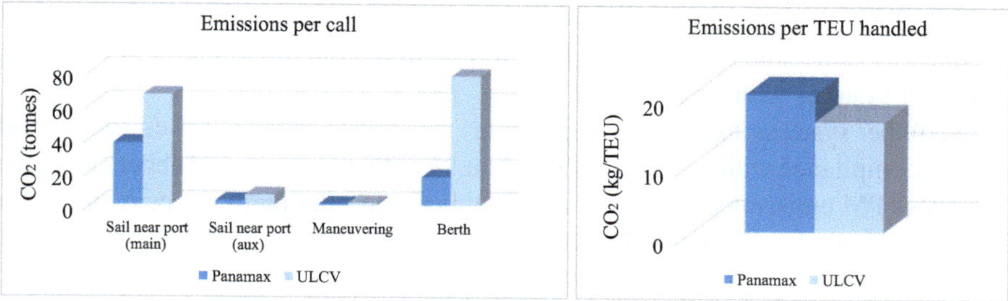

Fig. 12.6 Emissions per call for the two different classes and emissions intensity per TEU

vessels (Cariou 2011). Considering the benefits of a port, the slight change in operational speeds will have some positive impacts near the coastline, the extent of which will depend on the geography of the port and whether there is an extended period of low sailing during approach/departure.

4.2 Technologies

In terms of use of technology to improve the environmental performance of the port, the majority revolves around the electrification of the various operations and the gradual replacement of the use of combustion engines. On the maritime side, this concerns the use of shore power or cold ironing that connects vessels at berth with an electricity source and allows switching off the auxiliary engines. Zis et al. (2014) discuss the effectiveness of cold ironing as an emissions reduction option and construct a quantitative framework that allows an economic evaluation of the technology. This section will present the current status of cold ironing globally.

In California, six ports are affected by the at-berth regulation (see Sect. 3.3): the ports of Los Angeles (POLA), Long Beach (POLB), Oakland, San Diego, San Francisco, and Hueneme. The Port of Gothenburg in Sweden has two ferry (Ro-Ro) terminals with cold ironing capabilities. Shore power is supplied by local surplus wind-generated power and is marketed as a zero-emissions solution. Ferries have in general lower electricity requirements compared to other types, mainly lighting and ventilation during loading/unloading of vehicles (Zis and Psaraftis 2017). Therefore, the Gothenburg electrification process is much simpler than OGV in Californian ports. The port of Antwerp has provision for seven onshore power connection points at one terminal, for barges. In Hamburg, LNG barges are deployed that provide power to vessels at berth, a solution that is practically substituting MGO of auxiliary engines with LNG combustion. Zis et al. (2014) estimate that cold ironing can result in local emissions savings between 48% and 70% for CO_2, 3–60% for SO_x, 40–60% for NO_x, and 57–70% for BC of a container terminals ship emissions inventory.

Zis (2019) note that the low-sulfur regulation may actually be a barrier for cold ironing, as some ship operators may opt to invest in a one-off solution of installing scrubber systems with similar costs.

Scrubber systems are a technology mainly targeted to reduce SOx emissions and secure compliance with the SECA regulation. Vessels running on scrubbers will also emit less PM emissions, and while the scrubbers are running in the proximity of a port, the local emissions will be reduced (Zis et al. 2016). For some vessels, the scrubber systems are also operating during berth, but in general most vessels are using MGO at berth regardless of regulation for their auxiliary engines. Some ship operators have started using dual fuel engines that are capable of burning LNG for propulsion. LNG engines are considered more fuel efficient with a lower emission factor than conventional bunker fuel (Schinas and Butler 2016). LNG fuel has the additional benefit of virtually zero SO_x emissions and lower emissions for PM and NO_x. There are however concerns on methane slip which is a far more potent GHG than CO_2. For ports, ships sailing using LNG in their proximity will result in improved air quality.

On yard operations, the main environmental benefits will come from the deployment of more efficient ship to shore cranes that will increase the number of moves per hour and thus reduce the total turnaround time of large polluting vessels. At the yard, replacing handling equipment running on diesel fuel with hybrid or electric machinery will greatly reduce emissions at the yard. Deployment of AGVs can also greatly improve efficiency of horizontal moves at the yard while also reduce the requirement for lighting during night-time operations. On the hinterland side, ITS can be used to reduce the formation of queues at the gates. In addition, the gradual renewal of truck fleet coupled with attempts to reduce idling times of drivers will also result in great reductions in emissions at the gate. Finally, in the future the introduction of autonomous freight vehicles and the practices of platooning can also improve the transportation system and increase the capacity of road links near the port.

4.3 *Port Initiatives*

A number of port authorities are publishing annual reports on their environmental efficiency. The Port of Felixstowe (2017) published its ninth annual environment report for 2016–2017 focusing on energy consumption of in-port equipment and operations. The Port of Los Angeles produces comprehensive annual emissions inventories which have been used in the literature to provide base emission factors per engine type and activity. The inventory includes ship activities which are shown to be the most contributing in SO_x, NO_x, CO_2, and PM emissions.

A number of port authorities have their own green agendas that seek to improve the air quality near the port. Ports are emphasizing different environmental challenges according to their priorities, and therefore, there are initiatives that target all port operations (maritime, yard, and hinterland). An indicative list of programmes that address port operations (maritime, in-port, gate) are presented in Table 12.2.

4.3.1 Vessel-Oriented Programmes

Many port authorities that are not bound by existing regulation have been rewarding vessel operators that follow green practices. For example, in Singapore reduced port fees are required for ships that are using low-sulfur fuel or have good scores in their EEDI. Other ports promote the use of technologies such as cold ironing and offer it at competitive prices as electricity prices are typically lower than low-sulfur bunker fuel. Prior to the SECAs, the Port of Gothenburg would reduce the port tariff for vessels using scrubber systems. An interesting example is the port of Stockholm which provides financial help for retrofitting ferries to use scrubbers, provided their operators commit to call at the port for at least 3 years. Some port authorities are considering investing to LNG bunkering facilities which will result in cleaner vessels calling at these ports. LNG-fuelled vessels pay lower tariffs in Singapore and Rotterdam, while there are plans of the European Commission to develop LNG bunkering services in all EU ports within the Trans European Core Network by 2020 (European Commission 2013). A very successful initiative has been the introduction of the Green Flag Programme (which is a VSRP) in POLA and POLB. These port authorities offer monetary incentives for vessels that reduce their sailing speed in the proximity of the port at 12 knots. Zis et al. (2014) were the first to examine the efficiency of the VSRP programme. They find that it results in significant local emissions reduction at important costs for ship operators (loss of time or speeding up outside the zone). They conclude that the programme could be optimized to be tailored to specific vessel types. Linder (2018) conducted a survey to understand why the VSRP has seen such popularity in recent years, despite the economic penalties associated with its operation.

4.3.2 Non-vessel Programmes and Investments

A number of port authorities are upgrading their cargo handling equipment with the introduction of faster and more efficient machinery. This has positive effects in the energy efficiency of the terminal and at the same time reduces the turnaround time of vessels at berth and thus the vessel emissions generated near the port. Investments in energy generation within the port have been considered in smaller ports where space is available (Shoreham) for the introduction of renewable energy sources.

Relieving bottlenecks caused at the port's gate through improved planning may also contribute to the turnaround time (Zhao and Goodchild 2010). Giuliano and O'Brien (2007) were among the first to examine the effectiveness of the POLA and POLB terminal gate appointment systems and concluded that there was no evidence that the system reduced queues and thus emissions, though lack of ex ante data could have played a role. Truck emissions at the terminal however can be reduced by an

Table 12.2 Examples of green programmes in port authorities around the world

	Port operations						
	Maritime					Yard	Gate
Port	Low-sulfur fuel or scrubbers	Speed	Berth	Ship design/operational aspects	LNG bunkering	Electrification, hinterland operations	
Singapore	Green Port Programme			Green Ship Programme, Green Technology Programme	Available		
POLA POLB		VSRP	AMP	Smoke Stack Reductions (maintenance, control, and alternative fuel)			Clean Truck Programme
							Green Port Gateway
Gothenburg	Reduced tariffs				Planned for 2015		
Stockholm			AMP and incentive for retrofits				
Antwerp			AMP	Rewards for clean vessels	Available		
Rotterdam				Reduced fees LNG vessels	Available	Investing in RES	
Port of Virginia						Invest in hybrid container shuttle carriers	
Port of Georgia						Yard crane electrification	

optimized truck arrival pattern (Chen et al. 2013), and additionally booking systems have been introduced. Port authorities also design schemes where trucks below certain efficiency standards are banned from the port (Clean Truck Programme in POLA) and the use of cleaner trucks is rewarded. To reduce hinterland emissions, port authorities could adapt some measures such as:

- Promoting the retirement of older vehicles
- Introducing penalties for delayed arrivals
- Educating campaigns on driving behavior (e.g., reducing engine idling times when waiting)

These simple methods can improve terminal efficiency and the port's environmental performance.

5 Conclusions and Topics for Further Research

This chapter presented an overview of the current status quo on port environmental management. The academic literature is relatively scarce in comparison with research on whole journey aspects, but in recent years, the field of green ports has seen a renewed interest. The chapter aimed to define a green port, as a port that has launched specific initiatives to improve its environmental performance. The environmental challenges that ports are facing nowadays were presented. The role of regulatory bodies in reducing emissions globally was analysed, and examples of how a regulation that is targeting a different area can improve the environmental performance of a port were given. With regard to emissions and energy consumption, the chapter analyzed the different port operations (maritime, yard, hinterland) and the operational practices and technologies that can assist in overcoming these challenges. In emission terms, the most important contributor is the vessel operations, with an important role for specific pollutant types attributed to hinterland road and rail operations.

The author is convinced that the field of green ports will see additional attention in academia and the industry in the coming years, particularly with the potential inclusion of the maritime sector in an emissions trading scheme (ETS). Technologies like scrubbers or LNG engines have seen increased attention following the lower sulfur limit, and interesting research questions will arise from the global sulfur cap from 2020 onward. Cold ironing has already seen an increased attention in academia, and a potential increase in fuel prices may prompt additional vessel operators to consider this option. Research-wise, the main questions revolve around the emerging environmental and economic trade-offs from emissions reduction actions. Considering that ports have limited resources, and the infamous quote that "when you have seen one port, you have seen one port," the main question is how to get the best value for money for environmental programmes. The answer will vary from port to port depending on the throughput handled, the visiting fleet, the position of the port, and many other parameters.

Psaraftis (2016) proposed the push down-pop up paradox, whereby an effort to reduce emissions in one area can result in additional emissions somewhere else. With regard to ports specifically, Zis (2015) proposed the action-reaction concept where an emissions reduction action in one port can lead to increased emissions globally or at other ports. There are many open questions on the arising economic and environmental trade-offs of port emissions reduction options that will require an answer in light of new regulatory pressure that is coming.

Acknowledgments Work presented in this chapter draws heavily from the author's PhD Thesis. The doctoral work was co-funded by the Greek State Scholarship Foundation to which the author is grateful. The author would like to thank Michael G.H. Bell, Kevin Cullinane, and Harilaos N. Psaraftis for fruitful discussions over the last years on the topic of green ports.

References

Alderton, P. (2013). *Port management and operations*. London: Informa.

Angeloudis, P., & Bell, M. G. (2010). An uncertainty-aware AGV assignment algorithm for automated container terminals. *Transportation Research Part E: Logistics and Transportation Review, 46*(3), 354–366.

Aregall, M. G., Bergqvist, R., & Monios, J. (2018). A global review of the hinterland dimension of green port strategies. *Transportation Research Part D: Transport and Environment, 59*, 23–34.

Berechman, J., & Tseng, P. H. (2012). Estimating the environmental costs of port related emissions: The case of Kaohsiung. *Transportation Research Part D: Transport and Environment, 17*(1), 35–38.

Buhaug, Ø., Corbett, J., Endresen, O., Eyring, V., Faber, J., Hanayama, S., Lee, D., Lindstad, H., Mjelde, A., Palsson, C., Wanquing, W., Winebrake, J., & Yoshida, K. (2009). *Second IMO greenhouse gas study*. London: International Maritime Organization.

California Air Resources Board. (2012). *Ocean-going vessels – fuel rule*. Available at: http://www.arb.ca.gov/ports/marinevess/ogv.htm. Accessed Jul 2017.

Cao, M., & Golias, M. M. (2013). *Evaluation of the effect of gate strategies in drayage related emissions. Final report*. National Center for Freight and Infrastructure Research and Education. CFIRE 04-19

Cariou, P. (2011). Is slow steaming a sustainable means of reducing CO2 emissions from container shipping? *Transportation Research Part D: Transport and Environment, 16*(3), 260–264.

Carlo, H. J., Vis, I. F., & Roodbergen, K. J. (2014). Storage yard operations in container terminals: Literature overview, trends, and research directions. *European Journal of Operational Research, 235*(2), 412–430.

Chen, G., Govindan, K., & Golias, M. M. (2013). Reducing truck emissions at container terminals in a low carbon economy: proposal of a queueing-based bi-objective model for optimizing truck arrival pattern. *Transportation Research Part E: Logistics and Transportation Review, 55*, 3–22.

Colombaroni, C., Fusco, G., Isaenko, N., Quadrifoglio, L. (2017). Optimization of container operations at inland intermodal terminals. In *Models and technologies for intelligent transportation systems (MT-ITS), 2017 5th IEEE International Conference on* (pp. 69–74). IEEE.

Cullinane, K., & Khanna, M. (2000). Economies of scale in large containerships: optimal size and geographical implications. *Journal of transport geography, 8*(3), 181–195.

Cullinane, K., Tseng, P. H., & Wilmsmeier, G. (2016). Estimation of container ship emissions at berth in Taiwan. *International Journal of Sustainable Transportation, 10*(5), 466–474.

Davarzani, H., Fahimnia, B., Bell, M., & Sarkis, J. (2016). Greening ports and maritime logistics: A review. *Transportation Research Part D: Transport and Environment, 48*, 473–487.

De Meyer, P., Maes, F., & Volckaert, A. (2008). Emissions from international shipping in the Belgian part of the North Sea and the Belgian seaports. *Atmospheric Environment, 42*(1), 196–206.

Dragović, B., Tzannatos, E., Tselentis, V., Meštrović, R., Škurić, M. (2018). Ship emissions and their externalities in cruise ports. *Transportation Research Part D: Transport and Environment, 61*, 289–300.

Drake, J. M., & Lodge, D. M. (2007). Hull fouling is a risk factor for intercontinental species exchange in aquatic ecosystems. *Aquatic Invasions, 2*(2), 121–131.

El-Naggar, M. E. (2010). Application of queuing theory to the container terminal at Alexandria seaport. *Journal of Soil Science and Environmental Management, 1*(4), 77–85.

ESPO. (2012). *European port performance dashboard*. Available at: http://pprism.espo.be/LinkClick.aspx?fileticket=sWk5lsdRFI8%3D&tabid=3555. Accessed Jan 2018.

European Commission. (2013). *Alternative fuels for transport: Parliament committee vote supports roll-out of refuelling infrastructure*. [Press release]. Retrieved from http://europa.eu/rapid/press-release_IP-13-1168_en.htm.

Giuliano, G., & O'Brien, T. (2007). Reducing port-related truck emissions: The terminal gate appointment system at the Ports of Los Angeles and Long Beach. *Transportation Research Part D: Transport and Environment, 12*(7), 460–473.

Golias, M., Boile, M., Theofanis, S., & Efstathiou, C. (2010). The berth-scheduling problem: Maximizing berth productivity and minimizing fuel consumption and emissions production. *Transportation Research Record: Journal of the Transportation Research Board, 2166*, 20–27.

Greene, D. L., & Wegener, M. (1997). Sustainable transport. *Journal of Transport Geography, 5*(3), 177–190.

Imai, A., Chen, H. C., Nishimura, E., & Papadimitriou, S. (2008). The simultaneous berth and quay crane allocation problem. *Transportation Research Part E: Logistics and Transportation Review, 44*(5), 900–920.

Liao, C. H., Tseng, P. H., & Lu, C. S. (2009). Comparing carbon dioxide emissions of trucking and intermodal container transport in Taiwan. *Transportation Research Part D: Transport and Environment, 14*(7), 493–496.

Linder, A. (2018). Explaining shipping company participation in voluntary vessel emission reduction programs. *Transportation Research Part D: Transport and Environment, 61*, 234–245.

Marr, I. L., Rosser, D. P., & Meneses, C. A. (2007). An air quality survey and emissions inventory at Aberdeen Harbour. *Atmospheric Environment, 41*(30), 6379–6395.

Mooney, H. A. (2005). *Invasive alien species: A new synthesis* (Vol. 63). Washington, DC: Island Press.

Nikitakos, N. (2012). Green logistics: The concept of zero emissions port. *FME Transactions, 40*(4), 201–206.

Ng, S. K., Loh, C., Lin, C., Booth, V., Chan, J. W., Yip, A. C., Li, Y., & Lau, A. K. (2013). Policy change driven by an AIS-assisted marine emission inventory in Hong Kong and the Pearl River Delta. *Atmospheric Environment, 76*, 102–112.

Passet, R. (1979). *L'économique et le vivant* (Vol. 23). Paris: Payot.

POLA. (2018). Port of Los Angeles Inventory of air emissions - 2017. Available at: https://kentico.portoflosangeles.org/getmedia/880bc597-84bc-4ae6-94e2-59a2e6027f42/2017_Air_Emissions_Inventory

Port of Los Angeles. (2017). *Inventory of air emissions for calendar year 2016*. Available at: https://www.portoflosangeles.org/environment/studies_reports.asp. Accessed Mar 2018

Psaraftis, H. N. (2016). Green maritime logistics: the quest for win-win solutions. *Transportation Research Procedia, 14*, 133–142.

Saxe, H., & Larsen, T. (2004). Air pollution from ships in three Danish ports. *Atmospheric Environment, 38*(24), 4057–4067.

Schinas, O., & Butler, M. (2016). Feasibility and commercial considerations of LNG-fueled ships. *Ocean Engineering, 122*, 84–96.

Smith, T. W. P., et al. (2014). *Third imo ghg study 2014*. London. http://www.iadc.org/wp-content/uploads/2014/02/MEPC-67-6-INF3-2014-Final-Report-complete.pdf: International Maritime Organization (IMO).

Steenken, D., Voß, S., & Stahlbock, R. (2004). Container terminal operation and operations research-a classification and literature review. *OR Spectrum, 26*(1), 3–49.

Song, S. (2014). Ship emissions inventory, social cost and eco-efficiency in Shanghai Yangshan port. *Atmospheric Environment, 82*, 288–297.

Stopford, M. (2009). *Maritime economics, Vol. 3*. London: Routledge.

Talley, W. K. (2009). *Port economics*. Abingdon: Routledge.

Tzannatos, E. (2010). Ship emissions and their externalities for the port of Piraeus–Greece. *Atmospheric Environment, 44*(3), 400–407.

UNCTAD (2017). Review of Maritime Transport 2017. New York and Geneva, October 2017.

Zeng, Q., & Yang, Z. (2009). Integrating simulation and optimization to schedule loading operations in container terminals. *Computers & Operations Research, 36*(6), 1935–1944.

Zhao, W., & Goodchild, A. V. (2010). The impact of truck arrival information on container terminal rehandling. *Transportation Research Part E: Logistics and Transportation Review, 46*(3), 327–343.

Zhou, P. F., & Kang, H. G. (2008). Study on berth and quay-crane allocation under stochastic environments in container terminal. *Systems Engineering-Theory & Practice, 28*(1), 161–169.

Zis, T. (2015). The implications and trade-offs of near-port ship emissions reduction policies. PhD Thesis

Zis, T. P. (2019). Prospects of cold ironing as an emissions reduction option. *Transportation Research Part A: Policy and Practice, 119*, 82–95.

Zis, T., & Psaraftis, H. N. (2017). The implications of the new sulphur limits on the European Ro-Ro sector. *Transportation Research Part D: Transport and Environment, 52*, 185–201.

Zis, T., North, R. J., Angeloudis, P., Ochieng, W. Y., & Bell, M. G. H. (2014). Evaluation of cold ironing and speed reduction policies to reduce ship emissions near and at ports. *Maritime Economics & Logistics, 16*(4), 371–398.

Zis, T., Angeloudis, P., Bell, M. G., & Psaraftis, H. N. (2016). Payback period for emissions abatement alternatives: Role of regulation and fuel prices. *Transportation Research Record: Journal of the Transportation Research Board, 2549*, 37–44.

Chapter 13
The Way Ahead

Harilaos N. Psaraftis and Panos Zachariadis

Abstract The purpose of this chapter is to attempt to make an assessment on what may lie ahead as regards sustainable shipping. The focus of the chapter is the April 2018 decision of the International Maritime Organization on the formulation of an Initial Strategy to reduce maritime greenhouse gas (GHG) emissions. In that context, an assessment of the prospects for alternative fuels, which figure centrally in the Initial Strategy, is also included.

Abbreviations

AER	Annual efficiency ratio
BAU	Business as usual
BIMCO	Baltic and International Maritime Council
BRI	Belt and road initiative
CBDR-RC	Common but differentiated responsibilities and respective capabilities
CH_4	Methane
CO_2	Carbon dioxide
DCS	Data collection system
ECSA	European Community Shipowners' Associations
EEDI	Energy efficiency design index
EESH	Energy efficiency per service hour
EGR	Exhaust gas recirculation
EPA	Environmental Protection Agency (US)
ESPO	European Seaports Organisation
ETS	Emissions Trading System

H. N. Psaraftis (✉)
DTU Management Engineering, Technical University of Denmark, Kongens Lyngby, Denmark
e-mail: hnpsar@dtu.dk

P. Zachariadis
Atlantic Bulk Carriers Management, Ltd., Piraeus, Greece

H. N. Psaraftis (ed.), *Sustainable Shipping*,
https://doi.org/10.1007/978-3-030-04330-8_13

EU	European Union
FORS	Fuel Oil Reduction Strategy
GHG	Greenhouse gas
GWP	Global warming potential
HFO	Heavy fuel oil
IAPH	International Association of Ports and Harbors
ICS	International Chamber of Shipping
IMO	International Maritime Organization
IPCC	Intergovernmental Panel on Climate Change
ISPI	Individual Ship Performance Indicator
ITCP	Integrated Technical Cooperation Programme
kwh	Kilowatt hour
LBSI	Lean burn spark ignition
LDC	Least developed country
LNG	Liquefied natural gas
LPDF	Low-pressure dual fuel
MBM	Market-based measure
MDO	Marine diesel oil
MEPC	Marine Environment Protection Committee
MGO	Marine gas oil
NG	Natural gas
NGO	Nongovernmental organization
NH_3	Ammonia
NO_x	Nitrogen oxide
OECD	Organisation for Economic Cooperation and Development
SCR	Single catalytic reduction
SDG	Sustainable Development Goal
SECA	Sulfur emission control area
SEEMP	Ship Energy Efficiency Management Plan
SO_x	Sulfur oxide
SSS	Short sea shipping
UNFCCC	United Nations Framework Convention on Climate Change
US	United States

1 Introduction

The purpose of this chapter is to attempt to make an assessment on what may lie ahead as regards sustainable shipping. As without any doubt the single most important recent event that may impact international shipping in the years ahead has been the April 2018 decision of the International Maritime Organization (IMO) at the 72nd session of the Marine Environment Protection Committee (MEPC 72) to adopt an Initial Strategy to reduce maritime greenhouse gas (GHG) emissions (IMO 2018), the exclusive focus of this chapter will be the above decision. This is said realizing that other aspects of sustainable shipping, for instance, ship recycling, oil pollution, ballast water, logistical measures, sulfur, green ports, and others, are also important. However, we feel that the treatment of these subjects in the previous chapters of the book is adequate, and nothing more on these subjects needs to be added in the book. By contrast, not much has been said thus far in this book on the April 2018 IMO/MEPC 72 decision; therefore we feel that some comments are definitely necessary to that effect.

The rest of this chapter is organized as follows. Section 2 highlights some of the main elements of the IMO/MEPC 72 decision. Section 3 discusses the issue of alternative (low carbon or zero carbon) fuels, very much central within the IMO Initial Strategy. Commentary on the way ahead is provided in Sect. 4. The IMO/MEPC 72 decision is presented in its entirety as an Appendix to the chapter.

The treatment of alternative fuels in this chapter rather than elsewhere in the book stems mainly from two reasons: (a) such fuels figure prominently within the IMO Initial Strategy and (b) such fuels can conceivably represent the future in shipping, provided of course they will prove technically and economically viable. Our own assessment on this score is that, as things stand, the road ahead is a long one.

2 The April 2018 IMO Decision

The IMO Initial Strategy to reduce maritime GHG emissions is in the form of *Resolution MEPC.304(72)* and includes, among others, the following elements: (a) the vision, (b) the levels of ambition, (c) the guiding principles, (d) a list of short-term, medium-term, and long-term candidate measures with a timeline, and (e) miscellaneous other elements, such as follow-up actions and others.

We briefly highlight some of these elements below.

2.1 Vision

IMO remains committed to reducing GHG emissions from international shipping and, as a matter of urgency, aims to phase them out as soon as possible in this century.

2.2 Levels of Ambition

The Initial Strategy identifies levels of ambition for the international shipping sector noting that technological innovation and the global introduction of alternative fuels and/or energy sources for international shipping will be integral to achieve the overall ambition. Reviews should take into account updated emission estimates, emissions reduction options for international shipping, and the reports of the Intergovernmental Panel on Climate Change (IPCC). Levels of ambition directing the Initial Strategy are as follows:

1. *Carbon intensity of the ship to decline through implementation of further phases of the energy efficiency design index (EEDI) for new ships*
   To review with the aim to strengthen the energy efficiency design requirements for ships with the percentage improvement for each phase to be determined for each ship type, as appropriate
2. *Carbon intensity of international shipping to decline*
   To reduce CO_2 emissions per transport work, as an average across international shipping, by at least 40% by 2030, pursuing efforts toward 70% by 2050, compared to 2008
3. *GHG emissions from international shipping to peak and decline*
   To peak GHG emissions from international shipping as soon as possible and to reduce the total annual GHG emissions by at least 50% by 2050 compared to 2008 while pursuing efforts toward phasing them out as called for in the Vision as a point on a pathway of CO_2 emission reduction consistent with the Paris Agreement temperature goals

2.3 Guiding Principles

The principles guiding the Initial Strategy include:

(a) The need to be cognizant of the principles enshrined in instruments already developed, such as:

- The principle of nondiscrimination and the principle of no more favorable treatment, enshrined in MARPOL and other IMO conventions
- The principle of common but differentiated responsibilities and respective capabilities, in the light of different national circumstances, enshrined in UNFCCC, its Kyoto Protocol, and the Paris Agreement

(b) The requirement for all ships to give full and complete effect, regardless of flag, to implementing mandatory measures to ensure the effective implementation of this strategy
(c) The need to consider the impacts of measures on States, including developing countries, in particular, on LDCs and SIDS as noted by MEPC 68 (MEPC 68/21, paragraphs 4.18–4.19) and their specific emerging needs, as recognized in the Organization's Strategic Plan (resolution A.1110(30))
(d) The need for evidence-based decision-making balanced with the precautionary approach as set out in resolution MEPC.67(37)

2.4 *List of Candidate Measures*

2.4.1 Short-Term Measures (Finalized and Agreed by the Committee Between 2018 and 2023)[1]

- Further improvement of the existing energy efficiency framework with a focus on EEDI and SEEMP, taking into account the outcome of the review of EEDI regulations.
- Develop technical and operational energy efficiency measures for both new and existing ships, including consideration of indicators in line with the three-step approach that can be utilized to indicate and enhance the energy efficiency performance of shipping, e.g., Annual Efficiency Ratio (AER), Energy Efficiency per Service Hour (EESH), Individual Ship Performance Indicator (ISPI), and Fuel Oil Reduction Strategy (FORS).
- Establishment of an Existing Fleet Improvement Programme.
- Consider and analyze the use of speed optimization and speed reduction as a measure, taking into account safety issues, distance traveled, distortion of the market or trade, and that such measure does not impact on shipping's capability to serve remote geographic areas.
- Consider and analyze measures to address emissions of methane and further enhance measures to address emissions of volatile organic compounds.
- Encourage the development and update of national action plans to develop policies and strategies to address GHG emissions from international shipping in accordance with guidelines to be developed by the organization, taking into account the need to avoid regional or unilateral measures.
- Continue and enhance technical cooperation and capacity-building activities under the ITCP.
- Consider and analyze measures to encourage port developments and activities globally to facilitate reduction of GHG emissions from shipping, including provision of ship and shoreside/onshore power supply from renewable sources, infrastructure to support supply of alternative low-carbon and zero-carbon fuels, and to further optimize the logistic chain and its planning, including ports.
- Initiate research and development activities addressing marine propulsion, alternative low-carbon and zero-carbon fuels, and innovative technologies to further enhance the energy efficiency of ships and establish an International Maritime Research Board to coordinate and oversee these R&D efforts.
- Incentives for first movers to develop and take up new technologies.
- Develop robust lifecycle GHG/carbon intensity guidelines for all types of fuels, in order to prepare for an implementation program for effective uptake of alternative low-carbon and zero-carbon fuels.

[1]The Initial Strategy is subject to revision based on fuel oil consumption data collected during 2019–2021 and does not prejudge any specific further measures that may be implemented in Phase 3 of the three-step approach.

- Actively promote the work of the organization to the international community, in particular, to highlight that the organization, since the 1990s, has developed and adopted technical and operational measures that have consistently provided a reduction of air emissions from ships and that measures could support the Sustainable Development Goals, including SDG 13 on Climate Change.
- Undertake additional GHG emission studies and consider other studies to inform policy decisions, including the updating of marginal abatement cost curves and alternative low-carbon and zero-carbon fuels.

2.4.2 Medium-Term Measures (Finalized and Agreed by the Committee Between 2023 and 2030)

- Implementation program for the effective uptake of alternative low-carbon and zero-carbon fuels, including update of national action plans to specifically consider such fuels.
- Operational energy efficiency measures for both new and existing ships including indicators in line with three-step approach that can be utilized to indicate and enhance the energy efficiency performance of ships.
- New/innovative emission reduction mechanism(s), possibly including Market-based measures (MBMs), to incentivize GHG emission reduction.
- Further continue and enhance technical cooperation and capacity-building activities such as under the ITCP.
- Development of a feedback mechanism to enable lessons learned on implementation of measures to be collated and shared through a possible information exchange on best practice.

2.4.3 Long-Term Measures (Finalized and Agreed by the Committee Beyond 2030)

- Pursue the development and provision of zero-carbon or fossil-free fuels to enable the shipping sector to assess and consider decarbonization in the second half of the century.
- Encourage and facilitate the general adoption of other possible new/innovative emission reduction mechanism(s).

2.5 Follow-Up Actions

These are listed in Table 13.1.

We note again that this section is not encyclopedic; the complete IMO/MEPC 72 decision is included as an Appendix to this chapter.

Table 13.1 Key stages for the adoption of a Revised IMO GHG Strategy in 2023 as set out in the road map

Spring 2018 (MEPC 72)	Adoption of the Initial Strategy including, inter alia, a list of candidate short-, mid-, and long-term further measures with possible timelines, to be revised as appropriate as additional information becomes available
January 2019	Start of phase 1: data collection (ships to collect data)
Spring 2019 (MEPC 74)	Initiation of fourth IMO GHG study using data from 2012 to 2018
Summer 2020	Data from 2019 to be reported to IMO
Autumn 2020 (MEPC 76)	Start of phase 2: data analysis (no later than autumn 2020) publication of fourth IMO GHG study for consideration by MEPC 76
Spring 2021 (MEPC 77)	Secretariat report summarizing the 2019 data pursuant to regulation 22A.10
	Initiation of work on adjustments on initial IMO strategy, based on data collection system (DCS) data
Summer 2021	Data for 2020 to be reported to IMO
Spring 2022 (MEPC 78)	Phase 3: decision step
	Secretariat report summarizing the 2020 data pursuant to regulation 22A.10
Summer 2022	Data for 2021 to be reported to IMO
Spring 2023 (MEPC 80)	Secretariat report summarizing the 2021 data pursuant to regulation 22A.10
	Adoption of revised IMO strategy, including short-, mid-, and long-term further measure(s), as required, with implementation schedules

3 Alternative Fuels: Looking at the Crystal Ball

3.1 Preamble

It is very clear from the Initial Strategy that alternative (low-carbon and zero-carbon) fuels are centrally placed within the IMO decision document. Reference to such fuels is made in (a) the "level of ambition" section, (b) four instances in the set of short-term measures, (c) one instance in the set of medium-term measures, and (d) one instance in the set of long term measures. If anything, this reflects the fact that the development and eventual use of such fuels is a matter of high priority for the maritime industry, particularly if the targets set in the IMO Initial Strategy stand any reasonable chance to be reached. Put another way, if fossil fuels continue to be used in a "business as usual" (BAU) fashion, the chances of reaching the GHG reduction targets look slim. In that sense, it is reasonable to expect that low-carbon and zero-carbon fuels might provide the necessary "quantum leap" that might give the quest for decarbonization enough momentum to reach the stated targets. The obligatory question of course is whether the above expectation has a reasonable chance to be realized.

It is fair to say that the subject of alternative fuels is vast, with many studies and projects examining the subject from various twists and angles and also with specific real-life demonstration projects whose purpose has been to test such fuels in actual ships under specific scenarios in terms of technical and economic viability. See, for instance, among others, the work of DNV GL (2018) in the area. For recent surveys on the decarbonization possibilities of such fuels and other technologies and measures, see Bouman et al. (2017) and OECD (2018), among others.

It is also fair to say that, for all such work, and even though an upbeat picture is typically being painted by, among others, various players who have a stake (commercial or other) in such technologies, much uncertainty and divided opinion still prevails on the subject. As a result, the future of such fuels is very much uncertain. This section is an attempt to highlight some of the issues that exist in this area, with the hope that these issues eventually find a way to be addressed. This section is purposely brief, as an exhaustive analysis of various alternative fuels would require much more than one chapter in a book or even a book by itself.

To that effect, this section will concentrate on the GHG effect of several proposed alternative fuels for marine use as well as on practical aspects of application. Most of these alternative fuels are alleged as being "clean burning." This may be true, in part, if one focuses on the emissions affecting human health (SO_X, NO_X, and particulate matter (PM)). However, a rather rosy picture has been painted for these fuels also as regards their GHG footprint, something that several recent studies are strongly challenging. It turns out the life cycle footprint of nearly all proposed alternative fuels is quite poor and, in most cases, worse than current conventional liquid fuels (marine gas oil (MGO), marine diesel oil (MDO), or desulfurized fuel oil). As such, their branding as "transitional" or "bridge" fuels toward decarbonization is seriously questioned.

In the following, a more extensive discussion is devoted to natural gas, since most of the other proposed alternative fuels are in many ways by-products of, or originate from, natural gas, something which may not be widely known. For a more detailed study, the reader may consult the references herein.

3.2 *Natural Gas (NG) and Liquefied Natural Gas (LNG)*

Natural gas (NG) after extraction needs to be treated, through a fractionation process, to remove impurities (CO_2, water, sulfur, etc.), as well as propane and butane (those being 1–4% of the extracted gas) which are used separately. The purified gas (pure methane and ethane) is then cooled in stages to -162 °C where it becomes liquid (LNG). During the cooling process, excess nitrogen is also removed. The liquefaction reduces the volume of natural gas by a factor of 600, allowing it to be stored in insulated tanks or transported by ship or tanker trucks. LNG is regasified at the destination by heating and transported as gas via pipelines to consumers. When used as fuel of ships (other than LNG carrier ships), it is bunkered as liquid from tanker trucks or barges and stored as liquid onboard. Obviously, purification and liquefaction require large amounts of energy. The CO_2 intensity of liquefaction alone is 0.2–0.4 kg CO_2 released for every kg LNG produced (Jaramillo et al. 2005).

NG is about 90% methane (CH_4), and LNG is about 95% CH_4. CH_4 is a GHG much worse than CO_2 in terms of effect on global warming. The global warming potential (GWP) of CH_4 is not constant but depends on the time horizon contemplated. Thus, for a time horizon of 5, 10, 20, and 100 years, CH_4 is, respectively, 116, 110, 86, and 34 times worse than CO_2 (Myhre et al. 2013; Howarth 2014; Allen 2014). It is clear that a time horizon of 100 years, being the lifetime of CO_2 in the atmosphere, is not very relevant in the climate change discussion, given the urgency of the situation. In that sense, more and more scientists are calling for the US Environmental Protection Agency (EPA), academia, and regulators to stop using the (misleading) 100-year time horizon in GHG studies. In fact Myhre et al. (2013) state that "There is no scientific argument for selecting 100 years compared with other choices. The choice of time horizon is a value judgment because it depends on the relative weight assigned to effects at different times."

The above is an important point because using the proper GWP factor of 86 (instead of 25) immediately reverses all claims of LNG having a better GHG footprint than conventional liquid fuels. We refer to the GHG effect when LNG/CH_4 escapes unburned to the atmosphere.

Indeed, when NG or LNG is burned in a ship's engine, it produces about 20–25% less CO_2 than a conventional liquid fuel (HFO, MDO, or MGO), due to its lower carbon content. However, this assumes perfect combustion, which exists only in chemistry text books. In real life a quantity of NG remains unburned and is emitted to the atmosphere together with the combustion exhausts. This escape, also known as methane slip, occurs in sufficient amounts in the vast majority of marine engines (four-stroke and two-stroke dual-fuel or Otto cycle spark plug engines). For all these engines, the methane slip – even according to the engine manufacturers' stated numbers – renders them immediately worse than conventional liquid fuel engines, even when a mild GWP factor of 25 is used (Corbett et al. 2015; EU 2016). For some latest-type high-pressure combustion engines, methane slip is stated to be very small, but this refers more to the engine laboratory shop-test conditions and not necessarily to measured values in actual ship operation. In any case, multiplying even very small quantities of methane slip by 86 brings these state-of-the-art engines also at par to existing conventional ones, as is also discussed below.

The vast majority of gas engines on ships operate using the Otto cycle (as opposed to the Diesel cycle). Such engines have a published methane leak rate of 3–6 g CH_4/kwh corresponding to a fuel loss of 2–3%. Actual measurements invariably show a larger fuel leak rate (e.g., 6–8 g/kwh in optimal operation and much higher multiples at low loads) (Corbett et al. 2015). These are LPDF (low pressure dual fuel) engines with pilot fuel and lean burn spark ignition (LBSI) engines. SINTEF (2017) has measured the methane leak of these engines at 6.9 g/kwh and 4.1 g/kwh, respectively, with much higher values reported throughout the literature especially for LBSI engines. At these methane slip levels, either published by engine makers or independently measured, the GHG footprint of combustion is quite higher than conventional marine diesel engines, even using a 100-year GWP of 25. When a GWP of 86 is used, the GHG performance of NG combustion in marine engines becomes substantially worse than diesel or heavy fuel combustion. Furthermore, there is a limiting trade-off in that if the engines are operated at a lower air to fuel ratio to reduce methane slip, then NO_x emissions increase exceeding Tier III limits. Already, exhaust aftertreatment options are being discussed among regulators, involving methane oxidation/capture devices at the stack.

The latest-type newly introduced gas engines (high pressure dual fuel (HPDF)) are promoted as zero slip (0.2% of consumption) by their makers (e.g., the ME-GI DF engine by Man B&W), although independent references (Corbett et al. 2015) give 0.7 g/kwh as more common (i.e., 0.4% of consumption). Although this is definitely a big improvement over the Otto cycle engines, we should remember that even just a 1 g/kwh methane slip represents 86 g/kwh of equivalent CO_2 and thus any GHG benefit over standard diesel engines is mostly lost. In addition, the NO_x reduction benefit of these engines is diminished, requiring a selective catalytic reactor (SCR) or exhaust gas recirculation (EGR) to comply with NO_x regulations.

The above discussion concerns only the LNG methane slip during combustion at the engine, which is only one of the areas of methane slip, perhaps the most minor one. Nevertheless we felt it had to be addressed in some detail since it is usually under-evaluated, especially when the (inappropriate) GWP of 25 is being used to assess the greenhouse effect of natural gas. To the leaks at combustion, we must add the methane leaks before the engine (pumps, piping), at the ship's LNG storage tanks (boil off) and during bunkering, which are found to be very substantial (Corbett et al. 2015). With regard to the size of LNG tanks needed onboard, these are three to four times larger than those of conventional liquid fuels (DNV GL 2015), which may impact either the cargo carrying capacity or the cruising range of the ship. Equipping a ship for LNG use as fuel increases its new building cost by about 30%.

Lastly, with regard to LNG's particulate matter (PM) emissions, there have been concerns expressed that, although its combustion produces less PM than conventional fuels, it does produce ultrafine PM which cannot be addressed (EU 2016). Ultrafine particles penetrate the respiratory system and are transported to other parts of the body via the blood. They also may play a role in atmospheric processes, dictating the amount and lifetime of clouds, which can influence the climate.

Of course, for a fuel such as LNG, its *life cycle* methane slip should be considered. That involves possible leaks during transportation to consumer and, most of all, leaks during its extraction from the ground, which all currently are either ignored or severely understated from a proper global warming assessment. These leaks invariably occur, and they can be considerable.

The bulk of recent (post 2013) scientific literature indicates that LNG's life cycle global warming effect is much worse than that of conventional liquid fuels (diesel, heavy fuel oil) and may even be worse than that of coal. Many experts agree that if methane slip from the whole LNG life cycle (extraction to combustion) exceeds 3%, then LNG becomes worse than coal in terms of global warming effect. And, unfortunately, detailed and undisputed measurements in the United States show an overall average escape to the atmosphere of 3.6–7.9% natural gas at the extraction fields (Utah Basin as high as 11% and Los Angeles Basin at 17%) (Howarth 2014). Arguably the US shale gas extraction process results in more CH_4 escape than conventional gas fields, which are estimated at 4% leaks (Howarth 2014). However, in other countries, the leaks downstream (i.e., during liquefaction, transportation, and pipe distribution) are recorded higher. It should be mentioned that gas producers claim that extraction leaks remain generally at 1–2%, a figure, however, that has been discredited time and again by researchers' actual measurements.

To the above measurements of average 5.8% US extraction field leaks (and 4% estimated worldwide), the leaks from the distribution pipes to consumers should be added, a pipe network that in most cities is aged. Although such estimates for life cycle NG methane leak vary greatly, a realistic figure seems to be 5–7% of the total NG production. Current NG production (2017) is about 3 billion tonnes (3670 billion cubic meters) (BP 2018). Even assuming just a 4% life cycle leak, instead of 5–7%, multiplied by a GWP of 86 results in over 10 billion tonnes of CO_2 equivalent yearly, which is currently unaccounted for or largely underestimated. We remind that the total man-made CO_2, which we are trying desperately to reduce, is estimated at 36 billion tonnes per year, of which international shipping produces about 0.8 billion (IMO 2014). Considering the above, LNG as ship fuel should be viewed only as a SO_x- and also – for non-high-pressure dual-fuel gas injection (HPDF GI)-type engines – a NO_x-compliant fuel. Following from the above, its GHG footprint does not allow it to be considered either as a transitional or a short-term solution. It clearly is more disadvantageous in terms of GWP than conventional liquid fuels.

CH_4 currently contributes 40% of the heat-trapping effect of all human-produced GHGs in the atmosphere (AES 2018) as calculated by using the 100-year GWP. However, at the 20-year forward timescale, total global emissions of methane are equivalent to over 80% of global CO_2 emissions. Furthermore, while CO_2 in the atmosphere has increased by 35% in the last 300 years, methane has increased by more than 150% since 1750 (Oliver and Oliver 2018). All this suggests that CH_4 represents a more serious problem for the planet than CO_2 and, instead of LNG being promoted as a transitional fuel, efforts should be exerted to at least contain CH_4 leaks to the atmosphere.

3.3 *Liquefied Petroleum Gas (LPG)*

Liquefied petroleum gas (LPG) is any mixture of propane and butane in liquid form. Propane and butane are the first light distillates during crude oil refining by an amount of about 4%. It is also contained in the NG fields in amounts of 1–5%, and thus it is collected as a by-product of NG extraction. LPG combustion results in no SO_x and about 15% less CO_2 than fuel oil. However its GWP is three to four times that of CO_2 when LPG slip occurs (DNV GL 2018). For two-stroke marine engines, NO_x Tier III requirements are not met; thus EGR or SCR equipment is needed.

Liquid propane (boiling point −42 °C) and butane (boiling point 0 to −10 °C) are relatively easy to store and transport in pressurized or semi-refrigerated containers. Overall, however, its GHG reduction potential is modest, while LPG's availability is limited to play a major role as shipping fuel, being practically a by-product of LNG extraction or oil refining.

3.4 Hydrogen

About 95% of the world's hydrogen production comes from reforming fossil fuels, mostly NG (Milne et al. 2006; Collodi 2010). Obviously then the GHG issues described in the previous section on NG and LNG are also applicable to hydrogen, not including the CO_2 emissions in producing it from NG. The most common method to produce hydrogen is by steam CH_4 reforming whereby steam under high pressure, and in the presence of a catalyst, produces hydrogen and carbon monoxide and, with further "water-gas shift reaction," CO_2 and more hydrogen are produced. One tonne of hydrogen produced in this process releases 9–12 tonnes of CO_2 (Collodi 2010).

Hydrogen when burned emits no CO_2, no SO_x, and small amounts of NO_x. Thus, it is potentially a viable alternative fuel, provided the CO_2 intensity of its production could be addressed and provided that several technological and safety hurdles could be overcome. The CO_2 intensity of hydrogen production could be reduced by capturing and storing the released CO_2. Nevertheless, its origin, being mostly CH_4, results in the potent GHG effects related to methane's life cycle (methane slip, etc.). Clearly the carbon footprint of hydrogen produced from NG is higher than that of HFO or MDO (DNV GL 2018).

An alternative way to produce hydrogen is from water electrolysis. Only about 4% of the world's hydrogen production uses that method. However, even this hydrogen cannot be considered as "green," since electrolysis requires large amounts of electricity which usually comes from the grid. Only if this electricity originates from renewable sources (solar, wind, hydro) or even nuclear could the hydrogen produced be considered carbon-free. About 55 kwh is required to produce 1 kg of hydrogen at an assumed efficiency of more than 60%[2,3]. One kwh of electricity, when produced from a coal-burning power plant, generates about 1 kg of CO_2.[4] The US average is about 0.69 kg of CO_2 per kwh (DNV GL 2015), while China and Russia produce most of their electricity from coal and most other countries from diesel oil. At an assumed worldwide average of 0.80 kg CO_2 per kwh, 55 kwh to produce 1 kg of hydrogen from electrolysis will emit 44 kg of CO_2.

[2]https://en.wikipedia.org/wiki/Hydrogen_economy (accessed Sep. 7, 2018).

[3]https://cleanenergypartnership.de/en/faq/hydrogen-production-and-storage/ (accessed Sep. 7, 2018).

[4]https://carbonpositivelife.com/co2-per-kwh-of-electricity/ (accessed Sep. 7, 2018).

As things stand, the technology required to enable the use of hydrogen as a marine fuel is still under development. Hydrogen is not an easy fuel to handle, transport, and store. The boiling point of liquid hydrogen is −253 °C (DNV GL 2018); thus superinsulated (cryogenic) pressure vessels are needed for storage. Depending on the pressure, the size of the tanks needs to be 10–15 times larger than those of conventional liquid fuels (DNV GL 2015). Also safety issues due to the volatility of hydrogen need to be resolved.

To sum up, even though burning hydrogen would result in zero GHG emissions, as things stand, a number of very serious issues need to be resolved in order to make it a viable zero-carbon fuel.

3.5 Methanol

Today methanol is mostly produced from NG, in a process similar to the production of hydrogen (see section above). Thus the aforementioned GHG issues (methane slip of NG at extraction and transportation, etc.) are also applicable here, including the large carbon intensity of steam CH_4 reforming. Methanol is only employed as a transportation fuel on a significant basis for cars in China, where it is inexpensive by being produced cheaply from coal but with a highly negative GHG impact (IMO 2016). Methanol is easier to handle than hydrogen, being liquid at atmospheric pressure, thus the renewed interest into its possibility as marine fuel. However it is toxic and has a low flash point of only 12 °C, and, as such, several safety barriers must be employed.

Although methanol produces negligible SO_x emissions, its NO_x emissions, although reduced, are not down to Tier III levels. However, the well-to-propeller GHG emissions of NG-derived methanol are higher than liquid fuels (HFO, MDO) (IMO 2016) and a lot higher when the proper GWP of 86 (20-year time frame) is used. For methanol to offer any substantial GHG reductions, it has to be produced from biomass using renewable energy (wind/solar), something that for the time being is not realistic.

3.6 Ammonia

Ammonia (NH_3) is produced from hydrogen by adding nitrogen. Nitrogen is obtained from the air through liquid air distillation or an oxidative process where air is burnt and the residual nitrogen is recovered.[5] As such, in addition to the GHG effects of hydrogen and NG (being the primary source of hydrogen), the CO_2/GHG

[5]https://ghgprotocol.org/sites/default/files/Calculating%20CO2%20Emissions%20from%20Ammonia%20Production_0.pdf Accessed Sep. 7, 2018.

effects of nitrogen synthesis must be added to ammonia's lifetime GHG effects. The advantage of ammonia over hydrogen is that it can be stored as liquid in a temperature easier to maintain (−34 °C), while, being a hydrogen carrier, its liquid form allows more hydrogen storage per cubic meter than liquid hydrogen itself (OECD 2018). Ammonia used as fuel could offer GHG reductions to the extent that it is processed using renewable energy and is sourced from hydrogen made from electrolysis using renewable sources.

3.7 Biofuels (Ethanol, Biodiesel, etc.)

Biofuels are fuels produced from organic materials such as biomass, plants, animal waste, etc. Although they may offer a good potential for reducing CO_2, they have several downsides. One is the requirement for large agricultural land potentially resulting in food supply reduction, deforestation, and other environmental damages (OECD 2018). One important aspect of this damage is the associated loss of plant and forest carbon sinks, as well as cost increases in food. These aspects have prompted the California legislature to stop considering corn ethanol as carbon neutral and start the process of repeal of incentives for the production of corn ethanol (May 2018). Depending on the assumptions and data used, studies vary widely on the effectiveness of biofuels to reduce GHG, and it is not clear whether the energy used in the day-to-day farming practices; production and application of fertilizers, pesticides, and herbicides; and the production of biofuels offsets their GHG combustion benefits. In addition, concerns of air quality exist as the combustion of biofuels produces toxic and carcinogenic chemicals such as formaldehyde and acetaldehyde (Rio de Janeiro, where cars running on ethanol are common, has 160% more formaldehyde and 260% more acetaldehyde in the air than Tokyo or other cities where ethanol fuels are not used[6]). For marine use, caution should be exercised since some biofuels have a tendency to oxidize and degrade, due to bacteria development when stored over a few months. The current biofuel supply is limited and could only cover about 15% of the total fuel demand (OECD 2018). Attempts to produce biofuels from engineered "crops," such as algae, have so far proved unsustainable, while any production of large-scale biofuels (e.g., bio-LNG or ethane from agricultural and animal waste) is not considered realistic for the near- to medium-term future.

3.8 Fuel Cells

As also mentioned in Chap. 2, fuel cells convert the chemical energy of compounds, through electrochemical oxidation, to electric power, without combustion involved,

[6] http://theearthproject.com/biofuel/ Accessed Sep. 25, 2018.

releasing thermal energy in the process. The most usual fuel used is hydrogen, the exhaust being water. LNG, methanol, and ammonia could also be used. There are several different fuel cell technologies available, but none is mature enough to be used for main propulsion units of typical ships. This is due mainly to their limited lifetime and the large size required for both the fuel cells and the fuel tanks. Their GHG reduction potential is directly related to whether the fuel used was produced using renewable energy sources.

3.9 Conclusions on Alternative Fuels

For all the emphasis put by the IMO Initial Strategy on alternative fuels, it is clear that an essential step to assess if these fuels can deliver meaningful GHG reductions is the consideration of the environmental life cycle impacts of each proposed alternative fuel, and not just its combustion emissions. It is also essential to ensure that wider implications of fuel switches are properly accounted for. Failure to take upstream emissions into account risks locking in the sector to carbon-intensive solutions. As regards biofuels, their impact associated with cultivation, land-use change, and fertilizer use must also be assessed (Gilbert et al. 2018). Many studies underestimate the true upstream and combustion GHG effects of most alternative fuels by, among others, ignoring real measured data on methane slip or using 100-year GWP factors. Obviously, there are no easy solutions toward GHG mitigation of marine fuels, and it would be unfortunate if regulators, in their urge to act, promote fuels with worse lifetime GHG effects than current conventional fuels.

To these authors, this leads to the conclusion that to drastically reduce GHG emissions we need new technologies that would provide the necessary "quantum leap" vis-à-vis BAU. These can be better batteries, synthetic fuels, synthetic biofuels, or others. Until and unless these technologies become technically and economically viable and as things stand, current conventional liquid fuels have – sometimes by far – the smallest GHG footprint of all the above alternative fuels.

4 Looking at the Crystal Ball Cont'd

The reception of the April 2018 IMO/MEPC 72 decision was almost universally laudatory. Industry associations including the International Chamber of Shipping (ICS), the Baltic and International Maritime Council (BIMCO), the European Community Shipowners' Associations (ECSA), the International Association of Ports and Harbors (IAPH), the European Seaports Organisation (ESPO), but also the European Commission, the Organisation of Economic Cooperation and Development (OECD), and several nongovernmental organizations (NGOs) hailed the result as an important first step toward the eventual full decarbonization of shipping. There were very few expressions of dissatisfaction. For instance, the United States, which has backed out of the Paris Agreement anyway, did not vote for the Resolution. So did Saudi Arabia. Some environmental NGOs expressed disappointment with the result, and so did some members of the European Parliament.

Realizing that we are currently at a crossroads and the track that will be followed from now on is subject to many uncertainties, below we make a cursory and non-encyclopedic attempt to comment on some additional issues that we think are important as international shipping moves toward 2050. In these comments, we make clear that we only express our sincere and honest personal opinion, and we make no attempt to sound or appear politically correct:

1. Seven years after the adoption of the Energy Efficiency Design Index (EEDI), which is still (together with the Ship Energy Efficiency Management Plan (SEEMP)) the only mandatory GHG emission reduction measure, there is no doubt that the April 2018 IMO/MEPC 72 decision was a landmark decision. Achieving GHG emissions in 2050 which are at least 50% lower than they were in 2008 is a substantial and ambitious target that has to be taken very seriously by all involved.
2. We profess ignorance on whether or not the above target is compatible with the goals of the Paris Agreement. Climate experts are more competent to comment on this point. However, we think that this issue is probably of lesser importance as compared to some of the other issues that are raised below.
3. Any hope that substantial GHG reductions can be achieved by improvements on EEDI is in our opinion grossly unsubstantiated. Aside from the considerations of Chap. 3 of this book, in a recent study conducted for Danish Shipping, Smith et al. (2016) showed, among other things, that the existence of EEDI vs a scenario in which there is no EEDI as we move to 2050 amounts to a GHG emissions difference of about 3%.
4. The two stated principles that are centrally included in the Initial Strategy (a) nondiscrimination/no more favorable treatment and (b) Common but Differentiated Responsibilities and Respective Capabilities (CBDR-RC) are in direct conflict with one another. The latter principle was included so as to please the group of developing countries (mainly Brazil, Saudi Arabia, India, and others) who stood and continue to stand firmly behind CBDR-RC. In our opinion however, if there is a *single major obstacle* for any progress on maritime GHG emission reduction, it is definitely CBDR-RC, and one will need to find a way to circumvent or even eliminate this principle altogether if any serious progress is to be made. We obviously realize that doing so may not be politically correct, and the risk is that the issue may destabilize an already rather very delicate process.
5. There is no sense of priority among the wide array of candidate measures, all of which are on the table. Market-based measures (MBMs) have been put into the medium-term class (to be agreed upon between 2023 and 2030) *but only as a possibility*, even though the Damocles sword of a European Union (EU) Emissions Trading System (ETS) is looming. There appears to be no sense of urgency for any MBM, not even for reopening the MBM discussion. Some

industry circles seem to favor a bunker levy, but no one dares propose it at this point in time.

6. Related to this, it is unclear at this time what the EU will do. As stated in Chap. 11, the European Parliament decided in November 2017 to align itself with the IMO process on GHGs but that the European Commission will monitor the IMO process very closely. Depending on the pace of the IMO process, and in particular if that pace is not deemed satisfactory, one could not rule out a scenario that the EU unilaterally moves on early, so as to include shipping within the EU ETS, or at least do this conditionally. Already the European Commission (DG-CLIMA) has issued a tender for a study to investigate possible global regulatory measures to reduce GHGs from ships, focusing on two sets of measures: (a) clean fuels and (b) MBMs. The study is expected to be concluded in the second half of 2019.
7. In the event that shipping is included in the EU ETS, which in our opinion would be unfortunate as it would create distortions, it would be interesting to see what the IMO would do. A plausible (in our opinion) scenario is for the IMO to reopen the MBM discussion soon, at least so as to preempt EU action on ETS. But the political will to do so seems at this point invisible.
8. There is no clear link between any of the targets and the respective measures. The question is, for any of the reduction targets that were chosen, how can one be reasonably confident that these targets can be met by the measures proposed (and which measures)?
9. That GHG emissions are to reach a peak as soon as possible is a laudable goal but raises the obligatory question, how soon. According to the third IMO GHG study (IMO 2014), in 2008, the baseline year as far as the comparison to target values is concerned, CO_2 emissions of international shipping were estimated at 920.9 million tonnes and declined to 795.9 million tonnes in 2012, even though they reached 849.5 million tonnes in 2011. As of yet, and pending the fourth IMO GHG study (which will be commissioned in 2019 and finalized in 2020), there is no consensus on GHG emission figures after 2012. And even the above figures are based on the "bottom-up" (activity based) method, whereas emission figures based on the "top-down" (fuel sales based) method are significantly lower (624.9 million tonnes in 2008 versus 648.9 million tonnes in 2011 – there was no top-down estimate for 2012). There should certainly be consensus on which method is used ("bottom-up" numbers are 30–50% higher than "top down"), plus consensus on when the GHG peak is expected to occur. Barring any major world trade slowdown, it seems self-evident that for any GHG peak to be reached, some measures will have to be implemented- no peak will happen by itself.
10. The same is true as regards consensus on how "transport work" figures are defined. These are important so as to check the target of at least 40% GHG emissions per transport work reduction in 2030 versus 2008 levels (and at least 70% by 2050).
11. (Mandatory) "Speed reduction" (or speed limits) is included as a potential short-term measure, even though the term "speed optimization" was added so

as to make the measure more palatable to Chile and Peru, who are concerned about carrying cherries to China (as per Chap. 10). Speed limits may seem at first glance like a reasonable measure; however they are plagued by various deficiencies and would create distortions and other problems (see again Chap. 10). Still, it is a victory for the speed limit lobbyists (Clean Shipping Coalition and others) that this measure is now on the table at the IMO, only a few years after the IMO previously rejected it (at MEPC 61 in 2010). We note that this measure is also being considered by the European Commission, among other possible measures to reduce maritime GHG emissions.

Lack of prioritization among measures being an observation, and in the quest to meet the 2030 and 2050 targets, is there any measure that should receive priority? In our opinion there is.

As also mentioned in Chap. 11, and irrespective of the fact that MBMs are not up for discussion at the IMO in the foreseeable future, one idea that might be worth considering would be to impose a *significant* bunker levy at a global level. By significant we mean not 10 or 20 USD per tonne, as is being occasionally contemplated by industry, but *at least one order of magnitude higher*.

To put it very simply, if society truly does not like fossil fuels, or any other fuel that produces GHGs (and this includes LNG), and cannot, for obvious reasons, mandate their outright ban, society should at least try to implement the "polluter pays" principle by internalizing (even partially) the external costs of GHGs. The only way to do so is by putting a significant price on the fuels that produce these GHGs. Conversely, and so long as these fuels are affordable, there is no doubt that they will be used. All the debate on LNG, hydrogen, and other alternative fuels (see Sect. 3 above) critically hinges upon the economic dimension: we would like to know not only how much GHGs these alternative fuels would avert but also what is the cost of producing and using them. Conversely, and barring a technological quantum leap, for as long as these alternative fuels are not viable economically, they will not be used.

An important parenthesis here is that, and in order to avoid modal shifts to land-based modes, such a levy should not be confined to the maritime mode, and care should be taken to prevent modal shifts which could increase overall GHG levels. This is particularly true not only for short sea shipping (SSS) scenarios but also for longer-distance deep-sea services, especially now that the Belt and Road Initiative (BRI) is being pursued by China so as to link Asia and Europe.

As mentioned in Chap. 11, a substantial bunker levy would induce technological changes in the long run and logistical measures (such as slow steaming) in the short run. In the long run, it would lead to changes in the global fleet toward vessels and technologies that are more energy efficient, more economically viable, and less dependent on fossil fuels than those today. A levy would also raise monies that could be used for "out-of-sector" GHG emission reductions, aid to least developed countries (LDCs) and Small Island Developing States (SIDS), and other purposes.

However, and for the reasons stated earlier (see again Chap. 11), we realize that the prospects of such a development are, as things stand, very slim. This is so mainly for political reasons. Aside from the fact that proposing such a measure would not sound politically correct, a frequent argument that is made in that regard is that there is no need for an MBM since fuel prices are expected to increase anyway in 2020 due to the global 0.50% sulfur cap. Advocates of such argument say that this fuel price increase would be tantamount to a global MBM; therefore there is no need to institute MBMs on their own right.

We respectfully consider such an argument as a poor excuse for not taking action on MBMs. First of all, the extent of the anticipated fuel price increase is largely unknown, and it is reminded that contrary to the "gloom and doom" predictions before the imposition of the 0.10% sulfur limit in European sulfur emission control areas (SECAs) in 2015, fuel prices actually dropped. Second, some ships will opt to install scrubbers and thus still burn the cheaper high-sulfur fuel. Third, and even if fuel prices go up in 2020 (as is a likely scenario), this would be no way to implement the "polluter pays" principle and internalize the external costs of GHGs. Society would get no benefits, and no monies would be collected for LDCs, SIDS, or out-of-sector GHG emission reductions. In that sense, we think that the 2020 sulfur cap will hardly institute an MBM.

In a recent paper that was published a few months *before* the April 2018 IMO decision (Psaraftis 2018), the following statement was made: "... in spite of much talk about the maritime industry's commitment toward serious GHG emissions reductions, it is fair to say that such reductions are, as things stand, only a wish at this point in time." Based on what we have seen since then, including IMO/MEPC 72, we see no significant reason to retract the above statement. It is true that the April 2018 IMO decision has opened a new door and maybe created some momentum. However, substance-wise and in order to guarantee significant GHG emissions reductions in the future, one would have to abandon the BAU stance that still seems to pervade much of what is done in the maritime industry today and not be afraid to take bolder steps, even if these entail some political cost.

Of course, it can be argued that in the quest for a substantial decarbonization of the shipping industry, an unfair burden has been placed on the shoulders of the shipowner. The potential unfairness stems from the argument that if ship designers, shipbuilders, engine manufacturers, fuel producers, and other technology developers somehow fail to produce the set of technologies that would make the 2050 50% target feasible and viable, why should shipowners be held responsible for the failure? Shipowners of course have a substantial role to play by choosing appropriate ships to meet their need or even influencing their design. However, their role is limited by what is available in the market. At the same time, shipowners do have a substantial role in the maritime regulatory process (IMO, EU, and others), and the aim is for that process, among other things, to provide technology developers a workable framework and substantial incentives to produce viable decarbonization technologies.

We note here that this situation is completely different from the automotive industry setting, where the main responsibility for emissions reduction is placed on the *vehicle manufacturer*, who actually has to meet emissions requirements *on a fleet level basis* and based *on a variable speed profile*. Why something similar is not done with the shipping industry and instead the shipowner is the main player responsible for emission reduction at an individual ship level, is, in our opinion, intriguing. A governance model that is closer to what the automotive industry is doing may be, in our opinion, worth looking at. But doing this would surely necessitate some radical changes and could not happen overnight.

In Psaraftis (2018), it was also argued that "... as things stand, the international scene for the decarbonization of maritime transport has been rendered way too complex and fragmented, as well as political. Unnecessary complexity and fragmentation, coupled with factors that are mostly within the political sphere, will not help a speedy resolution of the issue. In fact they will definitely hinder prospects for substantial progress in the years ahead. Conversely, a necessary condition for substantial progress on the GHG front is the removal, or at least alleviation, of such political obstacles."

We see no reason to retract this last statement either. However, we sincerely hope that things move in the right direction, and the international shipping community finds a credible way to remove the above obstacles.

We also hope that the contents of this book may be of some help toward the above goal.

Appendix

Resolution MEPC.304(72) (Adopted on 13 April 2018)

Initial IMO Strategy on Reduction of GHG Emissions from Ships

The Marine Environment Protection Committee

RECALLING Article 38(e) of the Convention on the International Maritime Organization (the Organization) concerning the functions of the Marine Environment Protection Committee (the Committee) conferred upon it by international conventions for the prevention and control of marine pollution from ships,

ACKNOWLEDGING that work to address greenhouse gas (GHG) emissions from ships has been undertaken by the Organization continuously since 1997, in particular, through adopting global mandatory technical and operational energy efficiency measures for ships under MARPOL Annex VI,

ACKNOWLEDGING ALSO the decision of the thirtieth session of the Assembly in December 2017 that adopted for the Organization a strategic direction entitled "Respond to Climate Change",

RECALLING the United Nations 2030 Agenda for Sustainable Development,

1. ADOPTS the Initial IMO Strategy on reduction of GHG emissions from ships (hereinafter the Initial Strategy) as set out in the annex to the present resolution;

2. INVITES the Secretary-General of the Organization to make adequate provisions in the Integrated Technical Cooperation Programme (ITCP) to support relevant follow-up actions of the Initial Strategy that may be further decided by the Committee and undertaken by developing countries, particularly least developed countries (LDCs) and small island developing States (SIDS);
3. AGREES to keep the Initial Strategy under review, with a view to adoption of a Revised IMO Strategy on reduction of GHG emissions from ships in 2023.

Annex

Initial IMO Strategy on Reduction of GHG Emissions from Ships

Contents

1 Introduction

1.1 The International Maritime Organization (IMO) is the United Nations specialized agency responsible for safe, secure and efficient shipping and the prevention of pollution from ships.

1.2 The Strategy represents the continuation of work of IMO as the appropriate international body to address greenhouse gas (GHG) emissions from international shipping. This work includes Assembly resolution A.963(23) on *IMO policies and practices related to the reduction of greenhouse gas emissions from ships*, adopted on 5 December 2003, urging the Marine Environment Protection Committee (MEPC) to identify and develop the mechanisms needed to achieve the limitation or reduction of GHG emissions from international shipping.

1.3 In response to the Assembly's request, work to address GHG emissions from ships has been undertaken, including inter alia:

1. MEPC 62 (July 2011) adopted resolution MEPC.203(62) on *Inclusion of regulations on energy efficiency for ships in MARPOL Annex VI* introducing

mandatory technical (EEDI) and operational (SEEMP) measures for the energy efficiency of ships. To date more than 2,700 new ships have been certified to the energy efficiency design requirement;

2. MEPC 65 (May 2013) adopted resolution MEPC.229(65) on *Promotion of technical co-operation and transfer of technology relating to the improvement of energy efficiency of ships*, which, among other things, requests IMO, through its various programmes (ITCP,[7] GloMEEP project,[8] MTCC network,[9] etc.), to provide technical assistance to Member States to enable cooperation in the transfer of energy efficient technologies, in particular to developing countries; and
3. MEPC 70 (October 2016) adopted, by resolution MEPC.278(70), amendments to MARPOL Annex VI to introduce the d*ata collection system for fuel oil consumption of ships*, containing mandatory requirements for ships to record and report their fuel oil consumption. Ships of 5,000 gross tonnage and above (representing approximately 85% of the total CO_2 emissions from international shipping) are required to collect consumption data for each type of fuel oil they use, as well as other, additional, specified data including proxies for "transport work".

1.4 This Initial Strategy is the first milestone set out in the *Roadmap for developing a comprehensive IMO Strategy on reduction of GHG emissions from ships* (the Roadmap) approved at MEPC 70. The Roadmap identifies that a revised Strategy is to be adopted in 2023.

Context

1.5 The Initial Strategy falls within a broader context including:

1. other existing instruments related to the law of the sea, including UNCLOS, and to climate change, including the UNFCCC and its related legal instruments, including the Paris Agreement;
2. the leading role of the Organization for the development, adoption and assistance in implementation of environmental regulations applicable to international shipping;
3. the decision of the thirtieth session of the Assembly in December 2017 that adopted for the Organization a Strategic Direction entitled "Respond to climate change"; and
4. the United Nations 2030 Agenda for Sustainable Development.

[7]Integrated Technical Cooperation Programme http://www.imo.org.

[8]Global Maritime Energy Efficiency Partnerships http://glomeep.imo.org/.

[9]Global Maritime Technology Cooperation Centres Network http://gmn.imo.org/.

Emissions and Emission Scenarios

1.6 The *Third IMO GHG Study 2014* has estimated that GHG emissions from international shipping in 2012 accounted for some 2.2% of anthropogenic CO_2 emissions and that such emissions could grow by between 50% and 250% by 2050. Future IMO GHG studies would help reduce the uncertainties associated with these emission estimates and scenarios.

Objectives of the Initial Strategy

1.7 The Initial Strategy is aimed at:

1. enhancing IMO's contribution to global efforts by addressing GHG emissions from international shipping. International efforts in addressing GHG emissions include the Paris Agreement and its goals and the United Nations 2030 Agenda for Sustainable Development and its SDG 13: "*Take urgent action to combat climate change and its impacts*";
2. identifying actions to be implemented by the international shipping sector, as appropriate, while addressing impacts on States and recognizing the critical role of international shipping in supporting the continued development of global trade and maritime transport services; and
3. identifying actions and measures, as appropriate, to help achieve the above objectives, including incentives for research and development and monitoring of GHG emissions from international shipping.

2 Vision

IMO remains committed to reducing GHG emissions from international shipping and, as a matter of urgency, aims to phase them out as soon as possible in this century.

3 Levels of Ambition and Guiding Principles

Levels of Ambition

3.1 Subject to amendment depending on reviews to be conducted by the Organization, the Initial Strategy identifies levels of ambition for the international shipping sector noting that technological innovation and the global introduction of alternative fuels and/or energy sources for international shipping will be integral to achieve the overall ambition. The reviews should take into account updated emission estimates, emissions reduction options for international shipping, and the reports of the Intergovernmental Panel on Climate Change

(IPCC), as relevant. Levels of ambition directing the Initial Strategy are as follows:

1. *carbon intensity of the ship to decline through implementation of further phases of the energy efficiency design index (EEDI) for new ships*
   to review with the aim to strengthen the energy efficiency design requirements for ships with the percentage improvement for each phase to be determined for each ship type, as appropriate;
2. *carbon intensity of international shipping to decline*
   to reduce CO_2 emissions per transport work, as an average across international shipping, by at least 40% by 2030, pursuing efforts towards 70% by 2050, compared to 2008; and
3. *GHG emissions from international shipping to peak and decline*
   to peak GHG emissions from international shipping as soon as possible and to reduce the total annual GHG emissions by at least 50% by 2050 compared to 2008 whilst pursuing efforts towards phasing them out as called for in the Vision as a point on a pathway of CO_2 emissions reduction consistent with the Paris Agreement temperature goals.

Guiding Principles

3.2 The principles guiding the Initial Strategy include:

1. the need to be cognizant of the principles enshrined in instruments already developed, such as:
   1. the principle of non-discrimination and the principle of no more favorable treatment, enshrined in MARPOL and other IMO conventions; and
   2. the principle of common but differentiated responsibilities and respective capabilities, in the light of different national circumstances, enshrined in UNFCCC, its Kyoto Protocol and the Paris Agreement;
2. the requirement for all ships to give full and complete effect, regardless of flag, to implementing mandatory measures to ensure the effective implementation of this strategy;
3. the need to consider the impacts of measures on States, including developing countries, in particular, on LDCs and SIDS as noted by MEPC 68 (MEPC 68/21, paragraphs 4.18 to 4.19) and their specific emerging needs, as recognized in the Organization's Strategic Plan (resolution A.1110(30)); and
4. the need for evidence-based decision-making balanced with the precautionary approach as set out in resolution MEPC.67(37).

4 List of Candidate Short-, Mid- and Long-Term Further Measures with Possible Timelines and Their Impacts on States

Timelines

4.1 Candidate measures set out in this Initial Strategy should be consistent with the following timelines:

1. possible short-term measures could be measures finalized and agreed by the Committee between 2018 and 2023. Dates of entry into force and when the measure can effectively start to reduce GHG emissions would be defined for each measure individually;
2. possible mid-term measures could be measures finalized and agreed by the Committee between 2023 and 2030. Dates of entry into force and when the measure can effectively start to reduce GHG emissions would be defined for each measure individually; and
3. possible long-term measures could be measures finalized and agreed by the Committee beyond 2030. Dates of entry into force and when the measure can effectively start to reduce GHG emissions would be defined for each measure individually.

4.2 In aiming for early action, the timeline for short-term measures should prioritize potential early measures that the Organization could develop, while recognizing those already adopted, including MARPOL Annex VI requirements relevant for climate change, with a view to achieve further reduction of GHG emissions from international shipping before 2023.

4.3 Certain mid- and long-term measures will require work to commence prior to 2023.

4.4 These timelines should be revised as appropriate as additional information becomes available.

4.5 Short-, mid- and long-term further measures to be included in the Revised IMO GHG Strategy should be accompanied by implementation schedules.

4.6 The list of candidate measures is non-exhaustive and is without prejudice to measures the Organization may further consider and adopt.

Candidate Short-Term Measures

4.7 Measures can be categorized as those the effect of which is to directly reduce GHG emissions from ships and those which support action to reduce GHG emissions from ships. All the following candidate measures[10] represent

[10]The Initial Strategy is subject to revision based on fuel oil consumption data collected during 2019-2021 and does not prejudge any specific further measures that may be implemented in Phase 3 of the three-step approach.

possible short-term further action of the Organization on matters related to the reduction of GHG emissions from ships:

1. further improvement of the existing energy efficiency framework with a focus on EEDI and SEEMP, taking into account the outcome of the review of EEDI regulations;
2. develop technical and operational energy efficiency measures for both new and existing ships, including consideration of indicators in line with the three-step approach that can be utilized to indicate and enhance the energy efficiency performance of shipping, e.g. Annual Efficiency Ratio (AER), Energy Efficiency per Service Hour (EESH), Individual Ship Performance Indicator (ISPI) and Fuel Oil Reduction Strategy (FORS);
3. establishment of an Existing Fleet Improvement Programme;
4. consider and analyse the use of speed optimization and speed reduction as a measure, taking into account safety issues, distance travelled, distortion of the market or trade and that such measure does not impact on shipping's capability to serve remote geographic areas;
5. consider and analyse measures to address emissions of methane and further enhance measures to address emissions of Volatile Organic Compounds;
6. encourage the development and update of national action plans to develop policies and strategies to address GHG emissions from international shipping in accordance with guidelines to be developed by the Organization, taking into account the need to avoid regional or unilateral measures;
7. continue and enhance technical cooperation and capacity-building activities under the ITCP;
8. consider and analyse measures to encourage port developments and activities globally to facilitate reduction of GHG emissions from shipping, including provision of ship and shoreside/onshore power supply from renewable sources, infrastructure to support supply of alternative low-carbon and zero-carbon fuels, and to further optimize the logistic chain and its planning, including ports;
9. initiate research and development activities addressing marine propulsion, alternative low-carbon and zero-carbon fuels, and innovative technologies to further enhance the energy efficiency of ships and establish an International Maritime Research Board to coordinate and oversee these R&D efforts;
10. incentives for first movers to develop and take up new technologies;
11. develop robust lifecycle GHG/carbon intensity guidelines for all types of fuels, in order to prepare for an implementation programme for effective uptake of alternative low-carbon and zero-carbon fuels;
12. actively promote the work of the Organization to the international community, in particular, to highlight that the Organization, since the 1990s, has developed and adopted technical and operational measures that have consistently provided a reduction of air emissions from ships, and that measures could support the Sustainable Development Goals, including SDG 13 on Climate Change; and

13. undertake additional GHG emission studies and consider other studies to inform policy decisions, including the updating of Marginal Abatement Cost Curves and alternative low-carbon and zero-carbon fuels.

Candidate Mid-Term Measures

4.8 Measures can be categorized as those the effect of which is to directly reduce GHG emissions from ships and those which support action to reduce GHG emissions from ships. All the following candidate measures represent possible mid-term further action of the Organization on matters related to the reduction of GHG emissions from ships:

1. implementation programme for the effective uptake of alternative low-carbon and zero-carbon fuels, including update of national actions plans to specifically consider such fuels;
2. operational energy efficiency measures for both new and existing ships including indicators in line with three-step approach that can be utilized to indicate and enhance the energy efficiency performance of ships;
3. new/innovative emission reduction mechanism(s), possibly including Market-based Measures (MBMs), to incentivize GHG emission reduction;
4. further continue and enhance technical cooperation and capacity-building activities such as under the ITCP; and
5. development of a feedback mechanism to enable lessons learned on implementation of measures to be collated and shared through a possible information exchange on best practice.

Candidate Long-Term Measures

4.9 All the following candidate measures represent possible long-term further action of the Organization on matters related to the reduction of GHG emissions from ships:

1. pursue the development and provision of zero-carbon or fossil-free fuels to enable the shipping sector to assess and consider decarbonization in the second half of the century; and
2. encourage and facilitate the general adoption of other possible new/innovative emission reduction mechanism(s).

Impacts on States

4.10 The impacts on States of a measure should be assessed and taken into account as appropriate before adoption of the measure. Particular attention should be paid to the needs of developing countries, especially small island developing States (SIDS) and least developed countries (LDCs).

4.11 When assessing impacts on States the impact of a measure should be considered, as appropriate, inter alia, in the following terms:

1. geographic remoteness of and connectivity to main markets;
2. cargo value and type;
3. transport dependency;
4. transport costs;
5. food security;
6. disaster response;
7. cost-effectiveness; and
8. socio-economic progress and development.

4.12 The specification for and agreement on the procedure for assessing and taking into account the impacts of measures related to international shipping on States should be undertaken as a matter of urgency as part of the follow-up actions.

4.13 Disproportionately negative impacts should be assessed and addressed, as appropriate.

5 Barriers and Supportive Measures; Capacity-Building and Technical Cooperation; R&D

5.1 The Committee recognizes that developing countries, in particular LDCs and SIDS, have special needs with regard to capacity-building and technical cooperation.

5.2 The Committee acknowledges that development and making globally available new energy sources that are safe for ships could be a specific barrier to the implementation of possible measures.

5.3 The Committee could assist the efforts to promote low-carbon technologies by facilitating public-private partnerships and information exchange.

5.4 The Committee should continue to provide mechanisms for facilitating information sharing, technology transfer, capacity-building and technical cooperation, taking into account resolution MEPC.229(65) on *Promotion of technical co-operation and transfer of technology relating to the improvement of energy efficiency of ships*.

5.5 The Organization is requested to assess periodically the provision of financial and technological resources and capacity-building to implement the Strategy through the ITCP and other initiatives including the GloMEEP project and the MTCC network.

6 Follow-Up Actions Towards the Development of the Revised Strategy

6.1 A programme of follow-up actions of the Initial Strategy should be developed.
6.2 The key stages for the adoption of a Revised IMO GHG Strategy in 2023 as set out in the Roadmap, are as follows:

Spring 2018 (MEPC 72)	Adoption of the Initial Strategy (Initial IMO Strategy is subject to revision based on DCS data during 2019-2021 and does not prejudge any specific further measures that may be implemented in Phase 3 of the three-step approach.) including, inter alia, a list of candidate short-, mid- and long-term further measures with possible timelines, to be revised as appropriate as additional information becomes available
January 2019	Start of Phase 1: Data collection (Ships to collect data)
Spring 2019 (MEPC 74)	Initiation of Fourth IMO GHG Study using data from 2012–2018
Summer 2020	Data from 2019 to be reported to IMO
Autumn 2020 (MEPC 76)	Start of Phase 2: data analysis (no later than autumn 2020) Publication of Fourth IMO GHG Study for consideration by MEPC 76
Spring 2021 (MEPC 77)	Secretariat report summarizing the 2019 data pursuant to regulation 22A.10 Initiation of work on adjustments on Initial IMO Strategy, based on Data Collection System (DCS) data
Summer 2021	Data for 2020 to be reported to IMO
Spring 2022 (MEPC 78)	Phase 3: Decision step Secretariat report summarizing the 2020 data pursuant to regulation 22A.10
Summer 2022	Data for 2021 to be reported to IMO
Spring 2023 (MEPC 80)	Secretariat report summarizing the 2021 data pursuant to regulation 22A.10 Adoption of Revised IMO Strategy, including short-, mid- and long-term further measure(s), as required, with implementation schedules

6.3 The Marginal Abatement Cost Curve (MACC) for each measure, as appropriate, should be ascertained and updated, and then evaluated on a regular basis.

7 Periodic Review of the Strategy

7.1 The Revised Strategy is to be adopted in Spring 2023.
7.2 The Revised Strategy should be subject to a review five years after its final adoption.
7.3 The Committee should undertake the review including defining the scope of the review and its terms of reference.

References

AES. (2018). *Methane emissions: This year's biggest energy challenge*. American Energy Society. https://www.energytoday.net/environmental-health-impact/methane-emissions-years-biggest-energy-challenge/.

Allen, D. T. (2014). Methane emissions from natural gas production and use: Reconciling bottom-up and top-down measurements. *Current Opinion in Chemical Engineering, 5*(1), 78–83.

Bouman, E. I., Lindstad, E., Rialland, A. I., & Strømman, A. H. (2017). State-of-the-art technologies, measures, and potential for reducing GHG emissions from shipping – a review. *Transportation Research Part D, 52*, 408–421.

BP. (2018). *Statistical review of world energy*. British Petroleum report.

Collodi, G. (2010). Hydrogen production via steam reforming with CO_2 capture. *Chemical Engineering Transactions*, 19, 37–42.

Corbett, J. J., Thomson, H., & Winebrake, J. J. (2015). *Methane emissions from natural gas bunkering operations in the marine sector: A total fuel cycle approach*, Report for US Maritime Administration, 21 November.

DNV GL. (2015). *The fuel trilemma: Next generation of marine fuels*. Strategic Research and Innovation position paper 03-2015.

DNV GL. (2018). *Assessment of selected alternative fuels and technologies*. Guidance paper 2018-05.

EU. (2016). *Methane emissions from LNG-powered ships higher than current marine fuel oils*. European Commission DG Environment News Alert Service. http://ec.europa.eu/environment/integration/research/research_alert_en.htm, 28 January.

Gilbert, P., Walsh, C., Traut, M., Kesieme, U., Pazouki, K., & Murphy, A. (2018). Assessment of full life-cycle air emissions of alternative shipping fuels. *Journal of Cleaner Production, 172*, 855–866.

Howarth, R. W. (2014). A bridge to nowhere: Methane emissions and the greenhouse gas footprint of natural gas. *Energy Science & Engineering, 2*(2), 47–60.

IMO. (2014). Third IMO GHG study 2014, Co-authored by Smith, T. W. P., Jalkanen, J. P., Anderson, B. A., Corbett, J. J., Faber, J., Hanayama, S., O'Keeffe, E., Parker, S., Johansson, L., Aldous, L., Raucci, C., Traut, M., Ettinger, S., Nelissen, D., Lee, D. S., Ng, S., Agrawal, A., Winebrake, J., Hoen, M., Chesworth, S., & Pandey, A. International Maritime Organization (IMO) London, UK, June.

IMO. (2016). Study of the use of methanol as marine fuel. IMO doc. MEPC 69/INF.10.

IMO. (2018). Resolution MEPC.304(72) (adopted on 13 April 2018), initial IMO strategy on reduction of GHG emissions from ships, IMO doc. MEPC 72/17/Add.1, Annex 1.

Jaramillo, P., Griffin, W. M., & Matthews, H. S. (2005). *Comparative life cycle carbon emissions of LNG versus coal and gas for electricity generation. Civil and environmental engineering*. Pittsburgh: Carnegie Mellon University.

Milne, T. A., Elam, C. C., & Evans, R. J. (2006). *Task 16, hydrogen from carbon-containing materials* (Report for the International Energy Agency, Agreement on the Production and Utilization of Hydrogen). Golden: National Renewable Energy Laboratory.

Myhre, G., Shindell, D., Bréon, F.-M., Collins, W., Fuglestvedt, J., Huang, J., Koch, D., Lamarque, J.-F., Lee, D., Mendoza, B., Nakajima, T., Robock, A., Stephens, G., Takemura, T., & Zhang, H. (2013). Anthropogenic and natural radiative forcing. In T. F. Stocker, D. Qin, G.-K. Plattner, M. Tignor, S. K. Allen, J. Boschung, A. Nauels, Y. Xia, V. Bex, & P. M. Midgley (Eds.), *Climate change 2013: The physical science basis* (Contribution of Working Group I to the Fifth Assessment Report of the Intergovernmental Panel on Climate Change). Cambridge, UK: Cambridge University Press.

OECD. (2018). *Decarbonising maritime transport. Pathways to zero-carbon shipping by 2035*. Report of the Organisation for Economic Cooperation and Development, International Transport Forum, Paris, France, March.

Oliver, C. D., & Oliver, F. A. (2018). *Global resources and the environment*. Cambridge, UK: Cambridge University Press.

Psaraftis, H. N. (2018). Decarbonization of maritime transport: To be or not to be? *Maritime Economics and Logistics*. https://doi.org/10.1057/s41278-018-0098-8.

SINTEF. (2017). *GHG and NOx emissions from gas fueled engines*, SINTEF Ocean AS Report.

Smith, T., Raucci, C., Haji Hosseinloo, S., Rojon, I., Calleya, J., Suárez de la Fuente, S., Wu, P., & Palmer, K. (2016). *CO_2emissions from international shipping. Possible reduction targets and their associated pathways*. Report prepared by UMAS, London, UK, October.

Index

H. N. Psaraftis (ed.), *Sustainable Shipping*,
https://doi.org/10.1007/978-3-030-04330-8

S

MIX
Papier aus verantwortungsvollen Quellen
Paper from responsible sources
FSC® C105338

FSC
www.fsc.org

If you have any concerns about our products,
you can contact us on
ProductSafety@springernature.com

In case Publisher is established outside the EU,
the EU authorized representative is:
Springer Nature Customer Service Center GmbH
Europaplatz 3, 69115 Heidelberg, Germany

Printed by Libri Plureos GmbH
in Hamburg, Germany